Evolutionary Ecology

Concepts and Case Studies

Edited by

CHARLES W. FOX, DEREK A. ROFF,
AND DAPHNE J. FAIRBAIRN

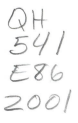

OXFORD

UNIVERSITY PRESS

2001

OXFORD

UNIVERSITY PRESS

Oxford   New York

Athens   Auckland   Bangkok   Bogotá   Buenos Aires   Cape Town
Chennai   Dar es Salaam   Delhi   Florence   Hong Kong   Istanbul   Karachi
Kolkata   Kuala Lumpur   Madrid   Melbourne   Mexico City   Mumbai   Nairobi
Paris   São Paulo   Shanghai   Singapore   Taipei   Tokyo   Toronto   Warsaw
and associated companies in
Berlin   Ibadan

Published by Oxford University Press, Inc.
198 Madison Avenue, New York, New York 10016

Oxford is a registered trademark of Oxford University Press.

Library of Congress Cataloging-in-Publication Data
Evolutionary ecology : concepts and case studies / edited by Charles W. Fox, Derek A.
Roff, and Daphne J. Fairbairn.
      p.   cm.
   Includes bibliographical references (p.   ).
   ISBN 0-19-513154-1; 0-19-513155-X (pbk.)
   1. Ecology.   2. Evolution (Biology)   I. Fox, Charles W.   II. Roff, Derek A., 1949–   .   III.
Fairbairn, Daphne J.

   QH541 .E86   2001
   577—dc21   00-053758

Cover art: A female broad-tailed hummingbird (*Selasphorus platycercus*) pollinating *Delphinium nuttallianum*.
Drawing by Mary V. Price, University of California, Riverside.

9 8 7 6 5 4 3 2 1
Printed in the United States of America
on acid-free paper

Evolutionary Ecology

Preface

Evolutionary biology and ecology share the goals of describing variation in natural systems and discovering its functional basis. Within this common framework, evolutionary biologists emphasize historical and lineage-dependent processes and hence often incorporate phylogenetic reconstructions and genetic models in their analyses. Ecologists, while cognizant of historical processes, tend to explain variation in terms of the contemporary effects of biotic and abiotic environmental factors. Evolutionary ecology spans these two disciplines and incorporates the full range of techniques and approaches from both. Evolutionary ecologists consider both historical and contemporary influences on patterns of variation and study variation at all levels, from within-individual variation (e.g., ontogenetic, behavioral) to variation among communities or major taxonomic groups. The overlap between evolutionary ecology and ecology is so broad that some previous treatments of the field (e.g., Pianka 1994) have little to distinguish them from standard ecology textbooks. However, recent advances in molecular genetics, quantitative genetics (e.g., multivariate models, analyses of quantitative trait loci), statistical methods for comparative analyses, and computer-intensive genetic modeling have enabled evolutionary ecologists to more explicitly incorporate lineage-dependent processes and constraints into their research programs. In modern evolutionary ecology, both the adaptive significance and the "evolvability" of traits are hypotheses to be tested, rather than a priori assumptions. As the chapters in this volume attest, contemporary evolutionary ecologists have assembled a very diverse and effective array of techniques and approaches to test these hypotheses.

Our primary objective in organizing this book was to provide a collection of readings, as an alternative to readings from the primary literature, that would serve as an introduction to contemporary research programs in evolutionary ecology. Having taught undergraduate and graduate courses in evolutionary ecology at our respective institutions, we recognized the need for such a volume and discovered that many of our colleagues, including some of the contributors to this volume, felt the same way. We hope that this book will fill this need and be suitable either as a textbook for evolutionary ecology courses offered at the graduate or advanced undergraduate level, or as a reader for graduate seminars on this same topic. We have asked authors to write their chapters for this audience. When writing the first part of the book, entitled "Recurring Themes," authors have assumed that students have the equivalent of at least one undergraduate course in ecology and one course in genetics, but they have not assumed any background in population or quantitative genetics, or in evolutionary theory. As indicated by the title, the concepts introduced in this section recur throughout the volume and are fundamental to most research in evolutionary ecology. For the succeeding chapters (parts II–V), we assume that students have a basic understanding of the evolutionary processes and concepts discussed in part I. Authors in all sections have also assumed that students have read the preceding chapters in the volume, allowing chapters to build on each other without repeatedly redefining terms or redeveloping basic concepts. However, we realize that some readers will select only a subset of chapters, and therefore when specific information from a previous chapter is necessary for understanding a concept or example, we have tried to ensure that a reference to the appropriate preceding chapter is included.

The chapters in this volume have each been written by a different author; all authors are leading researchers in their field. Chapters thus represent the current stage of evolutionary ecology better than any single-authored textbook could, and the diversity of authors introduces students to the diversity of ideas, approaches, and opinions that are the nature of science. However, a multiauthored textbook presents special challenges for students, just as team-taught lecture course presents challenges. Authors vary in the level at which they present their material and in the amount of background that they expect students to have when reading their chapter. Authors also vary in their writing styles and vary somewhat in the way that they organize their chapters. We have attempted to minimize the variation among chapters by providing guidelines to authors, by asking authors to communicate with each other while writing their chapters, and by aggressively editing and revising chapters as needed. We have also tried to minimize overlap among chapters and to ensure that chapters build on one another. Perhaps our major challenge as editors was to keep the volume to a reasonable length, given 28 independently written chapters. Each author was asked to contribute no more than 8000 words of text, using no more than six figures, plus tables, and 30 references. These restrictions precluded comprehensive reviews and forced each contributor to select only a few key references. Nevertheless, each chapter does serve as a good introduction to the research area by providing leading references to other reviews, books, and seminal papers. As editors, we take full responsibility for the resulting (and necessary) omission of many additional references, perhaps equally appropriate as examples or case studies. We hope that

readers will be inspired to delve more fully into at least some of the research areas and will thus have the opportunity to discover the vast and detailed literature that we have been unable to include.

Although this volume is intended to be suitable as a text for advanced undergraduates or graduate students, we also had a second objective—to produce a volume that is valuable to all researchers in ecology, evolution, and genetics. It is largely for this reason that we opted for a multiauthored volume rather than a traditional textbook style. This volume is a collection of chapters that describe the modern state of evolutionary ecology, including informed and thoughtful insights into where this field is, or perhaps should be, going. Researchers should find the chapters dedicated to their areas of expertise interesting food for thought, while the chapters covering more disparate areas should provide effective updates and insights that will, we hope, encourage cross-fertilization.

Evolutionary ecology is a very broad and diverse field that includes much of modern ecology and evolutionary biology. Unfortunately, we have only one volume within which to cover the field. We have tried to include as many topics as possible in the space provided, but of necessity, some topics had to be covered only briefly or omitted altogether. Thus, we have made numerous editorial decisions concerning content; we hope that you, as readers, will agree with most of these. The most substantial decisions involved what topics should be left out of the book. Undoubtedly, our personal interests and biases have influenced some of these decisions, but most omissions are for practical reasons. For example, we have opted not to include chapters on speciation because numerous edited volumes have been dedicated to the topic and it is generally well covered in general evolution textbooks. We have also limited our coverage of statistical and analytical techniques to introductions and brief descriptions within individual chapters. Readers will not find specific chapters dedicated to methodology such as chapters on molecular methods, methods of phylogenetic analysis, or methods for measuring genetic variance components. These are important techniques for us all to understand but are best acquired from specialized volumes (e.g., Brooks and McLennan 1991; Harvey and Pagel 1991; Avise 1994; Roff 1997). Instead, we focus on conceptual problems and case studies that may illustrate why and when particular methods and techniques are useful in evolutionary ecology, but we do not provide detailed recipes for application of the methods.

In closing, we express our gratitude to all of the authors contributing to this volume. Writing a book chapter is often a thankless task, and our stringent requirements have made the task especially difficult. We have been uniformly impressed not only by the very high quality of the contributions, but also by each author's cheerful willingness to respond to our requests for revision. Al-most all of these requests were made to standardize the style of the chapters and increase the cohesiveness of the volume as a whole. To the extent that we have succeeded in this, and in our overall goal of providing a state-of-the-art introduction to evolutionary ecology, we must thank the individual chapter authors.

Charles W. Fox
Derek A. Roff
Daphne J. Fairbairn

Contents

Contributors

Abrams, Peter A. Department of Zoology, University of Toronto, 25 Harbord St., Toronto, Ontario M5S 3G5 Canada.

Berenbaum, May. Department of Entomology, University of Illinois, 505 S. Goodwin Ave., Urbana, Illinois 61801-3795 USA.

Bronstein, Judith L. Department of Ecology and Evolutionary Biology, University of Arizona, Tucson, Arizona 85721 USA.

Crews, David. Section of Integrative Biology, University of Texas, Austin, Texas 78712 USA.

Damuth, John. Department of Ecology, Evolution and Marine Biology, University of California, Santa Barbara, California 93106 USA.

Dingle, Hugh. Department of Entomology, University of California, Davis, California 95616 USA.

Fairbairn, Daphne J. Department of Biology, University of California, Riverside, California 92521 USA.

Fox, Charles W. Department of Entomology, S-225 Agricultural Science Center North, University of Kentucky, Lexington, Kentucky 40546-0091 USA.

Futuyma, Douglas J. Department of Ecology and Evolutionary Biology, State University of New York, Stony Brook, New York 11794-5245 USA.

Hedrick, Philip W. Department of Biology, Arizona State University, Tempe, Arizona 85287-1501 USA.

Holyoak, Marcel. Department of Environmental Science and Policy, University of California, Davis, California 95616 USA.

Lively, Curtis M. Department of Biology, Indiana University, Bloomington, Indiana 47405 USA.

Kramer, Donald L. Department of Biology, McGill University, 1205 Avenue Docteur Penfield, Montreal, Quebec H3A 1B1 Canada.

Mazer, Susan J. Department of Ecology, Evolution and Marine Biology, University of California, Santa Barbara, California 93106 USA.

McKenzie, John A. Center for Environmental Stress and Adaptation Research, Department of Genetics, University of Melbourne, Parkville, VIC 3052 Australia.

Messina, Frank J. Department of Biology, Utah State University, Logan UT 84322-5305 USA.

Myers, Judith H. Department of Zoology, University of British Columbia, Vancouver, British Columbia, V6T 1Z4 Canada.

Nunney, Leonard. Department of Biology, University of California, Riverside, California 92521 USA.

Orzack, Steven Hecht. Fresh Pond Research Institute, 64 Fairfield St., Cambridge, Massachusetts 02140 USA.

Pechenik, Jan A. Department of Biology, Tufts University, Medford, Massachusetts 02155 USA.

Pigliucci, Massimo. Department of Botany, University of Tennessee, Knoxville, Tennessee 37996-1100 USA.

Reeve, Jeff P. Department of Biology, Concordia University, 1455 de Maisonneuve Blvd. West, Montreal, Quebec, H3G 1M8 Canada.

Reznick, David. Department of Biology, University of California, Riverside, California 92521 USA.

Rhen, Turk. Laboratory of Signal Transduction, National Institute of Environmental Health Sciences, National Institute of Health, Research Triangle Park, North Carolina 27709 USA.

Roff, Derek A. Department of Biology, University of California, Riverside, California 92521 USA.

Sakai, Ann K. Department of Ecology and Evolutionary Biology, University of California, Irvine, California, 92697 USA.

Savalli, Udo M. Department of Entomology, University of Kentucky, Lexington, Kentucky 40546-0091 USA.

Schluter, Dolph. Department of Zoology, University of British Columbia, 6270 University Blvd., Vancouver, British Columbia, V6T 1Z4 Canada.

Tatar, Marc. Department of Ecology and Evolutionary Biology, Brown University, Box G-W, Providence, Rhode Island 02912 USA.

Thompson, John N. Department of Ecology and Evolutionary Biology, Earth and Marine Sciences Building, University of California, Santa Cruz, California 96064 USA.

Travis, Joseph. Department of Biological Sciences, Florida State University, Tallahassee, Florida 32306 USA.

Waser, Nickolas M. Department of Biology, University of California, Riverside, California 92521 USA.

Westneat, David F. Center for Ecology, Evolution and Behavior, School of Biological Sciences, 101 Morgan Building, University of Kentucky, Lexington, Kentucky 40506-0225 USA.

Williams, Charles F. Department of Biological Sciences, Idaho State University, Pocatello, Idaho 83209 USA.

Wilson, David Sloan. Department of Biological Sciences, State University of New York, Binghamton, New York 13902-6000 USA.

PART I

RECURRING THEMES

1

Nature and Causes of Variation

SUSAN J. MAZER
JOHN DAMUTH

The field of evolutionary ecology is at its core the study of variation within individuals, among individuals, among populations, and among species. For several reasons, evolutionary ecologists need to know the causes and the effects of variation in traits that influence the performance, behavior, longevity, and fertility of individuals in their natural habitats. First, to determine whether the conditions for evolution by natural selection of traits of interest are fulfilled, we need to know the degree to which the phenotype of a trait is determined by the genetic constitution (or genotype) of an individual and by the environment in which an individual is raised. Second, to predict whether and how natural selection will cause the mean phenotype of a trait in a population to change from one generation to the next, we must understand the ways in which an individual's phenotype (for this trait) influences its genetic contribution to future generations (i.e., its fitness). Third, to understand why the phenotype of a given trait influences an individual's fitness, we need to know how the trait affects an individual's ability to garner resources for growth or reproduction, to avoid predation, to find mates, and to reproduce successfully. Finally, to evaluate whether the phenotypic differences we observe among populations and species may represent the long-term outcome of evolution by natural selection, we must understand how different phenotypes perform under different environmental conditions. In sum, with an understanding of the causes and consequences of phenotypic variation within and among populations, we can detect evolutionary processes operating at a variety of ecological levels: within random-mating populations; within and among subpopulations distributed over a species' geographic range; and even among multispecies associations. These goals, however, require a clear understanding of the nature of phenotypic variation.

The aim of this chapter and the next is to illustrate that the richness of evolutionary ecology has increased in direct proportion to our understanding of the multiple causes of intraspecific phenotypic variation. Before reviewing these sources of variation, it is worth considering briefly a fundamental question: What kind of variation is evolutionarily significant?

Any trait whose phenotype is reliably transmitted from parents to offspring over multiple generations has the potential to evolve. The rules of Mendelian genetics tell us that traits whose phenotypes are determined by nuclear genes operating in an additive manner (i.e., alleles whose effects are independent of the genetic background in which they are expressed) are most likely to fulfill this criterion. Indeed, the importance of this kind of inheritance has been considered so great that the proportion of total phenotypic variance in a trait that is due to the additive effects of nuclear genes is given a special term: *heritability*.

But what about traits that are partly or largely

3

influenced by nonnuclear genes, nonadditive interactions among alleles or loci, the maternal environment, an individual's current environment, the interaction between an individual's genotype and its environment, or the age or developmental stage of the organism exhibiting them? Over the last decade, it has become clear not only that such traits are ubiquitous in natural populations, but that they can evolve as well. Unlike Mendelian traits, however, the evolutionary trajectory of traits subject to these effects can be difficult to predict. The rate or direction of their evolution can depend on the degree and nature of population structure, interactions among individuals, and nonrandom mating. In addition, genetic drift can take on special importance in promoting the differentiation of populations when nonadditive sources of variation are prevalent.

Consequently, for evolutionary ecologists interested in predicting evolutionary change in a particular trait in a given population, it is important not only to determine whether traits are transmitted from parents to offspring, but whether they are transmitted in a predictable fashion. An understanding of all potential sources of phenotypic variation in a trait helps to achieve this goal. This chapter reviews the kinds of variation that interest evolutionary ecologists and notes their relevance to particular evolutionary questions. In addition, causes of variation within individuals are introduced. Chapter 2 considers components of variation among individuals. Together, chapters 1 and 2 consider the causes and evolutionary consequences of variation in both unstructured and structured populations, and we highlight our view that new insights into the potential for natural selection to cause phenotypic change in wild species will come from the study of subdivided populations in which mating is anything but random (see also Nunney, this volume).

Modes of Expression of Variation

Predicting the outcome of natural selection on ecologically important traits depends on being able to determine the quantitative relationships between phenotype, genotype, and fitness. If the phenotypic variation in a trait is genetically based and correlated with the fitness of individuals in a way that can be expressed mathematically, then it is possible to predict the direction in which the trait should evolve (Fairbairn and Reeve, this volume). The pattern of variation expressed by a trait, however, has a strong influence on the quantitative and experimental methods used to detect and to measure this relationship.

Discrete Traits

Traits whose phenotypes can be assorted into distinct, nonoverlapping classes exhibit discrete variation. The phenotypic frequency distributions of such categorical or qualitative traits are usually depicted as histograms, where the phenotypic categories are indicated on the x-axis, and the number or proportion of sampled individuals identified in each category is indicated on the y-axis (figure 1.1). Often, such traits are simple Mendelian traits controlled by a single locus. While the color and surface texture of Mendel's peas and the wing color of the peppered moth *(Biston betularia)* provide excellent if time-worn examples for introductory biology students, many other discretely inherited traits provide evidence for the potential for (or the limitations of) natural selection to mold genetic variation.

When the frequencies of multiple morphs are high enough that their abundances cannot be accounted for by mutation alone, this is identified as a polymorphism of considerable evolutionary interest. In such cases, it would appear that natural selection may not be effective at eliminating an inferior genotype. Alternatively, either the morphs may be identical with respect to both survivorship and reproduction, or the morphs may enjoy equal fitnesses because where one, say, has an advantage in fertility, another has an advantage in survivorship. Other possibilities are that each morph enjoys a fitness advantage in a particular microenvironment (where the population occupies a heterogeneous habitat), that the relative performance of the morphs varies over time, or that the performance of a given morph is gender-specific. Detecting the processes responsible for the maintenance of such polymorphisms in natural populations is a challenge for evolutionary ecologists, and numerous studies have aimed to do so.

In many animals, for example, body color is a discrete trait that varies among individuals and affects their vulnerability to their predators. For example, the adder, *Viperus berus,* exhibits two dorsal color patterns: black and zig-zag. In a 6-year mark-recapture study of island and mainland pop-

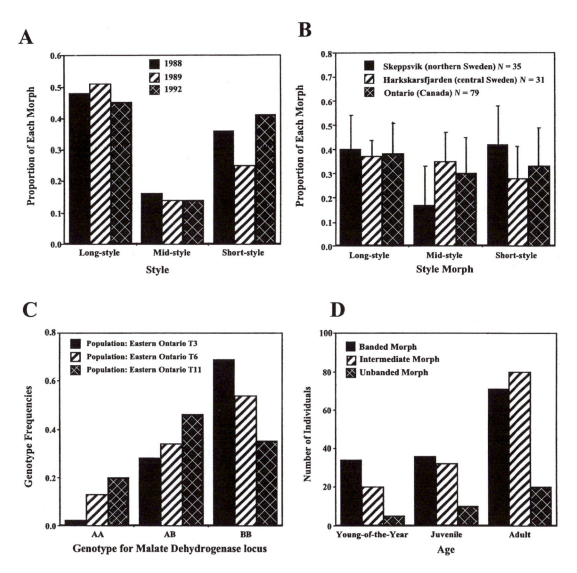

Figure 1.1 Examples of frequency distributions of discrete traits found in natural populations. (A) Frequencies of alternative style morphs in one eastern Ontario population of the tristylous perennial plant, *Decodon verticillatus* (Lythraceae). This population exhibits a marked deficiency of mid-style morphs that is persistent over time (Eckert and Barrett 1995). (B) Mean frequencies of style morphs in the tristylous perennial herb *Lythrum salicaria* (purple loosestrife; Lythraceae) sampled from populations in northern and central Sweden, where it is native, and in Ontario, where it has been introduced. The regional differences in morph frequencies are thought to be the result of both fitness differences between the morphs and historical factors (Ågren and Ericson 1996). *N* represents the number of populations sampled from each region. (C) Diploid genotype frequencies for the malate dehydrogenase locus sampled in three Eastern Ontario populations of *Decodon verticillatus* (Eckert and Barrett 1993). The genotypic frequencies do not differ significantly from Hardy-Weinberg expectations. (D) Frequencies of banded, intermediate, and unbanded body patterns in Lake Erie water snakes, *Nerodia sipedon insularum,* in different age classes. The relative abundances of the different morphs were not found to differ significantly among age classes, suggesting that there is no strong selection on body color patterns in this sample (King 1987).

ulations in and near Sweden, Forsman (1995) found that the relative performance of these two morphs differed between males and females. In male snakes, the black form suffered lower survival (apparently due to increased predation) than the zig-zag form, but in females, the pattern was reversed. Whether this pattern of gender-specific relative fitness accounts for the maintenance of the polymorphism is not certain. However, the relationship between body color and survivorship is clearly gender-specific, suggesting that the fitness of a given morph depends on the behavior of the individual exhibiting it. Individuals of the Lake Erie water snake, *Nerodia sipedon insularum,* also exhibit qualitative variation in body color pattern (King 1987). Differences in the relative abundances of banded and unbanded morphs between mainland and island populations, and among young-of-the-year, juvenile, and adult individuals, suggest that the probability of predation is both environment-specific and morph-specific.

Electrophoretic variation is less convenient to assess than visible polymorphisms, but several studies suggest that selection acts on allozymes[1] (or on closely linked genes) that appear to influence fitness through their physiological effects. Carter (1997) found that among populations of gilled tiger salamanders *(Ambystoma tigrinum nebulosum)* living in ephemeral ponds, there is a positive correlation between the frequency of homozygotes for the "slow" allele at the alcohol dehydrogenase locus (*Adh*-SS genotypes) and oxygen availability in the ponds. This geographic pattern is consistent with environment-specific selection under controlled conditions and with temporal patterns of selection within ponds: relative to the *Adh*-FF (Fast/Fast homozygotes) and *Adh*-FS (Fast/Slow heterozygotes) genotypes, *Adh*-SS genotypes are selected against under low-oxygen conditions.

Discrete traits controlled by one or a few loci are of special interest to evolutionists because population genetics theory allows the derivation of precise predictions of allele frequency changes from generation to generation. If the genetic basis of the phenotypic categories is well understood, allele frequencies can be precisely measured and tracked, providing a powerful tool for testing evolutionary hypotheses.

Quantitative Traits

In contrast to discrete traits are those for which the phenotype varies along a continuum and is deter-mined by alleles at multiple loci. The frequency distribution of such quantitative or "metric" traits is often illustrated as a histogram (as in the case of discrete traits), but here the x-axis is arbitrarily divided into convenient categories that, in sum, illustrate the shape of the distribution (figure 1.2). (Indeed, the width of the categories can have a very strong effect on the shape of the resulting distribution.) The y-axis shows the proportion or number of all sampled individuals whose "phenotypic value" for the trait falls within the boundaries of each category. Within populations, such polygenic traits often exhibit a normally distributed frequency distribution, although for many traits the raw values of the individuals' phenotypes must be transformed (e.g., log- or arcsine-transformed) to provide a scale that will generate a bell-shaped curve. Where the value of a trait must be expressed as an integer (e.g., the number of anthers per flower), the boundaries between the categories are less arbitrary, but the trait may nevertheless behave as a quantitative trait and be controlled by multiple loci.

The alleles and loci that influence a quantitative trait may each contribute additively to phenotype, whereby the change in the phenotypic value of a trait caused by an allelic replacement at a locus is independent of both the other allele at that locus and the genotypes expressed at other loci. Alternatively, alleles and loci may interact so that the effect on phenotype of an allelic change at a locus depends on the alleles or genotypes at this or other loci (i.e., dominance and epistasis; Mazer and Damuth, chapter 2, this volume). Only in the absence of dominance and epistasis, and where mating is random, can precise predictions be made concerning the similarity between parents and offspring or the phenotypic response to selection.

Many evolutionary ecologists focus exclusively on the evolution of quantitative traits simply because so many traits of known ecological importance and with strong effects on fitness are continuously distributed or known to behave as quantitative genetic traits. These include life-history traits such as germination time, juvenile growth rate, age of first reproduction, clutch size, number of reproductive bouts, longevity, fecundity, and fitness itself; behavioral traits such as running speed, foraging rate, and the rate of resource acquisition; physiological traits, such as photosynthetic rate or metabolic efficiency; and fitness-related morphological traits, such as size at birth and adulthood, bill size

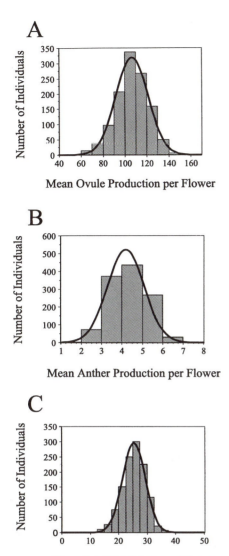

A

B

C

Figure 1.2 Frequency distributions of quantitative traits found in a greenhouse-raised experimental population of the annual salt marsh plant *Spergularia marina* (Caryophyllaceae) (Mazer et al. 1999). (A) The frequency distribution of the mean number of ovules produced per flower. (B) Distribution of the mean number of developmentally normal anthers produced per flower. (C) Frequency distribution of the mean area of all petals produced by a flower. $N = 1179$ individuals for all traits. The normal distribution corresponding to the mean and variance of each trait is superimposed on the histogram of the actual frequency distribution of phenotypic values.

in birds, wing length relative to body size, and the expression of secondary sexual characters such as tail length, flower size and color, and the size of color spots on bodies, wings, or petals.

One appealing attribute of normally distributed traits is that their statistical properties are well known (figure 1.3; Falconer and MacKay 1996). This means that it is a simple matter to ask whether population or species means differ significantly, potentially reflecting the direct outcome of evolution by natural selection (figure 1.4; Mazer and Lebuhn 1999; Reznick and Travis, this volume). It is also possible to conduct controlled breeding experiments from which to estimate the proportion of total phenotypic variance that is due to environmental versus genetic sources (e.g., Reznick and Travis, this volume). In addition, statistical methods have been derived to predict accurately the response to artificial and natural selection on normally distributed traits (Fairbairn and Reeve, this volume). Currently, agriculturalists and evolutionary ecologists alike use these methods both to estimate genetic and environmental causes of phenotypic variance and to predict how the mean phenotype of populations will (or should) change from generation to generation in response to artificial or natural selection.

The study and description of quantitative traits require some familiarity with the statistical parameters that can be measured given a sample of data representing a continuous variable. Any set of observations of a quantitative trait can be summarized by its mean (or average), variance, standard deviation, standard error of the mean, and coefficient of variation (among others) (figure 1.3). The mean and standard deviation are parameters reported in the units in which they are measured; the variance is a function of the square of these units and so is generally reported simply as a number. One problem with using these parameters to characterize a trait emerges when one aims to compare the variability of two or more traits. This is often a first step when attempting to predict which traits may most easily respond to natural selection. All else being equal, traits exhibiting high levels of phenotypic and genetic variation have a higher potential to undergo evolutionary change than those that do not. However, given that the units in which variation is measured are often trait-specific (with mass reported in grams, length in linear units, volume in cubed units, color in wavelengths, etc.), it is often meaningless to use, say, the standard devi-

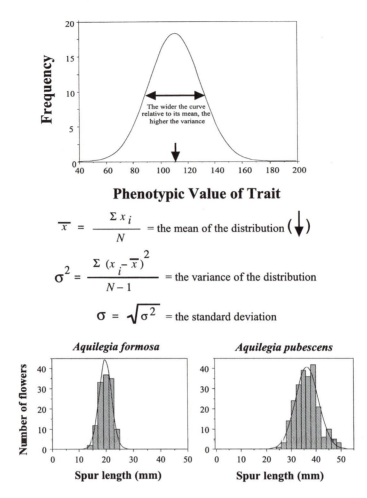

$$\overline{x} = \frac{\Sigma\, x_i}{N} = \text{the mean of the distribution } (\downarrow)$$

$$\sigma^2 = \frac{\Sigma\, (x_i - \overline{x})^2}{N-1} = \text{the variance of the distribution}$$

$$\sigma = \sqrt{\sigma^2} = \text{the standard deviation}$$

Figure 1.3 Statistical properties of quantitative traits. Top: The shape of a normal distribution, for a hypothetical trait whose values range between 40 and 180 units. The following parameters can be estimated for all quantitative traits measured on a sample of individuals representing a laboratory or field population: the sample mean, variance, and standard deviation. Bottom: Frequency distributions of floral spur length for two species of *Aquilegia* collected from field populations in the Bishop Creek Drainage (Inyo County, California). Flowers of *Aquilegia formosa* ($N = 129$) were sampled between 1950 and 2780 m in elevation; flowers of *A. pubescence* ($N = 236$) were collected at elevations of between 3400 and 3950 m. One flower per plant was sampled (Scott Hodges, unpublished data). The mean spur lengths of the two species differ significantly.

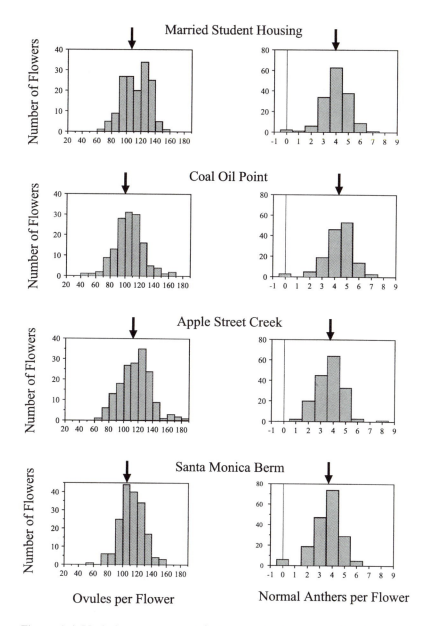

Figure 1.4 Variation among populations in the frequency distribution of two quantitative traits. The frequency distributions of ovule production and developmentally normal anthers per flower are shown for each of four greenhouse-raised populations of *Spergularia marina*. Each population was derived from seeds collected from a distinct wild population. There is significant variation among populations with respect to the mean number of ovules and anthers produced per flower under greenhouse conditions (Mazer and Delesalle 1996). Arrows indicate the phenotypic mean of the trait for each frequency distribution.

ation to compare the variability of two or more traits. Various solutions have been proposed to solve this problem, including the use of dimensionless parameters such as the coefficient of variation.

Because the statistical and mathematical properties of normal distributions are well known and tractable, theoretical models of the evolution of quantitative traits (for which a normal distribution is assumed) have been well developed. The success of these models in predicting the response of a trait to selection, based on its heritability and on the strength of selection, affirms the appropriateness of a quantitative genetic model of inheritance for many continuously distributed traits (Falconer and MacKay 1996). A summary of the statistical methods used to estimate the heritability of quantitative traits and to predict their evolutionary trajectories is beyond the scope of this chapter but is available in several recent volumes (Falconer and MacKay 1996; Roff 1997; Lynch and Walsh 1998). It is worth mentioning, however, what is perhaps the most useful theoretical and practical contribution to evolutionary ecologists by quantitative geneticists—the analysis of variance. The statistical analyses now conducted routinely to detect the causes of variation in quantitative traits and their effects on fitness and on each other, and to identify patterns of temporal and geographic variation would not be possible without Sir Ronald Fisher's invention of the analysis of variance (Fisher 1925).

Threshold Traits

A special case of discrete inheritance is represented by threshold traits. These are traits for which the phenotypes can be assigned to one of two or more distinct classes, but that are determined by alleles at multiple loci. The loci affecting a threshold trait each have a relatively small effect on some underlying trait that varies continuously, such as the concentration of a chemical product, the rate of development, or metabolic efficiency. Genotypes expressing less than some critical (or "threshold") value of this underlying trait will exhibit one phenotypic value, while those expressing more than the critical value will exhibit an alternative phenotypic state. In other words, there are discontinuities in the phenotypic expression of a continuous underlying variable.

Two-class (dimorphic) threshold traits may exhibit the expected (3:1) Mendelian ratios in the F2 generation produced by crosses among the F1 progeny of parents representing the two phenotypic classes, but the expected ratios do not appear when conducting backcrosses. The underlying trait on which the threshold is based is inherited as a quantitative trait and is termed the *liability*. The heritability of a discrete threshold trait is therefore estimated as the heritability of its underlying liability.

The relationships between a threshold trait and quantitative traits (such as size, fecundity, and fitness) are often nonlinear, and the use of highly controlled breeding designs and selection experiments to evaluate these relationships can strengthen conclusions concerning their inheritance and covariation (Roff et al. 1999). For example, Roff et al. (1999) used both approaches to examine the effects of a wing dimorphism (a threshold trait) on fecundity (a quantitative trait) in female sand crickets (*Gryllus firmus*). They found that short-winged flightless females have smaller flight muscles but higher fecundity than the long-winged morph, and that both wing morph and fecundity have a quantitative genetic basis. Moreover, artificial selection experiments confirmed the interpretation that there is an intrinsic negative correlation between wing length and fecundity. Selection favoring individuals with high (or low) fecundity resulted in a direct increase (or decrease) in fecundity and a correlated increase (or decrease) in the proportion of flightless females. Both wing morphs persist in natural populations because spatial and temporal heterogeneity in habitat persistence continually shifts the balance between selection for movement among patches (flight) and rapid population growth within patches (flightlessness).

As in the case of quantitative traits, dimorphisms or polyphenisms (where multiple phenotypic states exist) that behave as threshold traits can be subject both to strong environmental and genetic influences. The beetle *Onthophagus taurus* provides an example of dietary effects on phenotype (Moczek 1998). Males of this species are dimorphic with respect to horn development; the presence or absence of horns is determined by the quality and quantity of the food they receive from their parents. In other species, dimorphisms are strongly associated with a highly heritable trait. Quantitative genetic analyses of juvenile hormone esterase in the crickets *Gryllus firmus* and *G. assimilis* indicate that this enzyme (which degrades juvenile hormone) is highly heritable, responds rapidly to selection, and is a strong determinant of wing morph (Fairbairn and Yadlowoski 1997; Roff et al. 1997; Zera et al. 1998).

Sexually Dimorphic Traits

In organisms with separate sexes, it is common to observe that many traits differ between males and females. Sexually dimorphic traits often play a role in attracting or competing for mates or in raising offspring. Where behavior and its concordant risks of mortality are highly gender-specific, one may expect traits that influence fitness to evolve differently in males and females. Such traits may include body color, body mass, pheromone production, and mating calls; the expression of secondary sexual traits such as physical ornaments, flower size, or nuptial gifts; and parental care (Andersson 1994; Fairbairn and Reeve, this volume; Savalli, this volume). Dimorphisms may also evolve where the sexes differ in other social behaviors or in habitat preferences. In either case, gender-specific traits are usually interpreted as being the direct or indirect result of gender-specific patterns of sexual or natural selection.

Traits favored in males due to their positive effects on mating success are not necessarily favored among females, and vice versa. Where female choice is a major component of male reproductive success, sexual selection favoring elaborate courtship behaviors or visually attractive traits will be restricted to males. Where the outcome of direct competition among males determines their reproductive success, sexual selection favoring large size or aggressive behavior may be stronger in males than in females.

Similarly, where there are differences in the behavior of males and females unrelated to mating, the optimum phenotype may differ between the sexes. For example, where males spend more time foraging than females, natural selection favoring cryptic coloration may be stronger among the former. Traits for which the phenotype favored in one sex is actually selected against in the other sex are termed *sexually antagonistic* characters (Rice 1984). When the direction of selection is gender-specific, if there is a genetic mechanism (such as X-linkage) that permits the expression of a trait to be restricted to one sex, natural selection can result in significant differences between the sexes in either discrete or quantitative traits.

Hundreds of cases of sexual dimorphism have been documented in animals and in plants. Recent studies provide evidence that the dimorphism is the result of gender-specific patterns of sexual or natural selection (e.g., Fairbairn and Reeve, this volume; Savalli, this volume). For example, Grether (1996) reports evidence that the red wing spots restricted to male rubyspot damselflies are the result of selection operating during competition among males for mating territories, rather than the result of female choice. Bisazza and Marin (1995) argue that the sexual size dimorphism observed in the eastern mosquitofish *Gambusia holbrooki,* in which the males are smaller than the females, is the result of the mating advantage enjoyed by relatively small males during most of the reproductive season. Gwynne and Jamieson (1998) suggest that the evolution of the huge mandibles in male alpine wetas (*Hemideina maori*, Orthoptera) of New Zealand represent "cephalic weaponry" that have evolved in response to male-male battles for access to females. Similarly, the body size dimorphism in marine iguanas *(Amblyrhynchus cristatus)* appears to be the result of sexual selection favoring larger size more strongly in males than in females (Wikelski and Trillmich 1997). Above a given body size, female marine iguanas allocate resources to additional egg production rather than to increased growth, although both sexes grow to be larger than the apparent naturally selected optimum. Balmford et al. (1994) provide comparative data suggesting that sexual dimorphism in wing length among 57 species of sexually dimorphic long-tailed birds is the result of natural selection occurring concurrently with sexual selection on tail length. They argue that the secondary evolution of the wing size dimorphism serves to offset the functional costs incurred by the sexually selected tails.

Evidence for dimorphisms due to sexual selection is not restricted to animals with complex social interactions. Dioecious plant species also exhibit marked sexual dimorphism in traits related to mating success (Delph et al. 1996). Gender-specific sexual selection may be expected in species in which male success in delivering pollen to females is mediated by pollinators. Among male plants, reproductive success (or at least pollen removal) often increases linearly with visitation by pollinators; males benefit from multiple visits per flower, as only a fraction of their pollen is removed by any single visit. By contrast, reproductive success by females is often limited not by pollen but by other resources; the result is that relatively few pollinator visits are required to achieve maximum seed set, and females do not benefit from investing in attractants beyond those necessary to achieve this maximum.

The evolutionary outcome of this disparity is that traits favored in males (highly conspicuous, relatively large, or profusely flowered inflorescences) differ from those favored in females (smaller- or fewer-flowered inflorescences). Accordingly, the smaller but more numerous flowers in the inflorescences of male relative to female *Silene latifolia* (Meagher 1992) may be the result of competition among males to attract pollinators. Similarly, the more numerous flowers of male relative to female plants or inflorescences of *Wurmbea dioica*, *Salix myrsinifolia-phylicifolia*, and *Ilex opaca* appear to be due to gender-specific selection. Analogous to Balmford et al.'s (1994) observations concerning the evolution of sexually dimorphic wing size in birds, the dimorphism in sexually selected traits in plants seems to result in the evolution of gender-specific life-history traits. Male and female plants have also been found to differ in growth rates, phenology, frequency of reproduction, and plant height. These differences may represent the outcome of selection in females, which usually exhibit a much higher absolute investment in reproduction (due to fruit and seed production) than do males.

Causes of Variation and Their Evolutionary Consequences

Regardless of the kind of variation exhibited by a trait, predicting its evolutionary trajectory requires knowledge of its environmental and genetic basis. In the remainder of this chapter, we consider the causes and evolutionary consequences of phenotypic variation within individuals. Then (in chapter 2), we consider the causes and consequences of variation among individuals in random-mating, unstructured populations, and we describe recent conceptual advances concerning the consequences of population structure. In structured populations, genotypes do not interact at random, and the relationship between genotype and fitness can be highly sensitive to the identity of the genotypes with which an individual interacts (Nunney, this volume; Wilson, this volume).

Variation within Individuals

Ontogenetic Variation   Ontogenetic variation is the component of phenotypic variation in a trait expressed as an individual ages or progresses through sequential developmental stages. Ontogenetic variation can be expressed in two ways. First are age-related changes in an individual's phenotype. Traits such as body size, growth rates, hormone production, pigmentation, metabolic rate, and behavior are usually age-dependent. Second are changes in the phenotype exhibited by sequentially produced or "modular" organs. For example, the size of sequentially produced leaves, flowers, or fruits may change over time. Similarly, the size or number of seeds or eggs produced in successive fruits or clutches may change over time.

For traits that exhibit either type of ontogenetic variation, genetic and temporal sources of variation will be confounded unless special measures are taken to separate their effects. If ontogenetic variation comprises a sufficiently large proportion of total phenotypic variance, the proportion of phenotypic variance that is genetically based may be obscured to the point of being undetectable unless ontogenetic sources of variation are taken into account.

Consider a population of individuals for which a trait's phenotype varies over time (figure 1.5). When sampling this population, the magnitude of phenotypic variance may depend on the age structure of the population sampled. If a cohort of individuals is measured repeatedly over time, and if all individuals change their phenotype in the same way at the same rate, then the phenotypic variance will not change over time (figure 1.5A). On the other hand, if individuals differ in their ontogenetic trajectories, the phenotypic variance exhibited by the cohort may either increase or decrease (figures 1.5B–D: phenotype versus time, with lines converging or diverging over time). In this case, the magnitude of phenotypic variation detected in a population will depend on both the pattern of ontogenetic change exhibited by its members and the ages of the individuals included in the sample. Where different genotypes exhibit different patterns of ontogenetic variation, the amount of intergenotypic variation may also vary over time (figures 1.5B–D).

An example of this phenomenon is observed in floral traits among successively produced flowers of a short-lived, self-fertilizing, annual species in

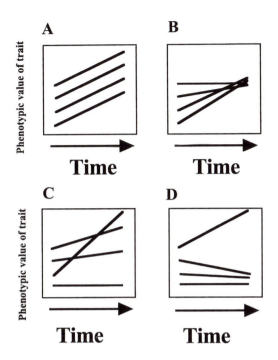

A

Phenotypic value of trait

Time

B

Time

C

Phenotypic value of trait

Time

D

Time

Figure 1.5 Alternative patterns of ontogenetic variation. Each line represents the phenotypic value of an individual (or the mean value for a genotype) as the individual ages or as successively produced organs are sampled. (A) All individuals (or genotypes) change in the same way over time. (B) Individuals (or genotypes) differ in their ontogenetic trajectories, the result being a temporal decrease in total phenotypic (or genotypic) variance. (C) Individuals (or genotypes) differ in their ontogenetic trajectories, the result being an increase in total phenotypic (or genotypic) variance over time. The crossing of lines represents a significant individual × time interaction, whereby the relative phenotypic values of individuals change over time. (D) Individuals (or genotypes) differ in their ontogenetic trajectories, but no changes in rank occur.

the Carnation family, *Spergularia marina* (Mazer and Delesalle 1996). Under uniform greenhouse conditions, offspring representing multiple maternal families from each of four wild populations produced flowers that were sampled over a 5-week period. The numbers of petals, anthers, and ovules in each flower were recorded to determine whether ontogenetic variation within individuals is so high that it masks genetically based variation among maternal lineages or among populations.

In each population, the mean phenotypes for all of these traits depended on the date on which they were sampled; ontogenetic changes in these traits contributed significantly to phenotypic variance. More important, analyses of variance found that the statistical significance of differences between populations was also varied among weeks (figure 1.6). As a result, conclusions as to whether the populations have differentiated with respect to the mean values of these traits depend on when flowers are sampled. In the pooling of data from all dates, populations differed with respect to the mean values of these floral traits, whether or not sampling date was controlled statistically. So, while ontogenetic variation in these traits is substantial, it does not completely mask genetic differences among populations.

When sampling a population that exhibits ontogenetic variation in a particular trait, the question of interest will dictate the kind of sampling protocol to use. If one aims to determine the absolute range of phenotypes expressed by a population, one should sample a wide range of individuals and times. To measure the genetic component of phenotypic variation among genotypes or populations, however, it is essential to control for the time at which individuals are sampled. This control can be done empirically, by sampling individuals at the same developmental stage, or statistically, by partitioning variance into components due to developmental stage and to family. Finally, to determine whether the ontogenetic trajectories are themselves genetically based, one must sample multiple individuals in multiple families (or genotypes) to determine whether there is a significant interaction between family membership (or genotype) and developmental stage. In other words, do families (or genotypes) differ with respect to the relationship between developmental stage (time) and phenotype, represented by the slopes or forms of the lines in figures 1.5 and 1.6?

Ontogenetic variation can be seen as a nuisance to studies of genetic variation, or it can itself be the subject of study. The observation that genotypes differ with respect to their pattern of ontogenetic variation suggests that the pattern itself may be subject to natural selection. One example of ontogenetic variation that appears to evolve by natural selection is the degree of abdominal spine elongation exhibited during ontogeny by dragonfly larvae (*Leucorrhinia dubia*; Arnqvist and Johansson 1998). Larvae exposed experimentally to cues

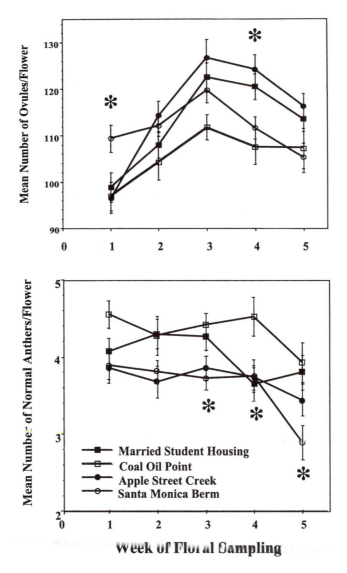

Figure 1.6 Ontogenetic variation in ovule and petal production per flower over a 5-week period exhibited by four populations of *Spergularia marina* near the University of California, Santa Barbara (Married Student Housing, Coal Oil Point, Apple Street Creek, and Santa Monica Berm; Mazer and Delesalle 1996). Each line represents the ontogenetic trajectory exhibited by the offspring of 7–10 maternal families sampled from a wild population and raised in the greenhouse. Asterisks indicate those weeks in which significant differences among populations could be detected. The ability to detect apparent genetically based differences among populations depends on the week during which flowers were sampled.

produced by fish predators produce harder and longer spines than control larvae. Consistent with this inducible defense, populations of larvae that are sympatric with fish predators exhibit either accelerated or longer-duration spine development than those in habitats free of predators. The onto-genetic trajectory for the degree of spine develop-ment appears to have evolved in response to selec-tive pressures imposed by predators. The evolution of developmental trajectories or "norms of reac-tion" is treated in detail by Schlichting and Pigli-ucci (1998; see also Pigliucci, this volume).

Somatic Mutations   In spite of the fact that all cells in an organism are derived from a single cell, so-matic mutation makes it possible for an adult to produce gametes with novel alleles not found in the zygote from which the adult developed. So-matic mutations are those that occur in an individ-ual's cells that possess a full complement of chro-mosomes. If these mutations occur in a cell lineage that participates in the production of the germ line, they can be passed on to the next generation.

Somatic mutations are of particular importance in long-lived organisms with indeterminate growth, such as colonial invertebrates and plants (Salo-monson 1996). For example, if a mutation occurs in meristematic tissue that develops into a lateral branch of a tree, gametes derived from the flowers produced by this branch will possess this mutation. Moreover, if the somatic mutation is beneficial and results in relatively high survivorship or growth rate of the tissue bearing it, the mutation may be carried by the vast majority of an individual's ga-metes. To illustrate, imagine that all the cells in a lateral branch express a somatic mutation that re-sults in the elevated production of a secondary chemical compound that deters herbivores. Leaf and flower production on this branch may be much higher than elsewhere on the plant, and the result may be a disproportionate production of ga-metes bearing the mutation.

Given that every shrub and tree supports doz-ens or hundreds of sites of rapidly dividing meriste-matic tissue (i.e., at every growing branch tip), it seems likely that the millions of gametes they pro-duce over a 100-year lifespan will include those from germ lines that carry somatic mutations. The gamete pool produced by such plants is itself a population that has evolved within a single genera-tion.

The studies cited in this introduction have by and large considered the evolutionary significance of phenotypic variation within individuals and populations that occupy or are considered to oc-cupy relatively homogeneous environments. In the following chapter, we shift the focus to variation among individuals and populations, and we con-sider the evolutionary consequences of environ-mental heterogeneity and of nonrandom interac-tions among genotypes.

Note

1. Enzymes that differ in electrophoretic mobil-ity as a result of allelic differences at a single locus.

2

Evolutionary Significance of Variation

SUSAN J. MAZER
JOHN DAMUTH

In this chapter, we consider the causes and evolutionary consequences of phenotypic variation among individuals in random-mating, unstructured populations. We focus on quantitative traits because they illustrate well the difficulties in determining genetic versus environmental causes of phenotypic variation. This is an important step when aiming to make precise predictions concerning phenotypic change in traits that influence individual longevity and reproduction. We also describe several recent conceptual advances concerning the evolutionary significance of population structure.

Variation among Individuals

Variation among individuals in quantitative traits is typically measured as total phenotypic variance (the variance estimated from all phenotypes measured in a population). This parameter has three convenient properties; first, total phenotypic variance can be partitioned into components that are themselves variances attributable to different causes, and second, these components are additive, summing to the total phenotypic variance (figure 2.1). These attributes allow one to identify and to compare the magnitudes of different sources of variance to determine their relative evolutionary and ecological importance. Third, even when total phenotypic variance in a trait is high, resulting in a high degree of overlap among the means of differ-

ent populations or genotypes, the statistical control of one or more variance components often permits the detection of significant differences among phenotypic means.

The proportion of total phenotypic variance accounted for by genotypic versus environmentally induced causes determines the degree of resemblance between parents and offspring or between other types of relatives (e.g., clonal replicates, siblings, half-siblings, maternal lineages). Given that a high degree of resemblance among relatives is a criterion for natural selection to cause evolutionary change, a major goal of evolutionary ecologists is to measure these variance components in wild populations.

Genetic Variation

The presence of genetic variation in quantitative traits is a prerequisite for evolutionary change, but genetic variation in most traits is a capricious parameter. Its expression depends on environmental conditions, and its magnitude changes as populations evolve, thereby influencing a trait's future potential to evolve. Another difficulty is that, in addition to being influenced by the expression of nuclear genes, the expression of quantitative traits is influenced by current environmental conditions, the condition of individuals in preceding generations, the degree of inbreeding, cytoplasmically inherited genes, the genetic composition of inter-

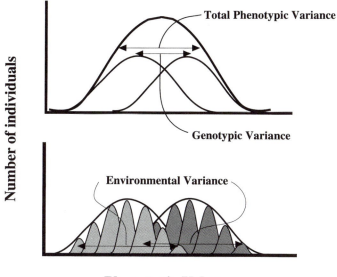

Phenotypic Value

Figure 2.1 Total phenotypic variation and some of its causes. A frequency distribution representing the phenotypic values of many individuals sampled from a population may itself comprise internal frequency distributions representing subsets of the sample. Here, the sample population includes representatives of two genotypes (each represented as its own frequency distribution in the top panel). Each genotype, in turn, is represented by multiple frequency distributions (bottom panel), where each of these represent the phenotypes measured in a distinct environment. The total phenotypic variance (represented schematically by the top-most arrow) is equal to the variance among genotypic means plus the variance among the means associated with the different environments (where environmental variance is estimated while controlling for genotype). A significant difference between genotype means may be more easily detected (in spite of the high overlap between the genotypic frequency distributions) when environmental variance is controlled statistically (e.g., through an analysis of variance).

acting individuals, and nonheritable interactions among alleles within and between loci. Consequently, knowing that a trait is under quantitative genetic control is not sufficient to predict its potential to evolve; one needs to be able to assess the degree to which phenotype (and its fitness effects) is transmitted to future generations.

The similarity between parents and offspring is measured by the *heritability* of a trait, which varies between 0 and 1. The heritability of a trait is of interest in evolutionary studies because its value is directly proportional to the magnitude of phenotypic change in the trait that is expected to occur in one generation in response to a given strength of selection. This principle is represented by a well-known equation, response to selection = heritability × selection differential:

$$R = h^2 S \qquad (2.1)$$

where the selection differential *[S]* is the difference between the mean of the population before selection and the mean of the population selected to contribute to the next generation, where the selected individuals represent one tail of the trait's frequency distribution. Consequently, if we know *S*, then the heritability of a trait allows us to predict the absolute amount of phenotypic change to expect (*R* represents the change in the phenotypic mean between generations). Estimating the heritability of a trait depends on being able to partition total phenotypic variance into its components. Heritability is most simply expressed as the ratio of additive genetic variance (see next section) to total phenotypic variance. Any source of variance in a trait other than additive genetic variance appears in the denominator of this fraction, reducing the trait's heritability, and decreasing the rate at which evolutionary change can occur due to natural or to artificial selection.

Phenotypic variance due to nuclear genetic differences among individuals can itself be divided into two components: additive and nonadditive variance, the latter being divided into dominance and epistatic variance. The transmission of cytoplasmic genes through either the maternal or the paternal lineage is a separate, non-Mendelian cause of similarity between parent and offspring. Above and beyond these sources of variation are recently discovered "epigenetic" phenomena, which include variation due to the behavior of nuclear genes that cannot be accounted for by Mendelian rules. In the following sections, we consider these components of variation in more detail.

Additive Genetic Variance   Additive genetic variance is the component of genotypic variance due to the additive effects of alleles on the phenotype of the individuals bearing them. One quantitative definition of additive genetic variance is as follows: If an individual mates with a number of randomly selected individuals in a population, its "breeding value" for a trait is defined as twice the average deviation of the individual's progeny from the population mean (the deviation must be multiplied by 2 because an individual contributes only half its genes to each offspring). The additive genetic variance in the trait is then defined as the variance in the breeding values of all members of a population. This definition underscores the fact that additive genetic variance is not difficult to measure directly in experimental populations (or in natural ones, if one can assume random mating *and* successfully raise the offspring under relatively natural conditions).

Nonadditive Genetic Variance   Nonadditive genetic variance can be defined as the component of phenotypic variance that cannot be predicted from the combined additive effects of a genotype's collective nuclear alleles. This component of genotypic variance can itself be divided into components due to dominance and epistatic variance.

Dominance variance   Dominance is identified when alleles at a single locus interact to produce a phenotype that would not be predicted on the basis of the average effects of these alleles when acting alone. For example, consider a pair of alleles (*A* and *a*) acting additively, where the *AA* genotype produces flowers with 6 petals and the *aa* genotype produces flowers with 4 petals. In a purely additive system, *Aa* individuals will produce 5 petals per flower. The mean petal number of a population composed of equal numbers of the three genotypes would be 5, and the variance among the three genotypic means would be 1.0. These values differ from the mean and variance of a genotypically identical population in which *A* is completely dominant over *a*. In the latter, where the three genotypes exist in equal proportions, the mean would be 5.33 petals ([6 + 6 + 4]/3) and the variance among the genotypic means would be 1.33. Two populations of identical genotypic composition will therefore exhibit different genetic variances if they differ in the degree of dominance. The difference between the two variances (0.33) is the variance due to dominance.

Where dominance variance exists, the similarity between parents and offspring is less reliable than where there is no dominance. In the example above, in an additive system, a parent with 4 petals *(aa)* might produce offspring with 4 or 5 petals, depending on the genotype of its mate. In contrast, when there is dominance, a parent with 4 petals might produce offspring with 4 or 6 petals.

Epistatic variance   Epistatic interactions are those in which the phenotype or fitness of a genotype at one locus depends on the genotype at one or more other loci (tables 2.1 and 2.2). Epistasis has the potential to constrain the rate at which natural selection eliminates allelic variation in populations when the fitness differences among two-locus genotypes are such that alternate alleles at each locus are unlikely to be purged. Although variance due to epistatic interactions can be measured in careful breeding experiments, most simple breeding designs used by evolutionary ecologists do not distinguish between dominance and epistasis as sources of nonadditive genetic variance.

Cytoplasmic (nonnuclear) variance   Cytoplasmic organelles and the genes they express are not inherited in a Mendelian fashion: Maternal and paternal cytoplasmic genes contribute unequally to an offspring's phenotype. Some traits are inherited through the maternal but not the paternal cytoplasm; the result is similarities between mother and offspring that greatly exceed those between father and offspring. Given that more plastids, mitochondria, and other organelles are generally transmitted

Table 2.1 Epistatic interactions, where the variance among genotypes at each locus depends on the allele frequencies at the alternate locus.[1]

		Locus 1		
	Genotypes	*AA*	*Aa*	*aa*
Locus 2	*BB*	11	10	9
	Bb	11	12	9
	bb	9	9	9

[1]The cells in this table show the phenotypic values of a hypothetical meristic trait (e.g., mean internode length, in mm) determined by diploid genotypes at two loci. When locus 1 expresses the *AA* genotype, locus 2 exhibits dominance, where the *B* allele is dominant over the *b* allele. When locus 1 is heterozygous, locus 2 exhibits overdominance, and the phenotypic value of the double heterozygote exceeds that of either homozygote. When locus 1 expresses the *aa* genotype, genotypic variation at locus 2 has no effect on phenotype. Similarly, when locus 2 is *BB*, locus 1 exhibits additivity. When locus 2 is *Bb*, locus 1 exhibits overdominance. And when locus 2 is *bb,* genetic variation at locus 1 has no effect on phenotype. The composition of a population at one locus will influence the apparent mode of genetic expression of alleles at the other locus.

through the cytoplasm of the egg than through that of pollen or sperm, this asymmetry is not unexpected. A general feature of cytoplasmic inheritance is that the direction in which a cross is made has a strong effect on the phenotype of the resulting offspring. Hybrid offspring produced by a cross in which the mother carries cytoplasm A and the father carries cytoplasm B may appear very different from those produced by the reciprocal cross. Although the opportunity for uniparental inheritance through the maternal genome is apparently greater than through the paternal lineage, uniparental inheritance is not limited to maternal cytoplasmic genes. Many cases of the inheritance of paternal nonnuclear genes have also been recognized, particularly in conifers.

Cytoplasmic genes (expressed by plastid DNA) have been found that influence a variety of ecologically and evolutionarily important traits, including chlorophyll production; heat and cold tolerance; herbicide, disease, and antibiotic resistance; ATP synthesis; and responses to phytotoxins (Mogensen

Table 2.2 Epistatic interactions, where alternate genotypic combinations at two loci generate genotypes of equal fitness.[1]

		Locus 1		
	Genotypes	*AA*	*Aa*	*aa*
Locus 2	*BB*	6	2	1
	Bb	2	6	2
	bb	1	2	6

[1]The cells in this table show the fitness values (e.g., the number of viable offspring produced) for the nine genotypes in a two-locus/two allele system. Here, the composition or allele frequencies at one locus will influence which allele is favored by selection at the other locus. The direction and strength of selection on alleles at one locus can be determined by the "genetic background" (i.e., the allelic frequencies at other loci) in which they are expressed.

1996). So it is not surprising that the uniparental inheritance of these genes can result in high similarity between one parent and its offspring with respect to fitness.

In sexually reproducing organisms, cytoplasmically inherited organelles that are uniparentally inherited behave as if they exhibit clonal reproduction because they are faithfully transmitted through the maternal (or paternal) lineage (usually with negligible rates of recombination, with the exception of plant mitochondria). If cytoplasmic genes strongly influence fitness, selection among maternal (or paternal) lineages can greatly increase the frequencies of those expressing high-fitness alleles.

Cytoplasmic inheritance of a trait can constrain the rate of evolution of nuclear genes that influence the same trait. As is the case for all sources of phenotypic variation, cytoplasmic genetic variation contributes to the denominator of the heritability equation. If the phenotypic variance in a trait is more strongly influenced by cytoplasmic genes than by additively expressed nuclear genes, then the heritability of the trait (defined as the proportion of genetic variation due to nuclear additive genetic variance) could be extremely low. This does not mean that the trait will not evolve; rather, it means that most of the evolutionary change will be mediated through cytoplasmic genes. Although cytoplasmic alleles are free of dominance effects because they are inherited as haploid genomes, their effects on phenotype are not necessarily fully additive. Cytoplasmic genes can interact with the additive and dominance effects of nuclear genes to contribute even more to phenotypic variance.

In plants, in addition to being affected by cytoplasmic genes, the growth and development of an embryo can be strongly influenced by the genetic constitution of its nutritive tissue, the endosperm, which includes multiple doses of the maternal genome. The expression of genes in the endosperm can therefore have similar effects on inheritance as cytoplasmic genes.

Epigenetic Variation: Genomic Imprinting and Epimutations   Genomic imprinting is the phenomenon in which an allele is differentially expressed depending on the parent from which it is transmitted (see Ohlsson et al. 1995 for review). The basis for this sex-specific expression is some kind of chemical "marking" or imprinting that determines the fate of the allele; usually, the function of the imprinted allele is disrupted (i.e., it is "silenced"),

although imprinting may also result in changes in the timing of gene expression. Imprinted genes are characterized by DNA methylation, where methyl groups are incorporated into the DNA of the imprinted gene.

Genomic imprinting is considered an "epigenetic" phenomenon because it occurs independently of the DNA sequence represented by the imprinted genes. This is not to imply that imprinting itself does not have a genetic basis, but it is not genetic variation in the traits themselves that causes their asymmetric inheritance.

In eutherian mammals, genomic imprinting has been observed to influence fetal growth (with paternally expressed genes enhancing and maternally expressed genes suppressing growth), muscle development, ectodermal tissue development, brain growth, and behavior. Epigenetic inheritance involving methylation has also recently been reported in plants (Cubas et al. 1999).

Our current lack of understanding of the genetic causes and inheritance of genomic imprinting has prevented the development of population genetics theory capable of offering clear predictions concerning the evolution of imprinted genes. The degree to which imprinted genes break Mendel's laws clearly provides a vexing challenge to theoreticians. The importance of imprinted genes in the expression of diseases promises to generate much empirical research on this mode of gene expression.

Interconversion of Nonadditive and Additive Genetic Variance   Although epistatic interactions may influence the rate and direction of response to selection on a trait (tables 2.1 and 2.2), the variance due to gene interactions plays little direct part in response to selection. Because of recombination, interactions among loci are broken up during reproduction and thus do not contribute to reliable, predictable similarities between parents and offspring—epistatic variance is nonadditive.

However, there are circumstances in which nonadditive variance can be converted into additive genetic variance and can contribute to evolutionary phenotypic change. Consider a drastic reduction in population size. Traditionally, it has been thought that a population passing through a size bottleneck should experience reduced evolutionary potential because of the loss of genetic variation. By chance, much allelic variation has been lost, and therefore, one would expect additive genetic variance to decrease for many traits. However, experiments show that passing through a bottleneck can substantially *increase* the total amount of additive genetic variance in a population (Bryant and Meffert 1993; Cheverud et al. 1999). The most likely explanation for this surprising result is that the loss of alleles—and drastic changes in allele frequencies—caused by the bottleneck has converted epistatic variance to additive genetic variance. In other words, although different genotypes at one locus may have had similar effects on fitness in the original genetic background, change in the genetic background due to a bottleneck may result in significant differences in fitness among these same genotypes.

A close look at table 2.3 illustrates how this may occur where strong epistatic interactions exist between two loci. In this case, when all combinations of genotypes at two loci are present in similar frequencies in the population, phenotypes may not be easily predictable from the genotypes at, say, the A locus—it is difficult to predict the effect on phenotype (and hence fitness) of bearing a particular allele at this locus. Considerable variance in the trait is nonadditive, and heritability is relatively low. However, if, owing to the population bottleneck (or to drift or selection), the b allele is lost from the population, all individuals will exhibit the BB genotype at the B locus, and genotype combinations (and their resulting phenotypes) in the lower two rows of the table will not occur. Now, the variation at the A locus has a highly predictable effect on fitness. Epistasis between these two loci has effectively disappeared, and instead, the A-locus alleles are behaving additively. The heritability of variation at the A locus has increased, even though the total genetic variation in the genome may have decreased (due to losing the b allele). Selection can now be effective in changing frequencies at the A locus, when it wasn't before.

Environmental Variation

For many traits, an individual's phenotype is highly sensitive to the quality of its current physical or biological environment; phenotypic changes exhibited by an individual or genotype in response to the environment are identified as phenotypic plasticity (Pigliucci, this volume). Different traits respond differently to environmental variation, so the evolutionary consequences of environmental variation are trait-specific. For environmentally sensitive traits, environmental heterogeneity inflates phenotypic, but not genetic, variance, thereby reducing the traits' heritabilities. By contrast, all else being equal (e.g., the magnitude of genetic variance and the strength of selection), traits whose expression is unaffected by environmental variation are more likely to exhibit an evolutionary response to selection.

Abiotic Environmental Variation   Numerous experiments have found that morphological and fitness-related traits are sensitive to physical (abiotic) conditions, including water and light availability, temperature, and substrate texture, and that traits differ in their sensitivity to these factors (Pigliucci, this volume). For example, a phenotypic response to shading is exhibited by *Iris pumila*, which re-

Table 2.3 Schematic example.[1]

| | | Locus 1 | | |
	Genotypes	*AA*	Aa	aa
Locus	*BB*	8.0	6.0	4.0
	Bb	2.0	4.0	4.0
	bb	1.0	1.0	3.0

[1]Epistatic contributions to nonadditive genetic variance are converted to contributions to additive genetic variance by loss of an allele at the epistatic locus. Each cell represents the phenotypic value of the combination of genotypes at two loci. When all genotypes at the B locus are present, there is high epistatic variance among genotypes at the A locus. Assuming that the phenotypic value is positively correlated with fitness, the effect of losing the b allele is that the A locus now behaves additively, and the result is strong selection favoring the A allele.

sponds to the low light levels in a woodland under-story (relative to an open dune site) by producing larger leaves. This type of response may be tentatively interpreted as the adaptive outcome of natural selection if it is clear that selection favors in each environment the phenotype that is usually expressed there. Phenotypic selection gradient analysis (Fairbairn and Reeve, this volume) is a useful tool for this task. For example, in *I. pumila,* the change in leaf size associated with light availability is consistent with the phenotypic selection gradients, which detected stronger selection favoring large leaves in the woodland than in the dune habitat (Tucic et al. 1998).

Developmental responses to temperature are also common. In two species of montane lizards of southeast Australia *(Bassiana duperreyi* and *Nannoscincus maccoyi),* the temperature at which eggs are incubated has a strong effect on a variety of traits expressed by newly hatched lizards. Eggs incubated at temperatures simulating the maternal uterus had higher viability than those cultivated at nest temperature. This result is interpreted as evidence for the selective advantage of vivipary, in which eggs are incubated within the uterus instead of externally in a nest.

Biotic Environmental Variation   One well-studied cause of biotically induced environmental variation is the presence of predators, damage that simulates predators, or experimentally manipulated cues (e.g., scents) that would in a natural situation indicate the proximity of a predator. For example, blue mussels *(Mytilus edulis)* cultivated in proximity to their starfish predators are smaller, have thicker shells, and have more meat per shell volume than mussels cultured in the absence of predators (Reimer and Tedengren 1996). In plants, attack by herbivores similarly induces the production of antiherbivore defenses (Karban and Baldwin 1997). Another major cause of environmental variation associated with the biotic environment is the amount and quality of food. For example, the amount of food received by nestlings affects juvenile condition (and subsequent survival) in the collared flycatcher *(Ficedula).* The population density of conspecifics often has a strong influence on life-history traits in plants and animals, apparently because population density influences per capita resource availability. For example, as population density is increased experimentally in wild radish *(Raphanus sativus),* mean survivorship, size at reproduction, plant biomass, ovules

per flower, mean individual seed mass, lifetime flower and fruit production, and lifetime maternal fecundity and yield decline, while mean petal area, pollen production, and pollen size remain constant (Mazer and Schick 1991a; Mazer and Wolfe 1998).

In animals, the apparent increase in stress that accompanies high population densities often promotes rapid development, as is seen among the larvae of the moth *Mamestra brassicae* (Goulson and Cory 1995). These larvae exhibit increased growth rates, a higher degree of melanization, smaller size at molting, and greater susceptibility to viral infection at high relative to intermediate densities. Interestingly, disease resistance is also low when larvae are raised singly, a fact suggesting that moderate levels of intraspecific interactions promote the development of disease resistance.

Maternal Environmental Effects   The condition of an offspring-bearing (i.e., maternal) individual can have a strong influence on the phenotype of her offspring, independent of the offspring's genotype. When a mother's phenotype, condition, or resource status is determined by her physical or biotic environment and influences offspring phenotype, this influence is called a *maternal effect.* Where there is variation among maternal individuals in the quality of their environment, at least some of the phenotypic differences among maternal families are often attributable to maternal effects. Maternal effects on offspring phenotype have been called *cross-generational phenotypic plasticity.* This description encapsulates the idea that a maternal individual responds to the environment, but the response is detected and measured in the next generation. That is, a mother's response to the environment is assayed by the phenotype of her offspring. Maternal effects therefore differ from phenotypic plasticity, which is defined as the phenotypic response of an individual to its current environment.

Unlike maternal effects on progeny phenotype caused by cytoplasmically inherited genes, environmentally induced maternal effects may either be independent of the mother's genotype or be the result of a genetically determined maternal choice of or ability to acquire a particular environment (Mousseau and Fox 1998). Where there is a nonrandom association between the genotype and the environment of maternal individuals, environmental and genetic maternal influences on offspring phenotype will be confounded.

Maternal effects often generate phenotypic similarities between parents and their offspring that appear to be heritable (e.g., see Reznick and Travis, this volume), but these similarities are deceptive when they are determined by extrinsic rather than genetic factors. The evolutionary significance of maternal effects depends on the distribution of environmental conditions among genotypes. If genotypes and environmental conditions are uncorrelated (i.e., there is no genotype-environment covariance), then even though maternal families produced in different environments will differ in phenotype, there will be no deterministic evolutionary consequences of these differences.

Deterministic, directional genotypic change through maternal environmental effects can occur, however, if there is a strong and persistent correlation between genotype and environment. In this case, the differences among maternal families can no longer be identified strictly as a "maternal environmental effect," as the condition that offspring phenotype is determined solely by the maternal environment does not apply. If a mother's genotype influences her ability to choose or to otherwise acquire environments that benefit her offspring, maternal genotypes associated with high-quality environments will be favored by natural selection.

Another way in which maternal individuals may exhibit different kinds of maternal effects concerns a maternal genotype's response to the environment. That is, for some genotypes, a mother's environment may greatly influence the phenotype of her offspring, while for other genotypes, offspring phenotype is independent of the maternal environment. When the expression of a maternal environmental effect differs among genotypes, the maternal effect has the potential to evolve.

Numerous examples of environmentally induced maternal effects on offspring phenotype have been documented (Mousseau and Fox 1998), particularly in domesticated species for which the causes of variation in offspring quality are of great economic interest. Among wild species, maternal effects are also well known; for example, in the common lizard, *Lacerta vivipara,* a female's feeding rate influences the maximum sprint speed of her offspring.

Maternal effects on seed size and quality are nearly ubiquitous in plants, and in some cases maternal effects influence traits expressed relatively late in life (Mazer 1987). In the annual wildflower *Nemophila menziesii,* maternal plants raised in competition with *Bromus diandrus* produce smaller seeds with increased dormancy and delayed germination relative to those raised in the absence of competition. In wild radish *(Raphanus sativus),* mean individual seed mass declines as maternal population density increases (Mazer and Wolfe 1998).

Maternal environmental effects on offspring phenotype can even persist into subsequent generations. In *Plantago lanceolata,* the temperature at which plants are raised can influence the adult traits of their grandchildren, even independently of the size of the seeds produced by grandparents (Case et al. 1996). These effects are also genotype-specific, indicating the presence of genetic variation in the expression of grandparental effects. This type of genetic variation is a prerequisite for natural selection to affect the evolution of maternal and grandmaternal environmental effects.

Genotype × Environment Interactions

Genotype-by-environment (or $G \times E$) interactions appear when the phenotype expressed by a particular genotype is sensitive to the conditions in which it is raised, and where genotypes differ in their responses to environmental conditions (e.g., Pigliucci, this volume, figure 5.1). Where strong $G \times E$ interactions affect a particular trait, the combination of the genotype and the environment generates a phenotype that cannot be predicted on the basis of what is independently known about each genotype and environment.

$G \times E$ interactions can be manifest in one of three ways (figure 2.2). In all cases, the direction of phenotypic change that occurs in response to an environmental condition or gradient differs among genotypes. There are several evolutionary consequences of strong $G \times E$ interactions that influence fitness-related traits. First, $G \times E$ interaction variance appears in the denominator of the heritability equation, thereby diminishing the heritability of traits subject to strong $G \times E$ interactions. Genotype × Environment interactions are therefore often considered to represent constraints to evolutionary change; all else being equal, where $G \times E$ variance is high, heritability will be low, reducing the potential response to selection. Second, assuming that differences between genotypic means reflect additive genetic variation and that other sources of variance are equal among environments, figure 2.2 B–D shows that the heritability of traits

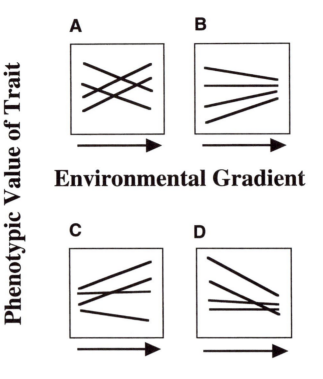

Phenotypic Value of Trait

Environmental Gradient

Environmental Gradient

Figure 2.2 Hypothetical relationships between phenotype and environment for four genotypes exhibiting a genotype × environment interaction *(G × E)* for the illustrated trait. In each panel, each line represents the phenotype exhibited by members of a particular genotype across an environmental gradient. (A) The relative phenotypic ranks of the genotypes change across environments. If the phenotype measured is fitness, this *G × E* would indicate that the genotype favored by selection would change across environments. (B) The phenotypic ranks of the genotypes remain constant across environments, but the magnitude of intergenotypic differences changes. Where intergenotypic differences reflect additive genetic variance, the heritability of the illustrated trait may differ across environments. (C and D) Both the genotypic ranks and the intergenotypic differences in phenotype vary across environments.

subject to strong *G × E* interactions may be highly environment-specific. *G × E* interactions can create situations where, in some environments, alleles will be neutral to selection while, in others, selection may be quite strong.

Genetic Variation in Phenotypic Plasticity   The expression of a *G × E* interaction indicates that there is genetic variation in phenotypic plasticity, and this will be detectable graphically as one of the patterns illustrated in figure 2.2. Detecting genetic variation in plasticity is of interest to evolutionary ecologists for two reasons. First, *G × E* interactions may contribute to the maintenance of genetic variation in fitness-related traits. If the optimum geno-type differs among environments and gene flow between environments is very low, then multiple genotypes may be maintained in a heterogeneous environment.

Second, the ability to detect *G × E* interactions raises the question of whether selection favors certain patterns of phenotypic plasticity over others. If different environments favor different phenotypes, then the genotype that expresses the optimum in each environment should be favored by selection, unless there is an overriding "cost" to phenotypic plasticity. If natural selection does act on genetically determined responses to the environment, one may be able to discover examples of adaptive phenotypic plasticity, where the nature of the response appears to be the evolutionary outcome of natural selection.

Differences among genotypes in phenotypic plasticity have been observed in a wide range of traits and species in both experimental and field conditions. In an experimental garden study of the wild radish *(Raphanus raphanistrum)*, $G \times E$ interactions in response to local population density characterized variation in both time to flowering and the volume of individual pollen grains (Mazer and Schick 1991a; figure 2.3) As a result, the heritability of these traits was density-dependent. In a laboratory study of ladybird beetle larvae, the expression of cannibalism in response to food availability differed among full sib families, and the heritability of cannibalism depended on food level (Wagner et al. 1999). Similarly, in a seed beetle *(Stator limbatus)*, genotypes differed with respect to changes in the size of eggs laid on two different host plants, and the heritability of egg size was host-dependent (Fox et al. 1999).

In plants, several cases of phenotypic plasticity in morphological or biochemical traits appear to be clearly adaptive. *Impatiens capensis* responds to shading (the proximal cue is a high ratio of red to far red wavelengths of light) by producing elongated stems with relatively few branches and accelerated flowering, which results in higher performance than in unelongated forms. This photomorphogenic response is interpreted as an adaptive response to avoid shade and is consistent with selection gradient analyses that have detected strong selection favoring tall plants at high density and short plants at low density. In *Nicotiana,* in response to attack by herbivores, plants produce nicotine, an herbivore deterrent.

On the other hand, temperature-dependent changes in leaf thickness and stomatal density in

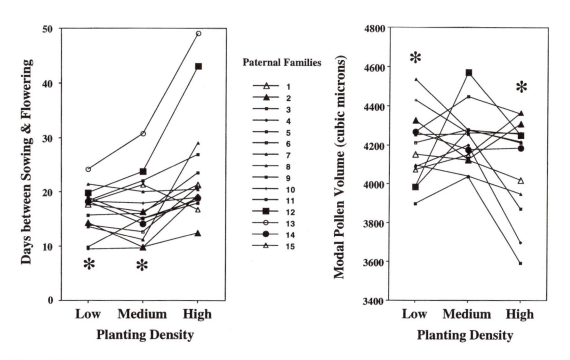

Figure 2.3 Genotype × environment interactions affecting flowering date and pollen grain volume in wild radish *(Raphanus raphanistrum)*. Seeds representing 13 paternal half-sib families were replicated across three planting densities and observed from the seedling stage through reproduction. Paternal families differed in their phenotypic responses to local population density with respect to several traits, including the number of days between sowing and flowering and the modal pollen of individual pollen grains. As a result of these $G \times E$ interactions, the magnitude of variation among paternal families is density-specific. Asterisks indicate planting densities in which the heritability of these traits was significantly greater than zero (Mazer and Schick 1991a).

Dicerandra linearifolia do not appear to be adaptive (Winn 1999). Based on results of controlled experiments, these temperature-sensitive traits differ as expected in summer and winter, but there is no evidence that selection favors this plastic response to temperature; in both seasons, selection on total plant biomass favors individuals with large, thick leaves, while stomatal density remains neutral to selection.

Interactions within and among species can also mediate the expression of $G \times E$ interactions. For example, the phenotypic plasticity of clonal replicates of *Lolium perenne* has been found to depend on whether or not they are infected with endophytic fungi. In some genotypes, the presence of endophytic fungi reduced phenotypic plasticity with respect to vegetative growth, while in others the fungi had no effect.

Variation in Structured Populations and Metapopulations

In many natural populations, individuals interact with conspecifics (especially close neighbors or kin) nonrandomly over space and time or form semi-isolated social groups or relatively small subpopulations. Populations within which individual interactions are spatially restricted are generally described as subdivided populations, structured populations, or structured demes (e.g., Wilson 1980 and this volume; Nunney, this volume). When the subpopulations are sufficiently spatially distinct that local extinction, migration, and recolonization are important, the overall population is often termed a *metapopulation*. When we consider selection in such situations, components of variation that are not directly involved in evolutionary responses to selection within a single, ecologically homogeneous, randomly mating population take on potential new significance. Also, the nature of the characters that are open to evolution in response to natural selection expands in scope.

Variation Underlying Multilevel Selection and Its Analogues

The complexity of ecological relationships and evolutionary forces in the context of metapopulations has permitted a wide variety of theoretical approaches and analytical techniques. To highlight a number of issues involving variation, we will concentrate on the perspective of modern multilevel (or group) selection theory (Nunney, this volume; Wilson, this volume). Although frequently the evolutionary forces generated by nonrandom interactions among individuals are modeled using alternative approaches (kin selection models, game theory, etc.; see Frank 1998), the fundamental "individual versus group" character can be discerned in all of them. Thus, much of what we discuss in the group selection context applies with suitable change of terminology to structured populations modeled from other points of view.

Phenotypic Traits and the Potential for Multilevel Selection From the perspective most relevant to evolutionary ecology, multilevel selection occurs in a structured population whenever an individual's fitness cannot be accounted for solely on the basis of its own phenotype, and information is required about properties of the group (or groups, subpopulations, etc.) of which it is a member (Heisler and Damuth 1987). In other words, an individual's fitness is context-specific; it depends on both the individual's phenotype and attributes of the group of which the individual is a member. There is thus some kind of "group effect" on individual fitness, independent of any fitness differences caused by individual phenotypic variation.

The selection gradient approach (Fairbairn and Reeve, this volume) has been extended to the multilevel case and can be used as an empirical analytical tool in the study of multilevel selection in nature (Heisler and Damuth 1987; Goodnight et al. 1992; Stevens et al. 1995). When focused more on behavioral interactions than on group structure, a complementary approach allows one to separate the effects of social interactions on fitness from selection directly on the individual phenotype (Wolf et al. 1999).

There will be no net force of selection deriving from group traits unless groups vary in those traits and thus, in most cases, vary in the distributions of individual phenotypes they contain. Any factor that promotes the variance among groups in their group-level characters and individual-level phenotypes thus enhances the effectiveness of multilevel selection. The limited dispersal inherent in structured populations is one source of interpopulation and intergroup variance. Others include inbreeding, habitat selection, tendency to associate with relatives or with similar phenotypes, modification of behavior to conform to that of group members (including the effects of learning and culture), and indirect genetic effects (see Nunney, this volume; Wasser and Williams, this volume; Wilson, this volume).

Group characters that are aggregates of individual phenotypes (such as group means of individual traits or frequencies of phenotypes) as well as characters that cannot be measured on individuals (such as population size) can potentially vary among groups and exert an effect on individual fitness. This means that many characters that cannot be the direct target of selection in homogeneous populations can be so in subdivided populations, including frequencies or proportions of different phenotypes, interactions among individuals, and the behaviors of nonreproductive individuals.

Interspecies Interactions   Interactions among individuals or among populations belonging to different species are of particular interest to ecologists. The possibility that relationships among species in multispecies associations (other than the simplest mutualisms) have evolved by a process of multilevel selection has been proposed but has remained almost unexplored empirically. The case of a simple mutualism, where an individual of one species does something that benefits an individual of another species and as a result receives a direct reward, presents no special problems for theory and has been studied extensively. However, when the reward returns as a benefit to the actor's population as a whole, this creates a situation analogous to the evolution of altruistic traits within a species (Wilson 1980; Frank 1994).

As in intraspecies multilevel selection, if communities are spatially structured, and the interactions among populations of different species and their fitness effects vary in space or across geographic "subcommunities," there should be circumstances under which such interspecific interactions can evolve by multilevel selection (Thompson, this volume). Experiments confirm that it is possible to obtain a response to selection directly on two-species interactions (Goodnight 1990). The significance of such evolutionary processes in the evolution and function of communities and ecosystems, and in attempts to manage them, remains almost completely unknown (Wilson 1997b).

Genetic Variation and Heritability in Multilevel Selection

Group characters under selection can be of many kinds, but in order for multilevel selection to re-

sult in evolution, there has to be a heritable basis for the characters that differ among groups and affect individual fitness. Sometimes the mapping between individual genotypes and group characters is straightforward, but in a multilevel context a variety of issues often make the precise prediction of evolutionary trajectories difficult.

First, nonadditive sources of genetic variance (such as epistatic variance) can contribute to a response to multilevel selection. Frequencies are characters that can differ consistently among populations and act as characters subject to multilevel selection. Models show that it is possible in some cases for multilevel selection to cause evolutionary change even when there is no additive genetic variance in the phenotypic traits that are involved (e.g., Wilson and Dugatkin 1997).

Second, in many cases the simple distinction between genetic and environmental sources of variance is no longer tenable. In subdivided populations, alleles and phenotypes of other individuals are often part of the "environment" experienced by a given individual. The environment thus evolves along with the organisms and has its own processes of inheritance (Moore et al. 1997; Wilson, this volume).

Third, the specific biological process for forming new daughter or colonizing groups, or for maintaining the ones that exist, may have considerable effects on the heritability of group traits (e.g., Wilson and Dugatkin 1997).

It is possible to estimate group-level heritabilities empirically by looking at the phenotypic relationships between "parent" and "offspring" groups. But in general, empirical estimation of relevant heritabilities and the partitioning of observed variation in metapopulations into components that are and are not involved in responses to multilevel selection would be a daunting task.

Epistasis, Population Differentiation, and Local Adaptation

We have seen how levels of epistasis and additive genetic variance for traits depend upon the allelic variation and allele frequencies at interacting loci (tables 2.1–2.3). This means that the additive genetic variance and the relationship between genotypes and fitness can depend upon the genetic background of the population.

This dependence leads to surprising results when there are high levels of epistasis in a metapopulational context (reviewed in Wade and Goodnight 1998). Where populations are small and epistasis is common, the significance of drift is that it can alter the genetic background that an allele experiences and thus can influence the rate and direction of genetic change in different populations. Drift and selection can convert nonadditive genetic variance to additive genetic variance, providing variation upon which selection can subsequently act. In fact, the effects of epistasis can result in some alleles being favored in some populations and being selected against in others, even if the ecological selective forces on the phenotypes do not differ (Goodnight 1995). Selection for local adaptation would necessarily favor different alleles in different backgrounds, leading to the formation of different adaptive allelic combinations in different populations (Goodnight 1995).

Thus, directional selection, which in a large panmictic population steadily removes genetic variation, can be a force promoting genetic differentiation in a metapopulation. Likewise, drift, which in a single-population context inhibits the effectiveness of selection, instead promotes local adaptation by converting existing nonadditive variation into an additive form on which selection can act.

Empirical studies of the genetics of metapopulations and the processes of population differentiation are necessarily challenging (e.g., Nunney, this volume), and the prevalence and importance of epistasis in natural populations is not yet well understood. Nevertheless, given the potential effects of gene interactions on evolutionary processes in metapopulations, this topic should increasingly gain the attention of evolutionary ecologists.

3

Natural Selection

DAPHNE J. FAIRBAIRN
JEFF P. REEVE

The theory that organisms become adapted to their environment through the process of natural selection has become so ingrained in modern biological thought, and more generally in Western culture of the late 20th century, that it is surely one of the great scientific paradigms of the present era. Evolution[1] and adaptation were both well-accepted concepts by the mid-19th century, at least among French and British natural philosophers. The theory of natural selection, developed by Wallace (1858) and Darwin (1859), provided a functional connection between the two processes. However, despite its logical consistency, natural selection was not accepted as a necessary or sufficient explanation for adaptation until the "evolutionary synthesis" of the mid-20th century, when knowledge from population and quantitative genetics, natural history (e.g., biogeography, ecology, behavior), systematics, and paleontology merged to form the unified theory of adaptive evolution known as *neo-Darwinism* (see Futuyma 1998 for a concise review of this history). Since that time, natural selection has been accepted as the universal mechanism leading to adaptation, and the two terms have become so closely associated as to be almost tautological. Adaptationist hypotheses are now fundamental to much of modern biology and are becoming increasingly apparent in more disparate fields, such as anthropology, medicine, biochemistry, and psychology (Futuyma 1999).

Nevertheless, there is much that natural selection cannot explain. For example, chance events may strongly influence macroevolutionary trends (i.e., the origin and extinction of species and higher taxa), some aspects of molecular evolution, and evolution within small or subdivided populations (Mazer and Damuth, this volume, chapter 2; Nunny, this volume). For this reason, adaptationist hypotheses should be viewed with skepticism until adequately tested (Reznick and Travis, this volume).

In this chapter, we carefully define natural selection and discuss methods of measuring selection in natural populations as a means of testing adaptationist hypotheses. These methods are most appropriate for testing hypotheses concerning the adaptive significance of contemporary trait distributions within and among populations ("microevolutionary" hypotheses) and thus have particular relevance for evolutionary ecologists. Readers will find many additional examples of these and other methods of testing microevolutionary adaptationist hypotheses throughout this volume, such tests being an essential component of most research programs in evolutionary ecology. Comparative methods more appropriate for testing evolutionary hypotheses on broader taxonomic or temporal scales ("macroevolutionary" hypotheses) are well summarized elsewhere (e.g., Endler 1986; Harvey and Pagel 1991).

Defining Natural Selection

Natural selection is notoriously difficult to define. In the broadest sense, the process of natural selection has been defined by the following deductive argument:

If there is:

  i. variation in some attribute or trait among biological entities *(phenotypic variation),*
  ii. a consistent relationship between the trait and fitness (a *fitness function*), and
  iii. descent with *heritability* for the trait (i.e., the variation in the trait must have a genetic component),

Then the trait distribution will change:

  I. within generations more than expected from ontogeny alone, and
  II. across generations "in a predictable way," until an equilibrium is reached.

This definition is true to Darwin's original description of natural selection, and was adopted by Endler (1986) in his review of selection in natural populations. However, in addition to being rather cumbersome, the deductive argument is flawed because conclusion (I) does not require premise *(iii)* and holds only for fitness differences caused by differences in survival (i.e., differences in fecundity alone will not cause within-generation changes in trait distributions).

In constructing a more concise and logically consistent definition of natural selection, most authors (e.g., Lande and Arnold 1983; Futuyma 1998) prefer to distinguish the process of natural selection occurring within generations (premises *i* and *ii*) from the evolutionary consequences of that selection (premise *iii* and conclusion II). This distinction follows from the classic mathematical models of population and quantitative genetics. In these models, within-generation selection parameters are used to predict change between generations, defined as response to selection. For example, in the standard one-locus, two-allele model of population genetics, with genotypes *AA*, *AB*, and *BB* in initial proportions p^2, $2pq$, q^2, selection is represented by *selection coefficients* associated with each genotype, s_{AA}, s_{AB}, and s_{BB}. This model yields the following general equation for change in allele frequency between generations (Δp) in response to selection:

$$\Delta p = \frac{pq[p(s_{AB} - s_{AA}) + q(s_{BB} - s_{AB})]}{1 - (s_{AA}p^2 + s_{AB}2pq + s_{BB}q^2)} \quad (3.1)$$

The analogous equation from quantitative genetics is

$$R = Sh^2 \quad (3.2)$$

where R is the change in the trait mean from one generation to the next (often shown as $\Delta \bar{z}$), h^2 is the narrow sense heritability[2] of the trait, and S is the *selection differential*. The selection differential is the within-generation change in the trait mean, estimated as $\bar{z}_s - \bar{z}_0$, where $\bar{z}_s$ and $\bar{z}_0$ are the trait means before and after selection, respectively. The standardized selection differential—calculated as S/σ, where σ is the phenotypic standard deviation—is called the *selection intensity, i*. If i is used, the response equation becomes:

$$R = i\sigma h^2 \quad (3.3)$$

In equations 3.1–3.3, change between generations (Δp or R) is modeled as response to selection, while selection is represented by a parameter quantifying change within generations (s_{ij}, S or i). In these mathematical representations, adaptive evolution occurs because of the interaction between selection within generations and some measure of the genetic basis (i.e., genotype or heritability) of the trait in question. Mathematical models of evolution therefore clearly distinguish between the process of natural selection, which occurs within generations, and evolution across generations in response natural selection, which occurs only if the traits in question are heritable.

Because of this legacy from evolutionary theory, we prefer the more restrictive definition of natural selection as a process occurring within generations, and we refer to any resulting change in phenotypic or genotypic distributions across generations as *response to selection*. We therefore adopt Futuyma's (1998, p. 349) concise definition of natural selection as *"any consistent difference in fitness among phenotypically different biological entities."*

What Is Fitness?

Given that we have defined natural selection as a consistent relationship between fitness and phenotype, it is essential that we develop a clear and practical definition of fitness. For simplicity, let us consider a definition appropriate for selection in which our "biological entities" are individual organisms, keeping in mind that this definition can be generalized to other levels of selection (see below). In its most general sense, fitness is success in contributing descendants to the next generation. If the contribution of descendants varies among individuals and is consistently related to some phenotypic trait, natural selection is acting on that trait.

This definition of fitness is very closely tied to net reproductive rate, R_0, defined as

$$R_0 = \int_0^\infty l_x m_x dx \qquad (3.4)$$

for continuous age distributions, and

$$R_0 = \sum_{x=1}^\infty l_x m_x \qquad (3.5)$$

for discrete age classes, where l_x is the probability of surviving to age x, and m_x is number of offspring[3] produced at age x. As a population parameter, R_0 is the expected or average lifetime reproductive success for that generation. When R_0 is estimated for individuals, l_x reduces to 0 or 1 for each age class, and R_0 becomes simply the number of offspring produced or lifetime reproductive success[4] (LRS) for each individual. If population size is relatively constant, net reproductive rate is generally the most appropriate measure of lifetime fitness.

Many natural populations would satisfy the assumption of stationary population size, at least to the extent that they show no sustained increases or decreases and fluctuations in population size are small relative to mean population size. However, for some organisms this assumption is clearly not valid. For example, species inhabiting ephemeral habitats exist in a mozaic of colonizations and extinctions so that population size can be expected to be either increasing or decreasing in any one generation and location. All parasites and "weedy" plants, and many insects (e.g., *Drosophila*, Roff, this volume) most certainly fall into this category. When population size is changing rapidly, the age schedule of reproduction becomes an important determinant of fitness. The appropriate measure of fitness is then r, the instantaneous rate of increase or "Malthusian parameter," which must be estimated by solution of the "characteristic equation":

$$\int_0^\infty e^{-rx} l_x m_x dx = 1 \qquad (3.6)$$

for continuous age distributions, and

$$\sum_{x=1}^\infty e^{-rx} l_x m_x = 1 \qquad (3.7)$$

for discrete age classes (Roff 1992 and this volume). Note that the characteristic equation reduces to $R_0 = 1$ when $r = 0$. More generally, $r \approx \ln(R_0)/\tau$ where τ is the mean generation time. This latter relationship illustrates the importance of the age schedule of reproduction in determining r. It also suggests that R_0 or $\ln(R_0)$ could be legitimately substituted for r as a measure of fitness if generation time can be presumed to be constant across all individuals being compared. This circumstance is not rare in natural populations. For example, many plants and insects have annual life cycles with a synchronizing season of dormancy. For such species, R_0 or LRS measured at the end of the reproductive season would be an appropriate measure of fitness even if population size were highly variable.

While both r and R_0 are used extensively in theoretical analyses of adaptive evolution, the difficulty of estimating either as a function of phenotype in wild populations has limited their application in empirical studies (but see Roff 1992 and this volume for examples). Much more commonly, researchers attempt to measure only certain components of fitness. This approach is generally much more feasible and is appropriate in many cases where the researcher is primarily interested in testing specific hypotheses about the adaptive significance of the trait in question (Reznick and Travis, this volume). For example, the hypothesis that the inducible crests in *Daphnia* (see figure 5.2 in Pigliucci, this volume) serve to protect their carriers from predation could be tested by measuring survival of *Daphnia* in the presence of predators as a function of crest size. Similarly, the hypothesis that an exaggerated secondary sexual characteristic in males is favored by sexual selection for that trait could be tested simply by measuring the correlation between mating success and trait value in a given population (e.g., Savalli, this volume, figure 15.2). In some cases, the fitness parameter may itself be a component of lifetime fitness (i.e., survival, fecundity, or mating success). However, in other cases, surrogate fitness measures can be used, on the assumption that they themselves correlate with lifetime fitness. For example, behavioral ecologists may use energy intake rate as a surrogate measure of fitness in studies comparing foraging strategies (Kramer, this volume).

The use of surrogate measures of fitness and studies of only certain components of lifetime fitness have proven to be very valuable for testing adaptationist hypotheses, as numerous examples in succeeding chapters will attest. However, the dynamics of trait evolution in response to selection (assuming $h^2 > 0$) are ultimately determined by the relationship between trait values and lifetime fitness. Trade-offs between the fitness effects of trait values during different phases of the life history often produce nonlinearities in the relationship between trait values and lifetime fitness that would not be obvious from studies limited to only a subset of life history stages (e.g., see the case study below; Roff, this volume). Studies that combine analyses of fitness components with estimates of lifetime fitness offer the best hope of both deducing the adaptive significance of the trait in question and understanding the evolutionary dynamics of the trait distribution.

Our discussion of fitness estimators has, to this point, implicitly assumed a constant environment and frequency-independence of fitness. In a variable environment, meaningful "global" estimates of fitness can be derived through judicious sampling, replication, and averaging of fitness estimates over space or time (Roff 1992). However, frequency-dependent selection poses particular problems and is most appropriately analyzed with the techniques of game theory (e.g., "evolutionarily stable strategies"), rather than by estimation of r, R_0, or analogous surrogate measures of fitness. A description of game theory and its associated methodologies is beyond the scope of this chapter, but concise introductions can be found in Roff (1992 and this volume), Futuyma (1998), and Kramer (this volume).

Levels of Selection

The use of the term *biological entities* in our definition of natural selection is purposely vague. Although we typically imagine natural selection acting on individual organisms and producing evolutionary change within population lineages, the process of natural selection occurs at all levels of biological organization, from molecules to species and higher

taxa (Williams 1992; Mazer and Damuth, this volume, chapter 2; Nunney, this volume; Wilson, this volume). The only requirement is a set of self-replicating entities on the same level of organization that satisfy premises *(i)* and *(ii)* above. For example, natural selection at the level of species would occur through differential rates of speciation and extinction within different species' lineages, if these rates were consistently related to species characteristics. From the perspective of evolutionary ecology, the key issue is not whether selection occurs at different levels of organization, but whether such selection has produced adaptation at those levels. A strong case can be made for selection at the level of the gene, and numerous examples of meiotic drive attest to its efficacy in producing adaptations that are detrimental at higher levels of organization. Similarly, altruistic behavior can be attributed to "kin selection."[5] Nevertheless, both theory and empirical evidence suggest that most organismal traits have evolved because they benefit individual organisms (Endler 1986; Williams 1992; Futuyma 1998). "Individual selection" is also sufficient to explain most emergent properties of higher levels of organization (e.g., geographic distributions, population growth rates, rates of speciation), although the possibility of reinforcing selection at several levels cannot be discounted (Williams 1992). Because of the preeminence of individual selection and the logistical difficulties of testing hypotheses concerning selection at other levels, we focus on studies of natural selection at the individual level.

Is Sexual Selection Distinct from Natural Selection?

Selection that arises from differences in mating or fertilization success rather than survival or fecundity is termed *sexual selection* (Savalli, this volume). Darwin clearly distinguished sexual from natural selection and regarded the two as often acting in opposition, sexual selection favoring traits such as bright coloration, exaggerated ornamentation and weaponry, that seem likely to reduce male survivorship. Despite this distinction, selection acting through differential mating success clearly satisfies the modern definition of natural selection. In fact, as our case study will illustrate, lifetime fitness functions cannot be estimated without inclusion of sexual selection and its trade-offs with other fitness components. Therefore, from the perspective of evolutionary biology, sexual selection is simply a form or subset of natural selection (Endler 1986; Futuyma 1998).

Is Opportunity for Selection a Measure of Selection?

The opportunity for selection, I, is simply the variance in relative fitness among individuals within a given episode of selection (Arnold and Wade 1984a; Walsh and Lynch 1998). Since natural selection is defined as the relationship between trait values and fitness (i.e., the covariance between phenotype and fitness), variance in fitness is necessary for selection to occur. In fact, the opportunity for selection defines the upper bound of the intensity of selection: if trait values are perfectly correlated with relative fitness, $i = I$. As a standard statistic, I can be useful in comparisons of the potential for selection in different episodes of selection, in different populations or organisms, or even at different levels of selection. Nevertheless, I is not a measure of selection because it includes only the variance in fitness and excludes the necessary covariance between trait values and fitness.

Measuring Selection in Natural Populations

Endler (1986) provides an excellent, comprehensive review of selection in natural populations in which he describes 10 major approaches to testing adaptive hypotheses for continuous and discrete traits and collates results demonstrating natural selection on 314 traits in 141 species. He discusses several categories of studies that we do not review in this chapter. The most important of these are studies of clinal variation; the use of null models such as the Hardy-Weinberg equilibrium; the use of models of selection such as optimality or ESS models; studies of population responses to perturbations; and comparisons among species. Readers will find many examples of these approaches in succeeding chapters of this volume, and Williams (1992) and Futuyma (1998) provide reviews.

In spite of this diversity of approaches, measurement of selection using S, i, or some form of selection gradient analysis (see below) has become the predominant method of quantifying natural selection within populations (table 3.1). For obvious logistical reasons, empirical studies show a clear bias toward studies of morphological traits in small,

Table 3.1 Organisms used and variables measured in studies designed to measure selection as S, i, or β in natural populations.

Organisms			Independent variables		Dependent variables[a]	
Taxon		n	Trait type	n	Trait	n
Plants[b]		55	Morphological	119	Fecundity[c]	63
Arthropods	Insects	38	Phenological	31	Mating success[d]	47
	Crustaceans	4	Behavioral	8	Lifespan or survival	44
Vertebrates	Birds	32	Physiological	6	Reproductive success[e]	15
	Reptiles	12	Life history	5	Flower or pollen production	7
	Fish	5			Others	9
	Amphibians	3				
	Mammals	3				

Source: Based on a survey of journals in ecology and evolutionary biology, January 1983–May 1999. The majority of these empirical examples were found in *Evolution* (101), *American Naturalist* (19), *Journal of Evolutionary Biology* (13), and *Ecology* (10).

Note: Variables in each column are listed in decreasing order of frequency. Number of studies $(n) = 188$.

[a]Measures of fitness.

[b]50 herbaceous flowering plants, 2 grasses, and 3 shrubs.

[c]Number or total mass of eggs, live births, seeds, or fruit.

[d]Includes 8 studies of pollination success in plants.

[e]Includes some aspect of offspring fitness, usually survival or growth.

relatively short-lived, terrestrial organisms (61% concern insects or herbaceous plants). Fecundity is the most common measure of fitness, followed by mating success and lifespan or survival. Quantification of fitness is still being used primarily to determine adaptive significance rather than to make future or retrograde predictions of evolutionary trajectories. This vast literature clearly cannot be effectively summarized in the restricted scope of this chapter. Instead, we concentrate on the statistical methodology for describing the relationship between fitness and trait values. The method that we describe in most detail, *selection gradient analysis,* has the advantage of simultaneously yielding both descriptions of current patterns of selection and quantitative estimates of the selection parameters needed for predictions of evolutionary trajectories.

Fitness Surfaces and Modes of Selection

The relationship between fitness and trait values that defines natural selection is presumed to be causative and is therefore properly called a *fitness function.* Unfortunately, validation of a statistical covariance between the trait and fitness is not sufficient evidence of a causative or functional relationship between the two. The latter must be inferred from other types of evidence and generally requires some form of manipulative experimentation (Endler 1986; Reznick and Travis, this volume). Statistical descriptions of fitness functions are therefore more correctly referred to as *fitness surfaces.*

Figure 3.1 illustrates four common shapes of fitness surfaces. To interpret these, we describe them using either a standard linear model, $w_i = a + \beta_1 z_i$, or a quadratic model, $w_i = a + \beta_1 z_i + \beta_2 z_i^2$, where w_i is the relative fitness[6] of individuals with trait value z_i. In figure 3.1A, the fitness surface is increasing and purely linear. This is classic directional selection favoring larger trait values. Figure 3.1B also illustrates directional selection favoring larger trait values, but in this case there is a nonlinear component to the fitness surface ($\beta_2 \neq 0$). In spite of this curvature, we use simple linear regression to estimate the net directional component of selection, which yields a slope of 0.22. Figures 3.1C, and D illustrate stabilizing selection, indicated by a fitness peak or maximum[7]. In figure 3.1C, the fitness peak is to the right of the popula-

tion mean phenotype and there is still a significant directional component to the fitness surface, indicated by $\beta_1 > 0$ in the simple linear regression. Pure stabilizing selection occurs only if, as in figure 3.1D, the population mean phenotype is at its "optimum" (i.e., the value that maximizes fitness). In such a case, β_2 is negative and simple linear regression does not yield a significant slope. By analogy, fitness surfaces with $\beta_1 < 0$ indicate directional selection favoring smaller trait values, while $\beta_2 > 0$ indicates concavity of the fitness surface. Concavity with an internal fitness minimum (the inverse of figures 3.1C and D) indicates disruptive selection.

Lande and Arnold (1983) coopted this regression approach to provide standardized descriptive statistics for fitness surfaces, called *selection gradients.* The univariate linear selection gradient, β_1 from the linear model, describes the component of directional selection or the average slope of the fitness surface. The univariate nonlinear or quadratic selection gradient, γ_1, describes the average curvature of the fitness surface and is given by the second derivative of the quadratic regression equation, $2\beta_2$. These are convenient summary statistics for describing the average geometry of the fitness surface (Walsh and Lynch 1998). When estimated from a regression of relative fitness on standardized trait values, as in our example, β_1 is mathematically equivalent to i, the selection intensity from equation (3.3), and so can be used to predict response to selection. When fitness cannot be defined in terms of 0 and 1 (e.g., mated vs. not mated; alive vs. dead), actually calculating i as the standardized difference between the mean of the population before and after selection requires weighting individual values according to their fitnesses. In practice, this is most easily accomplished by estimating i as β_1, which is, in fact, the covariance between relative fitness and the standardized trait value. The regression method thus provides an intuitively appealing and computationally tractable connection between the empirical estimation of fitness surfaces in natural populations and evolutionary quantitative genetics.

Multivariate Analysis of Selection

It is a truism in biology that organisms evolve and develop as integrated units, and that any given trait, be it morphological, behavioral, or physiological, will be correlated with many other traits.

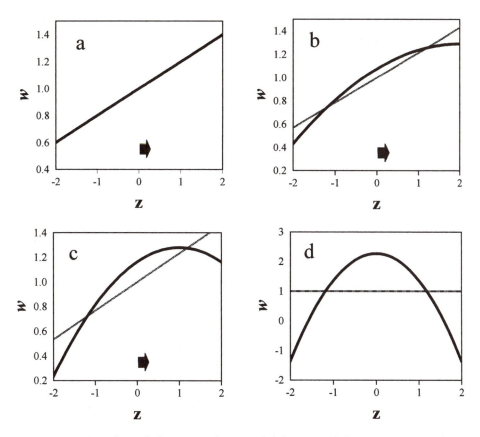

Figure 3.1 Hypothetical fitness surfaces (solid lines) and linear regression lines (hatched) for relative fitness *(w)* as a function of standardized trait values *(z)*. In each panel, mean fitness is 1 and the mean trait value before selection is 0. Arrows indicate the within-generation shift in the mean trait value expected under each selection regime. (A) Fitness is a linear function of trait values: $w = 1.0 + 0.20z$. (B) Fitness is a convex function of trait values: $w = 1.08 + 0.22z - 0.05z^2$. However, linear regression still accounts for the majority of the variance: $w = 1.0 + 0.22z$, $R^2 = 0.94$. (C) Fitness is a convex function of trait values: $w = 1.16 + 0.23z - 0.12z^2$, with an internal maximum at $z = 1$. However, linear regression still accounts for the majority of the variance: $w = 1.0 + 0.23z$, $R^2 = 0.78$. (D) Fitness is a convex function of trait values: $w = 2.27 - 0.91z^2$, with an internal maximum at $z = 0$. Linear regression does not account for any of the variance.

Phenotypic correlations among traits influence our perception of patterns of selection in the following manner. Consider two traits that covary within a given generation so that an individual with a high value of A is also likely to have a high value of B, and vice versa. Trait A strongly influences fitness so that higher values of trait A result in greater LRS (perhaps high A confers high fecundity or survivorship or both). Trait B is selectively neutral and has no influence on fitness. If we measure selection on trait A, using *S, i,* or β_1, we will detect directional selection favoring increased A. However, because individuals with high A also have high B, a similar analysis for trait B is also likely to indicate directional selection, even though B is selectively neutral. Our univariate methods of measuring selection cannot distinguish selection acting directly on a trait (trait A in our example) from *indirect selection* on correlated traits (trait B). As we shall see, distinguishing between direct and indirect selection is a necessary prerequisite for both testing adaptive hypotheses and predicting evolutionary response to selection. Multivariate statistical methods allow us to identify the traits subject to direct selection (i.e., the *targets of selection*) and to quantify the intensity of selection acting on each trait independently of its phenotypic correlations with other measured traits.

Calculating Multivariate Selection Gradients For more than one trait, the average slope of the fitness surface (β_1 above) becomes β, a vector of multivariate linear selection gradients[8] and is equivalent to the inverse of the phenotypic variance-covariance matrix multiplied by the vector of selection differentials, $\mathbf{P}^{-1}\mathbf{s}$. In practice, the coefficients of the vector β are calculated through a multiple regression of relative fitness on the trait values. For instance, with two traits, the regression model would be

$$w = a + \beta_{3A}z_A + \beta_{3B}z_B + e \qquad (3.8)$$

where the subscripts on the z's refer to traits A and B, a is a constant, and e is the error term. To avoid confusion with the univariate selection gradients described previously, we designate the multivariate linear gradients as β_{3i}, with the second subscript referring to the trait. Each coefficient is the partial regression of relative fitness on that trait and hence can be interpreted as the effect of the trait on relative fitness, when all the other traits in the model are held constant. Therefore, the multivariate selection gradients estimate linear selection acting directly on the traits, while the univariate estimators (s, i, or β_1) estimate net directional selection on the trait, including both direct selection and selection acting indirectly through selection on phenotypically correlated traits.

The multivariate linear gradients can be used to predict response to selection by means of the multivariate response equation (Lande 1979), which is the multivariate analogue of equation 3.2:

$$\Delta \bar{z} = \mathbf{GP}^{-1}\mathbf{s} = G\beta \qquad (3.9)$$

where the column vector $\Delta \bar{z}$ represents the change in trait means across generations (= response to selection), $\mathbf{G}$ is the additive genetic variance-covariance matrix, and β is vector of multivariate linear selection gradients.

The multivariate quadratic selection gradients are estimated from a multiple regression model that includes all squared and cross-product terms

in addition to the linear terms. For two traits, the regression model would be:

$$w = a + \beta_{3A}z_A + \beta_{3B}z_B + \beta_{4A}z_A^2 + \beta_{4B}z_B^2 + \beta_{5(AB)}z_Az_B + e$$

$$(3.10)$$

The quadratic selection gradients are conveniently represented as a matrix γ, where the diagonal elements, γ_{11} and γ_{22} (= $2\beta_{4A}$ and $2\beta_{4B}$, respectively) can be either positive (fitness surface is concave) or negative (fitness surface is convex). The off-diagonal elements $\gamma_{12} = \gamma_{21}$ (= $\beta_{5(AB)}$) describe selection for increased positive or negative covariance between the two traits (correlational selection). Because the linear coefficients obtained from equations 3.8–3.10 will not be the same, β is always estimated from the linear model (equation 3.8), while the full model (equation 3.10) is only used to estimate γ. Interpretations are simplified if traits are standardized to a mean of 0 and a standard deviation of 1, as in our univariate example (figure 3.1).

Episodes of Selection: Estimating the Components of Lifetime Fitness Most selection studies focus on single episodes of selection, and ask specific questions such as "Does wing length influence survival?" or "Does flower size affect pollinator visitation?" Such studies are valid as initial tests of hypotheses concerning the adaptive significance of the traits of interest. However, they tell us little about the evolutionary dynamics of these traits because they fail to set the episodic selection estimates in the context of lifetime selection. Conversely, estimates of lifetime fitness may shed little light on the functional basis of selection because the dependent variable necessarily represents the net effects of selection acting throughout the various life stages of the organism. Arnold and Wade (1984a,b) described a method for partitioning lifetime selection gradients into selection gradients for component episodes in a manner that permits assessment of the relative contribution of each episode to lifetime fitness. The method requires only that lifetime absolute fitness can be expressed as the product of the absolute fitnesses in several component episodes. For instance, male lifetime fitness might be measured as the product of preproductive survival (0, 1), number of matings (0, 1, 2, . . .), and number of fertilized eggs per mating (0, 1, 2, . . .). If the fitness components are multiplicative in this way, the episodic selection gradients will be additive (i.e., for k episodes, $\beta_{Life\text{-}time} = \Sigma\beta_k$) if all are calculated using the distribution of phenotypes present at the beginning of the first episode ($\mathbf{P}_0$). This "additive method" has the advantage of quantifying the relative contributions of the different episodes of selection to lifetime fitness. However, it gives a poor estimate of the fitness surfaces within episodes (other than the first) because the "cumulative fitness" from previous episodes is carried over (i.e., individuals with zero fitness at the start of the episode remain in the analysis). The use of $\mathbf{P}_0$ for every episode also means that episodic gradients cannot be calculated by standard multiple regression techniques. The additive episodic values are also sensitive to the ordering of episodes and so can only be unambiguously assigned when the episodes are strictly sequential.

To address some of these problems, Koenig and Albano (1987) introduced the "independent method," in which the gradients for each episode, β'_k, are calculated by the use of only the individuals with nonzero fitness at the start of the episode. These independent gradients are calculated by standard multiple regression, are insensitive to the ordering of episodes, and give an accurate description of the statistical relationship between fitness and trait values within each episode. However, the independent gradients for different episodes do not sum to lifetime gradients and so cannot be compared for their relative contributions to lifetime fitness. An alternative method for calculating gradients using adjusted $\mathbf{P}$ matrices (β_k^*; Arnold and Wade 1984a) combines the disadvantages of β_k and β'_k (i.e., it is nonadditive, cannot be calculated as a standard regression, and is sensitive to the ordering of episodes) and is best avoided.

Problems and Limitations of Multivariate Selection Gradient Analysis
Correlated characters Multiple regression coefficients are sensitive to the presence or absence of correlated traits in the model; the magnitudes and signs of the coefficients may change depending on which traits are included in the analysis. This sensitivity is proportional to the degree of correlation among the traits, and as correlations approach ±1, individual coefficients become very unstable (this is the problem of "multicollinearity"; Fairbairn and Preziosi 1996). The models used to estimate selection gradients should include all phenotypically correlated traits under selection. However, if traits are very highly correlated, they may profitably be reduced to a single variable, with little loss of information. This reduction improves the stability of the estimates and increases the statistical power of the analysis (Lande and Arnold 1983).

Environmental covariances Estimated selection gradients may be biased if environmental factors contribute to the covariance between trait values and fitness. For example, imagine a plant population in which fitness is determined primarily by concentration of a limiting soil nutrient. If high nutrient concentrations also lead to darker leaf pigmentation, a selection analysis would indicate positive selection for leaf pigmentation, even if the phenotypic covariance between pigmentation and fitness were entirely environmental. If environmental covariance is recognized, it can be removed by inclusion of either an indicator trait or a direct measure of the environmental factor in the analysis (Mitchell-Olds and Shaw 1987). An alternative is to estimate "genetic selection gradients" based on covariances among relatives (Rausher 1992). These are independent of environmental covariance and correct for even unsuspected environmental influences. Unfortunately, calculation of genetic selection gradients requires very large samples and specially designed breeding experiments not feasible in most natural populations.

Significance testing Significance tests for selection gradients follow the standard methodology for multiple regression coefficients. However, fitness data often violate the assumptions of parametric regression (Mitchell-Olds and Shaw 1987), and either transformation or computer-intensive resampling techniques such as jackknifing or randomization tests are frequently employed for hypothesis testing (Walsh and Lynch 1998). In the special case of dichotomous fitness values (e.g., alive or dead, mated or not mated), logistic regression can be used for both significance testing (Fairbairn and Preziosi 1996) and parameter estimation (Janzen and Stern 1998).

Whatever the method used, estimation of multivariate selection gradients is severely limited by the problem of correcting for experimentwise Type I error because each regression coefficient (β_{ij} in equation 3.10) requires an independent statistical test. For example, estimation of β and γ for a set of seven independent variables, as in our case study below, requires 35 separate statistical tests! In this example, a simple Bonferroni correction gives a critical alpha of 0.0014, making it very difficult to detect "significant" selection. To preserve power, some authors abandon estimation of selection gradients in favor of more general regression analyses (Ferguson and Fairbairn 2000), while others reduce the number of traits examined or test the significance of only a partial set of selection gradients (Preziosi and Fairbairn 2000). Other possible methods of boosting power include averaging over multiple replicates (Fairbairn and Preziosi 1996) or incorporating manipulative experiments into the analyses (Anholt 1991; Juenger and Bergelson 1998).

Applications and Extensions of Selection Gradient Analysis   Selection gradient analysis has been applied to a wide variety of morphological, behavioral, and physiological traits, and even to nontraditional characters such as fluctuating asymmetry, quantitative trait loci (QTLs), phenotypic plasticity, and growth trajectories. Studies of species interactions (competition, predation, parasitism, etc.) and of spatial and temporal heterogeneity in fitness surfaces are becoming increasingly common (e.g., Juenger and Bergelson 1998; Ferguson and Fairbairn 2000). A very small number of papers have estimated both β and G in order to make predictions using equation 3.9. However, to our knowledge, only one of these has compared these predictions with the changes that actually occurred in the following generation (Grant and Grant 1995). Several studies have attempted retrograde selection analysis (Lande 1979), in which the G matrix is combined with trait differences between populations to estimate the selection gradients responsible for the evolutionary divergence in trait values (e.g., Dudley 1996). In addition to these empirical applications, the basic theory has been considerably extended. For example, methods now exist for estimating selection when generations overlap (Stratton 1992), when selection acts at the level of the group (Queller 1992b), and in hierarchically structured populations (Heisler and Damuth 1987).

Related Methods of Analyzing Selection

Path Analysis   Path analysis begins with construction of a path diagram representing hypothesized causal relationships among a set of traits. Multiple regression or partial correlation analysis is then used to estimate "path coefficients" for each path. These partial regression or correlation coefficients quantify the relative importance of each path in explaining the variance in the target or ultimate trait.

The statistical objective of path analysis is to discover the path structure that best explains the covariance among a set of traits, rather than to derive a predictive equation as in regression analysis. However, the two approaches have much in common and are often combined. Path analysis permits analysis of complex webs of interactions among traits, with explicit estimation of direct, indirect and hierarchical effects. In principle, it should provide a more realistic description of the biological relationships than strict selection gradient analysis. Different paths within the diagram can be compared quantitatively, as can different path diagrams (i.e., different causal hypotheses). Path analysis is a general and highly flexible statistical approach, and there is as yet no single standard methodology for its use in analyses of selection. Kingsolver and Schemske (1991) provide a concise and lucid introduction to the literature in this area.

Visualizing Fitness Surfaces   As we have seen, fitness surfaces can be represented graphically by quadratic fitness functions (figure 3.1). However, the imposition of the quadratic function constrains the shape and complexity of the estimated surface. The true fitness surface may be considerably more complicated, with asymmetries, discontinuities, or multiple peaks. Schluter (1988) discusses the problem of more accurately visualizing these complexities and recommends the use of the cubic spline, a flexible curve-fitting method that describes a line of best fit through the data with no a priori mathematical function. Univariate splines are relatively easy to calculate and are useful adjuncts to gradient analysis (e.g., Preziosi and Fairbairn 2000). Unfortunately, graphical representation is less straightforward when multiple traits are included in the analysis, and an intuitive understanding of the shape of the fitness surface may be virtually impossible. Phillips and Arnold (1989) and Schluter and Nychka (1994) provide alternative approaches for tackling this problem, but both approaches are complex and neither has been widely adopted.

Case Study

Measuring Selection on Adult Body Size in the Waterstrider, Aquarius remigis

Preziosi and Fairbairn (2000) measured selection on adult body size in *Aquarius remigis*, a semiaquatic bug found on the surfaces of streams and small rivers throughout much of North America. Their goal was to understand the adaptive significance of sexual size dimorphism (SSD), and the study provides a good example of the application of selection gradient analysis to wild populations. Four characteristics of *A. remigis* in the study population at Mont St-Hilaire, Quebec, facilitated estimation of fitness for males and females over the entire adult life stage:

1. An annual or univoltine life cycle with non-overlapping generations.
2. Synchronization of reproduction in spring caused by adult, prereproductive diapause over the winter.
3. High catchability of all adults on the spatially limited, open habitat of the stream surface.
4. Negligible confounding of mortality and emigration due to low mobility of individuals within and among streams (adults lack wings in this population).

The only major limitation of this system was the absence of estimates of selection prior to the adult stage. Fortunately, SSD for total length is negligible until the adult stage, and the sexes do not differ in development time, survival to the final molt (eclosion), or date of eclosion. Thus, adult SSD is unlikely to be a correlated response to selection acting during earlier life stages. Preziosi and Fairbairn therefore assumed that adult SSD can be explained by differences between the sexes in adult fitness functions.

Experimental Design and Analytical Methods   Two components of fitness, survival from eclosion[9] to the reproductive season and reproductive longevity, were obtained from mark-recapture data. For 2 years (generations), all newly eclosed adults were captured, individually marked, and photographed (precise size measurements were obtained from the photographic negatives). Survival/longevity was then assessed by twice-weekly recaptures of the marked adults. The third component of fitness, daily reproductive success, was assayed as number of matings per day for males and number of eggs laid per day for females. Both parameters were measured in artificial enclosures placed in the stream and were assayed in 3- or 4-day bouts, two or three times during the reproductive season.

These assays were highly repeatable within individuals and showed no detectable seasonal variation (Preziosi and Fairbairn 1996, 1997), a fortunate result that allowed Preziosi and Fairbairn to collapse the information into single estimates of mean daily reproductive success for each marked adult. The resulting three fitness components are thus multiplicative: net adult fitness = [prereproductive survival (0 or 1)] × [reproductive longevity (days)] × [daily reproductive success].

Standardized selection gradients were calculated for net adult fitness and for each of the episodes, by using Koenig's independent method for the latter. For prereproductive survival, a 0/1 trait, gradients were estimated by the standard least squared regression techniques, but statistical significance was tested by logistic regression.[10] Patterns of net selection on total length were discerned from univariate selection gradients (β_1, γ_1) and cubic spline estimates of fitness surfaces (Schluter 1988).

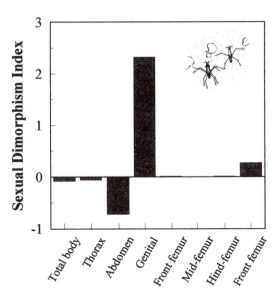

Figure 3.2 Sexual size dimorphism in *Aquarius remigis*. The *y*-axis is an index of sexual dimorphism calculated as the ratio of the larger to the smaller measurement, minus 1.0. If females are the larger sex, the index is arbitrarily made negative. The first seven variables are length measurements (i.e., total body length, genital length, etc.); the final variable is width of the front femur. The cartoon shows a single female and a mating pair (male mounted on the female) foraging on the stream surface.

Direct selection on body components was identified by use of multivariate selection gradients for seven correlated body components (figure 3.2). To preserve the overall power of the multivariate analyses, only linear gradients (β_{3i}) were estimated within episodes, and experimentwise error was controlled through a sequentially rejective multiple test procedure rather than simple Bonferroni corrections.

Results The univariate analyses indicated net stabilizing selection on total length in both sexes (table 3.2, figure 3.3). This interpretation is supported by statistically significant negative quadratic gradients combined with nonsignificant linear gradients (e.g., figure 3.1D; Phillips and Arnold 1989) in both generations for females and in generation 1 for males. Bootstrap analyses of the cubic spline approximations (Schluter 1988) confirmed the presence of an intermediate optimum body size, even for males in generation 2. In both generations the estimated optimum size for males (13.2 mm and 13.4 mm) was smaller than for females (13.4 mm and 13.8 mm), a finding supporting the hypothesis that female-biased SSD is favored by stabilizing selection acting independently on the two sexes. Further, the observed mean sizes indicate that both sexes were near their optimal body sizes in the generations studied (figure 3.3).

The analyses of episodes of selection and of selection on components of total length (tables 3.2, 3.3) suggest functional hypotheses to explain the net stabilizing selection on total length. No consistent pattern emerged for prereproductive survival; the statistically significant stabilizing selection in generation 2 males was not seen in the preceding generation and was not confirmed by bootstrap analysis of the cubic spline estimator (Preziosi and Fairbairn 2000). In contrast, body size was under strong and consistent selection during the reproductive season, particularly in females. Larger females had shorter reproductive lives in both generations (table 3.2), apparently because selection favored females with smaller abdomens (table 3.3). In contrast, larger females had higher daily reproductive success, again because strong selection acted on abdomen length (tables 3.2, 3.3). Partial regression analysis of the effects of fecundity on longevity, with size held constant, supported the hypothesis that the reduced longevity of larger females is a true cost of reproduction (Preziosi and Fairbairn 1997). The net stabilizing selection on adult body size in females thus appears to result from a trade-off between daily fecundity and reproductive longevity, selection in both cases targeting abdomen length.

Males show a similar pattern in generation 1, larger males gaining greater daily mating success but having shorter reproductive lives (table 3.2). However, longevity selection in males does not target any specific component of size, while daily reproductive success is clearly positively associated with external genital length (table 3.3). The selection for longer external genitalia is also significant in generation 2 and has since been confirmed in two additional generations (Ferguson and Fairbairn 2000), suggesting that it is a strong and consistent pattern of selection on males.[11] Additional analyses also confirm sexual selection for smaller pregenital bodies (thorax + abdomen), indicating that selection actually favors smaller males with longer genitalia (Preziosi and Fairbairn 1996; Fer-

Table 3.2 Univariate selection gradients for total length in male and female *Aquarius remigis*.

| | \multicolumn{6}{c}{Males} | \multicolumn{6}{c}{Females} |
| | \multicolumn{3}{c}{Generation 1} | \multicolumn{3}{c}{Generation 2} | \multicolumn{3}{c}{Generation 1} | \multicolumn{3}{c}{Generation 2} |
	n	β_1	γ_1	n	β_1	γ_1	n	β_1	γ_1	n	β_1	γ_1
Prereproductive survival	519	−0.02	0.06	456	0.10	**−0.34**	601	0.06	−0.02	560	−0.12	−0.06
Reproductive longevity	105	**−0.17**	−0.18	78	0.22	0.06	114	**−0.35**	0.14	100	**−0.26**	−0.08
Daily reproductive success	36	**0.38**	−0.12	35	0.02	−0.04	46	**0.24**	−0.12	34	**0.17**	0.08
Net adult fitness	450	0.13	**−0.42**	413	0.27	**−0.26**	533	0.00	**−0.28**	494	**−0.28**	**−0.42**

Note: Values in boldface are statistically significant ($\alpha = 0.05$) after correction for multiple comparisons. β_1, univariate linear gradient; γ_1, univariate quadratic gradient; n, sample size. After Preziosi and Fairbairn (2000).

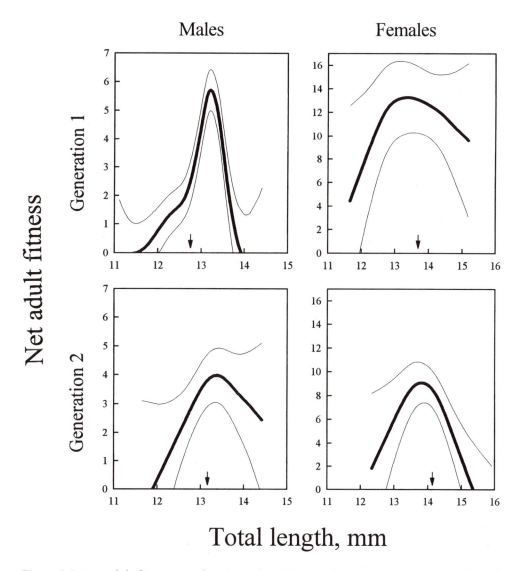

Figure 3.3 Net adult fitness as a function of total length for two generations of male and female *Aquarius remigis*. The *y*-axis shows absolute fitness in units of total number of matings for males and total number of eggs laid for females. Estimates include zero values for adults not surviving to reproduce (see text). The thick lines are fitted cubic splines, and the thin lines are ± 1 SE. The vertical arrows indicate the mean adult total length in the population prior to selection on adult size. (Adapted from Preziosi and Fairbairn 2000.)

guson and Fairbairn 2000). Therefore, net stabilizing selection on male total length appears to result from sexual selection favoring longer genitalia, combined with sexual selection for shorter pregenital segments and selection favoring smaller total length through differential reproductive longevity.

The pattern of multivariate selection on body components is remarkably consistent with the pattern of SSD for these traits (figure 3.2). The most dimorphic traits, abdomen and genital length, show the strongest and most consistent selection, and in both cases, selection for large size occurs in the larger sex and is associated with daily reproductive success. In comparison, thorax length shows little size dimorphism and is not the target of selection in any episode. This concurrence between patterns

Table 3.3 Multivariate linear selection gradients[a] (β_{3i}; see text) for the three components of adult lifetime fitness in male and female *Aquarius remigis*.

		Males		Females	
	Trait	Generation 1	Generation 2	Generation 1	Generation 2
Prereproductive survival	Thorax length	−0.01	−0.06	−0.10	−0.29
	Abdomen length	−0.18	−0.18	0.02	0.00
	Genital length	0.15	−0.15	**0.17**	−0.03
Reproductive longevity	Thorax length	0.22	−0.21	−0.01	−0.13
	Abdomen length	−0.16	0.08	**−0.41**	**−0.36**
	Genital length	0.34	0.21	0.05	−0.13
Daily reproductive success	Thorax length	0.18	0.12	−0.21	−0.02
	Abdomen length	0.14	−0.13	**0.24**	**0.20**
	Genital length	**0.36**	**0.37**	0.01	−0.05

Note: Values in boldface are statistically significant ($\alpha = 0.05$) after correction for multiple comparisons.[b] Sample sizes as in Table 3.2. After Preziosi and Fairbairn (2000).
[a]For clarity we do not present the gradients for the four leg measurements (lengths of the forefemur, midfemur, and hind femur, plus width of the forefemur). None were statistically significant for females, and only 2 of 24 were significant for males. No general patterns were evident in either sex.
[b]Corrected for multivariate estimates including all seven traits.

of SSD and patterns of selection supports Preziosi and Fairbairn's initial hypothesis that adult SSD is determined by selection acting in the adult stage.

This study is a good example of estimation of fitness surfaces for the purpose of understanding the adaptive significance of trait distributions in natural populations. It illustrates longitudinal analysis (selection measured by following individuals over time) of multiple episodes of selection, and it reveals complex interactions among fitness components, including clear examples of life history trade-offs. Further, it combines univariate and multivariate selection gradient analysis with graphical techniques for a more complete understanding of the interactions that lead to fitness differences. The study is also notable because all conclusions were validated by temporal replication across two, and sometimes four (see Ferguson and Fairbairn 2000), generations. However, the lack of estimates of selection on juveniles precludes use of the estimated selection gradients to predict response to selection (i.e., equation 3.9), and the study thus also serves as a reminder of the logistical difficulties of obtaining the requisite data for estimates of lifetime selection in natural populations.

Future Directions

Like many areas of research in evolutionary ecology, studies of selection in natural populations are constrained primarily by the logistical difficulties of obtaining adequate data, and by statistical conundrums associated with multivariate hypothesis testing and parameter estimation. The emergence of standardized selection gradients as widely accepted descriptive statistics for documenting selection is a welcome advance that facilitates consolidation of the field and comparisons among studies. Nevertheless, most studies of selection in natural populations still consider only single fitness components (e.g., sexual selection) and include little or no spatial or temporal replication. The success of these studies in testing or generating adaptive hypotheses for observed character states is laudatory, but they tell us little about more complex evolutionary processes such as the influence of life history trade-offs and genetic or developmental constraints on long-term evolutionary trajectories. For example, we know very little about the impact of ontogenetic effects: to what extent are characters expressed in one life history stage influenced by selection acting in previous or subsequent stages? Studies of natural selection will impact on these broader issues only if we can accumulate more examples of studies of lifetime fitness, multivariate selection estimates, and estimates from multiple episodes of selection. From a more ecological perspective, we have little quantitative appreciation of the spatial and temporal scales of adaptation, or of the relative importance of infrequent events and marginal or rare habitats in adaptive evolution. The logistical challenges of amassing data sets appropriate for addressing these questions are not trivial, particularly for long-lived organisms. Nevertheless, the greatest challenge may lie in deriving tractable statistical methods with sufficient power to detect multivariate selection within these complex data sets, particularly for large or uncommon organisms where sample sizes are likely to be low.

Notes

1. "Evolution" was accepted only in the sense of change in the earth's biota over past eons, not in the Darwinian sense of descent with modification from a single common ancestor.

2. Heritability in the narrow sense is defined as the proportion of the phenotypic variance in the trait that can be attributed to additive genetic variance, and it represents the degree to which offspring can be expected to resemble their parents. Throughout this volume, the term *heritability* refers to narrow-sense heritability unless stated otherwise. Broad-sense heritability, usually symbolized by H^2 rather than h^2, refers to the proportion of the phenotypic variance in the trait that can be attributed to genetic variance from all sources (additive plus nonadditive; see Mazer and Damuth, this volume, chapter 2).

3. In population ecology, this equation usually refers to number of female offspring per female parent and has an expected value of one for stationary populations. When used as a measure of individual fitness, total offspring produced is generally more appropriate because genes are passed on through both male and female offspring.

4. Note that some authors define LRS to include offspring survival to independence or even to first reproduction. Such estimates confound parental fitness with offspring fitness and are not appropriate measures of individual fitness.

5. For kin selection, the definition of fitness must be altered to include indirect descendants (e.g., siblings, nephews, nieces) in whom shared genes may be transmitted to the next generation. The contribution of indirect descendants (nonoffspring) is weighted according to the degree of relatedness or proportion of shared genes, and the resulting fitness measure is known as *inclusive fitness*.

6. In these analyses, relative fitness is estimated as absolute fitness divided by the population mean absolute fitness and thus has a mean of 1.

7. There has been some controversy concerning the interpretation of the quadratic term (β_2). Lande and Arnold (1983) originally suggested that $\beta_2 > 0$ and $\beta_2 < 0$ were sufficient to indicate stabilizing and disruptive selection, respectively. Later authors (e.g., Schluter 1988; Phillips and Arnold 1989) have argued that a local maximum or minimum must also be present. We follow the latter convention.

8. We follow Arnold and Wade (1984a,b) in referring to the coefficients of the vector β and the matrix γ as selection gradients. Some authors reserve the term *selection gradient* for the entire vector or matrix and refer to the individual coefficients as selection coefficients (e.g., Lande and Arnold 1983; Walsh and Lynch 1998). We reserve the term *selection coefficient* for the genotype-specific coefficients of Mendelian population genetic models (i.e., equation 3.1).

9. Terminal molt to the adult stage.

10. The probabilities determined by these two methods generally agreed to the third decimal place. Such agreement is typical unless one of the two categories occurs in very low frequency (Fairbairn and Preziosi 1996).

11. Male *A. remigis* remain in copula with their mates for several hours, during which time the genitalia are everted and inflated. Since males do not use their genitalia for display or combat, the consistent presence of strong sexual selection suggests some functional significance of large genital size during mating.

4

Adaptation

DAVID REZNICK
JOSEPH TRAVIS

When Charles Darwin and Alfred Russell Wallace proposed their theory of evolution by natural selection, the concepts of evolution and speciation were not new. Darwin introduced *The Origin* with "An Historical Sketch," in which he summarized the work of 34 previous authors who had speculated on evolution and the origin of species. What was new about Darwin and Wallace's proposition was natural selection as the mechanism of evolutionary change. Darwin further proposed that natural selection was a unifying process that accounts for adaptation, for speciation, and hence for the diversity of life on earth.

Darwin and Wallace proposed natural selection as a process that caused evolution. Adaptations are features of organisms that were shaped by this process. The modern version of Darwin and Wallace's theory allows for other agents of evolution, such as genetic drift, migration, and mutation, but adaptation remains a product of natural selection alone. The virtue of their proposal is that it allows us to develop testable hypotheses about cause-and-effect relationships between features of the environment and presumed adaptations.

Natural selection immediately became a source of controversy, although the nature of the controversy has shifted over time. First, there has been considerable debate about the definition of adaptation (e.g., Reeve and Sherman 1993). We do not wish to add to or summarize this debate because we feel that Darwin got it right the first time. Be-

sides defining a cause-and-effect relationship between selection and adaptation, Darwin emphasized that we should not expect organisms to be perfectly adapted to their environment. In fact, this emphasis was a large component of his argument against divine creation. For example, Darwin recognized, through his experience with artificial selection, that different aspects of morphology were in some way "tied" to one another so that selection on one trait would cause correlated changes in others that were not necessarily adaptive. He also recognized that organisms were subject to constraints that might limit their ability to adapt. Finally, he argued that how organisms evolved was a function of their history, so that the response to selection on the same trait would vary among lineages.

A more telling criticism considers the application of cause-and-effect reasoning to the interpretation of features of organisms as adaptations, and hence to the empirical study of adaptation. Gould and Lewontin (1979) argued that not all features of an organism are interpretable as adaptive because some are simply the by-products of other adaptations. Gould and Vrba (1982) argued that an association between a trait and a function does not demonstrate that the trait evolved for that function. Instead, traits could be coopted for new functions without further adaptation. Neither of these arguments represents a radical departure from Darwin's original concept of adaptation. They apply instead to the difficulty of understanding cause and effect in retrospect.

An adaptation is thus a feature of an organism that evolved in response to an identifiable form of natural selection. The term *adaptation* is actually used in two fashions. First, it is the process of change in response to natural selection. Second, it is the change that is seen in organisms as a result of selection. In either case, the use of the word implies that natural selection has acted on or is currently acting on the trait. The goal of our empirical study of adaptation is thus to focus on defining these cause-and-effect relationships.

Empirical Approaches to Studying Natural Selection

Early Historical Development

One might expect that natural selection immediately became the object of intense empirical study. However, if we trace the history of our current approaches to studying adaptation, they would lead us to the 1940s. Why was there a 90-year delay between the promulgation of these concepts and their empirical investigation?

There was a well-developed empirical program of study that began in the 1880s and was officially recognized as the "Committee on Evolution," funded by the Royal Society of London (Provine 1971). W. F. R. Weldon and Karl Pearson played major roles in establishing the committee and the associated research programs. Because Darwin emphasized the importance of heritable variation as the fuel for evolution, they focused on quantifying variation in natural populations. They and their colleagues concentrated on quantifying morphology in enormous samples of organisms, then evaluating the statistical distributions and correlations among different features of the phenotype.

One of their methods for evaluating natural selection was to compare the statistical distribution of characters in juveniles and adults in the same population. Differences between cohorts were assumed to provide a measure of selective death during development. One common observation was that the variance of traits in adults was smaller than the variance in juveniles, suggesting the selective death of extreme phenotypes, or stabilizing selection. Weldon argued that stabilizing selection would be the most commonly observed form of

natural selection. Directional selection, which causes a shift in the mean value of the population, would be relatively rare and would quickly run to completion, so it was less likely to be observed. He was right.

Weldon (1899) reported one study in which he did observe what appeared to be directional selection. Weldon and Herbert Thompson observed a narrowing of the relative frontal breadth of crab shells over time in a population from Plymouth Sound. This narrowing spanned samples collected before and after the completion of a breakwater that caused the accumulation of fine silt in the bay. Weldon hypothesized that silt coated the gills and interfered with respiration. The narrower shell was thus an adaptation that could limit the influx of sediment into the gill chamber. Weldon demonstrated that individuals with narrower frontal breadths were less likely to die when exposed to silty water in an experimental chamber. One prominent weakness of the work was his failure to show that this variation was heritable. In fact, results of another experiment suggested that this shape variation was environmentally induced. Nevertheless, it represented a promising beginning.

Why should a study program with such a strong start disappear? We infer disappearance from the nearly complete absence of references to this work in the postmodern synthesis reemergence of empirical studies of adaptation. The only references that we see today are to some of Pearson's statistical tests, such as the Pearson product-moment correlation. The primary reason was that this approach died as an innocent bystander to a conflict over the mechanism of inheritance. It was the misfortune of these scientists, later known as the biometricians, to oppose those who rediscovered Mendelian inheritance and, instead, to champion a mode of inheritance that proved to be wrong. More generally, by the early 1900s, natural selection was not considered an important factor in evolution (Provine 1971). If natural selection was considered unimportant, then there was little motivation to study adaptation. This situation did not begin to change until the 1930s and the publication of the first works of the modern synthesis (see Fairbairn and Reeve, this volume).

Roots of Current Approaches

The number of approaches used to study natural selection is too large to review here. We instead focus on those we have used in our own research. Other methods will be added in the course of presenting our "case studies." Reznick and Travis (1996) offer a more comprehensive review of methods. See Fairbairn and Reeve (this volume) for methods of measuring natural selection.

Classical Ecological Genetics   Ecological genetics considers the dynamics of discrete genotypes (genetics) in the field (ecology). Investigators monitored the numerical dynamics of discrete phenotypes that exhibited simple Mendelian patterns of inheritance. The changes in frequency of alternative phenotypes through the life cycle and the consistency of those patterns across generations were the signatures of natural selection. Investigators also frequently compared populations from different environments and sought correlations between the relative frequencies of phenotypes and variation in environmental factors. Significant correlations were interpreted as evidence that spatial variation in phenotype frequency was an adaptive response to the variation in that environmental factor.

The earliest case studies focused on visible polymorphisms in color or external markings of a trait (see Reznick and Travis 1996; Travis and Reznick 1998; Mazer and Dalmuth, this volume, chapter 1). These included industrial melanism, spot patterns in butterflies, banding patterns in snails, and chromosome inversion types in *Drosophila* (Ford 1971). The best-studied cases taught investigators a number of lessons, including that natural selection and adaptation were capable of being observed, quantified, and documented.

This approach had two disadvantages. First, its reliance on visible polymorphisms created a potential bias in assessing the nature of selection. The genes controlling these polymorphisms are a nonrandom sample of all genes, and there was no way to discern if they were under especially strong selection because of their visible effects on the phenotype. Second, the visible polymorphisms were often merely markers of genetic variation; the actual target of selection frequently included pleiotropic effects of those genes. Pleiotropy occurs when an individual gene influences multiple features of an organism; *pleiotropic effects* thus refers to effects of genes other than those on the visible polymorphism. These discoveries did not compromise the conclusion that selection was acting forcefully;

they did indicate that the actual targets of selection and its "typical" strength remained to be quantified rigorously. A lasting benefit was the advent of mark-recapture methodologies, which enabled investigators to evaluate the fates of individuals or phenotypes.

These same principles have been applied to biochemical polymorphisms, such as those seen at loci controlling enzymes and structural proteins. The strength of selection acting upon such variation may be weaker, but the various types of selective pressures and the methods used to understand those pressures are no different than those employed in the study of visible polymorphisms.

Studies of Continuous Trait Variation   An alternative approach is to focus on traits that are continuously distributed, or quantitative traits (see Mazer and Dalmuth, this volume, chapters 1 and 2; Fairbairn and Reeve, this volume). Such traits are often represented by quantities, such as length, mass, or shape. Their expression is usually controlled by many loci, although the loci involved are almost always unknown. The genetic basis of such traits is evaluated with quantitative genetics, which yields inferences about genetics from patterns of phenotypic similarity among relatives. The advantage of this approach is that it makes it possible to study the adaptive significance of virtually any feature of an organism, rather than being restricted to discrete polymorphisms. These are the types of traits that Darwin considered the most important in evolution because individual variation in them is ubiquitous and they are readily modified by artificial selection.

The current focus on continuous traits began when Russell Lande and colleagues developed a unified treatment of multivariate selection and quantitative genetic variation in evolving populations (reviewed in Roff 1997; see also Reznick and Travis 1996; Fairbairn and Reeve, this volume). Lande's treatment focused attention on two points. First, phenotypic correlations among traits allow selective changes in one to be translated into correlated changes in others (this effect is called *indirect selection*). Second, pleiotropic effects of the alleles controlling continuously distributed traits create genetically based covariances among them that channel the short-term direction of multivariate evolution. Taken together, these points suggested that studies of continuous traits needed to focus on a suite of developmentally and genetically associated characters in order to clarify how selection acts directly or indirectly on any one of them.

Case Studies

The methods chosen to study adaptation differ among studies. The choices are guided by the existing knowledge of the system and practical considerations—which questions about adaptation are operational and which methods can be applied to answer those questions. We illustrate these points by reviewing selected features of three studies of variation that we and our colleagues have pursued over a period of decades. Our studies include life history variation in guppies *(Poecilia reticulata)*, body size and life history variation in sailfin mollies *(Poecilia latipinna)*, and variation in life history and response to acute thermal stress in least killifish *(Heterandria formosa)*. The work on guppies and sailfin mollies has been reviewed in more detail elsewhere (Reznick 1996; Reznick and Travis 1996).

Guppies

Life History Theory and Guppies   Life history represents a specific form of adaptation that has been a focus of research since the late 1940s (see part II, this volume). The "life history" is the composite of variables that contribute to how organisms propagate themselves. Important variables include the age at maturity (Roff, this volume), frequency of reproduction, number and size of offspring (Messina and Fox, this volume), and "reproductive effort," or the proportion of available resources devoted to reproduction, as opposed to growth or maintenance. Important features of the theory are that it incorporates a quantitative definition of fitness and predicts how evolutionary changes in life history variables can maximize fitness (Roff, this volume).

One branch of life history theory predicts how life histories will evolve in response to a change in mortality rate. One prediction is that an increase in adult mortality rates will select for individuals that mature at an earlier age and devote more of their available resources to reproduction. An increase in the juvenile mortality rate will select for

delayed maturity and a reduction in the rate of investment in reproduction.

Our guppy research was initiated to evaluate these predictions in natural populations. Caryl Haskins and colleagues (1961) initiated guppy research in Trinidad during the 1950s. They characterized the distribution of guppies and guppy predators in the streams that drained the rainforests of the Northern Range Mountains. We have concentrated on two types of communities. "High-predation" communities are those where guppies co-occur with cichlids and tetras that frequently prey on guppies. "Low-predation" communities are those where guppies cooccur with only the killifish *Rivulus hartii*, which occasionally eats smaller size classes of guppies. Low-predation communities are found in the same streams as high-predation communities, but above barrier waterfalls that exclude the larger species of fish, a contrast that is repeated in several drainages throughout the Northern Range. Each stream presents a potentially independent example of adaptation to different predation environments. If living with predators means having a higher adult mortality rate, then life history theory predicts that guppies from high-predation localities should mature at an earlier age and have higher reproductive efforts than their counterparts from low-predation localities.

Mortality Rate Estimations   One goal was to establish whether there were really differences in the mortality rates of guppies from high- versus low-predation environments (Reznick et al. 1996). While it may seem obvious that the presence of predators will result in higher mortality rates, there are many factors that can influence a predator's choice of prey and many factors other than predators that can influence mortality rate. Since it is overall mortality rates that are the presumed agent of selection, it was necessary to measure them directly. To do so, we employed mark-recapture methods.

Many streams are divided into discrete pools with a low flow rate separated by riffles, or steeper, rockier portions of stream with higher flow rates. Guppies tend to congregate in pools and to have limited movement from one pool to another. We collected every guppy in a pool and marked it with a subcutaneous injection of acrylic latex paint diluted with physiological saline. We released the guppies in their native pool, then recollected them 12 days later. The probability of recapture is a very accurate estimate of the probability of survival for 12 days. We interpreted recapture probabilities in this fashion because we also evaluated the probability that guppies were present in the release pool but not caught and the extent to which guppies emigrated from the release pool; we also demonstrated that the marks did not influence their susceptibility to predation.

We repeated the mark-recapture study in seven high- and seven low-predation pools distributed among three streams of each type. We found that the probability of recapture of guppies from high-predation localities was significantly lower than for guppies from low-predation localities, or that they suffered a significantly higher mortality rate (figure 4.1). This result demonstrates that living with predators is correlated with higher mortality rates.

Comparative Studies   We used two types of comparative studies to compare the life histories of guppies from these different environments. Comparative studies are a classic first step for evaluating a presumed adaptation because they focus on populations of a target organism that are found in localities that differ in one or a small number of factors. In our case, the target organism was guppies, the environmental factor was predation, and the dependent variables were life history traits. The goal was to see if there is an association between the life histories of guppies and predation. Such an association is presumptive evidence that predation caused the guppies to evolve the observed life history patterns. First, we collected and preserved samples of guppies from a series of high- and low-predation environments; then we quantified variables that characterize the life history. We refer to

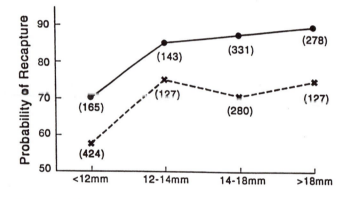

Figure 4.1 Recapture probabilities of juveniles, mature males, and females from seven high-predation *(Crenicichla)* and seven low-predation *(Rivulus)* localities. For purposes of analyses, we divided the fish into four size classes. The smallest two (< 12mm and 12–14 mm) consist of immature individuals. The 14–18-mm size class includes > 95% of mature males and all females reproducing for the first time. The > 18-mm-size class includes females that were producing their second or subsequent litters. The number of individuals marked and released in each size class are indicated in parentheses next to each data point. The probability values are derived from a log-linear analysis with survival, size, and predator as independent variables. The probability of recapture of guppies from *Crenicichla* localities was significantly lower than for *Rivulus* localities (*p* < .001). The probability of recapture from both types of localities was significantly less in the < 12-mm-size class than in the larger size classes (*p* < .001).

these descriptions as *life history phenotypes* because they are influenced by the genotype of the fish and the environment in which the fish developed. We then collected adult females from the field and returned them to the laboratory. Female guppies store sperm, often from multiple sires. Offspring from each female were reared to maturity in a common laboratory environment, then mated to

individuals from a different litter to produce the second generation of laboratory-reared offspring. These second-generation offspring were reared individually on controlled levels of food. We estimated their age and size at maturity, the number and size of their offspring, and the percentage of consumed resources that were devoted to reproduction as an index of reproductive effort. We re-

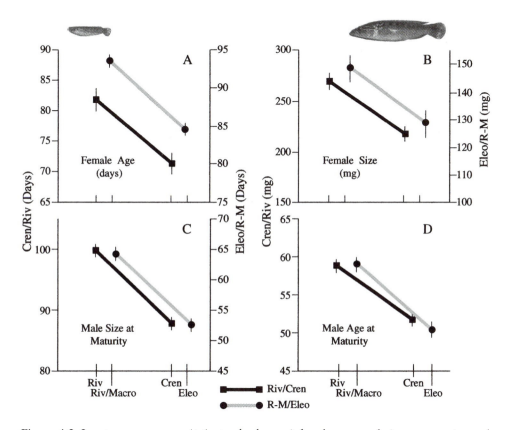

Figure 4.2 Least square means (± 1 standard error) for the age and size at maturity and size at first parturition in females from high versus low predation environments. These data are based on the comparison of fish from two high- and two low-predation sites on the north and south slopes, for a total of eight localities. The stippled lines represent the means for the high (Eleo) versus low (Riv/Macro) predation sites on the north slope. The solid lines are the corresponding values for the high (Cren) and low (Riv) predation sites on the south slope. We used lower levels of food availability in the experiment on north-slope guppies. A consequence of lower food is that these fish tended to mature at a later age and smaller size than the south-slope guppies. To facilitate comparisons among the two experiments, we scaled the y-axes differently so that the proportional differences between high- and low-predation fish remained the same. All differences between the high- and low-predation sites were significant ($p < .05$). (A) Female age at first parturition (days). (B) Female size at first parturition (wet weight in mg). (C) Male size at maturity (wet weight in mg). (D) Male age at maturity (days).

fer to this characterization as the *life history genotype,* since it is very likely that there is a genetic basis to any differences in life histories among guppies from different localities that persist for two generations in a common environment.

The life-history-phenotype approach has the benefit of being relatively easy, so it can be applied to a large number of localities over a wide area. The disadvantages of this approach are that the pheno-

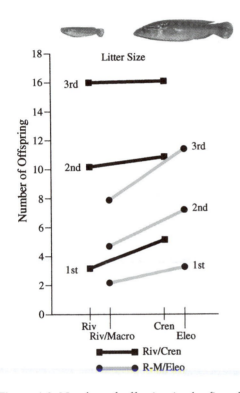

type includes genetic and environmental effects and that it is not possible to quantify some of the variables of interest, such as age at maturity. The life-history-genotype approach removes potential environmental effects from the comparison of guppies from different localities. We can also quantify all of the variables of interest. The disadvantage is that it is time-consuming and expensive, so we cannot evaluate as many localities. Our intent was to apply both approaches and exploit their combined advantages, which include wide geographical coverage and a deep description of the genetic differences among populations.

The two approaches yielded similar results (Reznick 1996). Guppies from high-predation environments mature at an earlier age and have higher rates of investment in reproduction than those from low-predation environments (figures 4.2–4.4). These results are consistent with theoretical predictions, suggesting that the differences in life histories are adaptations to the mortality rates experienced by guppies in high- versus low-predation environments. Our results also show how animals can evolve differences in reproductive effort. Guppies from high-predation environments tend to have shorter intervals of time between litters. They also tend to devote more resources to each litter, where the amount of resources is the product of the number and size of their offspring. They produce many more offspring per litter, but the individual babies are smaller.

Introduction Experiments   While these comparative studies produced a pleasing correlation between predation and life histories, we were interested in studying evolution from an experimental perspective. To do so, we continued a tradition, initiated by John Endler (1980), of experimentally manipulating the distribution of guppies and guppy predators. Manipulations allowed us to modify the guppies' environment in a specific fashion and hence to evaluate how they evolved in response to this modification. Manipulations make it possible to define with greater precision the factors that cause life history traits to evolve and hence to recognize those changes as adaptations to a specific type of selection. Our main experiments were done in streams that had waterfalls that barred upstream dispersal of all fish except the killifish *Rivulus hartii,* which is the key predator on guppies in low-predation communities. *Rivulus* are often found above barriers that exclude all other species of fish

Figure 4.3 Number of offspring in the first three litters for guppies from high- versus low-predation sites on both slopes of the Northern Range. These data are from the same experiments as summarized in figure 4.2. The stippled lines represent the results for the north-slope comparisons. In this case, guppies from the high-predation localities produced significantly more offspring, after correcting for female size differences, than their counterparts from the low-predation localities. The solid lines represent the results from the south-slope comparisons. Guppies from high-predation localities produced significantly more offspring in the first litter, but not in the second or third litters. Note that the fecundity differences between the north and south slopes are caused primarily by the lower levels of food availability in the north-slope study.

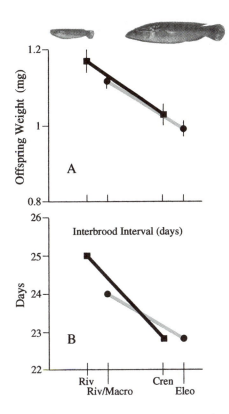

Figure 4.4 Least square means (± 1 SE) for high versus low predation sites for offspring weight (third litter only) and interbrood interval. These data are from the same experiments as summarized in figure 4.2. The stippled lines represent results from the north-slope comparisons, while the solid lines represent results from the south-slope comparisons. Guppies from either high-predation locality produced smaller offspring and had shorter intervals between broods than their counterparts from low-predation localities.

because they can disperse overland on rainy nights. In two such streams, guppies were collected from a high-predation locality below the barrier waterfall and then introduced into the guppy-free, low-predation locality above the waterfall. John Ender made the first such introduction to a tributary to the Aripo River in 1976. The second introduction was made by one of the authors (DR) on a tributary to the El Cedro River in 1981. Such introductions are an artificial range extension of fish within their own drainage. As such, they fall within the realm of naturally occurring events and are the

minimum possible perturbation that would allow us to test an evolutionary theory in nature.

Our principal prediction was that the lower mortality rate associated with this introduction would select for guppies that had delayed maturity and lower levels of reproductive effort. We also predicted that selection would favor a reduction in fecundity and an increase in offspring size because these differences characterize the usual comparison between high- and low-predation environments. We evaluated these predictions by comparing guppies from the introduction site to the "downstream control," which was the high-predation locality from which the introduced guppies were derived. We would conclude that the introduced populations had evolved if they differed significantly from the control population in their life history genotype, which involved a comparison of the second generation of laboratory-born fish.

We compared the experimental and control populations from the El Cedro River 4, 7, and 9 years after the introduction. We evaluated the Aripo Tributary 11 years after the introduction. The life histories of the guppies evolved as predicted; the differences between experimental and control treatments resembled those between low- and high-predation environments. Male age and size at maturity evolved far more rapidly than did female life history traits. For example, after only 4 years, the males from the introduction site on the El Cedro River differed from the males at the control site by an amount typical of comparisons between any pair of low- and high-predation environments, while the females exhibited no differences. Females from the introduction site began to show significantly delayed maturity by 7 years but were not different in any other life history trait. On the Aripo River after 11 years, there were significant differences in male age and size at maturity, female age and size at first parturition, offspring number, offspring size, and early life reproductive effort. All of these differences were consistent with our predictions (table 4.1).

Selection Gradient Analysis   The introduction experiments lend themselves to a more general evaluation of how evolution works in natural populations through the application of selection gradient analysis (Reznick et al. 1997; Fairbairn and Reeve, this volume). A common application of selection gradient analysis is a comparison of the phenotypes of survivors and nonsurvivors. Multiple re-

Table 4.1 Results of two introduction experiments in which guppies were introduced from a high-predation locality (i.e., predation by *Crenicichla*) into a low-predation locality (i.e., predation by *Rivulus*) that previously had no guppies.[c]

| Life history trait[a] | Introduction experiments | | | | | |
| | Reznick and Bryga (1987) | | Reznick et al. (1990) | | Reznick (1982) | |
	Control *(Crenicichla)*	Introduction *(Rivulus)*	Control *(Crenicichla)*	Introduction *(Rivulus)*	*Crenicichla*	*Rivulus*
Male age at maturity (days)	60.6 (1.8)	72.7 (1.8)**	48.5 (1.2)	58.2 (1.4)**	51.8 (1.1)	58.8 (1.0)**
Male size at maturity (mg-wet)	56.0 (1.4)	62.4 (1.5)**	67.5 (1.2)	76.1 (1.9)**	87.7 (2.8)	99.7 (2.5)**
Female age at first parturition (days)	94.1 (1.8)	95.5 (1.8)	85.7 (2.2)	92.3 (2.6)*	71.5 (2.0)	81.9 (1.9)**
Female size at first parturition (mg-wet)	116.5 (3.7)	118.9 (3.7)	161.5 (6.4)	185.6 (7.5)**	218.0 (8.4)	270.0 (8.2)**
Brood size, litter 1	2.5 (0.2)	3.0 (0.2)	4.5 (0.4)	3.3 (0.4)*	5.2 (0.4)	3.2 (0.5)**
Brood size, litter 2	6.3 (0.3)	7.0 (0.3)†	8.1 (0.6)	7.5 (0.7)	10.9 (0.6)	10.2 (0.8)
Brood size, litter 3	—[b]	—[b]	11.4 (0.8)	11.5 (0.9)	16.1 (0.9)	16.0 (1.1)
Offspring size (mg-dry), litter 1	0.91 (0.02)	0.87 (0.02)	0.87 (0.02)	0.95 (0.02)†	0.84 (0.02)	0.99 (0.03)**
Offspring size, litter 2	0.93 (0.02)	0.86 (0.02)*	0.90 (0.03)	1.02 (0.04)*	0.95 (0.02)	1.05 (0.03)*
Offspring size, litter 3	—[b]	—[b]	1.10 (0.03)	1.17 (0.04)	1.03 (0.03)	1.17 (0.04)**
Interbrood interval (days)	24.9 (0.4)	24.89 (0.4)	24.5 (0.3)	25.2 (0.3)	22.8 (0.3)	25.0 (0.03)**
Reproductive effort (%)	4.0 (0.1)	3.9 (0.1)	22.0 (1.8)	18.5 (2.1)	25.1 (1.6)	19.2 (1.5)*

Note: All of these results represent "common garden" comparisons of the second-generation, lab-reared offspring from wild caught females. The Reznick and Bryga (1987) data represent a comparison of the control (high-predation) and experimental (low-predation) populations for guppies introduced over a barrier waterfall on the El Cedro River. This assay represents a comparison of fish collected 4 years after the introduction, which is equal to approximately 6.9 generations. At this time, only male age and size at maturity had changed. In a subsequent assay, based on fish collected 7.5 years after the introduction, or approximately 12.7 generations, both males and females from the introduction site had delayed maturity. The results from Reznick et al. 1990 are based on a duplicate experiment on a tributary to the Aripo River. In this case, the parental generation was collected 11 years, or approximately 18.1 generations, after the introduction. Most aspects of the life history had evolved in the predicted fashion in this study. The Reznick (1982) data are from a comparison of two *Rivulus* and two *Crenicichla* localities and are included here to provide a frame of reference. Note that differences in mean values among experiments are largely attributable to differences in food availability. The Reznick (1982) study had the highest levels, Reznick et al. (1990) had intermediate levels, and Reznick and Bryga (1987) had the lowest levels of food availability. Details on these three references can be found in Reznick and Travis (1996).
[a]Differences in mean values among experiments are attributable to differences in food availability. Reznick (1982) had the highest levels, Reznick et al. (1990) was intermediate, and Reznick and Bryga (1987) had the lowest levels.
[b]Fish were kept only until they produced two litters of young.
[c]Comparisons between control and introduced fish were not significant unless indicated: **P < .01, *P < .05, †P < .10.

gression analysis of such data allows one to evaluate the degree to which individual traits predict survival (direct selection) and the extent to which changes in traits are caused by correlations with other traits that contribute to survivorship (indirect selection). This analysis of selection is not necessarily synonymous with evolution because it does not account for the genetic basis of phenotypic variation and thus the extent to which the mean phenotype changes in subsequent generations.

We analyzed an episode of evolution by comparing age and size at maturity in guppies of both genders from the control and experimental populations. We calculated a net "response to selection," which was the difference in mean values between the two populations in the life history genotype assay. We independently estimated the genetic variance-covariance matrix for the life history traits, which is a measure of the degree to which phenotypic variation within a population is genetically controlled. This matrix also estimates the extent to which correlations between phenotypic traits are genetically based. For example, we often found a high genetic correlation between age and size at maturity, which implies that many of the genes that control the age at maturity also control size at maturity. We used these two sets of information to infer how selection acted to produce the evolutionary changes that we saw in the introduction sites.

In both studies we found males to have far more genetic variation for age and size at maturity than females. This result implies that males evolved faster than females primarily because they had more genetic variation for selection to act on. The analysis also revealed that selection acted primarily on age at maturity and that size tended to change because it was genetically correlated with age. These general patterns were the same in both replicates of the experiment.

We used these results to quantify the rate of evolution with a metric called the *darwin*, which was invented by Haldane (1949). The darwin measures the proportional rate of change in a trait per unit time by dividing the difference between the natural log-transformed values of the trait at the beginning and end of the measurement period by the amount of time over which the change took place. The darwin can be applied to a diversity of data, including the fossil record, so it is possible to compare rates of evolution from different data sets. The striking feature of our results is that the apparently subtle changes over the period of a decade are very rapid when placed on the same scale as rates inferred from the fossil record. The rate of evolution was the same order of magnitude as is often observed in artificial selection and 10,000 to 10 million times faster than rates in the fossil record! Similar results have been obtained in other studies that have estimated rates of evolution through episodes of directional selection, the most famous of which are the studies of the Galapagos finches by Peter and Rosemary Grant and their colleagues (e.g., Grant and Grant 1995). Although we do not have the space to argue the point here, we feel that the differences in perceived rate occur because the fossil record is biased toward the perception of evolution being far slower than it really is. Such

results also support the argument that natural selection could be the primary mechanism of evolution throughout the history of life (see Reznick et al. 1997).

Complications   These results indicate that the differences among populations of guppies in life history traits are local adaptations caused by differences in mortality rate. We will not argue that predation is the sole cause of these patterns. Other results suggest that two other effects contribute to life history evolution in guppies: environmental factors that are correlated with predation and some indirect consequences of predation. Guppies from high-predation localities encounter higher levels of resource availability, in part because their streams are larger and have higher light levels and in part because predation produces lower guppy population densities. These differences among sites can serve as independent agents of selection or can interact with predation to select on the life history.

There are other data that indicate that mortality from predators is not the sole agent of selection on guppy life histories, so we cannot claim to understand fully the evolution of those life histories. This suggestion of multiple, interacting agents of selection puts our work on common ground with every other well-studied example of evolution by natural selection in natural populations and is probably a universal feature of the process, as discussed below.

Sailfin Mollies and Least Killifish

The guppy research began with the a priori prediction that predation is an agent of selection. An alternative approach is to document the pattern of variation among populations in phenotype distribution, to separate genetic from environment sources of phenotypic variation, and then to develop hypotheses about the agents that cause adaptive differences among populations. The best methods are dictated by the nature of the study system.

Comparing Patterns of Variation   The traditional approach of ecological genetics begins with surveying natural populations for variation in phenotypes and associations between that variation and environmental factors. A newer approach is to compare the spatial pattern of phenotypic variation to that expected if traits were diverging as a result of nonadaptive processes (e.g., genetic drift; see Mazer and Dalmuth, this volume, chapter 2; Nunney, this volume). We suggest that neither approach is foolproof and that each merits consideration; indeed, both approaches might be advisable, as we illustrate here.

While sailfin mollies and least killifish inhabit different worlds, the diversity of habitats each occupies and the diversity of other species with which each must contend makes them promising candidates for the study of local adaptation. Sailfin mollies inhabit salt marshes, tidal creeks, and calcium-rich freshwater habitats along the coast of the southeastern United States and northeastern Mexico plus the inland waters of peninsular Florida. Least killifish occupy a variety of freshwater habitats, including riverine floodplains, hard water springs, soft water lakes and swamps, and even brackish creeks in the coastal plain of the southeastern United States. Surveys of both species reveal extensive variation among populations in body size, a variety of life history attributes, and physiological properties such as tolerance to acute thermal stress (see Reznick and Travis 1996 for mollies; see Leips and Travis 1999; Travis et al. 1999; Baer and Travis 2000 for least killifish).

It is possible that this variation has no adaptive significance. It may have been created by the historic pattern of range formation; it may also result from ongoing effects of genetic drift in different populations and sporadic gene flow among them. If either nonadaptive explanation were correct, then the spatial pattern of trait variation ought to match the pattern of genetic variation at neutral markers. While a match between the two patterns is inconclusive, a mismatch suggests that historic factors cannot account for the trait distribution and that a hypothesis of local adaptation might apply.

The spatial genetic structures of the two species have broad similarities. Both species show geographic variation in allele frequencies with a broad pattern of increasing genetic differentiation between populations with increasing spatial separation (Trexler 1988; Baer 1998). But there is a key difference between the species. Sailfin mollies have a spatial genetic structure consistent with extensive gene flow among local populations, a conclusion reinforced by observations of apparent gene flow. On the other hand, the spatial genetic structure of least killifish populations is consistent with much lower rates of local exchange and much more genetic independence among local populations.

The spatial genetic structure in mollies offers a clear null hypothesis for quantitative traits. Variation among local populations in male and female body size, life history traits, and male morphology and behavior is large in magnitude and unrelated to the proximity of populations to one another. This discrepancy implicates local adaptation in these traits. Unfortunately, multivariate analyses of the ecological variables that vary among populations offer no direct insight into the potential agents of any such adaptation (Travis and Trexler 1987). Thus, while the traditional avenue of ecological genetics fails to implicate local adaptation in this case, the comparison to the pattern of neutral genetic structure succeeds.

The situation is reversed for least killifish. While these populations exhibit considerable local and geographic variation in a variety of quantitative traits, it is difficult to show that the pattern in this variation is incompatible with the expectation from neutral genetic markers. Least killifish exhibit a weak spatial pattern in neutral markers, and in some areas, the populations have not achieved an equilibrium between migration and drift. This result allows a variety of null hypotheses to be constructed, and consequently, it is difficult to falsify all of them and embrace the idea of local adaptation. But least killifish exhibit stronger associations between phenotypic variation and specific environmental variables than do mollies (see below).

Separating Genetic from Environmental Variation

While correlations between trait variation and environmental variation can reflect adaptive effects, these correlations could reflect environmental effects on trait expression. A variety of methods can separate genetic from environmental sources of trait variation, and choosing among them depends on the level of genetic detail of interest as well as the practicality of each method (Reznick and Travis 1996; Travis and Reznick 1998). The importance of this choice is revealed by our discovery of how nongenetic maternal effects (see Mazer and Dalmuth, this volume, chapter 2) can mimic genetic ones in the response to acute thermal stress in the least killifish.

Phenotypic patterns in this trait suggested that fish from a constant-temperature spring had a smaller range of thermal tolerance than fish from a shallow lake with widely fluctuating temperatures, which suggests an adaptive differentiation (Travis et al. 1999). The argument would be that the population in the fluctuating thermal environment requires wider tolerance to persist in that location than the population from the more constant thermal environment. Further, because maintaining the machinery for wide tolerance is expensive, the population in the more constant environment, which does not need as much of that machinery, ought to divert the energy that might be used for that purpose to other demands.

Two experimental results suggested a more subtle explanation. First, Forster-Blouin (1989) demonstrated experimentally that the thermal environment experienced by a gravid female affects her offspring's tolerance to acute thermal stress as adults. Second, genetic crosses and selection experiments with stocks from different populations indicated that the response to acute thermal stress had, at most, very low heritability (Baer and Travis 2000). In subsequent modeling, we found that it would be very easy for nongenetic maternal effects to generate different ranges of thermal tolerance in populations with different thermal regimes and for those differences to masquerade as adaptive genetic distinctions (Travis et al. 1999).

Developing and Testing Hypotheses about Adaptation   If nature were simple, we would find it easy to identify some agent of selection that favored one range of trait values in one location and a different range in another. However, as the guppy studies have shown, it is more likely that several ecological factors will combine to generate the forces of selection. Our studies of body size variation in sailfin mollies illustrate this point. We have identified four agents of selection on male body size (see Reznick and Travis 1996 and Travis and Reznick 1998). First, wading bird predation selects against larger individuals. Second, chronic cold stress in winter selects against smaller individuals. Third, sexual selection, through intermale competition and female choice, selects against smaller individuals. Fourth, the strong covariance of development time and body size causes fertility selection to work against larger body sizes as an indirect effect of selection for early maturity in populations growing in numerical size in spring. While the third and fourth agents appear to act in all populations, the first two are quite variable among populations and among years in the same population. This variability has made it very difficult to demonstrate the precise set of differences that maintains distinctions between any two populations of mollies.

Our studies of the least killifish have allowed more progress in isolating potential causes of adaptive differentiation. The challenge is deciding which among these agents actually act to maintain trait differences among populations. To help make this decision, we have used three types of experiments in mesocosms: artificial selection (Baer and Travis 2000), numerical dynamics in populations of experimentally constructed genotypes (Leips et al. 2000), and numerical dynamics in environments that re-create specific ecological differences among locations (Leips and Travis, unpublished data). Results from the latter experiments illustrate how difficult it can be to define the critical agent of selection.

There are substantial differences in water chemistry between the alkaline springs and acidic lakes and ponds in which least killifish are found (Leips and Travis 1999). Four lines of evidence suggested that such differences might generate adaptive differentiation. First, these chemical differences create distinct physiological challenges that might dictate different allocations of resources (Dunson and Travis 1991). Second, the distribution of poeciliid and fundulid species in the Florida peninsula are correlated with differences in water chemistry, a finding that suggests that such differences influence the ability of populations to persist in specific habitats (Dunson et al. 1997). Third, experimental work with fish and larval amphibians had shown that variation in water chemistry can alter development patterns and competitive interactions between species (Warner et al. 1993). Fourth, early screening of populations from each habitat suggested that fish in alkaline springs attained higher densities but had lower fecundity with larger individual offspring.

To evaluate the hypothesis, we studied experimental populations in 16 mesocosms, 8 with slightly alkaline water and 8 with acidic water. We introduced fish from two local populations from each habitat type into each condition, with two replicates per population per habitat type (see Warner et al. 1993 for methods). In each replicate, we introduced 10 adults of each gender. This experimental design crossed the fixed effect of "treatment" (alkaline or acidic water) with the fixed effect of "origin" (stock derived from a population from each habitat) with the random effect of local population nested within origin. If local populations were adapted to their water chemistry, populations from acidic habitats ought to perform better in acidic conditions and populations from alkaline habitats ought to perform better in alkaline conditions.

The initial results of this experiment, population numbers at the end of the first reproductive season, did not support this hypothesis (table 4.2). These numbers indicate each experimental population's capacity for increase from low initial density, which is a reasonable estimator of fitness for density-independent dynamics. Experimental populations derived from acidic origins attained numbers in alkaline conditions that were twice as high as those they attained in acidic conditions; experimental populations derived from alkaline conditions attained comparable numbers in the two conditions. Analysis of variance on the log-transformed abundances revealed no main effect of treatment but a strong effect of original habitat type (tested over the nested effect of population within habitat type, $F_{1,2} = 34.93$, $P < .05$); this effect was driven by the extraordinary response of the fish from acidic origins to alkaline conditions. The interaction of treatment and origin was not significant ($F_{1,10} = 3.42$, $.10 > P > .05$), but this lack of significance may be a function of our limited replication level. Nonetheless, it does not appear that water chemistry is an agent of direct selection promoting local adaptation.

We considered whether divergent natural densities were agents of selection for divergent life histories (Leips and Travis 1999). We constructed genetic stocks with different proportions of alleles from a high-density and a low-density population and examined the numerical dynamics of those stocks in artificial ponds (Leips et al. 2000). The phenotypic differences in life history traits that we had seen between the source populations persisted in the experimental stocks. Moreover, the expression of trait variation among stocks, particularly in offspring number and size, matched the predictions of density-dependent selection theory.

These results do not mean that water chemistry plays no role. Acidic habitats have lower densities of least killifish and higher relative abundances of predators (Leips and Travis 1999). The correlation of water chemistry with population density led us to a reasonable but incorrect hypothesis about the specific agent of selection. Fortunately, the differences among locations in the putative selective agents on least killifish have been very consistent across years (Leips and Travis 1999; Travis and Leips, unpublished additional data), in contrast to the situation with mollies. Which of these scenarios is more typical for investigations of local adaptation remains to be seen.

Conclusions

An implicit message from our "case studies" is that there is not a single formula for studying adaptation. Instead, one might employ a diversity of methods to evaluate the adaptive significance of traits. What these methods have in common is the goal of defining the cause-and-effect relationship between the trait and fitness and, as much as possible, of demonstrating that the trait evolved for a specific function. A key element of studying adaptation is thus to consider the element of history, or the features of the environment that caused the average properties of the population to change over time. We have learned three important lessons in our efforts to define cause and effect:

Adaptation is not unifactorial   Our scientific training encourages us to think in terms of null and alternative hypotheses and then to distinguish these with experiments. While this method is the most powerful approach to science, we might often be forced to evaluate multiple alternative hypotheses. Adaptations are shaped by all of the aspects of the environment that influence fitness, the genetic correlations within the phenotype, and the organism's history. For guppies, both predation and resource availability appear to be important, and the two may interact. For mollies, predation, winter survival, mating success, and development rate affect the fitness of males of different sizes. Popula-

Table 4.2   Average numbers of individuals in experimental populations (± 1 standard error) at the cessation of the reproductive season as a function of water chemistry treatment and type of habitat from which the stock originated.

Treatment	Original habitat type	
	Acidic water	Alkaline water
Acidic water	288 ± 18	349 ± 77
Alkaline water	603 ± 108	289 ± 81

tion processes such as gene flow can further complicate our ability to evaluate the fit between local phenotype and local environment (Nunney, this volume). Virtually every classic study in ecological genetics began with the evaluation of a simple proposition, such as "Melanic moths increased in frequency because they were less likely to be eaten by predators when seen against the darkened trunks of trees." All have ended with the realization that even simple traits are shaped by competing influences (Jones et al. 1977; Grant 1999).

Adaptation requires compromise Because adaptation involves so many factors, what we actually see is often shaped by trade-offs among competing influences. For example, the color and banding pattern of *Cepaea* snail shells affects the snails' vulnerability to visually oriented predators, but color also affects the thermal properties of the shell. Colors that are best for the prevailing thermal environment do not always provide the lowest risk of predation (Jones et al. 1977).

Fitness is context-specific The most fit phenotype (or genotype) is a specific function of which other phenotypes (or genotypes) are present. Adaptations arise as a consequence of the differential success among alternative phenotypes available in a population. For example, melanic moths form a higher percentage of samples from industrialized than from nonindustrialized regions. They also increased in frequency in some locations as the local environments changed through the effects of industrialization. The fitness of melanic moths varies between industrialized and rural regions, as does that of the peppered moth. The rank order of each

morph's fitness changes with environment, and this change in ranking determines which phenotype prevails. This point may seem obvious, but it is part of a powerful argument against the claim that "adaptive evolution is tautological" (e.g., Peters 1976).

The substance of the tautology argument is that fitness is judged in retrospect; the phenotype that increases in frequency is assumed to have the highest fitness. The existence of differential fitness among phenotypes is in fact often hypothesized from a correlation between the phenotype and an environmental variable. However, this is a hypothesis that is falsifiable and therefore scientifically valid. The biologist can falsify the hypothesis of adaptive evolution by estimating the average fitness of each phenotype in each environment and ascertaining if the ordering of those fitnesses corresponds with the correlation between phenotype relative frequency and environment. The application of life history theory to the guppy introduction experiments represents an example of this research program.

An important feature of the guppy introduction experiment and others like it is that natural selection can be shown to act very rapidly and hence be amenable to experiments performed during the lifetime of the investigator. While we laud Darwin for the success of his ideas, we are pleased that he was wrong in his characterization of natural selection as a slow, continuous process that would, only after great intervals of time, cause significant changes in organisms. We would not have succeeded in our work had Darwin been correct on this point.

5

Phenotypic Plasticity

MASSIMO PIGLIUCCI

Phenotypic plasticity is the property of a genotype to produce different phenotypes in response to different environmental conditions (Bradshaw 1965; Mazer and Damuth, this volume, chapter 2). Simply put, students of phenotypic plasticity deal with the way nature (genes) and nurture (environment) interact to yield the anatomy, morphology, and behavior of living organisms. Of course, not all genotypes respond differentially to changes in the environment, and not all environmental changes elicit a different phenotype given a particular genotype. Furthermore, while the distinction between genotype and phenotype is in principle very clear, several complicating factors immediately ensue. For example, the genotype can be modified by environmental action, as in the case of DNA methylation patterns (e.g., Sano et al. 1990; Mazer and Damuth, this volume, chapter 2). More intuitively, since environments are constantly changed by the organisms that live in them, the genetic constitution of a population influences the environment itself.

Perhaps the most intuitive way to visualize phenotypic plasticity is through what is termed a *norm of reaction* (figure 5.1). This genotype-specific function relates the phenotypes produced to the environments in which they are produced. The figure presents a simple example with a population made of three different genotypes experiencing a series of environmental conditions. Genotype 1 yields a low phenotypic value toward the left end of the environmental continuum (say, an insect with small wings at low temperature) but a high phenotypic value at the opposite environmental extreme (say, large wings at high temperature). Genotype 3, however, does the exact opposite, while genotype 2 is unresponsive to environmental changes, always producing the same phenotype regardless of the conditions (within the range of environments considered).

Even though the case presented in figure 5.1 is very simple (notice, for example, that the reaction norms are linear, which is unlikely in real situations), several general principles are readily understood following a closer analysis:

1. Let us consider the relationship between *phenotypic plasticity* and *reaction norms*. While the two terms are often used as synonyms, they are clearly not. A reaction norm is the trajectory in environment-phenotype space that is typical of a given genotype; plasticity is the degree to which that reaction norm deviates from a flat line parallel to the environmental axis. In the case of linear reaction norms, it is easy to see that genotype 2 has no plasticity, while genotypes 1 and 3 are plastic.

2. Also evident is that *the rank of different genotypes varies with the environment*. So, for example, genotype 1 yields the highest phenotypic value at one end of the environmental range, but the lowest value at the other end. If the phenotype we are studying has fitness consequences, this rank shift is important from an ecological and evolu-

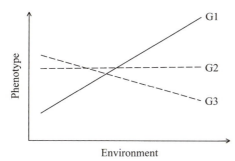

Figure 5.1 Hypothetical reaction norms for three genotypes (G1, G2, G3) drawn in environment-phenotype space. Notice that in some environments, the reaction norms almost converge to the same phenotype, while at the two extremes of the diagram, they diverge dramatically. Also, the phenotypic rank of the three genotypes is reversed between the extreme environments.

tionary perspective. If the highest fitness is associated with the highest phenotypic value, then genotype 1 would be favored in one set of environments and selected against in another. Furthermore, there is a range of environmental conditions for which all genotypes produce essentially the same phenotype, thereby behaving as neutral variants in respect to natural selection.

3. An important point related to the previous one is that unless no reaction norm crosses any other one, *no genotype will produce the highest- or lowest-ranking phenotype under all environments*. If there are different optimal phenotypes across environments, this means that selection may favor some genotypes in some environments and other genotypes in other environments. Depending on other conditions, including the frequency of the environments, the kind of selection, and the presence of costs associated with being plastic, this may explain maintenance of genetic variation in natural populations experiencing heterogeneous environments (Via and Lande 1985).

4. Last, *the concept of heritability needs to be reexamined in the light of phenotypic plasticity* (Lewontin 1974). Heritability is usually defined as the ratio between the genetic variance for a trait and the total phenotypic variance for the same trait:

$$H^2 = \sigma_G^2/\sigma_P^2 \qquad (5.1)$$

where H^2 is the heritability of a phenotypic trait, σ_G^2 is the (additive or total, depending on the experimental design) genetic variance, and σ_P^2 is the total phenotypic variance.[1] Figure 5.1 shows that the heritability of our trait (assuming equal phenotypic variances) is very high at the extremes of the environmental gradient (where the reaction norms diverge), because the genotypes produce widely different phenotypes, but is almost nil toward the center of the gradient (where the reaction norms converge), because the phenotypic values of all genotypes overlap. Therefore, a change in the environment can profoundly alter the heritability of a trait (e.g., Mazer and Schick 1991b).

Throughout the rest of this chapter we will briefly examine the history of the field of phenotypic plasticity studies, to see how we have arrived at the modern concepts of reaction norms and phenotypic plasticity. We will then discuss some recent theoretical and empirical advances in this area of research, with particular regard to the implications of plasticity for ecology and evolutionary theory. A series of specific examples of current studies on plasticity will be examined in some detail to give a flavor of what ongoing plasticity research is about, and the chapter will close with a brief discussion of some possible future lines of inquiry that are currently being explored and may become major foci of interest during the next decade or so. For a more expanded treatment of all these topics, see Pigliucci (2001).

Historical Overview: Genotype, Phenotype, and Dialectical Biology

The study of phenotypic plasticity is the modern incarnation of the ancient philosophical debate about the relative roles of nature and nurture. John Locke (1632–1704) proposed the idea of a tabula rasa, suggesting that humans are born as blank slates on which the environment writes their character. This is an early representation of the nurturistic position. Thomas Hobbes (1588–1679), on the other hand, was a pioneer of the application of mechanistic principles to explain human motivation and social organization and was therefore a forerunner of modern genetic determinism (the naturistic school). In a sense, the resolution of the philosophical debate about nature versus nurture has been achieved by the scientific synthesis provided by the study of phenotypic plasticity: Genotypes and environments interact with each other in a dialectical process throughout an organism's life.

The scientific study of genotype-environment interactions began with the introduction of the concept of reaction norm by Woltereck (1909), at the beginning of the twentieth century, to refer to a peculiar phenomenon observed in the crustacean *Daphnia*, known today as cyclomorphosis (figure 5.2). Essentially, when laboratory lines of this organism are exposed to the presence of a predator, they respond by altering the shape of their body to produce a "helmet," which is effective in reducing predation pressure. Woltereck erroneously believed that the existence of reaction norms falsified the hypothesis of the fundamental distinction between genotype and phenotype proposed a few years earlier by Johannsen. The reasoning was that if the environment could alter the phenotype produced by different genotypes, then the genotype itself was not distinct from the phenotype it produced. Johannsen (1911) published a masterful rebuttal in which he showed why the idea of reaction norm does not contradict the genotype-phenotype dichotomy. On the contrary, it allows us to understand how genotypes and environments interact to yield a given phenotype. It is ironic that to this day Woltereck's confusion crops up in discussions of phenotypic plasticity in academic circles.

Reaction norms and phenotypic plasticity did not play a prominent role during the neo-Darwinian synthesis of the 1930s and 1940s. However, a milestone in the field was published in 1947 (and translated into English in 1949) by Schmalhausen under the title "Factors of Evolution." In it, the author outlined a general theory of the ecology and evolution of phenotypic plasticity, elements of which are still being investigated today (Schlichting and Pigliucci 1998).

Nevertheless, it was not until Bradshaw published a very influential review in 1965 that research on phenotypic plasticity was brought out of the realm of applied science (crop scientists in particular have always been interested in the "unwelcome" effects of the environment on their plants) and into the main stage of evolutionary theory. Bradshaw was the first to clearly state two fundamental concepts concerning our understanding of plasticity. First, plasticity is a character in its own

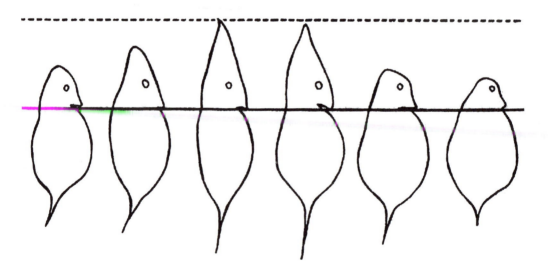

Figure 5.2 Cyclomorphosis in *Daphnia*, the first published example of adaptive phenotypic plasticity. The development of the "helmet" or crest (middle individuals) is catalyzed by the presence of a predator and can be induced simply by the release of specific chemicals in the water, even without the actual physical presence of the predator. The helmet reduces predation rate but probably carries a cost, either metabolic or in terms of swimming efficiency, which is presumably why *Daphnia* that are not exposed to the cue do not develop the helmet. (From Dodson, 1989, adapted from Woltereck's original.)

right, genetically controlled, and can therefore evolve somewhat independently of other aspects of the phenotype. Second, plasticity is not a property of an entire genotype; it needs to be studied in reference to specific environments and traits: A given genotype can be plastic for one trait in response to one set of environmental conditions but not to another set, or it can be plastic for some traits but not others in response to the same set of conditions.

The modern era of plasticity research was opened by a theoretical work and a series of conceptual reviews that not only summarized the state of empirical research, but also actively explored new ideas in the field. The theoretical work was Via and Lande's 1985 paper, which incorporated the concept of reaction norms within the new framework of evolutionary quantitative genetics. Following earlier suggestions, they treated reaction norms as interenvironment genetic correlations. Via and Lande considered the expression of one character in a particular environment as a trait distinct from the expression of the same character in another environment. Therefore, they could calculate a genetic correlation between the two "traits" and use it as an indicator of the degree to which the environment alters genetic expression (if the genetic correlation is significantly less than unity, some genes must act on the trait in one environment, but not in the other—but see Schlichting and Pigliucci 1995 for a critique of this approach).

I do not have enough space here to discuss the conceptual reviews that punctuated the field between the late 1980s and the early 1990s, but the interested reader will find there plenty of challenging ideas and summaries of the best classical empirical studies concerning the evolution of complex phenotypes (Schlichting 1986), plasticity as an adaptation (Sultan 1987), the macroevolutionary consequences of plasticity (West-Eberhard 1989), and the genetics (Scheiner 1993) and molecular biology of plasticity (Pigliucci 1996).

Modern Concepts in Plasticity Research

The Adaptive Plasticity Hypothesis

Perhaps one of the most controversial and difficult areas of study in plasticity research concerns the possibility that plasticity may be an adaptive char-acter directly targeted by natural selection. Much effort has been directed toward carrying out empirical tests of the adaptive plasticity hypothesis, in particular by using appropriate phenotypic manipulations of either the environmental conditions or the genotypes of interest (Schmitt et al. 1999).

Let us consider a simple case of two environments with different phenotypic optima. An adaptive plasticity hypothesis can be considered validated when a plastic genotype performs better than a less plastic one across conditions. While a specialist (nonplastic) genotype might perform as well as or even better than a plastic one under one condition, it will do significantly worse under the other condition. Depending on the frequencies of the two environments, the plastic genotype may be favored over the long run. Notice that we are talking about the plasticity of a crucial trait related to fitness, not of fitness itself. One would expect the reaction norm for fitness to show as little plasticity as possible across environments for a genotype to be successful.

An example of testing the adaptive plasticity hypothesis is the work of Dudley and Schmitt (1996) on shade avoidance in plants. This is a type of plasticity in which an individual can perceive the presence of other plants (and therefore of ensuing competition) by means of detecting changes in the spectral quality of incident light. The reason is that red light is absorbed by neighboring plants since it is photosynthetically active, while far red is reflected because it is of too low energy to be used in plants' metabolism. The optimal phenotype of a plant under sunlight is different from the one of a plant under shade, in that in the latter case it is advantageous to grow taller, reduce branching, and accelerate the time of flowering. Dudley and Schmitt manipulated the environment in which individuals of *Impatiens capensis* were growing in order to produce short, sun-adapted phenotypes and elongated, shade-adapted ones. They then put both types of plants under conditions of competition (high density of neighbors) and noncompetition (low neighbor density). The results were spectacular and accorded neatly with the expectations of the shade avoidance hypothesis (figure 5.3). Elongated plants did well under competition, but worse under the noncompetitive regime, and vice versa: Nonelongated plants did well under noncompetition but worse when surrounded by neighbors. *If* the two environments are experienced frequently enough under natural conditions, *then*

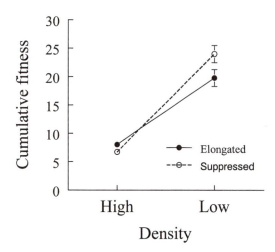

Figure 5.3 Dependency of cumulative fitness in *Impatiens capensis* on neighbor's density. Suppressed plants do better than elongated ones at low density, where they can exploit the lack of competition while being structurally more sturdy, but worse at high density, where they cannot outgrow their competitors in search of light. (From Dudley and Schmitt 1996.)

more plastic genotypes will fare better on average than less plastic ones.

Costs of Plasticity

A second crucial concept concerning modern theory and empirical research on phenotypic plasticity is the idea that plasticity does not come free of charge. It is reasonable to ask if the ability to monitor environmental conditions and to alter the developmental program in response to changes in such conditions carries costs that might somewhat reduce the fitness of a plastic genotype. In fact, to some extent, this must be so, otherwise one would expect only adaptively plastic genotypes to exist whenever environmental heterogeneity occurs (that is, in virtually every circumstance, outside of some computer models).

Incorporation of costs of plasticity into models of the evolution of reaction norms has been attempted by van Tienderen (1991), and a complete theoretical framework for the study of costs and limits to plasticity has been set out by DeWitt et al. (1998). The classification proposed by DeWitt et al. includes five types of possible costs:

Costs of Plasticity

Maintenance: energetic costs of sensory and regulatory mechanisms
Production: excess cost of producing structures plastically (when compared to the same structures produced through fixed genetic responses)
Information acquisition: energy expenditures for sampling the environment, including energy/time not used for other activities (e.g., mating, foraging)
Developmental instability: plasticity may imply reduced canalization of development within each environment, or developmental "imprecision"
Genetic: due to deleterious effects of plasticity genes through linkage, pleiotropy, epistasis with other genes

To these, one can add four types of limits:

Limits of Plasticity

Information reliability: The environmental cues may be unreliable or changing too rapidly.
Lag time: The response may start too late compared to the time schedule of the environmental change, leading to maladaptive plasticity.
Developmental range: Plastic genotypes may not be able to express a range of phenotypes equivalent to that typical of a polytypic population of specialists.
Epiphenotype problem: The plastic response could have evolved very recently and function more like an "add-on" to the basic developmental machinery than an integrated unit. As such, its performance may be reduced.

The main problem with this classification of costs and limits is that many of them are very difficult to address empirically and could easily be lumped with other categories in any reasonably realistic situation. This is clear if one considers the fact that so far no attempt to empirically demonstrate the existence of costs to plasticity has succeeded. Students of phenotypic evolution will surely find this area of research a particularly challenging and potentially rewarding one.

Plasticity Genes and the Molecular Basis of Phenotypic Plasticity

A third area of intense activity in plasticity research concerns the molecular basis of adaptive plastic responses. Schlichting and I have proposed the idea that some genes can be considered "plasticity genes" because their main function (and presumably the reason they were naturally selected) is to detect environmental changes and initiate an appropriate cascade of developmental events leading to an adaptive phenotype. The latest definition of plasticity genes conceptualizes them as "regulatory loci that directly respond to a specific environmental stimulus by triggering a specific series of morphogenic changes" (Pigliucci 1996, p. 169). Notice that this is not to say that only this category of genes influences genetic variation for plasticity in natural populations. On the contrary, many different kinds of genes, including transduction signal genes and internal receptors (e.g., responding to hormone signaling), may contribute. But these other genes are more difficult to pin down as primarily involved with plasticity (and therefore as having been selected for that purpose) since they usually carry out a variety of other functions as well.

How do we find plasticity genes? So far, two major strategies have been proposed. On the one hand, one can identify "candidate genes" based on previous knowledge of the function of gene products specified by sequences identified by mutagenesis or reverse genetics. On the other hand, the increasingly sophisticated technique of QTL (quantitative trait loci[2]) mapping could be used to identify genes that underlie differences in phenotypic plasticity in natural populations; this mapping could eventually lead to identifying the exact function of such genes and therefore to determining if they fall within the definition of plasticity genes.

A good example of plasticity genes is provided by the phytochromes that underlie the shade avoidance response discussed above, to which I will return in the next section of this chapter. A direct use of QTL mapping to uncover plasticity genes has been published, for example, by Wu (1998). Wu examined QTLs affecting four characters in trees obtained from an interspecific cross of *Populus trichocarpa* and *P. deltoides*. His idea was to test for the presence of genes affecting phenotypic plasticity through allelic sensitivity or by gene regulation.

Allelic sensitivity is a situation in which the same genes control the expression of a trait in each environment, but the behavior of their gene products is altered by the environmental conditions in a more-or-less proportional manner. For example, temperature and pH are known to affect the kinetic properties of enzymes, thereby altering biochemical processes at the subcellular level. These alterations can in turn resonate at the morphological level as whole-organism plasticity. Plasticity by gene regulation, on the other hand, implies that some regulatory switches are turned on or off by changes in the environment. This turning on or off will cause batteries of genes to be active only in a specific subset of environmental conditions and not in others.

Wu reckoned that QTLs that are active in only one environment may constitute a good example of regulatory plasticity, while QTLs that are active across environments, but whose contribution to the phenotype changes in an environment-specific manner, may be good candidates for allelic sensitivity. Indeed, he found that most of the QTLs uncovered by his analyses fell into the regulatory plasticity category, with some evidence for allelic sensitivity. These results are not as straightforward as they may sound, because QTL analysis is still largely a statistical approach and does not reveal anything directly about gene action. Furthermore, the environment-specific QTLs identified in this manner must be part of the sets of genes being turned on or off, but they are unlikely to be the primary switches themselves, which are presumably active across all environments.

These difficulties notwithstanding, the study of plasticity genes has direct relevance for evolutionary questions concerning reaction norms and how they change. While it is possible to describe the evolution of phenotypic plasticity in purely statistical terms by using quantitative genetics theory, such an approach is necessarily limited to the short- or medium-term evolutionary time frame for which the assumptions underlying quantitative genetics approximately hold. The existence of plasticity genes, however, opens up the possibility of investigating the macroevolution of plastic responses, including their very origins. For example, shade avoidance plasticity is determined by phytochromes (see below), which are members of an ancient gene family whose elements are common throughout the flowering plants and green algae and possibly include cyanobacteria (Mathews and Sharrock 1997). It is likely that these proteins' initial function was quite different from what it has come to be in flowering plants, possibly involving accumulation of pigments as a protection from light. A combination of modern techniques in phylogenetics and physiology will make it possible to trace the origin of different types of plasticity through gene duplication and differentiation of new functions.

Plasticity of Character Correlations: The Whole-Organism Phenotype

A fourth field of investigation of phenotypic plasticity that has been lurking in the literature for some time and has seen varied attempts at a better understanding is the plasticity of phenotypic integration, that is, how environments and genotypes interact in altering the relationship among several characters to shape the whole-organism phenotype (Schlichting 1989). The importance of this line of research is made obvious once one reflects on the fact that organisms are not just a collection of independent characters that evolve separately from each other. Organismal traits are linked to each other to a varying degree by the existence of genetic correlations among the traits themselves. In fact, the understanding of how sets of correlated characters are assembled and disassembled by natural selection is one of the most promising avenues for the study of the appearance of phenotypic novelties.

A genetic correlation is a statistical measure that is assumed to reflect the overlap in the genetic control of two given traits. If most of the same genes affect both traits, the genetic correlation will be close to unity (positive or negative, depending on the type of effect the genes have on the two traits). Otherwise, the genetic correlation will be closer to zero. This is a very important parameter in quantitative genetics because it estimates the degree to which independent evolution of distinct characters is actually possible. Notice that, from a mechanistic perspective, at least two different causal mechanisms can originate a genetic correlation: Either largely the same set of genes affects both characters (pleiotropy) or the genes are distinct but are physically located so close to each other on the same chromosome that they are almost invariably inherited as a single unit, thereby behaving as a "supergene" (linkage).

Stearns et al. (1991) have made clear exactly how the environment can alter the genetic correlation between two characters. They started out by suggesting that genetic covariances between traits should not change sign when the traits in question are tightly integrated developmentally or physiologically. Under these conditions, one would expect natural selection to "canalize" the pattern of genetic covariance. However, Stearns and collaborators were able to propose a simple model in which the alleles at one locus act additively on two distinct traits across a series of environments. While the allelic effects on the reaction norms of the resulting genotypes are linear, they lead to quadratic functions of the genetic variances and covariances. The covariance quadratic function can then be readily shown to assume negative values toward the middle of the environmental range and positive values for both extremes of the gradient. It remains to be seen how Stearns et al.'s model can be generalized to more complex and biologically realistic cases, but the value of this example is in demonstrating that one can obtain apparently complex patterns of genotype-environment interaction on trait covariances even within the framework of a very simple genetic system.

Clear experimental evidence of the fact that the environment can indeed alter genetic correlations has been published, for example, by Windig (1994), who studied the wing patterns of the tropical butterfly *Bicyclus anynana* (a model system to which I will return later). A principal components analysis (figure 5.4) of the phenotype-environment relationships in these animals shows that both temperature (on PC2) and season (dry/wet, on PC1) dramatically alter the phenotype of these insects. In the diagram, the vectors represent individual phenotypic variables, and the angles among them are proportional to the correlations among traits. So, for example, there is a strong positive correlation between "pupil" and "outer ring," while these are both negatively correlated with "marbling" (whose corresponding vector lies in the opposite direction); similarly, all three characters are independent from "wing size," since this vector is positioned perpendicularly to the first three—and so on for the remaining characters. What the analysis shows is that the seasonal and thermal forms are effected by shifts in the phenotypic space defined by the correlations among characters. Windig found that several of these genetic correlations remained constant across environments, but two

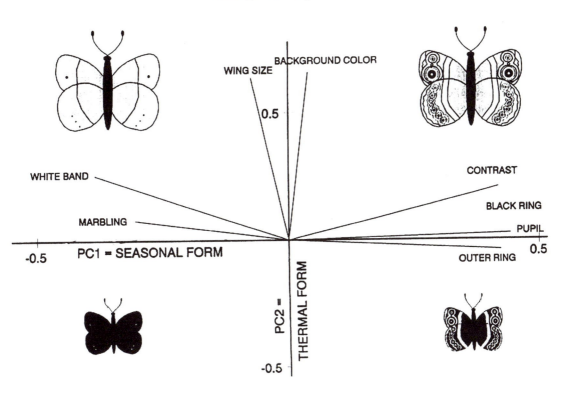

Figure 5.4 Phenotypic variation induced by genotype-environment interaction in butterflies of the species *Bicyclus anynana*. The variation is summarized by principal components axes, the horizontal one differentiating the seasonal forms, the vertical ones separating the thermal forms. (From Windig 1994.)

changed in a particularly revealing fashion. The relationship between larval developmental time and an index of seasonal form was significant in the wet season, probably because of a trade-off due to the energetic cost of producing the wet pattern. This relationship, however, disappeared during the dry season because the corresponding wing pattern is likely to be less costly to make. This is an example of a physiological mechanism potentially underlying a change in genetic correlations. The second variable correlation was between the indices of thermal and seasonal form, which was negative at low and high temperatures, but not significant at intermediate temperatures. Winding suggests that this might reflect an environment-dependent functional trade-off; for example, at high temperatures either the butterflies develop into good fliers (small and dark, better for foraging males) or into better motionless insects (larger and pale, better for egg-laying females). In the first case, a visible eyespot would interfere with the thermal requirements (because of its light color); in the second

case, it might contribute to minimizing damage due to predation. Conversely, in the dry season, the butterflies might develop either a phenotype adapted for survival (large and pale) or better suited for reproduction (small and dark). These and several other details make Windig's work a particularly good and rare example of phenotypic integration within the framework of a well-studied system.

The problem with plasticity of phenotypic integration is that if offers formidable challenges from empirical, statistical, and theoretical viewpoints. Empirically, it is cumbersome to accurately measure a large number of traits on a large number of individuals under a variety of environmental conditions. Statistically, it is not immediately obvious how best to conduct tests for the significance of environmentally induced changes in the multivariate phenotype, or how to reduce the multidimensionality of the problem to something that can be intuitively but accurately grasped by the human mind (principal components are one way to do this, but they only occasionally result in something that can be readily interpreted biologically, as in the case of figure 5.4). It is difficult to clearly formulate expectations for the hypothesis of adaptive plasticity when multiple characters are considered simultaneously. More often than not, studies of phenotypic integration result in the observation of complex, but not necessarily meaningful, patterns; this is a fascinating challenge for the next generation of students interested in the plasticity of complex phenotypes.

Case Studies: Plant Canopies and Butterfly Eyespots

Many of the concepts discussed in the preceding parts of this chapter can be seen interwoven with each other in two particularly clear case studies of phenotypic plasticity: the adaptive reaction of plants to shade induced by neighboring individuals, known as the *shade avoidance response,* and the seasonal variation of the eyespot on the wings of some butterflies. I will therefore examine these two cases in some more detail here as illustrative of general principles of research on phenotypic plasticity.

Molecular Biology and Ecology of Shade Avoidance

Plant physiologists and ecologists have long been aware of the fact that some plants respond in a seemingly adaptive way to canopy shade and that this response is mediated by one or more specific photoreceptors. However, the two ends (ecological and molecular) of this field of research had rarely met until very recently.

As mentioned above, shade avoidance is best described as a syndrome comprising a coordinated alteration of morphological as well as life history characteristics whenever the light spectral quality, measured as the ratio of red to far red, decreases below the normal sunlight level of about 1:1. From a functional ecological standpoint, the plant can follow one of two shade-avoiding strategies. If it belongs to a species capable of actually overtopping the canopy, a dramatic reduction of lateral branching, increased apical dominance, and accelerated vertical growth will increase its chances to gain access to a spot literally under the sun. If, on the other hand, the plant is a poor competitor un-

der a canopy, the best option will be to accelerate the life cycle and flower early enough to avoid most of the direct competition, producing seeds before an actual canopy closure; after all, some reproductive fitness is better than none.

From the viewpoint of molecular physiology, shade avoidance is made possible by the action of specific molecules termed *phytochromes.* These are quantum light counters particularly sensitive to red and far red light. Five phytochromes have been described in flowering plants (A, B, C, D, and E), and the shade avoidance syndrome has been attributed mostly to the action of phytochrome B, although phytochrome A and possibly at least also phytochrome C play a role in this complex morphological response.

While morphological shade avoidance is well known in understory plants that can take advantage of canopy gaps, its molecular biology has mostly been studied in the opportunistic weed *Arabidopsis thaliana,* which shortens its life cycle instead of overtopping its competitors. Nevertheless, *A. thaliana* does have a more varied ecology than it is usually credited with (Pigliucci 1998), and while its shade avoidance response is less intense than that of some of its close relatives, this species is found in a variety of habitats in which both inter- and intraspecific competition for light does occur.

Given all of the above, shade avoidance in *A. thaliana* provides the ecological geneticist with a rare opportunity to examine directly the molecular basis of ecologically relevant characters. Johanna Schmitt and I took advantage of this opportunity by studying the reaction norms of single and double mutants affected in key components of the shade avoidance reception and transduction systems. Figure 5.5, for example, shows that eliminating phytochrome B (the slightly sloped broken line) greatly reduces (but does not eliminate) a crucial component of the shade avoidance response: leaf production during the vegetative phase of the life cycle. In *A. thaliana* there is a strong genetic correlation between leaf production and flowering time, so that often the only way to flower earlier (a requirement of the shade avoidance strategy) is to arrest prematurely the production of rosette leaves. Our study also demonstrated two other important facts: First, the shade avoidance response is characterized by molecular redundancy, since only the elimination of *all* five phytochromes (the lowest broken line) completely flattens the reaction norm

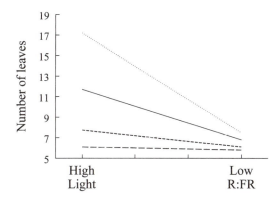

Figure 5.5 Reaction norms of single-gene mutants affected in their photoreception in *Arabidopsis thaliana*. The trait considered here is leaf production during the vegetative phase of the life cycle in response to sunlight or a low R:FR ratio simulating foliar shade from neighbors. Notice the opposite response (relative to the wild type, in high light) of a blue receptor mutant (dotted line) compared to a series of phytochrome mutants (broken lines). The solid line represents the wild type. (Modified from Pigliucci and Schmitt 1999.)

yielding a nonplastic genotype. Second, under high light, the blue receptor actually acts in opposition to the phytochromes (dotted line) in respect to the wild type (solid line), so that eliminating its functionality actually prolongs the vegetative phase in that environment. Additional analyses revealed the effects of these mutations on reproductive fitness, and a follow-up study focused on the epistatic effects of these loci when the genotypes are subjected to a variety of ecologically relevant conditions.

It is important to understand that the point of this approach is not to investigate the genetic basis of natural variation for plasticity, an issue that can be best addressed by QTL analysis of natural populations. What one is after in this case is rather the functional molecular ecology of well-characterized types of plasticity. The two are distinct areas of research, which may converge at one or two levels: First, it may turn out that some of the key regulatory loci identified by molecular developmental biologists are also the same that produce at least part of the observable phenotypic variation in natural populations, as some preliminary research has already shown. Second, the study of these regulatory loci will enable evolutionary biologists at least to ask questions concerning the long-term evolution

of plastic responses, including how they actually originated in deep phylogenetic time.

Eyespots and Seasonal Forms of Bicyclus anynana

Our second paradigmatic example of recent plasticity research involves the ecology and molecular developmental biology of eyespots in the African butterfly *Bicyclus anynana* and is mostly based on the work of Paul Brakefield and collaborators (e.g., Roskam and Brakefield 1999). These animals experience a regular alternation of dry and wet seasons, which brings about a completely different ecology in this species. During the dry season, conditions are such that the butterfly rarely moves about, being mostly intent on surviving until conditions improve. In this case, the eyespot becomes hardly visible, so the animal can mimic the dull environmental background and avoid predation (figure 5.6 *left*). When the wet season begins, the conditions are completely different. This is when the butterflies are most actively in search of mates, and their defense strategy changes accordingly (figure 5.6 *right*). The eyespot is now very prominent and has the potential to benefit the insect by attracting predatory birds toward nonvital parts of the body.

Brakefield and coworkers have amassed an impressive amount of information about this system, characterizing the basic ecology and selective pres-

Figure 5.6 Dry season (left) and wet season (right) forms of the tropical butterfly *Bicyclus anynana*. The two forms are the result of phenotypic plasticity of the complex molecular machinery controlling the development of the eyespot and are characterized by markedly distinct ecologies. (Pictures kindly provided by P. Brakefield—Leiden University, Netherlands.)

sures, as well as investigating its molecular and developmental biology. For example, by using mutants that alter the pattern of formation of the eyespot as well as transplants of eyespot developmental organizers, these researchers were able to demonstrate that the features of the eyespot can be altered in a completely independent fashion from other body structures, including different wing-pattern elements. This allows a high degree of evolutionary flexibility to the eyespot, enhanced by the inherent developmental plasticity described above.

Furthermore, Brakefield and colleagues were able to demonstrate that eyespot patterns can evolve dramatically over a very short period of time. This evolution was evident from a series of selection experiments, as well as from comparisons of closely related species of *Bicyclus* whose evolutionary relatedness is known. The outcome showed that most of the response to selection can be accounted for by changing allele frequencies at one or very few loci characterized by major regulatory action. This again points toward a potential convergence of the molecular biology of regulatory genes and the population genetics of evolutionarily relevant loci that was unexpected until a few years ago.

Perspectives for the Future: Where is Plasticity Research Going?

It is always dangerous to predict the future, especially in rapidly moving fields of inquiry such as phenotypic plasticity. However, one can reasonably speculate on what some of the most promising current lines of research will develop into within the next 5–10 years. At best, this prediction might provide graduate students and active researchers with some hint about possible research programs. At worst, it will do no harm.

As the preceding section on shade avoidance and butterfly eyespots should have made clear, we are likely to see an increased convergence of lines of inquiry from the thus far mostly disparate fields of molecular developmental genetics and population and evolutionary ecology. For example, research is currently under way to study variation for shade avoidance in natural populations of *Arabidopsis thaliana* as a preliminary step toward conducting QTL surveys of relevant genes (J. Schmitt, M. Purugganan, and T. Mackay, pers. comm.). So

far, QTL analyses have been carried out in a mostly haphazard way, by the crossing of genotypes characterized by extreme phenotypes and the examination of a limited number of such crosses. The increasing sophistication of both the molecular and the statistical methods involved in these studies is making it possible to couple extensive investigations of natural populations (representative of variation encountered in the field) with a mechanistic look at the genetics of evolutionary change.

The kinds of questions one would like to answer by these methods deal with the role of regulatory genes in evolution. For a long time, these genes have been considered too important and the consequences of their alteration too far reaching for there to be any allelic variation at these loci. Recently, however, not only have such genes be implicated in major evolutionary transitions (e.g., Carroll et al. 1995), but we have increasing evidence of within-species sequence variation that could be exploited by natural selection to produce alternative phenotypes (e.g., Purugganan and Suddith 1999). Should these preliminary reports become widespread, they will force us to rethink large components of our current understanding of phenotypic evolution (Schlichting and Pigliucci 1998).

Another area of plasticity research that has largely been neglected because of logistic as well as methodological difficulties is the comparative study of the evolution of reaction norms. When we think of the evolution of plasticity, we should compare extant reaction norms with the likely patterns of plasticity in a group's immediate ancestors, not with a flat, no-plasticity null hypothesis. While the latter may make more sense within a framework of functional ecological and optimality studies, it hardly represents a good guess of real historical patterns. However, while several plasticity studies do compare different species, such comparisons are usually limited to a small number of taxa and lack an explicit phylogenetic hypothesis. The reasons are good. First, plasticity experiments have the unfriendly tendency to grow in size in a multiplicative fashion if the researcher wishes to include more environments or taxa; second, reliable low-level phylogenetic hypotheses are not easy to come by and are still not generated by systematists mostly interested in macroevolutionary patterns.

This situation, however, is slowly changing, and examples of explicitly comparative studies of the evolution of phenotypic plasticity are starting to appear (e.g., Roskam and Brakefield 1996; Pollard et al. 2000). Many open questions will need to be addressed in the near future. For example, how often does plasticity significantly change throughout the evolutionary history of a given clade? Do the patterns of change follow mostly phylogenetic inertia, thus implying strong genetic constraints, or do they match ecological expectations, according to the adaptive plasticity hypothesis? When one considers larger phylogenetic groups, how often does a potentially adaptive plastic response arise de novo as opposed to simply being a modification of a preexisting one?

Finally, and perhaps ironically, we really need to devote more effort to understanding the ecology of phenotypic plasticity. After the gargantuan pioneering field studies of Clausen, Keck, and Hiesey in the middle of the 20th century, we have seen a plethora of experimental research under controlled conditions, but very few studies of phenotypic plasticity have been set on an ecologically realistic stage. Since the introduction of modern quantitative techniques allowing us to quantify the intensity and type of natural selection under field conditions (see Fairbairn and Reeve, this volume) we still refer to classical examples of adaptive phenotypic plasticity, such as heterophylly in plants and conditional metamorphosis in salamanders, with hardly any evidence that they actually *are* under selection. Once again, the logistics of these experiments are indeed daunting, and they represent formidable challenges from the point of view of statistical analyses. Yet, examples have appeared in the literature (Weis and Gorman 1990), and the way is marked clearly enough that we can expect major progress in this area during the next decade or so.

The study of phenotypic plasticity is a fundamental component of both basic and applied science. In basic research, it represents the interface between genes and environments, "nature and nurture," and it allows evolutionary biologists to link population genetics with ecological questions, one of the main missing links of the modern neo-Darwinian synthesis. From the standpoint of applied science, plasticity has historically affected the way we select new varieties of plants and animals, and the new merging of molecular and organismal research programs will continue to direct our efforts in this area. Furthermore, the study of genotype-environment interactions provides a solid theoretical framework for a better understanding of human nature itself, as well as of the limitations of our research into its biological basis.

Notes

1. "Narrow-sense heritability," the ratio of additive genetic to phenotypic variances (σ_A^2/σ_P^2), is generally symbolized by h^2. "Broad-sense heritability," the ratio of total genetic to phenotypic variances (σ_G^2/σ_P^2), is symbolized by H^2.

2. A quantitative trait locus is a region of a chromosome, defined by linkage to a marker locus, that has a significant effect on a quantitative trait. The phenotypic effects of QTLs are typically detected through crosses between inbred lines differing in average expression of the quantitative trait.

6

Population Structure

LEONARD NUNNEY

Population structure is a ubiquitous feature of natural populations that has an important influence on evolutionary change. In the real world, populations are not homogenous units; instead, they develop an internal structure, created by the physical properties of the environment and the biological characteristics of the species (such as dispersal ability). However, our basic ecological and population genetic models generally ignore population structure and focus on randomly mating (panmictic) populations.

Such structure can profoundly change the evolution of a population. In fact, the myriad of influences that population structure exerts can only be hinted at in a single chapter. Since an exhaustive review is not possible, I will focus on presenting the conceptual issues linking mathematical models of population structure to empirical studies. To do this, it is useful to recognize two different kinds of population structure that both reflect and influence evolutionary change. The first is genetic structure. This is defined as the nonrandom distribution of genotypes in space and time. Thus, genetic structure reflects the genetic differences that develop among the different components of one or more populations. The second is what I will call *proximity structure,* defined by the size and composition of the group of neighbors that influence an individual's fitness. Fitness is commonly influenced by local intraspecific interactions. Perhaps the most obvious example is competition. When individuals compete for some resource, they don't usually compete equally with every other member of the population; in general, they compete only with a few of the most proximate individuals.

These two forms of population structure, genetic structure and proximity structure, provide a foundation for understanding why we have shifted away from viewing populations as homogenous units. For good reason, this is a theme that is explored in many of the other chapters in this book.

Genetic Structure

Genetic structure can develop within a population over a single generation, generally either as a result of local family associations or as a result of spatial variation in selection. For example, limited seed dispersal results in genetic correlations among neighbors even in the face of long-distance pollen movement, due to the clustering of maternal half sibs. Similarly, strong selection of the kind seen in the classic example of heavy-metal tolerance in grasses results in local adaptation within small subgroups of populations even when gene flow is high. However, structure of this kind will be limited, unless there is some degree of reproductive isolation between subgroups. Reproductive isolation reinforces the effects of limited dispersal and selection.

Partial reproductive isolation is common in natural populations. Generally, it is the physical distance separating individuals that precludes complete random mating and promotes the development of spatial genetic structure. However, temporal genetic structure can also develop when subgroups of a population become separated in time. One of the most extreme examples of this effect is seen in the Pacific pink salmon. Odd- and even-year cohorts show marked genetic differences within the same river. These semelparous[1] fish take exactly 2 years to mature, and hence, alternate-year cohorts are reproductively isolated.

The evolution of a population can be profoundly influenced by its genetic structure, since it affects local adaptation, coadaptation, and, over the longer term, speciation. All of these are important evolutionary processes. First, understanding local adaptation can be crucial to understanding the ecology of any spatially structured population. Spatial genetic structure both facilitates and is part of this process. The reason is that selection for genotypes with high local fitness is most effective when the local group is already genetically isolated. In addition, such selection acts to further enhance genetic structure, at least at the loci involved in the adaptive change. Second, a related effect is the development of genetic coadaptation. Coadaptation is due to the nonadditive interactions among genes that can develop within isolated gene pools (demes). When individuals within different demes adapt to their respective local environments (regardless of whether these environments are very similar or very different), they may do so in different ways, each recruiting a set of alleles that interact in a complex (nonadditive) way to determine fitness. If so, any mixing of the different sets through hybridization can result in lowered fitness (coadaptive breakdown), particularly in the second (F_2) and subsequent hybrid generations. Genetic coadaptation is an essential precursor to Wright's shifting balance model of evolution, which will be discussed later. It can also contribute, along with local adaptation, to the phenomenon of outbreeding depression (see Waser and Williams, this volume). This is indicated when the offspring of parents from different subgroups of a genetically structured population have a lowered fitness.

The link between genetic structure and outbreeding depression has long been recognized as central in most models of speciation. More generally, speciation almost always depends on genetic structure (with the exception of chromosomal speciation). Models of allopatric (geographic) speciation rely on spatial structure created by physical barriers to dispersal (and consequently to random mating), and models of sympatric (nongeographic) speciation often depend on small-scale spatial structure or temporal structure (e.g., differences in the timing of reproduction) to drive assortative mating.

The cornerstone of population genetics is its mathematical models. Modeling of genetic structure requires a simplification of the complexities of the real world. We generally use one of three different models for the analysis of genetic structure: the island model, the isolation-by-distance model, and the stepping-stone model.

Proximity Structure

The region within which intraspecific interactions (such as competition) determine fitness defines proximity structure. The interactions of an individual may be restricted to a few of its neighbors (its proximity group), so proximity structure often occurs on a much smaller scale than genetic structure. Wilson introduced the term *trait group* to describe the special situation where such interactions take place within discrete groups and showed that such interactions between an individual and its immediate neighbors can have an important influence on natural selection (see Wilson, this volume).

When the genotype of an individual and the genetic structure of its proximity group are correlated, the evolutionary outcome can be particularly interesting. Specifically, there is the potential for successful group selection. This concept was introduced to explain the evolution of traits that appeared to be good for the group but that had no obvious individual benefit. Possible examples ranged from the sterility of worker bees to population self-regulation to community structure. In general, traits do not evolve for the "good of the species," contrary to the belief of many early naturalists, and "group selection arguments" must be critically evaluated. However, the importance of group selection in nature still remains controversial.

Concepts: A Historical Perspective

The Island Model and Random Genetic Sampling

Sewell Wright was convinced that, in nature, adaptive evolution was most likely to occur in genetically structured populations. Wright's conviction led him to develop the basic models on which we still rely. Of these, the simplest and most frequently used is the island model. In this model, the population is divided into a series of identical islands, with some fixed proportion of individuals dispersing each generation. Dispersing individuals are randomly distributed across all islands, with no regard to distance. Thus, there is no tendency for migrants to favor nearby over distant islands.

An important result derived from the island model was the expected gene frequency distribution across islands, given a large population subdivided into many identical islands of size N. For a single polymorphic locus with alleles A and B at a frequency of p and q locally (within an island) and P and Q globally, the distribution is:

$$\phi(q) = C \; e^{4Nsq+2Ntq^2} \; q^{4N(mQ+v)-1} \; p^{4N(mP+u)-1} \quad (6.1)$$

where the (weak) local selection acting against allele A is defined by the genotype fitnesses w(AA) = 1, w(AB) = 1 + s, and w(BB) = 1 + 2s + t. The immigration rate is m, and the mutation rates from A to B and vice versa are u and v, respectively. The constant C normalizes the distribution. Equation 6.1 (despite its complexity) is important because it explicitly recognizes that population structure develops and evolves as a result of the interaction of local random genetic sampling, local selection, mutation, and immigration. But to understand the specific effects of selection, migration, and mutation on population structure, we need to examine the effect of random genetic sampling in the absence of these factors. The effect of random sampling alone provides us with a null hypothesis.

When N adults in an isolated subpopulation produce the offspring that become the N adults of the next generation, we might expect gene frequencies to remain the same, provided no other processes (such as selection) are influencing their change. In an infinite population, this is indeed true. However, in a finite population, the random sampling of the 2N parental gene copies means that some gene copies will be chosen more than once and others not at all. This process can be viewed in two ways, either as a change in gene frequency (random genetic drift) or as an increase in the identity by descent of gene copies (inbreeding) (see figure 6.1). In any event, the effect of random sampling is to create genetic differences among isolated subpopulations, even though they are subject to identical conditions.

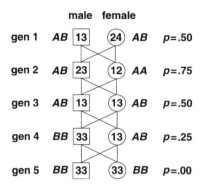

Figure 6.1 Five hypothetical generations of full-sib mating illustrate inbreeding and genetic drift. (A) Inbreeding. In the initial pair of heterozygotes, each of the gene copies is uniquely labeled. The first autozygous individual—that is, with both of his copies identical by descent (IBD)—is seen in generation 4. By generation 5, inbreeding at this locus is complete, since all copies originate from a single ancestral copy (copy 3). (B) Random genetic drift. The frequency of allele A (= p) fluctuates randomly until fixation occurs for allele B in generation 5.

Genetic drift reflects the effect of random sampling within a single isolated subpopulation, integrated over time. Similarly, inbreeding reflects the gradual increase in identity by descent (IBD) of gene copies within each subpopulation over time. IBD defines the probability of an individual being autozygous, that is, the probability of carrying two copies of a gene that share a common ancestor (figure 6.1). However, we must define IBD carefully. If we trace ancestry back far enough, IBD is universal. The phylogenetic reconstruction of human mtDNA and of the human Y chromosome back to single, ancestral copies (the so-called mitochondrial Eve and Y-chromosome Adam) is the pattern expected of any genetic element in a finite population. Thus, IBD must be measured relative to some higher level of structure, and Wright recognized that this property could be exploited to measure population structure.

The link between inbreeding and structure is not immediately obvious. We generally think of inbreeding in terms of mating with close relatives, and typical examples would be full sib mating and selfing. However, a mating pair of close relatives (or indeed a selfing individual) can be viewed as a very small unit of population structure. As a result, in a population with a high frequency of full sib mating, the inbreeding is reflected in the high genetic correlation among mating pairs relative to randomly chosen pairs. It is also reflected in the high genetic variance among mating pairs (and in the lowered variance within pairs). When mating pairs are transient, this type of spatial structure is temporary. However, it is the link between inbreeding within population subunits and the genetic correlation within (and genetic variance among) those subunits that Wright exploited to develop his hierarchical method describing genetic structure. It is now routine to describe population differentiation in terms of Wright's hierarchical coefficient F_{ST}, which reflects the variance in gene frequency of subpopulations *(S)* contained within the total *(T)* population (see Waser and Williams, this volume, box 7.1).

As we all know, deviations from expectation due to random sampling can be large if the sample size is small. In exactly the same way, the rate of change in gene frequency due to genetic drift and the rate of increase in inbreeding are both inversely proportional to population (or subpopulation) size, N. This size dependence raises a crucial question: What is the appropriate measure of population size? Wright (1931) recognized the need for a standardized measure relevant to evolutionary change, and for this purpose, he defined the "effective population size," N_e. Wright's measure works by comparing the effect of random sampling (usually the rate of genetic drift or inbreeding) of any real population to his "ideal" population. Wright's ideal population is isolated and of constant size N, and the individuals mate at random, are hermaphrodite, and have nonoverlapping generations. The rate of inbreeding per generation in such a population is $1/(2N)$; hence, the rate of inbreeding in any real population is, by definition, $1/(2N_e)$.

Isolation-by-Distance and Stepping-Stone Models

Wright's island model with its idealized island subpopulations is not appropriate when genetic structure develops within a homogenous landscape due to limited dispersal. To analyze this scenario, Wright (1943) developed his isolation-by-distance model. In this model, limited dispersal creates a spatial autocorrelation, that is, the genetic correlation be-

tween two individuals declines as the distance between them increases. Wright defined population structure in the model using his "genetic neighborhood." He defined the neighborhood of an individual as the region surrounding that individual within which its parents can be considered to have been picked at random (see below).

Kimura (see Kimura and Weiss 1964) introduced a third model of population structure, the stepping-stone model, that fits conceptually between the island model and the isolation by distance model. As in the island model, the population is divided into many island subpopulations lacking internal genetic structure, but as in the isolation-by-distance model, dispersal is restricted. Specifically, dispersal is constrained, usually to a single "step," so that individuals can disperse only to an immediately adjacent island.

Measuring Genetic Structure

Hierarchical F Statistics Wright's hierarchical inbreeding coefficients are an important tool for measuring genetic structure. The simplest level of organization is the diploid individual (generally labeled as level *I*), since each individual can be viewed as a small group of two gene copies. Above this level, we can hierarchically structure a population into larger and larger units, each corresponding to a different level of structure. In this framework, the generalized coefficient F_{XY} can be defined as the correlation of pairs of gene copies within level X, relative to the correlation of pairs of gene copies within the higher level Y:

$$F_{XY} = (F_{XZ} - F_{YZ})/(1 - F_{YZ}) \qquad (6.2)$$

where Z defines the highest level. Obviously, if the groups that make up level X are simply a random sample of the Y unit that contains them, then the correlation is 0. On the other hand, if level X is highly inbred, then the correlation will approach its maximum value of 1. Note that these correlations are also inbreeding coefficients, since the probability of IBD is equivalent to the correlation between two copies (see also Waser and Williams, this volume).

The most common partition used to define genetic structure is F_{ST}, which is the correlation of gene copies within a subpopulation *(S)* relative to some larger population *(T)*. F_{ST} can be evaluated by comparing the correlation of all pairs of gene copies from within each island, to the correlation of all possible pairs of gene copies within the larger unit.

There has been some confusion in the literature over the best way to estimate these hierarchical parameters, and how to define them. Much of this confusion arises because Cockerham (1973) introduced a new set of hierarchical coefficients that, like Wright's, are based on genetic correlations. Cockerham's measures have useful statistical properties, but only Wright's measures link directly to population genetic theory. Cockerham's hierarchical coefficients F'_{XY} can be defined as the correlation of pairs of gene copies within level X, relative to the correlation of pairs of gene copies *from different units of level X* within the higher level Y. The italics highlight where Cockerham's definition differs from that of Wright.

These two definitions have led to various alternative formulations for F_{ST} and related measures (see table 6.1). A detailed evaluation of the relative merits of these measures is beyond the scope of this chapter, but a brief comparison of F_{ST} and one of Cockerham's statistics, θ_S, serves to illustrate the problem created by these two parallel systems.

θ_S is usually defined nonhierarchically as the correlation of gene copies from different individuals within an island and is often considered identical to F_{ST}. The two parameters are related as follows:

$$F_{ST} = [\theta_S - F_{TM} + (F_{IM} - \theta_S)/(2N - 1)]/(1 - F_{TM})$$

(6.3)

where I, S, T, and M define hierarchical levels from the individual to the (often hypothetical) metapopulation (i.e., collection of populations), and N is the size of each subpopulation (Nunney 1999). If coancestry (i.e., inbreeding) within the total population can be ignored, then F_{TM} is zero and the two coefficients are indeed essentially identical:

$$F_{ST} = \theta_S + (F_{IT} - \theta_S)/(2N - 1) \approx \theta_S \quad (6.4)$$

However, it cannot be generally assumed that the two approaches yield similar results if more than two levels are included in the analysis, that is, if F_{TM} is nonzero. This problem persists even when the θ''s are corrected relative to the next level (as F'_{XY}) if the number of sample units (e.g., subpopula-

Table 6.1 Hierarchical measures of population structure that compare subpopulations *(S)* to total populations *(T)*, based on Wright's definitions of F_{ST} versus Cockerham's F'_{ST} (see text).

	Wright's F_{ST}	Cockerham's F'_{ST}
Definitions[a]	$F_{IT} = (f_0 - f)/(1 - f)$ $F_{ST} = 1 - t_0/t$	$F'_{ST} = (\theta_S - \theta_T)/(1 - \theta_T)$
Other (biased) forms:[a,b]	$F_{ST} = Var(p)/pq$ (2 alleles) $F_{ST} = G_{ST} = 1 - H_S/H_T$ (Nei 1977)	$F'_{ST} = \theta_S$
DNA sequence data:[a]	$K_{ST} = 1 - K_S/K_T$ (Hudson et al. 1992)	$N_{ST} = 1 - v_w/v_b$ (Lynch and Crease 1990)
Microsatellite data:[a]	$R_{ST} = 1 - S_W/S_T$ (Slatkin 1995)	

Note: In general, $F_{ST} \neq F'_{ST}$

[a]Each of the following measures relate to the average over all pairs of gene copies chosen from the unit specified:

f_0, f: probability of IBD within subpopulation or total population.

θ_S, θ_T: probability of IBD within subpopulation or total population, but comparing only different lower-level units; e.g.,θ_T excludes comparisons within the same subpopulation.

t_0, t: mean coalescence time within subpopulation or total population.

H_S, H_T: expected heterozygosoty (assuming Hardy-Weinberg ratios) within subpopulation or total population.

K_S, K_T: mean number of substitutions per nucleotide site within subpopulation or total population.

v_w, v_b: mean number of substitutions per nucleotide site within subpopulation or total population, but comparing only different lower-level units.

S_w, S_T: mean square size difference within subpopulation or total population.

[b] p,q, $Var(p)$: frequency of the two alleles in the total population (p, $q = 1 - p$) and the variance among subpopulations.

tions) is small (see Nunney 1999). Under these conditions, care must be taken over the population genetic interpretation of the Cockerham coefficients.

Wright's Neighborhood   Wright introduced his isolation-by-distance model to investigate the development of genetic structure within a continuously distributed population. Even under ideal conditions, in which only genetic drift is influencing overall gene frequencies within the population, local genetic structure can develop as a result of limited gene flow.

Gene flow reflects the successful dispersal of individuals or gametes and can be expressed as the mean distance moved by gene copies per generation. However, Wright (1943) showed that a more useful measure is the mean square of distance moved (σ^2). In his isolation-by-distance model, Wright defined the genetic neighborhood as his unit of population structure. As noted in the more precise definition given earlier, the neighborhood of an individual at birth is the region around its birthplace that is expected to encompass the birthplaces of the individual's parents. Clearly, as dispersal distances increase, the neighborhood must enlarge. Given normally distributed dispersal distances (equal in all directions) of individuals from their place of birth to their place of reproduction, the area of the neighborhood in a two-dimensional habitat is $4\pi\sigma^2$. Note that σ^2 is also the variance of the dispersal distribution, when it is viewed as a symmetrical distribution around the point of origin. Thus, if d is the density of individuals in the population, then the number of individuals in neighborhood is

$$N_b = 4\pi\sigma^2 d \qquad (6.5)$$

Neighborhood structure can be detected by means of Wright's hierarchical F statistics. If island subpopulations show internal genetic structure because of isolation by distance within them, then F_{IS} will be positive due to an excess of homozygotes. This excess is an example of the Wahlund effect, which occurs whenever random mating subsamples differing in gene frequency are pooled and tested for their fit to Hardy-Weinberg ratios. (This same principle underlies the definition of G_{ST} in table 6.1). The effect can be factored into a distinct level of genetic structure (let us say, level D), whereby F_{DS} would quantify the islands' internal genetic structure in terms of the variance in gene frequency among sample sites within each island.

Gene Flow

Measures of F_{ST} and neighborhood size quantify the genetic structure of populations. As such they are useful for comparing populations. However to understand the biological reason for a particular level of structure, these parameters must be linked to patterns of dispersal. The direct link between genetic neighborhood and the mean square of dispersal distance was noted above (equation 6.5), and this relationship has been refined further to include such factors as nonnormal dispersal and pollen movement.

Neighborhood size is usually estimated by measuring current patterns of dispersal. However, genetic structure reflects an integration of these patterns over time. In particular, if the present level of dispersal is less than the historical norm, then genetic structure may be much less than that predicted from current dispersal data.

One way to estimate dispersal from genetic structure is to use the relationship derived from the island model that links F_{ST} to the migration rate among islands. Wright (1951) showed that at equilibrium:

$$F_{ST} = 1/(1 + 4N_e m) \qquad (6.6)$$

where N_e is the effective size of islands, m is the immigration rate (as a proportion of the total), and assuming $m \ll 1$. This relationship ignores the effects of mutation and selection; however, it is generally little affected by either realistic levels of mutation or weak selection (i.e., selection coefficients less than m). However, it must be noted that if there is strong selection that varies spatially, then F_{ST} measures the effects of selection and equation 6.6 does not apply. But with this proviso, equation 6.6 provides a simple method for estimating the effective number of migrants per generation $(N_e m)$. Furthermore, simulations suggest that analogues of F_{ST} (see table 6.1) can be used to estimate $N_e m$ from DNA data (e.g., Slatkin 1995).

It can be seen from equation 6.6, that if $4N_e m$ increases above 1, then F_{ST} decreases rapidly. This relationship leads to the general rule that if $4N_e m \gg 1$ (often expressed simply as $N_e m > 1$) then genetic differentiation among islands will be minimal. On the other hand, substantial genetic differentiation is expected if $4N_e m \ll 1$. These conditions show that, in the absence of selection, little more than one effective migrant per generation is needed to overcome the diversifying effect of genetic drift, and that this result is independent of the size of the islands.

Ecological Genetics

While Wright and others were developing their theory, ecological geneticists were exploring spatial structure in nature. Thanks to the pioneering work of such figures as E. B. Ford and Theodosius Dobzhansky, the ubiquity of genetic structure in nature became clear. Ecological genetics provided us with the classic examples of evolutionary change that illustrate the interplay of selection, migration, and genetic drift. As noted earlier, heavy metal tolerance in several grass species was studied by Bradshaw and coworkers in areas of mining pollution in Wales, and this work remains one of the best examples of local adaptation. It shows clearly the role of gene flow (in this case, primarily wind-blown pollen) in maintaining a cline of tolerance downwind of the mine. The famous work of Kettlewell on industrial melanism in the peppered moth, *Biston betularia,* showed that, as a result of bird predation, the typical form was favored in unpolluted rural areas of northern England, whereas in polluted industrial areas, the melanic form had spread to near fixation. Similarly, the work of Cain and Sheppard on the snail *Cepaea nemoralis* showed a role for bird predation in maintaining the color and banding polymorphism between beechwood and grassland habitats. But their work also revealed "area effects" where localized patterns of color and banding did not correlate with the environment, suggesting a possible role for genetic drift.

In the 1940s and 1950s, Dobzhansky showed that genetic structure in nature was not limited to morphology. He demonstrated that the inversion polymorphisms of *Drosophila pseudoobscura* showed considerable geographical variation across the southwestern United States. More surprisingly, he found that the flies exhibited marked genetic structure along an altitudinal gradient of just 60 miles in the Sierra Nevada of California. Since Dobzhansky's work with inversions, many molecular tools have become available for detecting genetic structure in natural populations. The allozyme studies of the 1970s revealed many cases of clinal variation, on both a large and a small geographical scale (reviewed by Mitton 1997). Examples include, on a large scale, the latitudinal gradient in the alcohol dehydrogenase polymorphism in *Drosophila melanogaster* and, on a small scale, the estuarine cline

in the aminopeptidase-I polymorphism in the mussel *Mytilis edulis.*

Studies of allozyme variation sparked a heated debate between the "neutralists" and the "selectionists." Kimura, as the foremost proponent of the neutral model, argued that the spatial patterns of allozyme variation seen in natural populations were consistent with the processes of random genetic drift, migration, and mutation (see, for example, Maruyama and Kimura 1980). He saw no evidence for natural selection maintaining polymorphism at this biochemical level. Others, particularly those with a background in ecological genetics (for example, Clarke 1975), were convinced that natural selection was responsible for patterns seen in nature. This debate was never fully resolved, in part, because both sides had a point. As Wright had noted some 40 years earlier (see equation 6.1), random genetic drift influences spatial genetic structure, but the extent of its influence depends on, among other things, the strength of selection. Random genetic drift is best seen as part of the null hypothesis, rather than as the alternative to selection.

Theory Linking Genetic Structure and Genetic Diversity

The genetic diversity of a population and how that diversity is structured in space (and sometimes in time) can have important implications for a population's evolution and, in some cases, its likelihood of extinction. For this reason, the maintenance of genetic diversity is a question central to conservation biology (see Hedrick, this volume).

Effective Population Size   The level of genetic diversity maintained in a population can be influenced by a large number of factors, but one parameter of particular importance is the effective population size, N_e (Nunney 2000). Most discussion of N_e focuses on the size of population subunits (e.g., the islands of the island model), but here we are concerned with how that substructure influences the effective size of the total population (e.g., the collection of islands considered together).

Consider a population (of size gN) divided into g island subpopulations, each of size N. Assuming an island model with each subpopulation contributing equally to the pool of migrants, the effective size of the whole population (or metapopulation) is

$$N_e = gN/(1 - F_{ST}) \tag{6.7}$$

(Wright 1943). Note that as $F_{ST} \to 1$, the effective size of the metapopulation approaches infinity, because low interisland dispersal blocks the effectiveness of genetic drift. As a result, the genetic variation of the metapopulation becomes distributed among islands (i.e., the islands become genetically different) while relatively little variation persists within islands (i.e., the islands become inbred). This pattern can have serious negative consequences for a metapopulation, even though, according to equation 6.7, it has a high effective size (see Nunney 2000). In particular, each island becomes, in effect, an independent evolutionary unit with little adaptive potential, since the genetic variation in other islands is not readily accessible. This scenario is particularly relevant to conservation biology, where a threatened species may be distributed among a few isolated reserves (see Hedrick, this volume).

The traditional island model assumes that each subpopulation is independently regulated and contributes equally to the migrant pool. This is probably a rather unusual situation in nature. More realistically, successful islands will produce more migrants than unsuccessful islands. This assumption underlies the interdemic island model. In the interdemic model, even random differences in female fecundity will result in differences in island productivity. These differences are sufficient to promote interdemic genetic drift, which has a marked effect on the effective size of the metapopulation. Following Wright's (1943) derivation of equation 6.7, but allowing Poisson differences in female fecundity to affect island productivity, gives

$$N_e = gN/(1 + F_{ST}) \tag{6.8}$$

(Nunney 1999). Equation 6.8 shows that under the interdemic model, increasing genetic structure ($F_{ST} \to 1$) reduces N_e, which is the precise opposite of its effect under the traditional island model (see equation 6.7).

Variance in reproductive success among individuals acts to decrease N_e in both a panmictic population and one subdivided into islands. This effect can be incorporated, together with the general effect of variation in island productivity, in a relationship that generalizes (6.7) and (6.8):

$$N_e = \frac{gN}{(\frac{1}{2} + V_i)(1 - F_{ST}) + 2NF_{ST}V_Sg/(g - 1)} \tag{6.9}$$

(Whitlock and Barton 1997; Nunney 1999) where V_i is the standardized variance in reproductive success among individuals within islands, V_S is the variance in island productivity, and N and g are the size and number of islands, respectively.

Under extreme conditions of extinction and recolonization, the variance in island productivity (V_S) dominates (equation 6.9). Whitlock and Barton (1997) showed that such variance can dramatically lower the effective size of a metapopulation and that the effect is most severe if the variance in island productivity is largely due to cycles of extinction and recolonization. They showed that:

$$N_e \approx g/[4(m + e)F_{ST}] \tag{6.10}$$

where e is the probability of island extinction and m is the migration rate. Under these conditions, the effective size is primarily dependent on the number of islands rather than the number of individuals. In practical terms, equation 6.10 illustrates that cycles of extinction and recolonization are very efficient at removing genetic variation even though the overall numerical size of a metapopulation may remain large. This is another result with particular relevance to conservation genetics (see Hedrick, this volume).

Local Adaptation   When environmental conditions vary spatially, then genetic diversity is generally enhanced. If patches are large relative to dispersal distances, local adaptation leads to the maintenance of genetic variation. However, most of this variation reflects differences between patches, with relatively little within-patch genetic variance. This is the pattern seen in the classic examples of ecological genetics mentioned earlier (e.g., industrial melanism). A more complex issue concerns whether genetic variation can be maintained in a patchy environment when the population is panmictic or nearly so. As yet, there is no definitive answer to this problem, and it remains uncertain whether or not environmental heterogeneity can maintain significant levels of polymorphism when there are few barriers to gene flow between habitat patches. However, a number of models demonstrate the possibility, provided certain assumptions hold.

Levene (1953) modeled a randomly mating population distributed over different types of habitat patch. Success in each patch was determined by a single locus, with alternate alleles favoring high fitness in alternate environments (i.e., there was a strong genotype × environment interaction). A stable polymorphism could be maintained, provided that each patch was independently regulated. Independent regulation generates a frequency-dependent effect that promotes stability; however, the conditions for polymorphism are still fairly restrictive. Recent models have reinforced these conclusions and have shown that limited migration, strong selection, and genotype × environment interaction are all important in promoting polymorphism.

Recently, a single-locus, two-patch model (Kisdi and Geritz 1999) incorporated some important additional features into the framework used by Levene (1953). In a quantitative genetic approach, the fitness of each genotype in a given patch is determined by its phenotypic distance from the optimum; and in addition, long-term evolution is included, during which time alleles can give rise to slightly different alleles by mutation. Kisdi and Geritz (1999) found that a strong selective difference between the patches promoted polymorphism, and that a polymorphic outcome commonly consisted of one of the homozygotes and the heterozygote adapted to the alternate patches, with the remaining homozygote having a uniformly low fitness.

Evolution in a Structured Population

As noted earlier, the local population regulation assumed in the "traditional" island model is probably extremely unusual in nature. More generally, regions of a population that are particularly successful, because of genetic or environmental factors, contribute disproportionately to the next generation. This differential productivity is one way of viewing how spatial structure can influence evolutionary change. Unfortunately, since local differences in productivity can be found in almost any spatially structured polymorphic population, observing differential productivity tells us nothing about the nature of the underlying natural selection. It therefore takes a more careful examination to determine which form of selection is acting (see table 6.2). We will examine each of these forms of selection in turn, after first confirming that group productivity differences occur in even the simplest model of selection in a structured population.

Hierarchical Selection and the Paradox of Productivity   The differential productivity of regions within a population is sometimes incorrectly used to suggest the occurrence of a special form of selection (group selection; see table 6.2). Certainly, the differential productivity of groups *sounds* as if it would result in group selection! Unfortunately, as we just noted, any form of selection acting in a structured population will generally lead to "differential productivity." This effect can be illustrated by using a powerful technique for partitioning hierarchical selection introduced by Price (1970). His method of partitioning applies generally to any selected character, and in particular, he showed that the change in mean gene frequency *(p)* at a single locus due to selection could be expressed as

$$\Delta p = [Cov(w_i, p_i) + E(w_i \Delta p_i)]/\overline{w} \qquad (6.11)$$

where $\overline{w}$ is the mean fitness, and the subscript i relates to units at a lower level within the population. In a population subdivided into islands, we can partition evolutionary change into two components: change due to the covariance *(Cov)* between the average fitness of an island and its gene frequency (the between-group "productivity" effect) and change due to the average effect of selection *(E)* that is occurring in each subpopulation (the within-group effect). The expectation term *(E)* can itself be expanded in the form of (6.11), showing the hierarchical nature of this approach.

In general, a region with a high frequency of high-fitness genotypes will be more productive than a region with a lower frequency of such genotypes. To understand this effect, consider a simple example of individual selection, in which a population of haploids is segregating two genetic types (A and B) at a frequency of p and q, with fitness 1 and $1 - s$, respectively. The population is divided into subpopulations of size N, with F_{ST} determined primarily by interisland dispersal. The deterministic change in gene frequency per generation is defined by

$$\Delta p = spq/\overline{w} \qquad (6.12)$$

(where $\overline{w} = 1 - sq$). Group structure has no effect on this rate of change. Partitioning equation 6.12 according to equation 6.11 gives (approximately)

$$Cov(w_i, p_i) = s\,Var(p_i) = spqF_{ST} \qquad (6.13a)$$

$$E(w_i \Delta p_i) = spq(1 - F_{ST}) \qquad (6.13b)$$

Thus, the influence of individual selection on gene frequency change through differential group pro-

Table 6.2 Natural selection in a structured population.

Type of selection	Is fitness dependent on interactions?	Individual selection?	Effect of spatial structure
Interdemic selection	No	Yes	Spatial differences in gene frequency create different distributions of individual genotypes within groups.
Proximity-dependent selection	Yes	Yes	An individual's fitness is affected by the genotypes of neighbors (the proximity group).
Group selection	Yes	No	Similar genotypes are positively associated. An individual's fitness is affected by this positive association.
Kin selection	Yes	No	Similar genotypes are positively associated due to close kinship. An individual's fitness is affected by this kin association

ductivity is approximately $spqF_{ST}/\overline{w}$ (using equation 6.13a). But we should also note that, for most purposes, evolution in this population is best understood through the unpartitioned equation 6.12, since population structure has absolutely no effect on the course of selection.

Wright's Shifting Balance Model and Interdemic Selection It should now be clear that Sewell Wright developed the basic methodology we use to understand the formation of genetic structure within a population. But he went further. Based on the interdemic version of his island model, Wright proposed a far-reaching hypothesis of adaptive evolution in nature. This was the shifting-balance model. Wright's genetic work on coat color in guinea pigs convinced him of the importance of nonadditive interactions between genes. As a result, he viewed adaptation very differently from one of the other great theoreticians of the time, R. A. Fisher. Fisher believed that the most successful alleles were the "good mixers" (i.e., those that had a more or less constant additive effect regardless of genetic background). To Fisher, adaptation was a continual process of improvement, complicated primarily by a changing environment that could alter the genetic optimum.

Wright's view was more complex, with adaptation represented by a multidimensional genetic landscape of adaptive peaks and maladaptive valleys. The problem of adaptation, as seen by Wright, was moving a population from a moderate adaptive peak to a higher one. This required moving the gene pool of the population through a valley of a lower average fitness. Wright's model relied on a "shifting balance" between selection and random genetic drift in a subdivided population. The shifting balance model invokes three "phases." In phase 1, the island subpopulations cluster around a particular adaptive peak, but given moderate levels of genetic isolation, their relatively small effective size would allow some of these islands to drift into a less adaptive valley (i.e., random genetic drift overcomes selection). In phase 2, the high fitness of genotypes close to a new adaptive peak influences those islands that drifted away from the original peak. This drives selection within these islands, and they evolve out of the "valley" to the new adaptive peak (i.e., selection overcomes drift). If this new strategy is superior to the old one, then phase 3 comes into play. This final phase is driven by island productivity, a process that Wright called "interdemic selection." He assumed that the best-adapted islands would produce the most migrants, and that the differences in productivity would pull all subpopulations to the new, higher adaptive peak.

The simplest scenario for Wright's shifting-balance model is the case of heterozygote disadvantage with two alleles, A and B. Let the fitness of the three genotypes be: $w_{AA} = 1$; $w_{AB} = 1 - s_1$; $w_{BB} = 1 + s_2$, and assume that initially the B allele is rare. Note how in this case, even though only one locus is involved, the alleles have a highly nonadditive interaction that defines the fitness of the heterozygote. In a large panmictic population, the B allele can never increase, even though BB is the fittest genotype. On the other hand, in a population structured into a number of relatively small demes, genetic drift could enable the B allele to become established in some demes. Interdemic selection would then promote the spread of the beneficial allele in the remaining demes.

Note that throughout the process of shifting balance, the fitnesses of the genotypes never change. The problem that Wright sought to resolve through population structure was the detrimental effect of recombination (or, in the preceding example, heterozygosity). In the presence of strong nonadditive gene interactions, recombination will tend to break up rare beneficial combinations of alleles into non-beneficial combinations. As a result, these alleles cannot increase in frequency in a large panmictic population. The division of a large population into several smaller demes makes it more likely that a good combination can get established through the action of shifting balance. As a result, evolution progresses differently in a subdivided versus an undivided population, in the presence of nonadditive gene interactions.

Wright's shifting-balance model remains an important theoretical cornerstone of the study of adaptive evolution in structured populations. Nevertheless, its practical importance is still debated (see Coyne et al. 1997).

The third phase of Wright's model invokes interdemic selection. This can be defined as the differential productivity of demes due to individual selection (table 6.2). The definition excludes the effects of differential productivity due to random sampling (interdemic genetic drift; Nunney 1999) and due to group selection. Interdemic selection has importance beyond the shifting balance model since it provides a general basis for modeling the dynamics of individual selection in structured populations and metapopulations.

Group Selection   In the early 1960s, an important evolutionary debate over the role of population structure was initiated. This centered on a disagreement over the importance of group selection (selection that favored the group over the individual). Wynne-Edwards (1962) proposed that many natural populations are regulated at densities too low to be explained by external factors such as food availability or predation. He suggested that natural selection favored populations exhibiting self-sacrificing social conventions that avoided overexploitation of their resources because it guaranteed the future survival of the group. Maynard Smith (1964) strongly criticized this view, stressing that such group selection is unlikely to succeed. He pointed out that a mutant that "cheated" by not conforming to the altruistic social norm of reproductive self-restraint would be able to exploit the available resources to enhance its fitness and consequently spread through the population by individual selection.

The main prerequisite for successful group selection to work in nature is that altruists become nonrandomly associated with each other and hence avoid some of the negative fitness effects of being with selfish nonaltruists. Thus, we can define group selection as the differential productivity of demes due to the fitness effects of the positive association of similar genotypes (Nunney 1985). But could the extinction and recolonization scenario proposed by Wynne-Edwards maintain such associations? Models of Maynard Smith (1964) and others showed that altruism was unlikely to evolve unless the islands (or groups) were very isolated. To maintain the necessary associations, dispersal between extant islands had to be very low, and yet, colonization of vacant habitat had to be effective. Such conditions are probably extremely unusual in nature.

Maynard Smith (1964) noted that one type of selection, identified by Hamilton (1964), facilitated the evolution of altruism. This was kin selection (see Wilson, this volume). Kin selection (table 6.2) depends on the interaction of close relatives, and most researchers now consider it a special case of group selection. As such, it may be the only widespread form of group selection.

Proximity-Dependent Selection   Much of our population genetic theory is based on the assumption of constant relative fitness; however, in nature this may be the exception rather than the rule. Intraspecific interactions are probably ubiquitous in determining, at least in part, an individual's fitness. Proximity structure defines the spatial scale at which these interactions act. This scale can vary widely, although in general it is probably quite small. It may include all of an individual's neighbors within an island subpopulation (or larger unit), as in the extinction-recolonization group selection models. More usually, the proximity group may consist of a just a few immediate neighbors, as in, for example, a plant's competitors for light.

Such competitive interactions are, of course, examples of individual selection, and proximity-dependent selection can be defined as a form of individual selection that depends on proximity-group interactions (table 6.2). Given such interactions, fitness becomes frequency-dependent, since it depends on the mix of genotypes in the proximity group. Indeed strong frequency-dependent interactions have frequently been observed in experimental studies of intraspecific competition in plants and of larval competition in *Drosophila*. The proximity group of individual plants varies because the genotypes of its immediate neighbors vary, and less obviously, *Drosophila* larvae also inhabit a structured environment in nature, made up of spatially separate resource patches (generally rotting fruit), and larvae in different patches will interact with a different mix of genotypes. Population structure of this kind is likely to be common.

It is well known that group (and kin) selection can profoundly alter the outcome of natural selection by favoring the spread of altruistic traits. It is less obvious that proximity-dependent selection can also lead to unexpected evolutionary optima. For example, we usually expect a 1:1 sex ratio to evolve; however, Hamilton (1967) showed that, under some circumstances, a temporary and random association of egg-laying females favors the evolution of a female-biased sex ratio (see Orzack, this volume). Parasitoid wasps often have a population structure that corresponds closely to Hamilton's (1967) model, and female-biased sex ratios are common.

Hamilton's (1967) model illustrated a very important point: Population structure can influence the outcome of individual selection. Wilson has shown that Hamilton's finding was quite general (see Wilson, this volume). When individuals interact and those interactions are localized within temporary, random trait groups, then novel evolutionary optima can be favored just as was seen with sex ratio. However, the requirement of isolated groups is an analytical convenience. The same con-

clusions apply whenever individuals interact exclusively with those in their immediate proximity.

Future Directions

Gene Flow

Genetic structure is crucially dependent on the levels of gene flow prevailing now and in the past, and prevailing at both small and large distances. One of the future challenges for population geneticists is to refine the methods for measuring gene flow at these different temporal and spatial scales. Prior to the widespread use of allozyme variation in the late 1960s, the main method of measuring gene flow was by direct observation, for example, by observing the movement of marked animals. This method is a valuable adjunct to genetic methods; however, rare long-distance movement of individuals or gametes can be important in molding genetic structure, and these events are very difficult to document. Allozyme variation provided researchers with estimates of F_{ST} that could be compared with the dispersal (and other life history) characteristics of species (see, for example, Hamrick and Godt 1996). However, the relatively low levels of variability found at allozyme loci have limited our power to analyze different temporal and spatial scales.

We have entered an era in which a wide range of genetic variation can be examined, different types of variation being appropriate for different scales. Microsatellite loci, because of their high levels of variability, can be used to estimate genetic correlations on a smaller spatial scale than is generally possible with allozymes. In addition, using microsatellites it is often feasible to unambiguously recognize individuals and their offspring and make direct estimates of current levels of gene flow. These estimates allow us to determine if current levels of dispersal are consistent with current genetic structure.

At the other extreme of spatial and temporal scales, phylogenetic methods detect long-term movement patterns. Phylogeography recently celebrated its 10th anniversary, and its success is based largely on the analysis of animal mtDNA variation. Such variation is effective at detecting intraspecific spatial structure that has developed over relatively long periods of time and over relatively large areas.

The method relies largely on deriving a robust phylogeny of the genetic variation and, in some cases, the application of the molecular clock. The challenge for the future is twofold. First, there is a need to incorporate in the analysis other types of genetic variation, such as nuclear sequence data. This is a particularly important issue for the study of plant phylogeography, where the lack of appropriate molecular variation has been a serious problem. Second, there is a need to interpret phylogeographic patterns in terms of gene flow (see Templeton 1998).

Coalescent techniques[2] are being increasingly used to develop theory that exploits the similarities and differences of allelic DNA sequences or microsatellite sizes to infer common ancestry. Recent work combines spatial and temporal scales in attempting to estimate not only "when" but also "where" the common ancestors existed (Epperson 1999). The goal of these approaches is to interpret genetic data from local, landscape, and geographical scales in terms of history and ecology. But a major problem remains. Ecological processes such as population dynamics and gene flow can be highly variable. Studies of extinction-recolonization dynamics (see Whitlock and Barton 1997) have shown that fluctuations in population size can profoundly influence genetic structure, and incorporating such variability remains a real challenge.

Ecological Genetics

The study of genetic structure at the DNA level and its interpretation in terms of gene flow provide only one facet of the influence of population structure on evolution. Specifically, adaptive evolution depends primarily on the interaction of gene flow with natural selection, and much of this interaction is driven by ecology. Indeed, one consequence of metapopulation dynamics is that the rate of gene flow itself is subject to natural selection (Olivieri and Gouyon 1997).

Local adaptation, particularly in plants, can be demonstrated through transplant experiments. These have often revealed the "home site advantage" consistent with local adaptation (see Waser and Williams, this volume). The scale of this adaptation can be as small as a few meters or less. This finding of small-scale local adaptation raises several important questions: First, what is the strength of local selection in natural populations relative to gene flow? Second, what environmental factors

drive this selection? And third, are the multiple patch models of Levene (1953) and others, discussed earlier, compatible with such adaptation?

Similarly, in animal populations, there is a need for studies of spatial structure that combine elements of both ecology and population genetics. The feasibility of such studies is enhanced by the increased availability of highly polymorphic genetic markers, but they still require a long-term commitment. However, such studies provide valuable information. For example, the study of spatial patterns of life history variation in the guppy, *Poecilia reticulata,* by Reznick and coworkers has demonstrated experimentally that local adaptation is strongly influenced by the presence or absence of specific predators (see Reznick and Travis, this volume). And while the metapopulation study of the butterfly *Melitaea cinxia* by Hanski and his collaborators (for example, Saccheri et al. 1998) does not address adaptation, it has shown how patterns of extinction and colonization both influence and are influenced by genetic structure.

Field studies of fine-scale population structure have been hampered by a lack of genetic markers. Thus, it is not surprising that there is a lack of field studies investigating the role of proximity structure on the evolution of natural populations. As noted earlier, its influence on sex ratio in parasitoid wasps in the wild is well documented (see Orzack, this volume). But there are still relatively few data on the importance of local competitive interactions in driving the evolution of natural populations. Stevens et al. (1995) provided one example. They showed that interactions among individuals of the plant *Impatiens capensis* have a frequency-dependent effect on fitness. Specifically, small individuals benefited disproportionately when surrounded by other small individuals. Similar competitive frequency-dependent effects have been documented many times in laboratory experiments using *Drosophila;* however, field evidence is lacking.

Ultimately population structure and its influence on evolution can be studied only in natural populations. The increased availability of DNA-based genetic markers will facilitate such studies in the future. While many of the advances in molecular genetics have prompted an increasingly reductionist approach to evolution, it is perhaps paradoxical that the same technology can also be used to reinvigorate the study of ecological genetics and spatial patterns of adaptation at the phenotypic level.

Notes

1. Semelparous organisms die after one bout of reproduction; in comparison, iteroparous organisms are capable of repeated bouts of reproduction.

2. Coalescent techniques are based on the assumption that all gene copies within a given lineage have a common genealogy and form a monophyletic lineage that can be traced back to a single common ancestor ("coalescence").

7

Inbreeding and Outbreeding

NICKOLAS M. WASER
CHARLES F. WILLIAMS

Contemplate the descent of a piece of DNA (or RNA in organisms using this as their genetic material). The DNA is copied, and copies are passed to descendants. If the copies were error-free we could rightly think of them as perfect clones that pass down indefinitely through the eons. This logic led Richard Dawkins to speak of immortal coils in his book on selfish genes; here, it instead brings up issues of the common ancestry of genes and of individuals, and of the definition and consequences of inbreeding and outbreeding, the subjects of this chapter.

When two individuals share one or more ancestor, they are relatives, both in common parlance and by technical definition in biology. The consequence of their mating is inbreeding, that is, the production of offspring receiving copies of a given gene through both mother and father that can be traced to the common ancestor(s). These gene copies are *identical by descent* (IBD; not to be confused with an acronym for inbreeding depression, see below), a shorthand for "identical by the fact of descending as copies of the same original piece of DNA" (see Nunney, this volume, figure 6.1). The probability that two gene copies are IBD in a diploid individual, or its *inbreeding coefficient*, symbolized by f, is a simple function of the genetic relatedness of its parents and the segregation of genes during meiosis and gametogenesis. Because the probability is one-half that two gametes from the same individual carry identical gene copies, fer-

tilization by self produces f of one-half, a brother-sister mating or parent-offspring mating produces f of one-quarter, a first-cousin mating produces f of one-sixteenth, and so on (see "Measurement of Inbreeding and Outbreeding," below). In these examples, we assume that neither common ancestor(s) nor parents themselves are inbred; such inbreeding reflects additional common ancestry and so inflates f. From all of this, a definition of *outbreeding* as "mating of nonrelatives" follows automatically.

As just defined, inbreeding and outbreeding rely on an *absolute* measure of relatedness. An alternative definition that may be of more value in real, finite populations (as opposed to ideal, infinite ones) is that inbreeding is mating with relatives more often than expected by chance, and outbreeding the opposite. By this *relative* measure, inbreeding yields an excess of homozygotes over random-mating expectation $(0 < f < 1)$, while outbreeding yields a deficit $(-1 < f < 0)$. This approach also leads to an interpretation of f as a *fixation index* that describes deviations of genotype frequencies from Hardy-Weinberg expectations, whether from nonrandom mating alone or from other causes.

These definitions are not without semantic and conceptual problems. Consider for example that all individuals of a given species, indeed all organisms on earth, share a single ancestor, implying that *all* matings are inbred! However, the implication is false in a practical sense, since inbreeding must always be measured implicitly or explicitly with reference to some more-or-less recent ancestral generation. A fixed reference generation (or reference population, which amounts to the same thing) provides a benchmark by which we can assess IBD of gene copies, allows us to estimate f, and yields a distinct boundary between inbreeding and outbreeding. It is critical to keep in mind that our dichotomy of inbreeding and outbreeding is relative and always refers to a certain ancestral generation! Furthermore, just as the degree of inbreeding can vary, so can that of outbreeding, ultimately because of mutation. It is mutation and the sorting of mutants by genetic drift and selection that lead to changes in a potentially immortal copy of DNA. New genetic variation is constantly generated and preserved, as Darwin argued without knowing the mechanism. Thus, all gene copies are not identical even though they ultimately descended from a distant common ancestor, and for this reason also, the implication that all matings are inbred is false.

These contemplations show that there exists a continuum of genetic similarity of individuals, from the closest kinship (individuals belonging to the same completely inbred line, $f = 1$), across the boundary into "outbreeding" ($f = 0$) and continuing through ever greater genetic dissimilarity to the boundary of the species itself, beyond which it is still conceivable, however, that the two individuals may mate and pass on gene copies. For flowering plants (angiosperms) and some animals, we can imagine a *bimodal* set of mating opportunities (figure 7.1). What range of inbreeding and outbreeding is exemplified in the matings that actually occur? One of the simplest classes of mating to demonstrate conclusively in the absence of sophisticated genetic information is a self-mating. As a result, and because selfing figures centrally in early theories of the evolution of mating systems (Sakai and Westneat, this volume), the study of selfing versus "outcrossing" has dominated discussion among those working on cosexual organisms (those in which both male and female sexes are expressed within the same individual), especially angiosperms. In truth, however, a range of less intense "biparental" inbreeding dominates the mating opportunities for many organisms, in which restricted dispersal of individuals and/or gametes leads to kinship structure within populations (Nunney, this volume). In angiosperms, inbreeding may be reduced by partial or complete genetic incompatibility mechanisms. Inbreeding may also be limited in plants by tempo-

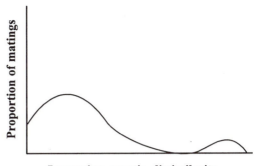

Increasing genetic dissimilarity

Figure 7.1 A bimodal distribution of potential matings may characterize populations of many flowering plants and some animals. The left-hand mode represents the inbred side of the continuum, ranging from selfing or close biparental inbreeding through to the greatest degree of genetic dissimilarity that occurs between conspecific mates, given the dispersal biology of the species. The right-hand mode captures the possibility of matings between individuals of subspecies or closely related species that encounter one another in sympatry (e.g., due to secondary contact). In most animals, the smaller second mode is not realized, and the conceivable matings between the two modes also are not realized because individuals of that intermediate genetic dissimilarity do not live in sympatry or within the distance of one another's dispersal.

ral and physical separation of male and female reproductive organs, and in other organisms by dispersal before reproduction of one or both sexes. Whether inbreeding avoidance is the primary source of selection in the evolution of these mechanisms is debated. To a first approximation, a moderate to large degree of inbreeding characterizes many sessile organisms, including angiosperms and marine invertebrates; much less inbreeding characterizes birds and mammals (see Thornhill 1993 for review).

Inbreeding and outbreeding carry with them predictable consequences for phenotypic expression. Inbreeding commonly leads to *inbreeding depression*, a decline in the mean value of characters, including components of fitness such as viability and fecundity (figure 7.2). Inbreeding depression can be severe: Progeny of selfed matings in angiosperms, for example, not uncommonly suffer a fitness loss >50% (e.g., Price and Waser 1979; Charlesworth and Charlesworth 1987). The converse of inbreed-

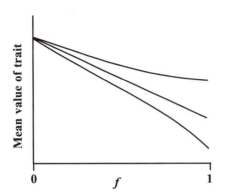

Figure 7.2 Decline in mean value of a character as a function of the inbreeding coefficient f. In the absence of interactions between loci affecting a character (i.e., epistasis), the mean value of the character under inbreeding (M_f), relative to the mean under outcrossing (M_0), is

$$M_f = M_0 - 2f \Sigma \, dpq$$

where d is the dominance deviation of the heterozygote relative to the mean of homozygotes, p and q are the frequencies of alternate alleles, and the sum is taken across all loci. The empirical observation is that d tends to be positive, so in the absence of epistasis we expect linear inbreeding depression as shown in the middle function graphed above (for a character such as development time, d is likely to be negative—heterozygotes will have shorter development time than the mean of the homozygotes—and the equation above will predict an *increase* rather than decrease with f, as we might expect). In contrast, if the dominance at one locus interacts with that at another (dominance × dominance epistasis), we expect nonlinear inbreeding depression, either a decline in character value that grows increasingly steep with f ("reinforcing" epistasis; lower function graphed above), or one that grows more shallow ("diminishing" epistasis; upper function graphed above). If the character is fitness and its mean value is expressed on a log scale, the slope of a linear decline estimates the genetic load, B.

ing depression is *heterosis,* an observation of increased character mean upon the crossing of inbred individuals. However, with crosses between more and more genetically dissimilar individuals, we expect at some point to encounter another decline, which logically can be called *outbreeding depression* (e.g., Price and Waser 1979). Our definition here is purposely vague in stating the critical

level of genetic dissimilarity, because this level is poorly understood at present. However, the critical level may correspond to surprisingly small scales of spatial, ecological, or taxonomic separation in some species. At these scales, variation in the genetic makeup of populations, and in environments experienced by offspring, is expected to lead to spatial and temporal variation in the expression of outbreeding depression (e.g., Waser et al. 2000).

In what follows, we discuss historical roots of the scientific study of inbreeding and outbreeding, explore how they can be measured, discuss mechanisms of inbreeding and outbreeding depression, and note some consequences for the evolution of mating systems. We then turn to our own studies of inbreeding and outbreeding and their fitness consequences in montane wildflowers. We conclude with suggestions for future work to help fill the gaps in our understanding of inbreeding and outbreeding and their genetic, ecological, and evolutionary mechanisms and consequences.

History

Charles Darwin married his first cousin, Emma Wedgewood; their children had $f = 1/16$ assuming no other shared ancestors. This is not an unusual marriage in human history or across cultures. Incest, or mating of close kin such as members of the same nuclear family, is a strong human taboo; the boundary line often hovers around the level of first cousins. Marriage rules have political and cultural foundations, but the incest taboo also may have biological roots. Although we do not know when inbreeding depression was first appreciated, humans may have recognized early some link between matings of kin and the incidence of birth defects and juvenile mortality. Such a recognition also may have come from animal husbandry, in which inbreeding is a necessity of a small herd, and inbreeding depression is expected (Darwin 1868).

Avoidance of incest requires knowledge of close kin, which was virtually automatic in tightly knit ancestral societies. In fact, the codification of kinship is another universal feature of human culture. The counterpoint is some universal definition of nonkin, which includes both those individuals with which friendly relationships such as marriage are allowed, and those who are so distinct as to be seen as alien. The expression of these constructs in language at some point served as a precursor to a scientific description of inbreeding and outbreeding.

As a scientist, Darwin made two major contributions to thinking on inbreeding and outbreeding. First, he compiled evidence from domestic animals and plants (Darwin 1868), and he also extended the early pollination experiments of Andrew Knight (Darwin 1876) to deduce that inbreeding depression is widespread, so that "nature abhors perpetual self-fertilisation." He also found that plant species that regularly self-fertilize suffer less inbreeding depression than nonselfers, and that some experimental plant lineages seem to purge themselves of inbreeding depression following enforced selfing. Second, in his broader discussion the origin of species, Darwin provided a modern framework for thinking about the partial and complete reproductive barriers to hybridization that had been studied a century earlier by Joseph Gottlieb Kölreuter and Carl Friedrich von Gärtner. As a means of investigating and demonstrating the fixity of species, these earlier workers began the exploration, still incomplete, of fitness consequences of wide outbreeding; Darwin placed this exploration within an evolutionary and phylogenetic framework.

Darwin's work on inbreeding and inbreeding depression was followed by a steady increase in similar studies with mice, rats, domestic stock, maize, and other plants (see Wright 1977; Lynch and Walsh 1998). The rediscovery of Mendel at the turn of the century allowed the consequences of inbreeding for genotype frequencies to be appreciated and linked with the empirical results of character expression. Indeed, Mendel (in the "lost" paper of 1865) appears to have been the first to show that the frequency of heterozygotes declines by one-half per generation under complete selfing. Experiments during the first two decades of the 20th century clearly documented both inbreeding depression and heterosis for *Drosophila* and maize and articulated the two major genetic explanations that prevail to this day (see "Inbreeding and Outbreeding Depression," below). But a quantitative measure of inbreeding remained elusive until Sewall Wright's (1922) invention of path analysis to solve for the correlation of uniting gametes (yet another definition of f) from pedigrees.

Wright's invention formed the foundation of all future discussion of inbreeding. It was followed by the recognition that "genetic viscosity" (i.e., restricted spatial dispersal of genes) characterizes many populations. This recognition, in turn, led to the development of methods for describing the spatial genetic structure of populations, including hierarchical versions of the fixation index. In particular, Wright developed a view of evolutionary change in which population substructure, inbreeding, and random genetic drift play central roles (see reviews in Wright 1977; Nunney, this volume).

Our modern perspective on inbreeding and outbreeding has other roots as well. One is the recognition of diverse mating behaviors in animals, such as birds and mammals, and the focus this brought on empirical studies of mating patterns, including the kinship they embody and their fitness consequences (Thornhill 1993; Sakai and Westneat, this volume). In parallel came the recognition of diverse mating systems in plants. More recently, the short-term adaptive value of sexual reproduction has become an issue in evolutionary biology, which raises the possibility that some of the benefits of outbreeding parallel those of engaging in sexual recombination (e.g., Lloyd 1980). From these roots, and fertilized by methods borrowed from other fields such as molecular biology, grow our current perspectives on inbreeding and outbreeding.

Measurement of Inbreeding and Outbreeding

Written language and other technological advances eventually allowed humans to keep pedigree records for themselves and for their animals. Pedigree information makes possible an unambiguous calculation of f, at least with reference to the earliest ancestral generation for which information is known (figure 7.3).

Unfortunately, long lifespan, uncertainty about paternity (and even maternity), and other factors usually combine to make pedigrees difficult or impossible to obtain in nature. Wright developed formulae for expected homozygosity in populations with regular systems of mating, and later for the effects of mutation, selection, migration, effective population size, and population substructure on f. Here f provides a single measure of the deviation in heterozygote frequency from Hardy-Weinberg expectations due to all causes (Wright 1977; box 7.1). More important for the measurement of inbreeding and outbreeding in natural populations was Wright's hierarchical subdivision of the fixation index into the so-called F statistics (box 7.1;

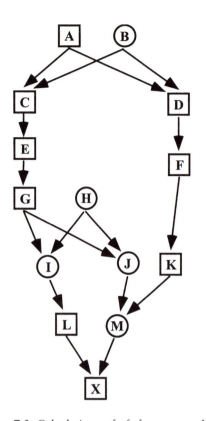

Figure 7.3 Calculation of f from a pedigree. Squares are males, circles are females, arrows show transmission of gene copies across generations. Individuals A, B, G, and H are common ancestors of X through both his father, L, and his mother, M, and thus are possible sources of alleles that are IBD. For simplicity, we omit individuals outside the lineages leading to these common ancestors (e.g., the father of L). The method of Wright (1922) traces each possible path that leads from an individual of interest through one parent back to a common ancestor and returns to the individual through the other parent. The contribution of that path to f is one-half raised to the power of the number of individuals along the path (not counting the individual of interest). Assuming no common ancestor is inbred, f is the sum of these contributions. In this example, there are four paths (LIGECADFKM, LIGECBDFKM, LIGJM, LIHJM), so the inbreeding coefficient of X is $(0.5)^{10} + (0.5)^{10} + (0.5)^5 + (0.5)^5 = 0.0645$. Note that most of this inbreeding coefficient is due to the mating of first cousins L and M, with little contribution from more distant common ancestors A and B. This is a real example: X is the paternal grandfather of NMW (see Waser, in Thornhill 1993).

Nunney, this volume). Keeping in mind that the correlation (standardized covariance) of alleles *within* individuals is equivalent to their standardized variance *among* individuals, the F statistics partition the variance among individuals in a total population (F_{IT}; equivalently the total deviation from expected Hardy-Weinberg heterozygosity) into components due to the average fixation (including inbreeding) of individuals within subpopulations (F_{IS}), and due to genetic differentiation among subpopulations relative to the panmictic expectation in the combined sample (F_{ST}). F_{IS} is often misrepresented as the "inbreeding coefficient" but, as described earlier for f, depends in addition on other evolutionary factors, including additional hierarchical levels of population subdivision. Inbreeding in nature has predictable consequences for patterns of gene fixation. For example, species of highly self-fertilizing plants on average maintain high levels of homozygosity within populations (high F_{IS}) and because pollen dispersal is restricted show greater differentiation among populations (high F_{ST}) than do more outcrossing species (Hamrick and Godt 1996).

With the advent of molecular markers several methods have been developed to calculate coefficients of relatedness, r, in the absence of pedigrees. These coefficients estimate the probability of IBD of two gene copies carried in different individuals; for example, r of full siblings is one-half. Note that r between individuals is twice the value of f for their offspring. Queller and Goodnight (1989) showed how information on allele frequencies from molecular markers can be combined across loci to generate virtually unbiased estimators of r between population subclasses or even individuals, and developed the correspondence with F statistics.

Another approach, developed explicitly for cosexual plants, is to estimate the *selfing rate, s* (or *outcrossing rate, t*) from the genotypes of progeny arrays. The mixed-mating model (Brown 1990), on which most outcrossing estimators are based, assumes that matings are a mixture of pure selfing and outcrossing. As already noted, some proportion of outcross matings will be with relatives in any situation of genetic viscosity. The selfing rate, $s = 1 - t$, therefore reflects both true selfing and all forms of *biparental inbreeding*. Multilocus estimates of s will be smaller than averages across estimates for single loci, in proportion to the contribution of biparental inbreeding, so that by comparing the two, one can estimate the contribution of true selfing versus other forms of inbreeding (Brown 1990).

A single-locus or multilocus t indicates outbreeding without characterizing how dissimilar the alleles are that are not IBD, and thus how genetically dissimilar the parents were. But as argued already, some substantial range of genetic dissimilarity of parents is likely to characterize matings in many species. Given that this is so, how can we characterize different "outbred" matings? Two possibilities present themselves. Several measures of *genetic distance* have been used to characterize the degree of genetic differentiation between subdivisions within species (e.g., populations, geographic races, subspecies) and between species and higher taxa. In a single taxon (such as *Drosophila* flies), genetic distance often appears to increase monotonically with taxonomic distinctness; hence, it is reasonable to use genetic distance to index a continuum of outbreeding. Genetic distance depends on the identity of alleles sampled from two parents or populations, and this dependence seems at first to return us to the problem that greater and greater genetic dissimilarity will be collapsed falsely into a single determination of nonidentity and classified all together under "outbreeding." However, because genetic distance is taken across multiple loci, there is scope for an ever greater fraction of loci to carry nonidentical alleles, and this feature allows for a great range of dissimilarity indeed! A more modern approach is to make use of the tremendous increase in resolution of genetic change that has become available with newer molecular methods such as protein and DNA sequencing, RAPDs (randomly amplified polymorphic DNA), and single- and multilocus VNTRs (variable number of tandem repeats = satellite DNA). These methods have transformed the field of systematics; with judicious choice of regions of proteins or of DNA, in which amino acid or nucleotide substitutions are occurring at appropriate rates, they also can be used to explore genetic differentiation on finer scales within species.

Inbreeding and Outbreeding Depression

Regardless of the method we might use to characterize the anticipated continuum of outbreeding, the connection between a phenotypic effect (such as outbreeding depression) and a measure of genetic distance or sequence divergence is an exercise in empiricism. In other words, extant theories of outbreeding depression are not explicitly couched in terms of any such measures (indeed, an analogous state of affairs holds for the relationship between inbreeding depression and f). This is not surprising, because there is no universal relationship between genomewide dissimilarity (or similarity) and the genetics of outbreeding (or inbreeding) depression. Emlen (1991) did derive a theoretical expectation for outbreeding depression that depends on the strength of local adaptation of parental populations, a form of genetic distance at those loci involved in adaptation to local conditions. However, local adaptation itself must be measured for each species under study, so we are left again without any universal relationship between genomewide genetic similarity and phenotypic consequences.

Recall that the phenotypic consequences are labeled *inbreeding* and *outbreeding depression*. What are their mechanisms? Two major explanations for inbreeding depression were proposed soon after its link to Mendelian inheritance. These explanations, respectively, invoke *overdominance*, in which heterozygotes are superior to homozygotes at single loci affecting a character (including fitness), and directional *partial dominance*, in which the recessive or partially recessive allele confers a lower phenotypic value. In each case, increased homozygosity following inbreeding leads to decline in the mean genetic value of the character, either because alleles that are complementary as heterozygotes segregate as homozygotes (overdominance hypothesis), or because rare, deleterious (partially) recessive alleles are expressed in homozygous combination (partial dominance hypothesis). Most evidence supports the partial dominance hypothesis (Wright 1977; Charlesworth and Charlesworth 1987; Lynch and Walsh 1998; Byers and Waller 2000), although compelling cases of overdominance do exist. Furthermore, if alternate alleles at a marker locus each are tightly linked to different recessive deleterious alleles at other loci, an excess of heterozygotes will arise at the marker locus (termed *associative overdominance*) even though there is no overdominance at this or the other loci. Such a mechanism could underlie some observed correlations between fitness and heterozygosity. Finally, note also that nonadditive effects across loci (i.e., epistasis), frequency-dependent selection, heterogenous selection

Box 7.1 Calculation of f and F statistics

The Hardy-Weinberg Equilibrium

The underlying basis for estimating gene fixation in natural populations is the Hardy-Weinberg (HW) equilibrium. Castle in 1903 and Hardy and Weinberg in 1908 showed that, given random mating (or random union of gametes) and no selection, mutation, migration, or random genetic drift, the frequencies of alleles and genotypes remain constant from one generation to the next. These expected frequencies provide the null hypothesis to test for the effects of one or more evolutionary factors, including inbreeding and outbreeding, on the genetic makeup of a population.

For diploid organisms, frequencies of genotypic classes can often be estimated directly from individual phenotypes. Allelic variants of enzymes (allozymes) provide a good phenotypic marker for underlying genetic variation because they are codominant; heterozygotes are readily distinguished from homozygotes because of their different mobilities through an electrophoretic gel (figure 7.4).

For example, 100 individuals from a population of plants were sampled for allozyme variation. At the *Pgm* enzyme locus, three genotypes were observed: homozygous fast (two copies of the faster migrating allele = *FF*), homozygous slow *(SS)*, and heterozygous *(FS)*. The observed *genotype frequencies* in the sample are the proportions of each allelic combination. For example, if there are 42 *FF*, 36 *FS*, and 22 *SS*, the frequency of fast homozygotes is *freq(FF)* = 42/100 = 0.42; *freq(FS)* = 0.36; and *freq(SS)* = 0.22. From these genotype frequencies the *allele frequencies* are then calculated:

$$p = freq(F) = freq(FF) + 0.5 \times freq(FS) = (0.42) + 0.5 \times (0.36) = 0.60$$

$$q = freq(S) = freq(SS) + 0.5 \times freq(FS) = (0.22) + 0.5 \times (0.36) = 0.40$$

where $p + q = 1$.

If mating is at random in a large population, then allele frequencies predict the proportional representation of alleles in gametes. The expected genotype frequencies in the next generation are then calculated as the probabilities of random union of different gametes to produce diploid offspring. The probability of producing an *FF* offspring = probability of an *F* sperm *(p)* uniting with a *F* egg *(p)* = $p \times p = p^2$. Likewise, the probability of an *SS* offspring = $q \times q = q^2$, and the probability of *FS* = $(p \times q) + (q \times p) = 2pq$.

In our example, the expected genotype frequencies in the next generation are therefore

$$freq(FF) = p^2 = (0.60)^2 = 0.36,$$

$$freq(FS) \quad 2pq = 2(0.60)(0.40) = 0.48,$$

$$freq(SS) = q^2 = (0.40)^2 = 0.16, \text{ and}$$

$$p^2 + 2pq + q^2 = 1.$$

The Fixation Index, f

The observed genotype frequencies in our hypothetical example were of *freq(FF)* = 0.42, *freq(FS)* = 0.36, and *freq(SS)* = 0.22, obviously different than those expected under HW equilibrium, due to an excess of homozygotes. The fixation index, f, measures this deviation between expected heterozygosity *($H_E = 2pq$)* and observed heterozygosity *(H_O)*:

$$f = (H_E - H_O)/H_E = (0.48 - 0.36)/0.48 = 0.25.$$

It can be seen that $0 < f < 1$ for observed heterozygote deficiency and $-1 < f < 0$ for heterozygote excess. When the population is in HW equilibrium, $f = 0$. One generation of random mating will establish HW when f is nonzero initially. If we observe nonzero f that remains constant from one generation to the next, then the expected allele and genotype frequencies also remain constant and can be expressed in terms of p, q, and f:

(continued)

$$[p^2 + pqf] + [2pq - 2pqf] + [q^2 + pqf] = 1.$$

Therefore, inbreeding $(f > 0)$ increases homozygote frequencies relative to HW by a factor of $2pqf$ and decreases heterozygote frequencies by $2pqf$. Outbreeding (e.g., negative assortative mating, $f < 0$) has the opposite effect.

F Statistics

F statistics are a set of fixation indices that measure deviations in heterozygote frequencies within and among different levels of a hierarchically structured population. In the simplest case, the entire population being sampled comprises two or more spatially discrete subdivisions or subpopulations (e.g., different populations of a species). The deviation of observed heterozygote frequencies among individuals within the total sample, from the expected heterozygote frequencies if the total sample were in HW equilibrium, is symbolized as F_{IT}. This total gene fixation can then be broken down into components due to fixation in individuals within each subpopulation (F_{IS}), and due to differences in allele frequencies among the different subpopulations relative to the HW expectation in the total sample (F_{ST}). The method of estimating F statistics developed by Weir and Cockerham (1984) is in fact analogous to a nested analysis of variance and provides a means of combining data from many loci.

Total gene fixation, F_{IT}, can therefore be thought of as due to the combined effects of inbreeding and other evolutionary factors within subpopulations, and differences in allele frequencies among subpopulations can be thought of as due to distant common ancestry, drift and selection within subpopulations, and migration among them. For example, consider three subpopulations differing in allele frequencies at an allozyme locus, *Pgi* (table 7.1). For the illustrated subpopulations, the heterozygosities associated with the three levels of the hierarchy are

H_I = mean observed heterozygosity in the three subpopulations = 0.303

H_S = mean expected HW heterozygosity in each subpopulation = 0.360

H_T = mean expected HW heterozygosity in the total sample = $2\bar{p}\bar{q}$ = 2(0.738)(0.262) = 0.387

The F statistics are then calculated as the deviation in heterozosity at one level of the hierarchy relative to HW expectations at a higher level:

$$F_{IS} = (H_S - H_I)/H_S = (0.360 - 0.303)/0.360 = 0.158$$

$$F_{ST} = (H_T - H_S)/H_T = (0.387 - 0.360)/0.387 = 0.070$$

$$F_{IT} = (H_T - H_I)/H_T = (0.387 - 0.303)/0.387 = 0.217.$$

The F statistics are mathematically related as $(1 - F_{IT}) = (1 - F_{IS})(1 - F_{ST})$. In this example, there is a heterozygote deficit $(H_O < H_E)$ in each subpopulation, leading to a moderate value of F_{IS}. This deficit suggests that inbreeding may be important within subpopulations, and indeed, F_{IS} is equivalent to the average fixation index, f, within subpopulations. Allele frequencies vary among subpopulations as indicated by F_{ST}. F_{ST} can also be defined in terms of the standardized variance in allele frequencies among subpopulations $(\sigma^2/\bar{p}\bar{q})$ and is a measure of genetic differentiation. The value of F_{ST} in this example suggests that moderate levels of gene flow occur among subpopulations, preventing their complete isolation, but not homogenizing them genetically. The total fixation in the population, F_{IT}, is therefore due primarily to evolutionary factors reducing heterozygosity within subpopulations, such as inbreeding, and to a lesser extent to moderate gene flow among subpopulations.

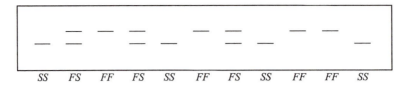

Figure 7.4 Schematic electrophoretic gel showing variation at an allozyme locus. Two alleles that differ in mobility through the gel are segregating in this sample. The faster-migrating allele is designated *F,* while the slower-migrating allele is *S.* The three genotypic classes, *FF, FS,* and *SS* can readily be distinguished.

across environments, and other mechanisms promoting heterozygosity may also in theory contribute to inbreeding depression.

If we assume that different loci contribute to fitness independently in a multiplicative fashion, the *log* of fitness should decline linearly with *f*. The slope of this relationship, *B,* is a measure of the *genetic load* of deleterious recessives, one measure of inbreeding depression (Charlesworth and Charlesworth 1987). But estimating the relatedness of mates in natural populations is problematic, so another approach with cosexual organisms such as angiosperms is to take the ratio of fitnesses of selfed to outcrossed progeny ($1 - w_i/w_o$) and refer to this as inbreeding depression, δ (Charlesworth and Charlesworth 1987; Byers and Waller 2000). This method will fail to detect nonlinearities (figure 7.2), and can generate misleading comparisons if outcrossing is defined differently across species. Ritland (1990) shows how to estimate inbreeding depression from changes in *f* across life history stages or generations, a measure, however, that shares some of the same limitations.

Under the partial-dominance mechanism, the inbreeding history of a population should influence the magnitude of inbreeding depression and the

evolution of mating systems. Lande and Schemske (1985) developed a model of mating system evolution in which a history of inbreeding will purge a population of its genetic load and thereby reduce the magnitude of inbreeding depression over time (notice the historical link to Darwin's experiments with plants). They concluded that most populations should therefore evolve either toward complete selfing or complete outcrossing. However, subsequent work has indicated that intermediate *mixed-mating* systems are common and can be evolutionarily stable (e.g., Uyenoyama et al. 1993; Byers and Waller 2000). The empirical evidence for purging is mixed as well, and Byers and Waller (2000) discuss several factors that may limit purging. Other ecological and genetic explanations for the maintenance of outcrossing mechanisms in plants and inbreeding avoidance in animals should therefore continue to be explored.

With regard to outbreeding depression, two general explanations are proposed (e.g., Price and Waser 1979; Templeton 1986; Campbell and Waser 1987). The first, which may be termed a *physiological mechanism,* invokes nonadditive gene action within or among loci that leads to a reduction in phenotypic expression in F_1 or later hybrid genera-

Table 7.1 Observed and expected heterozygosities *(H_O and H_E)* for three hypothetical subpopulations differing in allele and genotype frequencies at the *Pgi* locus.

Subpopulation	Genotype frequency			Allele frequency		Heterozygosity	
	FF	*FS*	*SS*	*p*	*q*	H_O	H_E
1	0.74	0.21	0.05	0.845	0.155	0.21	0.262
2	0.65	0.28	0.07	0.790	0.210	0.28	0.332
3	0.37	0.42	0.21	0.580	0.420	0.42	0.487
Mean	0.587	0.303	0.110	0.738	0.262	0.303	0.360

tions. For example, Lynch and Walsh (1998) predicted the phenotypes of hybrids largely in terms of *crossbreeding effects* involving various forms of two-locus epistasis (these crossbreeding effects capture the degree and form of genetic dissimilarity of parental populations). In Lynch's model, both single-locus dominance and epistasis contribute to F_1 heterosis or outbreeding depression (see also Schierup and Christiansen 1996). Outbreeding depression is generally anticipated to be greater in F_2 and later hybrid generations, as coadapted complexes of alleles that interact favorably are broken apart by meiosis and recombination, and Burton (1987) and Fenster and Galloway (2000) report just this trend.

The second mechanism of outbreeding depression may be termed *environmental* and proposes that outbreeding produces hybrid progeny ill suited to either parental environment. As noted above, Emlen (1991) modeled a version in which the genome contains structural gene loci with alternate alleles conferring adaptation to alternate ends of an environmental cline, and predicted outbreeding depression quantitatively as a function of the average strength of adaptation of the parents to their native environments (measured with reciprocal transplant experiments). This prediction was upheld quite closely for salmon. As summarized by Waser (1993), local adaptation over short physical distances appears to be common in flowering plants, so that an environmental mechanism may contribute to relatively fine-scale outbreeding depression reported for a number of plant species.

Case Studies with Montane Plants

The Nelson's or Nuttall's larkspur *Delphinium nuttallianum* Pritzel (= *D. nelsonii* Greene) is a small perennial herbaceous wildflower found in subalpine regions of the western United States. Near the Rocky Mountain Biological Laboratory (RMBL) in Colorado, its blue flowers are pollinated about equally by hummingbirds (figure 7.5) and queen bumblebees. Our work over several decades on the reproductive biology of this species gives us insight into inbreeding and outbreeding in *D. nuttallianum* and the role of animal pollinators in affecting these matings, and into the consequences for offspring fitness in the form of inbreeding and outbreeding depression.

We can derive an expectation of the mating system of *D. nuttallianum* from direct observations of pollinator behavior, from knowledge of the spatial kinship structure of populations, and from knowledge of the performance of matings over increasing physical distances, assumed to correspond on average to increasing genetic dissimilarity of parents. The flowers are cosexual and protandrous, expressing female sexual receptivity only after most pollen is shed. Flowers mature sequentially from the base of the flowering stalk (inflorescence). Bumblebees tend to begin foraging at the bottom of the inflorescence and to move upward, thus promoting outcrossing by first visiting older female- and then younger male-phase flowers before flying to another plant. In contrast, hummingbirds, which predominate as visitors during the first few weeks of flowering each summer, exhibit no such directionality. Because seed dispersal is restricted and most pollen dispersal is as well, we expect local kinship structure to develop within populations (Nunney, this volume) and biparental inbreeding in pollinations between neighboring plants. Finally, hand pollinations using pollen from single individual donors indicate that *D. nuttallianum* is partly self-compatible, setting approximately 35% fewer seeds after selfing as after outcrossing over a distance of 10 m (Price and Waser 1979).

Taken together, these studies predict a range of matings in natural populations from selfing through various degrees of biparental inbreeding and into a substantial degree of outbreeding. On the whole, we might therefore anticipate finding some intermediate value of the outcrossing rate, *t*. Multilocus estimates of *t* in four populations in 1991 were quite high, ranging from 0.88 to 0.97 (Williams et al. 2001), but subsequent estimates from 1997 and 1998 were lower, averaging 0.75. These results suggest substantial year-to year and site-to-site variation in mating patterns (F. Saavedra and C. F. Williams, unpublished).

An interesting contrast with *D. nuttallianum* is the sympatric Barbey's larkspur, *D. barbeyi,* which lives in similar habitats and has similar floral morphology, pollinators, and seed dispersal. The main difference is the number and size of inflorescences; *D. nuttallianum* usually has a single stalk with 5–15 total flowers (3–5 open per day), whereas *D. barbeyi* produces 50–1500 flowers (with an average of 170 open per day) on multiple (up to 50) stalks. This difference should lead to more within-plant selfing (geitongamy) in *D. barbeyi*. Indeed,

Figure 7.5 A female broad-tailed hummingbird *(Selasphorus platycercus)* pollinating *Delphinium nuttallianum*. Scale bar is 1 cm. (Drawing by Mary V. Price.)

our estimate of the multilocus outcrossing rate for *D. barbeyi* is only 0.55 (Williams et al. 2001).

These differences in the mating system of the two congeners appear to have both ecological and genetic causes. Differences in pollinator behavior based on inflorescence size and architecture influence the "primary" outcrossing rates and the genetic make-up of arriving pollen, with more geitonogamy in *D. barbeyi*. The high values of multilocus *t* in *D. nuttallianum* also suggest that actual pollen dispersal in our *D. nuttallianum* populations is greater than indicated by direct observations of pollinator behavior. Indeed, a thorough electrophoretic survey of spatial genetic variation on a hierarchy of scales ranging from centimeters to kilometers failed to reveal much structure in *D. nuttallianum*, suggesting substantial long-distance pollen dispersal (Williams and Waser 1999). In contrast, the more selfing *D. barbeyi* had pronounced genetic differentiation among population subdivisions, >10 times that of *D. nuttallianum* at all spatial scales (Williams et al. 2001). The reduction in pollen flow due to increased selfing in *D. barbeyi* and the selective elimination of the seed progeny of more inbred matings in *D. nuttallianum* (which may occur less often in *D. barbeyi;* Williams et al. 2001) appear to lead to these differences in spatial genetic structure between species.

Schulke and Waser (2001) independently used experimental populations of *D. nuttallianum* plants in flower pots to discover that rare long-distance flights by pollinators can deliver surprising amounts of pollen over distances up to several hundred meters, so long as the intervening landscape contains few rewarding flowers of any species. Hummingbirds might be especially important mediators of this long-distance pollen flow among isolates, since separate studies suggest that they lose relatively little pollen from their bodies as they fly, compared to bumblebees. Hence, although hummingbirds may affect more within-plant delivery of self-pollen than bumblebees, they may also affect wider outcrossing. These possibilities remain to be explored.

Not only do we know something about the average degree of inbreeding and outbreeding in *D. nuttallianum,* but we also know about the fitness consequences of inbred and outbred matings (Waser and Price 1994 and references therein). By hand-pollinating many hundreds of flowers over 9 years of experiments, we determined that seed production declines on average by about 35% in self-pollinations, and by about 25% in pollinations between plants that are near neighbors (which are related roughly at the level of first cousins), relative to seed set in pollinations between plants separated by 10 m. Conversely, pollinations over distances of 100–1000 m produced on average about 25% fewer seeds than pollinations over 10 m. These differences in seed set appear to match quite well differences in prezygotic pollen tube growth, suggesting that stylar discrimination based on maternal-paternal interaction is at play, rather than inbreeding and outbreeding depression (which are characteristics of inbred or outbred progeny and hence are postzygotic phenomena).

On the other hand, inbreeding and outbreeding depression are clearly manifest in the survival and reproduction of seed progeny from such crosses. In early experiments (Price and Waser 1979), we planted seeds in a natural environment distinct from either maternal or paternal environments, and we recorded substantial inbreeding and outbreeding depression in progeny survival, suggesting a contribution from a physiological mechanism. In later experiments, we planted seeds near their maternal parents, where natural seed dispersal would have deposited them, and recorded their life histories with annual censuses for up to 11 years after planting. These experiments showed that progeny of 10-m matings again fared best in terms of survival, year of first flowering, and number of flowers produced, suggesting a contribution to outbreeding depression from an environmental mechanism. Specifically, the progeny of matings over 1 m achieved only about 50% the estimated long-term fitness of

the progeny of matings over 10 m, and progeny of matings over 30 m achieved only about 15%! This level of outbreeding depression over short distances is very surprising and is greater than that predicted by Emlen's (1991) model of an environmental mechanism, given the strength of local adaptation measured in *D. nuttallianum.* These results again suggest that a physiological mechanism, in addition to an environmental mechanism, contributes to outbreeding depression in this species, or that the environmental mechanism differs from the one Emlen considered.

We earlier pointed out the lack of detectable genetic structure at marker loci in *D. nuttallianum,* which might appear to contradict the conclusion that local adaptation (i.e., an environmental mechanism) contributes to outbreeding depression. However, lack of structure at loci that are selectively neutral or nearly so does not preclude spatial differentiation at other loci that are under local selection, that is, that confer adaptation to local conditions (Waser 1993; Williams and Waser 1999).

In summary, matings in *D. nuttallianum* span a range from close inbreeding (selfing) to outcrossing over distance of hundreds of meters or more. Genetic analysis of the mating system and spatial genetic structure of populations suggest that the realized mating structure is much less inbred than direct observations would at first indicate. Both ecological (floral morphology and pollinator behavior) and genetic (inbreeding and outbreeding depression) factors influence patterns of realized gene flow. How physiological (genetic compatibility) and environmental (local adaptation) factors contribute remains elusive. As our ability to unambiguously identify parentage of seeds in natural populations increases with new molecular methods, we hope to approach a more complete description of inbreeding and outbreeding and its consequences in this and other species.

Future Directions and Unsolved Issues

The study of inbreeding and outbreeding is a fascinating domain with strong conceptual links to the study of mating system evolution and ecology, and of the genetic structure of populations. There is much still to learn, and new empirical and theoretical tools with which to proceed. Computer power and molecular methods make it an achievable goal to move beyond traditional oversimplifications of mating opportunities (e.g., "selfing vs. outcrossing") to explicitly characterize inbreeding and outbreeding in natural populations. Once we have described the mating system as *who mates with whom to produce offspring*, the door is open for more direct connections with animal behavior, dispersal biology, and the like. Similarly, there is much still to do in measuring inbreeding and outbreeding depression in nature, and in exploring their genetic mechanisms.

Inbreeding depression and outbreeding depression are not static phenomena but themselves coevolve with the mating system (e.g., Uyenoyama et al. 1993) and with the spatial genetic structure of the population (Campbell and Waser 1987; also see Nunney, this volume; Sakai and Westneat, this volume). This dynamical interaction has important implications, which are still being explored, for equilibrium patterns of mating and of fitness in natural populations. For example, the mating systems of angiosperms are astoundingly diverse, and an understanding of their flexibility on microevolutionary scales, shedding light on their diversity across species, will be improved by understanding the fitness consequences of possible matings. Similarly, systematic studies of the fitness effects of matings across a range from extreme inbreeding to wide outbreeding within single species remain rare. Furthermore, it would be desirable to couple empirical results from such studies with theoretical models that allow three-way simultaneous coevolution of the mating system, fitness effects, and spatial population structure, in order to explore dynamics and possible equilibrium states (includ-

ing alternative equilibria). Recent modeling efforts have brought us closer to this level of realism, but we are not there yet!

There are many reasons to engage in these efforts. Some questions lie squarely within the realm of microevolution, with implications for ecology, population biology, and animal behavior over short time spans and spatial scales. Other questions lie within the realm of macroevolution. Of particular importance is an understanding of the mechanisms by which outbreeding depression (aka partial reproductive isolation) evolves. The study of partial reproductive crossing barriers within species was very active only a few decades ago, especially with angiosperms, and deserves to be reinvigorated. The grand prize is a closer understanding of the genetic changes that occur during the development of complete reproductive isolation between higher taxa. These genetic changes may be of kinds that Mendel could have understood or may often involve phenomena (such as inter- and intragenomic conflict) that have entered our consciousness much more recently. Furthermore, the evolution of reproductive isolation may not occur in lockstep with the phenotypic changes by which we usually identify distinct species (Waser 1998). But whatever mysteries are ultimately revealed, reproductive isolation is the sine qua non of macroevolution, and so this is a grand prize well worth striving for.

Acknowledgments   For discussion and advice, we thank Maria Bosch, Daphne Fairbairn, Alan Fix, Mary Price, and Jessica Ruvinsky. Our field studies of *Delphinium* were supported in part by the U.S. National Science Foundation.

PART II

LIFE HISTORIES

8

Age and Size at Maturity

DEREK A. ROFF

Age and size at maturity have been an object of interest to humans since the domestication of animals and plants, for one of the objectives of domestication was to produce an organism that grew fast and matured early at a large size. Selection was also practiced to produce animals that could be used for such purposes as hunting and portaging, and to produce products for pleasure alone, as seen in the many ornamental varieties of dogs, cats, goldfish, pigeons, and plants. All of these instances demonstrate that age and size at maturity are traits that are relatively easily molded by artificial selection and, by extension, natural selection.

Historically, artificial selection experiments were concerned not with the evolution of age and size at maturity in natural populations but with the production of economically more valuable plants and animals. Recently, there has been a substantial increase in the quantitative genetic analysis of nondomesticated organisms, which has shown that, with respect to morphological traits such as adult size, there is typically abundant additive genetic variance, with heritabilities averaging approximately 0.4 (reviewed in Roff 1997). Life history traits, such as the age at maturity, show, on average, lower heritabilities (approx. 0.26) but still enough for rapid evolutionary change.

Quantitative genetic analyses have shown that age and size at maturity can evolve, but the most significant advances in our understanding of the factors favoring particular age at maturity/body size combinations are due to mathematical models predicated upon the assumption that selection maximizes some fitness measure such as the rate of increase, r. In a paper entitled "Adaptive Significance of Large Size and Long Life of the Chaetognath *Sagitta elegans* in the Arctic," McLaren (1966) produced a seminal analysis in which he incorporated all the important elements that have appeared in subsequent analyses of the evolution of age and size at maturity. Specifically, McLaren attempted to take into account the trade-offs produced by increased fecundity being bought at the expense of delayed maturity and increased mortality. In this chapter, I shall primarily consider analyses that have followed in McLaren's footsteps. More recently, the question of the evolution of life history traits and particularly body size or its components has been addressed by measurement of selection in wild, free-ranging populations; this important advance will not be dealt with here as it is covered by Fairbairn and Reeve (this volume).

Concepts

An Overview of Modeling Approaches

There are two broad modeling approaches to the analysis of evolutionary problems, phenotypic models, and genetic models. In the first approach, no attempt is made to model the genetic basis of traits; it is simply assumed that sufficient genetic variation exists so that evolution is not constrained by genetic architecture. While this assumption is reasonable with respect to the equilibrium conditions, genetic architecture can considerably influence evolutionary trajectories, and hence, the phenotypic approach is contingent on the state's having attained equilibrium. The method seeks to construct the fitness surface and hence the optimal combination of trait values. Phenotypic models can be conveniently divided into the optimality method and game-theoretic analysis.

The concept of trade-offs is central to present theories of the evolution of life history traits, for it is trade-offs that limit the scope of variation. For example, selection will typically favor an increased fecundity, but this may require an increased body size, which entails and increased development time, which itself reduces fitness by decreasing *r*. Alternatively, an increased fecundity might be achieved by a decreased propagule size, which may decrease offspring survival. Thus, an increased fecundity may be prevented because of a trade-off with development time or propagule size. Within the set of possible combinations, there will be at least one combination that exceeds all others in fitness. Optimality analysis assumes that natural selection will drive the organism to that particular set. To initiate an analysis using the principle of optimality we must designate some parameter to be optimized. In the present case, we assume that there is some measure of fitness that is maximized by natural selection. The second step is to construct a set of rules that defines the life history pattern of the organism, hypothetical or real, under study. Within these rules, there will exist a variety of possible life histories; the optimal life history is one that maximizes fitness.

Game theory is really a subset of optimality modeling; it is appropriate when interactions are frequency-dependent. The approach comprises two essential elements: First, it is assumed that particular patterns of behavior will persist in a population provided no mutant adopting an alternate behavior can invade. Such stable combinations are termed evolutionarily stable strategies (ESSs). The concept of the ESS is not unique to game theory; the maximization of fitness measures in optimality models are all ESSs within the context in which they are appropriate. Second, for each type there must be

an assigned gain or loss in fitness when this type interacts with another individual. From this payoff matrix, we compute the expected payoff for each behavior. For two behaviors to be evolutionarily stable, their fitnesses must be equal. With respect to age and size at maturity, game theory has been primarily applied to explain the occurrence of male dimorphisms such as are found in Atlantic salmon. This topic will not be dealt with in this chapter (for reviews see Roff 1992, 1996).

Genetic modeling proceeds by defining the genetic mechanism determining the phenotypic trait and the fitness value (e.g., survival or fecundity) of each phenotype. By use of the appropriate mathematical machinery (e.g., explicit Mendelian genetics in the case of simple genetic architectures and quantitative genetic methods for polygenic traits), the equilibrium trait values can then be determined. The critical problems are the correct definition of the genetic architecture and the estimation of the selection gradients. Age and size at maturity are traits that typically show continuous variation and hence the appropriate genetic models are polygenic, although considerable insight can be gained by using even single-locus models. The general formulation of the quantitative genetic model $\Delta \bar{z} = G\beta$, where $\bar{z}$ is a vector of trait means, G is the genetic variance-covariance matrix, and β is a vector of selection gradients for each character (see Fairbairn and Reeve, this volume). For two traits, such as age and size at maturity, the model can be written as $\Delta z_1 = \beta_1 h_1^2 + \beta_2 r_A h_1 h_2$, $\Delta z_2 = \beta_1 h_2^2 + \beta_1 r_A h_1 h_2$, where h_i^2 is the heritability of the ith trait, r_A is the genetic correlation between the two traits, and Δz_i and β_i are in standardized units. The above formulation emphasizes the fact that the evolutionary trajectory is governed by the genetic correlation, which is the genetic parameter that causes the phenotypic trade-off to generate a change in the phenotypic value of the offspring of the selected parents. However, except under very specific circumstances, the genetic correlation will not constrain the system from moving to any particular combination. The trade-offs in the phenotypic models assume that the organism can only move along the trade-off line, which is implicitly assuming that the genetic correlation is ±1. This difference between the two approaches remains a problem to be satisfactorily solved. It remains true, however, that in order to be evolutionarily significant, the trade-offs in the phenotypic models should have a genetic basis, meaning that there exists a nonzero genetic correlation of the appropriate sign between the two traits.

Trade-Offs in the Age and Size at Maturity

Trade-offs are manifested at the phenotypic and genetic levels. At the phenotypic level we observe a phenotypic correlation between traits so that a change in one trait that itself increases fitness is opposed by the antagonistic effect of a change in another trait. For example, because it increases the rate of increase, a decreased age to maturity will increase fitness, all other things being held constant. However, because of the reduced time for growth, a decreased age at maturity might result in a decreased body size. Because in many species fecundity and fighting ability are correlated with body size, a reduced age at maturity could result in a loss of fecundity in females and reduced matings in males (Roff 1992).

When the organism is not at its phenotypic equilibrium, a phenotypic trade-off will produce a selection differential but will not produce a corresponding evolutionary change unless the trade-off is genetically determined, as indicated by a genetic correlation of the same sign as the phenotypic correlation (Fairbairn and Reeve, this volume; Reznick and Travis, this volume). Although trade-offs are typically denoted by negative genetic correlations, this need not be so: In the above example, the correlation between age to maturity and body size is positive, but it represents a trade-off by virtue of the corresponding negative genetic correlation between age at maturity and fecundity.

An increased body size can be achieved either by an increased development time or an increased rate of growth. In the latter case, there may be no resulting correlation between development time and adult body size. On the other hand, such an increased growth rate could be selected against because of a trade-off between growth rate and mortality rate (Roff 1992; Arendt 1997). The analysis of trade-offs between growth and reproductive success (fecundity, number of matings) must therefore consider the interactions among growth rate (i.e., the growth trajectory), development time, adult size, immature survival, and fecundity.

For any given individual, only two of the three traits, growth rate (g_{rate}), development time (d_{time}),

and adult body size (a_{size}) are independent; for example, suppose that the growth trajectory is linear (or can be linearized with an appropriate transformation), $a_{size} = c_0 + g_{rate}d_{time}$, where c_0 is the initial size, which I shall assume is very small relative to the adult size and hence can be ignored (i.e., $c_0 = 0$). In a variety of animal and plant species, adult size appears to be determined by a size threshold for maturation (Roff 2000), in which case, development time would be the dependent variable and the growth relationship is best represented by the equation $d_{time} = \frac{a_{size}}{g_{rate}}$. By taking logs, this equation can be converted into the simple additive model $D_{time} = A_{size} - G_{rate}$, where the uppercase letters designate log-transformed values. If adult size and growth rate are genetically uncorrelated, then there will be a positive genetic correlation between development time and adult size, regardless of the genetic variation in growth rate. So, in this instance, large adult size can only be attained at the expense of an increased development time; thus, for the reasons given previously, development time and adult size would represent a trade-off. If adult size and growth rate are genetically correlated, then the covariances between development time and the two "independent" traits are

$$Cov_A(D_{time}, A_{size}) = V_A(A_{size}) - Cov_A(A_{size}, G_{rate})$$

$$Cov_A(D_{time}, G_{rate}) = Cov_A(A_{size}, G_{rate}) - V_A(G_{rate}) \quad (8.1)$$

where $V_A(\bullet)$ is the additive genetic variance and $Cov_A(\bullet)$ is the additive genetic covariance. If there is a negative covariance between adult size and growth rate (i.e., large adult size means a slow growth rate), then the genetic correlation between development time and adult size will again be positive. However, if there is a positive genetic correlation between adult size and growth rate, then it is possible for there to be a negative genetic correlation between development time and adult size. Similarly, the genetic correlation between development time and growth rate can be positive or negative depending on the relative values of the covariance between adult size and growth rate and the additive genetic variance of growth rate. Thus, when both growth rate and adult size are genetically variable and correlated, there is no longer necessarily a trade-off between size and development or between growth rate and development time.

A survey of the literature showed that 26% of estimated genetic correlations between development time and adult size were negative, though not necessarily statistically significant (6/23 species; Roff 2000). Thus, in the majority of cases, there was a trade-off between adult size and development time. In at least three cases, the negative genetic correlations were significantly smaller than zero, and hence, the lack of a trade-off in these cases cannot be ascribed to low statistical power. The variation in sign the genetic correlation between development time and fecundity is consistent with the observed sign of the genetic correlation between growth rate and adult size and growth rate and development time, though only two species were available for this comparison (Roff 2000).

Does Large Size or Delayed Age at Maturity Increase Fecundity?

Large adult size per se is of no evolutionary importance unless it affects the fitness of an organism. For females the most likely effect is an increased fecundity, for which there are abundant phenotypic data (Roff 1992). Among 11 different species, there are 10 positive genetic correlations and 1 negative, which is significantly different from the null hypothesis of no association ($P = .0059$, one-sided binomial test; Roff 2000). Thus, overall, there is evidence for a positive genetic correlation between body size and fecundity. Given that there is also a strong tendency for development time and body size to be positively genetically correlated, can we conclude that, in general, there will be a trade-off between development time and fecundity? The, perhaps surprising, answer is no. This can be shown by the following argument.

Typically, there is a linear relationship between fecundity and body size on a logarithmic scale, $F = C_0 + C_1 A_{size}$, where F is log-transformed fecundity and C_0, C_1 are constants. For simplicity I shall assume that C_1 is equal to 1 and there is no genetic covariation between C_0 and F; the covariance between fecundity and development time can then be shown to be

$$Cov_A(D_{time}, F) = \frac{1}{2}\{V_A(C_0) + V_A(F) + V_A(A_{size})$$
$$- 2Cov_A(D_{time}, G_{rate})\}$$
$$= \frac{1}{2}\{V_A(C_0) + V_A(F) - V_A(A_{size})$$
$$+ 2Cov_A(D_{time}, A_{size})\} \qquad (8.2)$$

The most striking message of the above equations is that it is possible to have a positive genetic correlation between development time and body size (i.e., $Cov_A(D_{time}, A_{size}) > 0$), but a negative correlation between fecundity and development time. Similarly, it is possible to have a negative correlation between development time and adult size but a positive correlation between fecundity and development time. Thus, it is not possible to conclude from the genetic correlation between development time and adult size if there is a trade-off between development time and fecundity, even if there is a positive genetic correlation between fecundity and adult size.

Unfortunately there are few (five species) estimates of the genetic correlation between development time and fecundity, and these tend to have very large standard errors. There are both significant positive and negative genetic correlations, but the data are too few to allow us to draw any general conclusions (Roff 2000). The question of whether there is generally a trade-off between development time and fecundity remains unresolved, despite its central place in many life history models.

Current Models

Table 8.1 shows a representative set of optimality analyses for the evolution of age and size at maturity. In some cases, the models explicitly examine reproductive effort rather than age or size at maturity; it is generally assumed that a decreased reproductive effort will be associated with an increased age at maturity, although there is no logical requirement for this. Almost all analyses and all those shown in table 8.1, do not explicitly incorporate density-dependent phenomena. Density dependence has typically been dealt with by simply assuming that r equals zero. However, the optimal strategy when food is scarce because population density is high may be quite different from when it is scarce, even when there are no competitors. In the former case, there may be selection for rapid foraging and eating to obtain food before one's competitors, whereas in the latter case, rapid feeding will not in itself increase the total ration that the immature can obtain.

The vast majority of models assumes that size increases with age and fecundity increases with size. Further, as a consequence of the assumed growth curve, the fecundity at age function, $m(x)$, is either concave down (e.g., models 4, 5, 6) or concave up (e.g., models 3, 7, 8, 9; figure 8.1), which has very significant effects on the fitness consequences of delayed maturity (see below).

In addition to the shape of the $m(x)$ function, present models can be categorized according to the environment and the type of mortality. The most common mortality pattern assumed is stage-specific in that one pattern is assumed for immature individuals and another for mature individuals, the latter pattern being established at maturation. Less frequently (model 1), an age-specific or size-specific pattern is assumed, in which case maturity itself does not signal a change in the mortality regime. Most analyses have focused on evolution in a constant, nonseasonal environment. For such an environment, the most appropriate fitness measure is the Malthusian parameter r (or R_0 if it is assumed that $r = 0$) given as $\int (x)m(x)e^{-rx}dx = 1$, where $l(x)$ is survival to age x and $m(x)$ is age-specific fecundity (female births). Four generalizations can be made (table 8.1):

1. It sometimes matters whether one maximizes r or R_0 (models 3, 4).

2. Increasing age-specific mortality of immatures (model 1) favors an increase in reproductive effort and hence by inference an increase in the age of maturity. Conversely, a mortality rate that increases in the older age groups favors a decrease in the age of maturity. Changes in the age and size at maturity of wild populations (both unmanipulated and manipulated by changing the predation pressure) of guppies appear to evolve under predation in the manner so predicted (Reznick et al. 1996; Reznick and Travis, this volume), although it has not been possible to demonstrate the required differences in size-specific mortality regimes in natural, unmanipulated populations (Reznick et al. 1996).

In life histories, in which mortality is stage-specific, the opposite trends hold (models 4, 5, 6, 7): Increasing mortality in the adult stage favors a delay in maturity. The apparently opposite result for model 2 is due to the assumption that animals always mature at age 1; this model predicts that reproductive effort should decrease if "adult" mortality decreases. As stated above, low reproductive effort is generally associated with a delayed maturity. Applying this principle here is not strictly possible because of the fixed age at maturity. According to the stage-specific assumption, an increase in the ratio of adult to juvenile survival (i.e., increased A/B) will favor a decrease in the optimal

age at maturity and an increased reproductive effort; this pattern has been observed in populations of brook trout (Hutchings 1993).

In the stage-specific models, a female incurs the mortality associated with the adult phase only by metamorphosing into the adult form. If adult mortality is high, selection favors females that delay entry into the adult stage, remain immature, and grow larger. Likewise, a high juvenile mortality favors rapid exit from this stage into the adult. The animal thus has the option of varying how long it is subjected to either mortality regime. If mortality is solely age-dependent, a female cannot escape the source of mortality by delaying reproduction; therefore, if mortality in later age groups increases, selection will favor females that reproduce before they are subjected to the increased mortality. If the cost of reproduction in terms of mortality declines with age, selection will act to delay maturation.

3. The effect of varying growth depends on the particular parameter of the growth function that is affected (e.g., model 4). Intuitively, it would seem likely that a delay in the age at maturity would be favored by an increased growth rate, because this delay gives an increased marginal value of fecundity relative to survival. This hypothesis is supported by models 3 and 7, in which an increase in the growth rates favors an increased age at maturity. An increase in the parameter c in models 4, 5, and 6 is in apparent contradiction to the hypothesis in that size at age increases with c; however, because growth is asymptotic (approaching a) the age at which marginal increases in fecundity no longer favor continued growth is reached earlier with an increased c. Thus, the importance of the growth function does depend on the growth rate, which itself depends on the shape of the function: Growth curves that are linear or accelerating will produce delayed reproduction with increased parameter values, whereas growth curves that are decelerating may favor earlier reproduction, contingent on the parameter altered (e.g., in model 4, earlier reproduction is favored by increased a or c, but delayed reproduction is favored by increased b). An important message here is that to test the effects of the growth environment, it is essential to measure the growth trajectories and determine which components of the growth function are changing. In *Drosophila* species and the fish *Gambusia holbrooki*, decreased rations lead to increased age at maturity and decreased body size (Berrigan and Koella 1994; Weeks and Meffe

Table 8.1 Some representative life history models for the analysis of age and size at maturity.

Model	Fecundity at maturity $m(\alpha) =$	$l(x)$ function, $e^{-f(x)}$ $f(x) =$	Parameter	Age at maturity r	Age at maturity R_o	Size at maturity r	Size at maturity R_o	Reference and notes
			Constant, nonseasonal environment					
1	Increases with α	Age-specific	Young	+		+		Law 1979: RE models[1]
			Old	−		−		
			All	0		0		
2	Constant or geometrically increasing with α	Immature = Constant, Mature = Ax	A	−		−		Schaffer 1994b. Maturity at age 1, RE models
3	$a\alpha^b$	Immature = Ax, Mature = Semelparous	a	+	0	+	0	Roff 1992
			b	+	+	+	+	
			A	0	−	0	−	
4	$a(1 - e^{-c\alpha})^b$	Stage specific mortality; Immature = Ax, Mature = Bx	a	−	0	−	0	Roff 1992
			b	+	+	+	+	
			c or A	−	−	+	+	
			B	+	0	+,−		
5	$a(1 - e^{-c\alpha})^b + d$	Immature = Ax; Mature = Constant	c			+	+	Stearns and Koella 1986 ("neg" means that mortality decreases with c)
		Constant neg with c	c				−	
		Constant neg with c	c			+	+	
		Constant neg with c	c			+,−		
6	$a(1 - e^{-c\alpha})^b + d$	Immature = Ax; Mature = Bx	c		−		+	Berrigan and Koella 1994
			A		−		−	
		$(0.05 + 0.5c)x$	c		−		+,−	
		$(0.01 - 0.05c + 2.5c^2)x$	c		−		−,+	
		$(-0.01 + 2c - 2.5c^2)x$	c		−		+	
		$0.4 - 0.95c$	c		−			
7	$a(d + c\alpha)^b$	Ax	A		−		−	Perrin and Rubin 1990. d is the initial size
			c		+		+	
		A and c positively correlated	A,c		+		+	
		A and c negatively correlated	A,c		+		+,−	

Seasonal environment with fixed length of favorable season, S

#	Model / Function	Parameter	Effect	Reference / Notes
8	No mortality except at end of season; ab^{α}	S	+	Cohen 1971
		a	−	
9	$(ab\alpha + d^{b})^{\frac{1}{b}}$; Ax; Cw^{-B}	S	+	Kozlowski and Weigert 1987; d is the initial size
		a	+	
		b	+	
		A	−	
		C	−	
10	Semelparous with $m(x)/l(x)$ a concave function of α	S	+,−	Roff 1980; sawtooth cline
11	Increasing function of age and size; Ax	S	+	Rowe and Ludwig 1991

Stochastic environment—effect of increasing temporal variation in specified parameter (note that most use RE)

#	Model / Function	Parameter	Effect	Reference / Notes
12	Semelparous, mean = a; Immature = Ax; Some fraction matures each x	a	+	Cohen 1966; no age structure.
13	Constant, mature at age 1; Immature A; Mature Bx	A	+	Schaffer 1974a; Charlesworth 1994, p. 222; RE model
		B	−	
14	Mortality rates included in $m(x)$; Iteroparous, $P(m) = \dfrac{k^{g}}{\Gamma(g)}\, m^{g-1} e^{-gm}$	m	+	Tuljapurkar 1990; $P(m)$ is the probability density function of fecundity

Stochastic environment—effect of spatial variation

#	Model / Function	Parameter	Effect	Reference / Notes
15	Immature = Ax; Mature = Ax; a^{b}	a	$-(0)$[2]	Kawecki and Stearns 1993
		A	$0(-)$[2]	

Note: The models consist of a function that determines the relationship between age and fecundity, $m(\alpha)$, and a function that specifies the probability of surviving to age x, $l(x) = e^{-f(x)}$, where the function shown in the table entries is the mortality function, $f(x)$. Shown is the change in the optimal age or size at maturity (given r or R_0 maximized) when the specified parameter is increased (+ = age or size increased; − = age or size decreased; 0 = no change; +,− = concave down function; −,+ = concave up function; no entry, change not determined). Lowercase letters are used for parameters of the fecundity function and uppercase letters for the survival function.

[1] RE models = models that are actually based on changes in reproductive effort, from which changes in size and age at maturity are inferred.

[2] Result in the absence of migration.

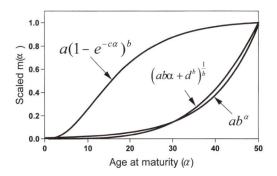

Figure 8.1 Examples of the two types of $m(\alpha)$ function shown in table 8.1. Curves have been scaled so that they have a value of 1 at an age of 50 (arbitrary units). Parameter values differ between curves.

1996); without knowing what parameters were changed by the ration (including survival), it is not possible to use such data to confirm or refute theoretical predictions.

4. Survival is frequently negatively correlated with reproductive effort, which is also negatively correlated with future growth and fecundity (Roff 1992). Such trade-offs can produce a wide range of changes in age and size at maturity (models 6, 7).

The analysis of evolution in a seasonal environment (of fixed length S) has revealed the same general patterns as for a nonseasonal environment, namely, that a delay in reproduction is favored by a low immature mortality rate or a high growth rate (models 8, 9). In all models, an increase in season length selects for an increased age at maturity unless the increase is sufficient for an increase in the number of generations to be favored (model 10). Increases in body size with season length and a "sawtooth" change in body size corresponding to a switch in voltinism have been observed in several species of Orthoptera (reviewed in Roff 1992) and Lepidoptera (Nylin and Svard 1991).

Environments may be variable in either time or space. In a temporally stochastic environment, the appropriate fitness measure is the geometric mean of r or its equivalent in an age-structured environment (Charlesworth 1994). If there is genetic variation, the realized trait values will fluctuate about the optimal combination but will always be somewhat displaced because the response to selection in the preceding generation will generally produce a phenotype that is not optimal for present conditions. Thus, in a stochastic environment, an optimal solution is impossible so long as genetic variation persists. The issue typically examined in the models shown in table 8.1 is whether increasing environmental variation alters reproductive effort, the general conclusion being that increased environmental variation favors decreased reproductive effort and hence a delayed age at maturity. Examination of reproductive efforts in a variety of fish species in relation to environmental variation showed no significant statistical correlation (Roff 1992). However, the occurrence of many cases in which only a portion of a population enters dormancy argues for the general importance of temporal uncertainty in modulating the life history.

The analyses entered in table 8.1 assume that the organism does not have any cue that can be used to predict the state of the subsequent environment. Several cues are, however, potentially available; for example, high density may indicate a poor environment to follow, or photoperiod can indicate that the period remaining for growth is declining. Many insects have indeed been observed to shorten their development time in relation to day length indicative of a late date (reviewed in Nylin et al. 1996). In such circumstances, an organism has two options: reduce development time by maturing at a small size, or increase growth rate. In the lepididopteran genus *Lasiommata,* the reduction in development time is brought about primarily by an increased growth rate (Nylin et al. 1996).

Only recently has the solution for predicting the optimal life history in a spatially variable habitat been developed (Kawecki and Stearns 1993). Suppose that the habitat is spatially heterogeneous and organisms disperse across the habitat. In this case, the appropriate fitness measure is the overall r, not the r that is characteristic of a particular environment. Thus, the characteristic equation must be expanded as $\int p(h) \int l(x, h) m(x, h) e^{-rx} dx dh = 1$, where $p(h)$ is the probability of habitat h. The optimal combination of trait values can be markedly different from the average set determined by considering each habitat separately (model 15).

Most environments are likely to show variation in both space and time. Such variation will favor either migration or some sort of "sit-tight" strategy such as dormancy. Selection for migration by active locomotion (as opposed to ballooning, etc.) will favor fast growth rate and large size, a prediction supported by both interspecific and intraspecific comparisons (Roff 1992).

Case Study: The Evolution of Age and Size at Maturity in *Drosophila*

Insects are particularly useful organisms for the analysis of the evolution of age and size at maturity. First, with few exceptions, insects do not grow after achieving adult size, which reduces the potential number of factors that must be incorporated into an analysis. Second, many have short generation times and are easily reared in large numbers in the laboratory, facilitating genetic analyses. The genus *Drosophila* and the species *D. melanogaster* in particular have been extensively studied and have provided models for the evolution of age and size at maturity under a variety of conditions.

Predicting the Size at Maturity in Drosophila Melanogaster *under Density-Independent Conditions*

Drosophillids typically feed on yeast growing in sap exudates, rotting fruits, and so on, habitats that are probably generally very ephemeral. As a consequence, *Drosophila* species in general can be considered colonizing species, though the possibility of reaching high population densities may occur in some species. Like most *Drosophila* species, in the wild *D. melanogaster* is primarily a colonizing species, though its ease of rearing under laboratory conditions has permitted an examination of its evolution under extremely high population densities. In this section, I shall examine the evolution of age and size at maturity of *D. melanogaster* under density-independent conditions, using *r* as a measure of fitness (Roff 1981).

Body size and development time are genetically correlated and can be related by the trade-off function (figure 8.2)

$$D(L) = d_0 L^{d_1} + d_2 \qquad (8.3)$$

where $D(L)$ is development time for an adult with thorax length L, and the d_i's are constants. As noted above, the evolutionary importance of large size resides in the increased fecundity associated with increased size (Roff 1992, 2000). The age-

schedule of fecundity in *D. melanogaster* is triangular in shape and can be described by the function (figure 8.2)

$$m(x) = m_0 \left(1 - e^{-m_1(x-m_2)}\right)e^{-m_3 x} \qquad (8.4)$$

where age x is measured from the day of eclosion, and the m_i's are constants. Fecundity measured over some given period is allometrically related to thorax length and hence can be most simply incorporated into the $m(x)$ function by replacing m_0 with $m_4 L^{m_5}$ (figure 8.2). Fecundity and adult size are genetically correlated, and hence, the above equation represents an evolutionarily significant trade-off. Thus, because of its correlation with fecundity, selection favors an increased body size, which in turn leads to a genetically correlated response in development time. A continuous increase in body size will be opposed by the increased development time decreasing r in two ways. First, an increase in mean generation time decreases r, and second, if larval mortality rate is constant the increased, development time will lead to an increased preadult mortality. Assuming constant mortality rates for the larval and adult periods the $l(x)$ function can be written as

$$l(x) = e^{-l_0 D(L)} e^{-l_1 x} \qquad (8.5)$$

where l_0 is the mortality rate in the larval stage (averaged over the egg, larval, and pupal stages) and l_1 is the adult mortality rate. The characteristic equation is thus given as

$$\sum_{x=m_2}^{\infty} l(x)m(x)e^{-r(x+D(L))} = 1 \qquad (8.6)$$

The above equation describes a curvilinear function between adult size *(L)* and *r* (figure 8.2), indicating stabilizing selection towards an intermediate optimum. The predicted optimal value falls within the observed range in body size of *D. melanogaster*. The values used to obtain the curve shown in figure 8.2 were drawn from a variety of laboratory experiments or crudely estimated for field conditions, and hence, the correspondence between prediction and observation suggests that the optimal body size is relatively insensitive to parameter values (Roff 1981). Pairwise variation of parameter values confirmed this, variation within reasonable ranges producing optimal thorax lengths ranging only between 0.8 mm and 1.1 mm (Roff 1992).

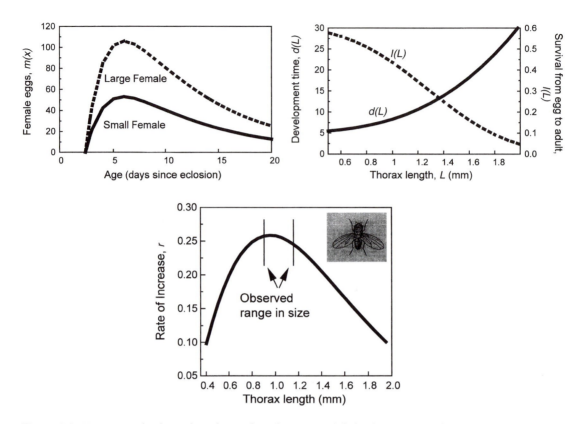

Figure 8.2 Upper panels show the relationships between adult body size and life history traits in *Drosophila melanogaster*. The upper left panel shows the age schedule of reproduction for a small and large female. The upper right panel shows the development time required to attain a specified adult size (indicated by thorax length) and the immature survival for this size. Lower panel shows the estimated relationship between fitness, as measured by the rate of increase and adult size in *D. melanogaster*. (After Roff 1992.)

Predicting the Size at Maturity in *other* Drosophila *Species*

The sensitivity analysis suggests that other species of *Drosophila* that have similar ecologies will have similar body sizes. North American *Drosophila* species range in size from approximately 0.7 to 1.5 mm, which are sizes that can be made the optimal size by relatively minor changes in one or several parameter values. However, some of the Hawaiian *Drosophila* are two or three times as large as *D. melanogaster*. Can such extreme variation be predicted from the above model, and if so, do the corresponding changes match what is observed in the Hawaiian species? According to the body-size–development-time trade-off function, the large sizes of the Hawaiian species should be accomplished by

a very extended development time. This is, in fact, what is observed (figure 8.3). Analysis of equation 8.4 shows that large adult size will be favored by high survival rates and a very low adult fecundity.

In *D. melanogaster*, fecundity is correlated with the number of ovarioles and hence can be used as an approximate index of fecundity in other *Drosophila* species. For the Hawaiian Drosophillidae *(Drosophila, Scaptomyza, Antopercus, Ateledrosophila)*, there is a highly significant regression between ovariole number and thorax length (log(ovariole number) = 0.73 + 1.9log(thorax length), $r = 0.68$, $P < .00001$), with the number of ovarioles being substantially less than found in *D. melanogaster* (figure 8.3). This is indirect evidence for the low predicted fecundity of the Hawaiian species. This conclusion is reinforced by fecundity data for

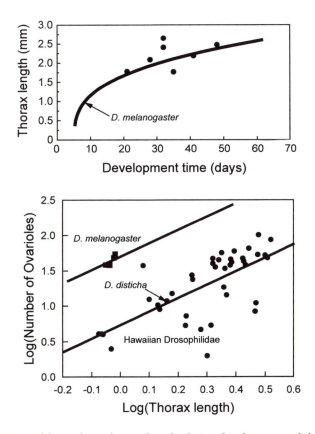

Figure 8.3 Upper panel: Solid lines show the predicted relationship between adult size and development time for *Drosophila melanogaster*. Symbols show the observed body size and development time for some Hawaiian *Drosophila*. (After Roff 1992.) Lower panel: Relationship between the number of ovarioles and body size in some Hawaiian Drosophilidae ($\bullet$, $\text{Log}_{10}Y = 0.73 + 1.9\text{Log}_{10}X$, $r = 0.68$, $n = 46$, $P < .00001$). Also shown ($\blacksquare$) are data for six populations of *D. melanogaster*. The line passing through these points is the regression line of the Hawaiian Drosophilidae moved upward. (Data from Capy et al. 1983 and Kambysellis and Heed 1971.)

the Hawaiian species, *D. disticha,* which indicates a fecundity in the laboratory of less than 0.5 eggs per day (Robertson et al. 1968) and an ovariole number typical of Hawaiian species and far below that of *D. melanogaster* (figure 8.4).

Age and Size at Maturity under Crowded Conditions

There have been two basic approaches to addressing this problem. In one approach, a reaction norm is constructed by varying the parameter or parameters that are hypothesized to be affected by the high density (Stearns and Koella 1986; Berrigan and Koella 1994); in the second approach, population dynamics are explicitly incorporated in the model (Mueller 1988). All the "reaction norm" models predict a negative relationship between adult size and development time, as is generally observed. In his analysis, Mueller (1988) considered two alternatives: Either larvae are able to modify the threshold size that must be reached to initiate metamorphosis, or they are able to adjust the efficiency with which they process food. The first model predicted the same response as the "reaction norm" models (i.e., a reduced size with increasing density), whereas the second model predicted an increasing body size.

Experimental evaluation of selection occurring in high-density populations of *D. melanogaster* has shown that all of the above analyses are flawed in ignoring several important trade-offs. Populations

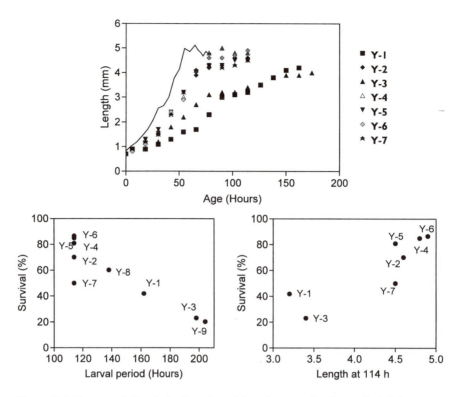

Figure 8.4 Top panel: Symbols show larval length at age for *D. mulleri* fed on pure cultures of yeast (Y-1..Y-7). (Data from Wagner 1944.) Solid line shows the growth of *D. melanogaster* fed on standard medium. (Data from Alpatov 1929.) Lower panels: Survival from hatching to adult emergence as a function of duration of the larval period and the length achieved 114 hr posthatch.

maintained at high density were found to evolve and at the high density achieved body sizes larger than flies from populations maintained at low density. However, this difference disappeared when flies were reared at low densities. The increased competitive ability came about because of an increased feeding rate of the K-selected larvae and a higher height at pupation (which increased their survival by reducing the probability of being submerged in the food medium by the activities of other larvae). All other things being equal, an increased feeding rate should be under directional selection. It is not under directional selection, because there is a negative correlation between feeding rate and the minimum size for pupation (Joshi and Mueller 1996). In an environment in which food is scarce because of factors other than the action of other larvae, selection for faster feeding will

be opposed by the increased food requirement, which might not be available.

Age and Size at Maturity in a Heterogenous Environment

Predicting the evolution of age and size at maturity in a heterogenous environment depends on the level of migration among habitat patches. Events that are temporally rare can have very important influence, whereas events that are spatially rare should typically be less important (Roff 1992). The potential importance of environmental heterogeneity can be illustrated with life history data collected by Wagner (1944) for the cactophilic *Drosophila* species *D. mulleri*. This species lays its eggs on cacti of the genus *Opuntia*. High numbers of flies are associated with the fruiting season of *Opuntia*,

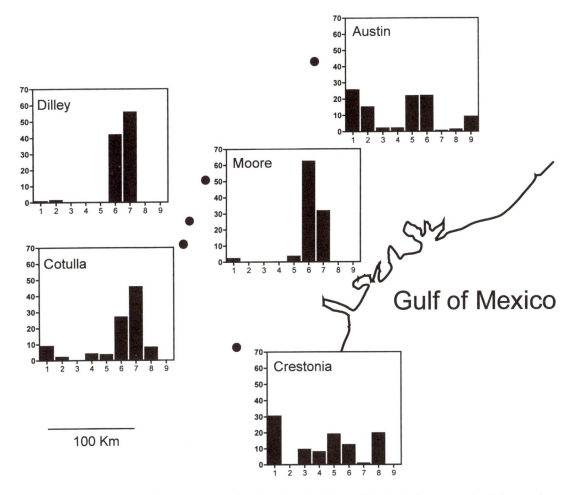

Figure 8.5 Distribution of yeast types at five locations in Texas. The basic data comprised the number of colonies obtained by plating diluted juice from the fruit of *Opuntia*. The results shown are the frequency of each yeast type averaged with respect to sample and site. Number of samples per site: 9 (Crestonio), 15 (Cotulla), 15 (Dilley), 6 (Moore), 20 (Austin).

and it is believed that the cactus fruit are the major or sole source of food for the larvae. Wagner identified nine different yeast types from *Opuntia* and measured the growth and survival of *D. mulleri* larvae on each. Growth rates for *D. mulleri* varied greatly among the nine yeast types, with exceedingly slow growth on Y-1, Y-3, and Y-9 and growth rates comparable to *D. melanogaster* on the other six (figure 8.4). There was a highly significant negative correlation between survival and larval period and between growth rate, as indexed by size achieved after 114 hr, and survival (figure 8.4).

We would expect to find the highest growth and survival rates on those yeast types that occur frequently both spatially and temporally in the habitat. Yeast types that occur very rarely will not exert a large adaptive pressure. The very low growth and survival rates on Y-3 and Y-9 suggest that these are such rare types, a hypothesis confirmed by their almost complete absence in *Opuntia* collected from five widely spaced sites (figure 8.5). In contrast, Y-6, the yeast to which *D. mulleri* appears well adapted (highest survival and growth rates), occurs at a relatively high frequency at all sites. The yeast Y-1 is of particular interest because *D. mulleri* grows very poorly on it but still has moderate survival; this yeast occurs at a very low frequency at three sites (Dilley, Cotulla, Moore) but is a significant component of the flora at Austin and Crestonia, suggesting that it does exert an adaptive pressure in some locations or at some times.

If the distribution of yeast types were as found at Dilley or Moore, selection could favor specialization on Y-6 and Y-7. However, a composition such as found in Crestonia or Austin would favor a less specialized genotype and the ability to cope with Y-1, which is rare in the other sites. The extent to which specialization can occur will depend on migration rates and temporal variation in floral composition. The present data do not permit an assessment of the importance of spatial and temporal variation in floral composition in shaping the age and size at maturity of *D. mulleri,* but they do show the potential for an important influence.

Future Directions

Theoretical analyses on the optimal age and size at maturity assume a trade-off between growth, survival, and fecundity. Despite the importance of these assumptions, there are surprisingly few data demonstrating the genetic bases for such trade-offs. Further, there is evidence that in some cases, adult size and development time are positively correlated, whereas in other cases, they are negatively correlated. There is a clear need for more quantitative genetic analysis of the covariation between life history traits directly or indirectly related to maturity. Such analyses should not be restricted to a laboratory setting but should take into account the ecological setting in which evolution has occurred.

Most analyses are predicated on the assumption that selection is density-independent. At present, there is insufficient theoretical or empirical investigation of the effects of density dependence on the evolution of age and size at maturity. The work of Mueller and his colleagues has shown that the situation may be more complex than initially supposed.

There is a growing recognition that the world is spatially and temporally heterogenous, and hence models must incorporate this factor. When cues exist on the state of future conditions, we expect that selection will favor the evolution of phenotypic plasticity. We now have the basic theoretical tools to tackle this problem, and considerable progress could be made in this area within the near future.

Because they are readily distinguished and are intimately tied to fitness, age and size at maturity are excellent model traits for the analysis of evolutionary change. The traits are common to all living organisms, and hence, an analysis in one organism is likely to have very general significance.

9

Offspring Size and Number

FRANK J. MESSINA
CHARLES W. FOX

If we look across all organisms, we find that some species produce only one or a few large offspring per reproductive bout (e.g., most birds and mammals), others produce 10s or 100s of intermediate-size offspring (e.g., most plants and insects), and yet others produce many 1000s of offspring (e.g., some marine invertebrates). How can we account for such broad variation? In this chapter, we review many of the environmental and demographic variables that influence the evolution of offspring size and number.

In the first section, we discuss how the trade-off between offspring size and number is an important determinant of offspring size. An individual's resources can be allocated to three basic functions— growth, somatic maintenance, or reproduction. Resources directed toward reproduction can in turn be used to produce either many small offspring or a few large offspring. Thus, for a fixed amount of resources available for reproduction, it necessarily follows that there is a trade-off between the number and size of offspring during a given bout of reproduction.

Trade-offs between offspring size and number during a single reproductive bout are a primary determinant of offspring size for most semelparous organisms, which reproduce once in their lifetime (e.g., salmon and century plants). For iteroparous organisms, however, lifetime reproduction is divided into many discrete bouts, with intervening periods of no reproduction. Evolutionary explanations for the number and size of offspring in these organisms must also consider how reproductive effort during any one period affects future survival and reproduction. The second part of our chapter considers the evolution of offspring number among long-lived, iteroparous organisms, especially vertebrates. We focus on the clutch sizes of birds that produce altricial[1] (nidicolous) young. Because each nestling requires much parental care, we expect strong selection toward producing the most appropriate number of offspring for a given environment.

The trade-off between current and future reproduction can also affect semelparous animals if offspring must be distributed among scattered resources. Many insects, for example, lay eggs on small, discrete hosts, and their sedentary offspring often cannot move between hosts. A female that places too many eggs on a host faces the same diminishing returns as a bird that produces more nestlings than it can provision. Because females encounter many potential hosts during their lifetimes, the optimal number of eggs to deposit at any one site will depend on the abundance, distribution, and quality of other sites. We conclude the chapter by examining the evolution of offspring number in insects whose hosts typically support the development of only a few offspring.

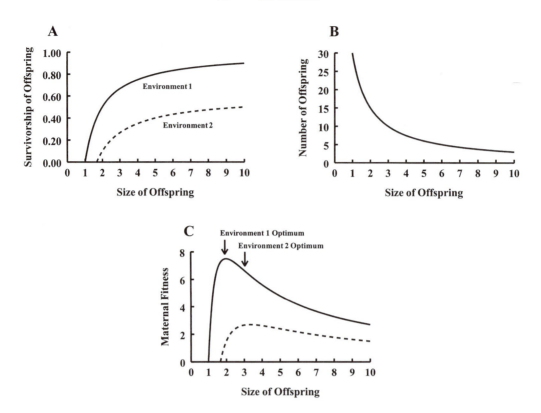

Figure 9.1 Graphical depiction of the Smith-Fretwell model of optimal offspring size (units are arbitrary). (A) The first assumption: Fitness of offspring increases with increasing offspring size (for simplicity, we assume that offspring size affects only offspring survival). (B) The second assumption: There is a trade-off between offspring size and number, so that the number of offspring produced necessarily decreases as investment per offspring increases. (C) The total number of adult offspring that a female will produce is maximized at an intermediate size of offspring. The shape of the curve simply reflects the product of the curves in A and B.

The Evolution of Offspring Size

The first formal analysis of optimal offspring size was offered by Smith and Fretwell (1974). They calculated the size of offspring that a female must produce to maximize her total number of grand-offspring. They started with two assumptions: that offspring fitness (W_{Young}) increases with greater parental investment per offspring (I_{Young}) (figure 9.1A), and that there is a trade-off between offspring size and the number of offspring produced; for any fixed amount of parental investment into reproduction (I_{Total}), the number of offspring necessarily decreases as the investment per offspring increases (figure 9.1B). A female can produce N offspring, where $N = I_{Total}/I_{Young}$. Any increase in N requires less investment per offspring

(lower I_{Young}) or a higher total investment (I_{Total}). To simplify their model, Smith and Fretwell assumed that I_{Total} is constant. Thus, maternal fitness, $W_{Parent} \propto N * (W_{Young}) = (I_{Total}/I_{Young}) * (W_{Young})$, the product of the number of offspring (fecundity) and the fitness of each offspring (figure 9.1C). The value of I_{Young} that results in the highest parental fitness is the value that maximizes $(I_{Total}/I_{Young}) * (W_{Young})$.

This model illustrates three points that have become the subject of much empirical and theoretical exploration. First, for any fixed amount of parental allocation to reproduction, offspring size is under balancing selection; fitness per offspring increases with offspring size, but maternal fecundity increases as the investment per offspring decreases (figure 9.1). Second, there is an evolutionary conflict of interest between parents and their offspring (Einum and Fleming 2000). Because offspring fitness increases with increasing investment per offspring (I_{Young}), the value of I_{Young} that maximizes offspring fitness (figure 9.1A) is larger than the value that maximizes parental fitness (figure 9.1C). Third, any environmental variable that changes the relationship between investment per offspring and offspring fitness should also change the optimal offspring size in a population (e.g., environment 1 vs. environment 2 in Figure 9.1). Since the original Smith-Fretwell model, more complex models have examined optimal offspring size under a range of conditions (review in Clutton-Brock 1991; Roff 1992; Fox and Czesak 2000), but most start with the same assumptions: that there is a trade-off between offspring size and number, and that offspring fitness increases with increasing parental investment per offspring.

Trade-offs between offspring size and number (some of which have been demonstrated to be genetically based) have been detected in many semelparous organisms that do not exhibit parental care but have been less frequently detected when females are iteroparous (e.g., many vertebrates, perennial plants) or provide parental care (reviews in Roff 1992; Fox and Czesak 2000). The reason for this discrepancy may be that a negative correlation between offspring size and number is expected only when the quantity of resources allocated to reproduction (I_{Total}) is fixed. In nature I_{Total} is often variable. For example, variation in juvenile or seedling growth can produce substantial variation in size at maturity, which is often correlated with individual reproductive effort. We may thus observe that some females produce both more *and* larger offspring, the result being a positive rather than negative correlation between offspring size and number. To detect a trade-off, we would need to consider parental size as a covariate. Other sources of variation in I_{Total} are less easily quantified or controlled; these include variation in resource acquisition during the adult stage and variation in the degree of parental care.

Many studies have examined the second assumption of the Smith-Fretwell model: a positive relationship between offspring size and offspring fitness. They often find that smaller eggs or seeds are less likely to hatch or germinate. Smaller hatchlings or seedlings generally have lower survival and are generally less able to withstand stress from competition, starvation, desiccation, low oxygen, low temperature, or environmental toxins (Fox and Czesak 2000). In most animal taxa, small offspring partially compensate for their small initial size by extending development time and maturing at a "normal" size, although it is occasionally observed that small eggs produce small adults.

There are, however, circumstances in which selection may favor smaller propagule size. For example, in some insects smaller eggs hatch sooner and are thus exposed to egg parasitoids for shorter durations. In plants, the optimal seed size often represents a balance between germination requirements and the need for seeds to disperse. Smaller seeds tend to disperse further and are often found among successional species and among species where competition between parents and offspring is likely.

Offspring Size in Variable Environments

Selection on offspring size is likely to vary across time and space, and even among offspring produced by a single female, but few studies have quantified this variation (Fox and Czesak 2000). Variable selection may explain much of the variation in offspring size observed among species or populations. Geographic clines in offspring size are commonly observed and suggest that some environmental variable that changes across the cline affects the intensity or direction of selection on offspring size. Clines have been argued to depend on gradients in temperature, salinity, growing season, food abundance, and other factors. In *Drosophila*, egg size increases with latitude, and experimental studies suggest that temperature is the driving variable (Azevedo et al. 1996). Yet the direction of a cline can vary even among related organisms. For example, some insects lay larger eggs in cold climates, others lay smaller eggs in cold climates. In many organisms, offspring size varies substantially across space but does not exhibit a cline (e.g., offspring size of fish and crustaceans in different lakes or ponds).

When the optimal offspring size consistently varies over small spatial scales, selection may favor larger offspring than predicted by the Smith-Fretwell model (e.g., Forbes 1991). The reason is that producing larger offspring reduces the variance in offspring survival (across environments, and thus across generations if each generation potentially experiences a different environment) and thus results in higher geometric mean[2] fitness. Alternatively, spatial variation in selection can favor increased variance in progeny size or adaptive plasticity in progeny size in which females are able to adjust the size of offspring in response to immediate environmental cues. For example, many cladocerans lay larger eggs when food availability is low, presumable because larger offspring survive periods of food stress better than small offspring. Similarly, some plants respond to the presence of competitors by producing larger seeds, so that seedlings can better compete for resources.

The size of offspring may also vary substantially within a family. In animals, much of this variation is correlated with maternal age. Many vertebrate females produce larger offspring as they age, whereas most insects produce progressively smaller offspring. Both adaptive and nonadaptive explanations have been proposed to explain these patterns (Roff 1992; Fox and Czesak 2000). Offspring size also varies within a single brood. For example, in some birds, the last egg laid in a clutch is especially small and survives only as "insurance" when an elder sibling dies or when the availability of food is especially high. Alternatively, we should recognize that there might be physiological limitations in the ability of females to make identically sized offspring.

Violations of Model Assumptions

Like most optimality models, the Smith-Fretwell model ignores much real-world biology. For example, the model does not consider how reproductive effort may evolve in response to selection on offspring size or number. The model also incorporates only one constraint, the trade-off between offspring size and number. It does not consider morphological or physiological constraints on the ability of females to make especially large or small offspring, although such constraints may be important. For example, the necessity for offspring to fit into the brood pouch of a female may constrain the evolution of larger offspring in *Daphnia*, even when larger offspring are favored in the local environment. Physical and physiological constraints on offspring size have also been identified in some vertebrates (Roff 1992) but are generally poorly known.

Smith and Fretwell (1974) also assumed that selection always favors increased fecundity. However, this is often not the case. For example, female animals that have very high fecundity may not always be able to lay all of their eggs before dying.

In many parasitic insects (e.g., herbivores and parasitoids), females may be limited by the ability to find hosts. This limit will relax selection for increased fecundity and can shift the optimal offspring size to a larger value than predicted by the Smith-Fretwell model.

Selection for high fecundity may be relaxed for other reasons. For example, in animals that exhibit parental care, large clutches may be less easily tended or defended than small clutches. Similarly, offspring within larger clutches may experience increased competition or conflict that decreases offspring fitness (Parker and Begon 1986). Thus, both parental care and sibling competition can select against large clutches, the result being a change in optimal offspring size without a change in the relationship between offspring size and offspring fitness. The influence of parental care and sibling competition for the evolution of clutch size is discussed in the next two sections of this chapter.

The Evolution of Offspring Number

Although the literature pertaining to the evolution of offspring number is vast, we identify three key developments. A starting point is Lack's (1947b) hypothesis concerning the number of eggs per nest in altricial birds. Lack noted that the number of offspring fledged from a nest is the product of the size of the clutch and the per capita survival of nestlings in a clutch of that size. If nestling survival declines with increasing clutch size (perhaps because of greater competition for food), there will be an intermediate clutch size (called the *Lack clutch size*) that produces the most fledglings, that is, maximizes the product of clutch size and survival (figure 9.2). Stabilizing selection should maintain clutch sizes close to this value until changing ecological conditions alter the relationship between egg number and offspring survival. Much subsequent discussion of clutch size evolution has considered factors that cause deviations from the Lack solution, mostly toward smaller values.

For organisms that distribute offspring among scattered resources, the search for suitable egg-laying sites resembles the search for suitable food items. A second historical development was the application of foraging theory (especially the marginal value theorem) to the analysis of offspring number. Charnov and Skinner (1984) argued that

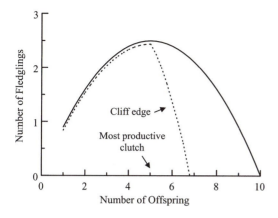

Figure 9.2 Lack's model (solid line) of the most productive clutch in an altricial bird as a product of the number of eggs (N_o) and the survival rate of offspring in different-sized clutches. The model assumes survival is a negative linear function of clutch size $(1 - cN_o)$, with slope c. Thus, the number of fledglings $= N_o(1 - cN_o)$. The dotted line represents a "cliff edge," in which there is a greater penalty associated with producing too many eggs than with producing too few. (After Stearns 1992.)

and will be expected to seek a disproportionate share of resources (with the severity of sibling competition depending on relatedness coefficients and inclusive-fitness costs). A focal offspring's fitness is usually higher if clutch size is small, but parental genetic interests may be best served by larger clutches and an equal division of resources among offspring. In other words, selection acting on genes expressed in offspring may be different from selection on genes expressed in parents. In the face of sibling competition, the parental optimum number of offspring may be lower than it would be without competition. One reason for this is that reduced sibling competition in small broods may raise offspring fitness enough to offset the initial production of fewer young. Thus far, models of parent-offspring conflict appear to exceed empirical tests, in part because it is difficult to identify unambiguous cases. For example, observations of siblicide and subsequent brood reduction cannot be taken as prima facie evidence of parent-offspring conflict or its resolution. Siblicide can increase parental fitness if brood size is adjusted toward the optimum set by the current environment.

a female visiting multiple resource patches may achieve maximum fitness by leaving fewer offspring than each patch can support. This result depends on a decline in the per capita fitness of offspring with increasing clutch size, so that the *rate* of fitness gain (i.e., the fitness increment per time spent laying or per egg laid) also decreases with each additional egg. A female should leave a host when the marginal rate of fitness gain drops to a point where she could gain fitness more rapidly on another host. The optimum number of eggs per host will therefore be a function of the costs of finding and handling additional hosts. Because a female's physiological state changes with successive visits to potential hosts, dynamic optimization models provide a useful refinement of the deterministic approach of Charnov and Skinner (Mangel 1987). These models predict that a female's clutch size will depend on several state variables, including her load of eggs and her rate of encountering new hosts.

A third development was the recognition of a potential conflict between parents and offspring with regard to the optimal number of offspring per breeding event (Mock and Parker 1998). Each offspring is more related to itself than to its siblings

Offspring Number of Iteroparous Organisms with Parental Care

Trade-offs within and between Reproductive Bouts Initial tests of Lack's hypothesis analyzed nestling survival in altricial birds. In many cases, observed clutch sizes were smaller than the Lack value (Roff 1992). Experimentally enlarged broods yielded more fledglings rather than fewer, and it appeared that parents were producing fewer young than they could rear successfully. The simplest explanation for this pattern is that the number of fledglings is merely one component of fitness. In particular, using the number of fledglings to estimate the optimal clutch size ignores both within- and between-generation trade-offs that are associated with high reproductive effort.

Intragenerational trade-offs occur when there is an inverse relationship between a parent's current reproductive effort and its future reproductive performance. This trade-off is thought to depend on the competitive allocation of resources to reproduction versus somatic maintenance. If above-average brood sizes are produced at the expense of somatic maintenance, parents with larger broods may be less able to withstand food shortages, demands of migration, extreme climates, or natural enemies during the interval between breeding events. There is continuing debate as to the most appropriate way to estimate this cost of reproduction (chapter 10, this volume), but several experiments have detected reductions in the survival, mass, and future breeding success of birds that were induced to rear more offspring than they originally produced (Murphy 2000).

Intergenerational trade-offs are evident when the postfledging success of offspring from enlarged broods is lower than that of offspring from small or unmanipulated broods. Experimental increases in offspring number have been shown to reduce offspring survival to breeding age and to delay the age of first reproduction (Humphries and Boutin 2000). An interesting intergenerational trade-off is a negative maternal effect for clutch size itself. Female pied flycatchers (*Ficedula hypoleuca*) with enlarged broods produced daughters with smaller than average broods (Schluter and Gustafsson 1993). It is becoming increasingly apparent that the conditions an individual experiences early in life have lasting effects on fitness among endotherms. In some cases, the negative effects of enlarged broods are evident within the first few days of juvenile development.

Some observations of clutch sizes below the Lack value may reflect inadequate experimental protocols rather than built-in trade-offs. Parents in brood manipulation studies do not always incur the full demands of producing a higher number of offspring. Among altricial birds, parents forced to rear extra nestlings are typically spared some of the costs of producing or incubating additional eggs. Production and incubation costs have been considered trivial relative to the costs of rearing, but accumulating evidence suggests otherwise (Monaghan and Nager 1997). The number of offspring that can be reared successfully will therefore be overestimated, and one may falsely conclude that observed clutch sizes are smaller than the Lack solution. Similar biases pertain to postnatal manipulations used to estimate optimal litter sizes of mammals (Sikes and Ylönen 1998).

Despite these problems, manipulations of offspring number are preferable to purely observational studies, which are even more likely to produce spurious results. For example, differences in resource acquisition before the breeding season will allow some birds to lay relatively large clutches, produce high-quality offspring, and yet incur no survival cost. Differences in resource acquisition can thus mask trade-offs that would be apparent if all parents had initially similar levels of resource acquisition.

Effects of Variable Environments Physiological trade-offs within or between generations are far from the only explanations for clutch sizes below the Lack prediction. The most productive clutch may also be less than optimal if the quality of the environment fluctuates among years, and parents cannot predict "good" and "bad" years before they reproduce. Under these conditions, the optimal clutch size may be one that minimizes the variance in fitness among breeding events rather than one that yields the highest fitness during any single event. Because fitness values are multiplicative across generations rather than additive, the *geometric mean* number of surviving offspring is a better measure of overall fitness than the arithmetic mean. The geometric mean is influenced by the variance in values and is especially sensitive to low values; when two sets of values have the same arithmetic mean, the set with the lowest variance will have a higher geometric mean. The *bad-years* or *bet-hedging*[3] *hypothesis* emphasizes that laying a smaller clutch is less rewarding in "good" years but is also less risky in "bad" years.

Boyce and Perrins (1987) analyzed long-term clutch-size variation in great tits, *Parus major,* and found that modal clutch sizes were below the Lack solution but close to a value that maximized geometric mean fitness. A problem with this analysis (noted by the authors) is that the geometric mean is most appropriate for populations with nonoverlapping generations and may not be the best measure of average fitness in age-structured populations. Parental survival will generally dampen the adverse effects of environmental fluctuations; increasing parental survival relative to offspring survival causes the optimal clutch size to increase and eventually converge toward the Lack solution, which maximizes arithmetic mean fitness (see also Pettifor et al. 2001).

Individual Optimization and Phenotypic Plasticity The foregoing discussion considered the optimal clutch size for an entire population of iteroparous organisms. This population-level optimum obviously need not apply to each individual within a population. Several studies have examined the degree to which clutch or litter sizes are adjusted to a parent's particular history or physiological condition. One brood-manipulation study suggested that each female great tit produces a clutch that is about one egg short of her individual Lack optimum (Pettifor et al. 1988). Thus, a female that laid 9 eggs was capable of rearing 10, whereas one that laid 5 eggs could produce only 6 fledglings. The *individual optimization hypothesis* provides yet another reason for average or modal clutch sizes below the Lack prediction, but evidence in support of this idea is equivocal for birds and mammals (Murphy 2000; Pettifor et al. 2001).

Individual optimization should be viewed as a special case of state-dependent life-history traits. The dependence of offspring number on an individual's condition can presumably be assessed by internal or external state variables, such as mass, fat reserves, parasite load, dominance rank, or territory quality. Effects of female condition on egg-laying decisions are rather easily demonstrated among insects, whose egg loads and informational states change almost continuously. Among vertebrates, however, it is not always clear which variables best describe an individual's state, and many variables are highly correlated.

Individual adjustment of offspring number requires some level of plasticity. We might expect natural selection to produce optimal reaction norms for clutch size as it does for other life history traits (Pigliucci, this volume). Yet we cannot assume that observed levels of plasticity in offspring number are adaptive. For example, laying a small clutch when food is scarce could be an adaptive response to future food availability to juveniles, or it may simply signal that the adult itself is in poor condition. An intriguing example of adaptive plasticity is the adjustment of offspring number (or size) according to mate quality. Recent evidence suggests that some vertebrate females produce fewer eggs (or smaller eggs) when paired with low-quality males (Reyer et al. 1999; Cunningham and Russell 2000). By reducing current reproductive investment, these females can increase their residual reproductive value for future matings with higher-quality males.

A high degree of plasticity can itself favor average clutch sizes below the Lack value, especially if it arises merely because females do not have precise control over offspring number. The *cliff edge hypothesis* postulates that costs associated with deviating from the Lack clutch size are asymmetrical; the penalty associated with exceeding this value may be much greater than the cost of falling below it (figure 9.2). Under this scenario, a genotype that produces a range of clutch sizes that average below the Lack value may outcompete an equally plastic genotype whose average clutch size equals or exceeds the Lack prediction. Such a cliff edge effect has been detected among litter sizes of mice (Morris 1998).

Brood Parasitism and Natural Enemies In many altricial birds, the number of offspring per brood is not under the strict control of the individual or pair that produces the brood. Conspecific brood parasitism occurs when a female lays at least one of her eggs in the nest of another female. This form of parasitism is especially likely to affect clutch size evolution when females cannot recognize parasitic eggs and are "determinate layers" (i.e., they cannot alter their production of eggs according to the number of eggs already in a nest). The *parasitism insurance hypothesis* posits that a female should lay fewer eggs than the Lack number because the risk of receiving a parasitic conspecific egg will occasionally lead to an overcrowded nest and high mortality of all nestlings. Empirical support for this hypothesis is still scant, in part because it is difficult to rule out other explanations.

An interesting twist on the parasitism insurance hypothesis is the suggestion that, in the case where parasites also maintain their own nests, conspecific brood parasitism should reduce the parasite's clutch size as well as the host's. Lack argued that a female should stop laying as soon as the expected increment in fitness from her next egg falls below zero. Similar reasoning suggests that a female with an opportunity to become parasitic should do so as soon as the fitness gain from adding an egg to her own nest drops below the expected gain from putting that egg in another female's nest. This switch to parasitism is expected to occur when the parasite's own clutch is still smaller than the Lack value. A study of American coots, *Fulica americana*, found that females tended to switch to parasitism at the predicted threshold, and that clutch sizes of parasitic females were lower than those of females that lacked the opportunity to become parasitic (Lyon 1998).

Predation has long been thought to influence the evolution of offspring number, but its importance among birds probably depends on nesting habits. Predation risk can favor smaller clutch sizes if larger broods are more apparent to predators or are exposed for a longer period. Predation risk can also interact with food limitation to determine optimal clutch sizes. If predators use parental activity to find nests, they may constrain the rate at which parents can deliver food (Martin et al. 2000). When there is a reasonably high probability of losing an entire brood to predation, selection will also favor females that "reserve" enough resources for a second nesting attempt.

Constraints on Offspring Number   We have discussed the evolution of optimal offspring number as though it were a simple outcome of adaptation to local environments. Yet several factors may constrain individuals from producing the locally optimum number of offspring in nature. For example, when animals and plants routinely migrate between habitats that differ in quality, gene flow may be sufficient to prevent a precise adjustment of either offspring number or size to local conditions.

Another potential constraint is a lack of genetic variation for offspring number. Studies of vertebrate populations have often estimated nonzero heritabilities for clutch or litter size, but at least some of these estimates are derived from experiments that do not eliminate nongenetic, parental (especially maternal) effects. Even cross-fostering experiments (in which eggs or neonates are switched among sets of parents) will overestimate heritabilities if differences in egg provisioning or other prenatal effects inflate resemblances between relatives. Moreover, nonzero heritabilities under controlled experimental conditions do not preclude the possibility that most clutch size variation in nature has environmental rather than genetic causes. Because the genetic architectures underlying life history traits are still poorly understood, we do not know how strongly the evolution of offspring number is constrained by antagonistic pleiotropy (i.e., by negative genetic correlations with other fitness components; Rose and Bradley 1998). Antagonistic pleiotropy may in fact provide the underlying mechanism for some of the inter- and intragenerational trade-offs discussed above.

A lack of available genetic variation can also account for persistent phylogenetic effects on offspring number. In some taxa, it appears that offspring number is not as evolutionarily labile as other life history traits. Böhning-Gaese and Oberrath (1999) analyzed a series of morphological, behavioral, and life history traits in 151 species of birds. Phylogenetic effects were stronger for clutch size than for 17 of the other 21 traits examined. The importance of phylogeny in determining avian clutch size may derive from an underlying positive correlation between clutch size and body size, which also shows a relatively strong phylogenetic signal among birds.

The role of phylogeny is especially apparent when entire taxa are invariant for offspring number. Among lizards, geckos nearly always produce a clutch size of two, and anoline lizards lay only one egg per clutch. Shine and Greer (1991) concluded that invariant or nearly invariant clutch sizes have evolved independently over 20 times among lizards. Their analysis suggests that invariant clutch sizes may be a by-product of small body size, as smaller lizards show less variation in both body size and clutch size. Roff (1992) reviews other factors that influence offspring number among squamate reptiles, including climbing requirements, foraging habits, and mode of reproduction. Constraints on offspring number among anoles, which produce eggs continuously and scatter them widely, are likely to be different from those affecting hole-nesting birds, which provide substantial parental care in one location (Stearns 1992). In some respects, anoles more closely resemble the short-lived insects discussed below.

Offspring Number among Foraging Insects

Larval Competition and the Single-Host Maximum   Many female insects scatter their eggs among multiple resource patches. Some herbivorous species attack small seeds, and parasitoid wasps typically lay eggs in or on other insects. A female must locate a suitable host and choose the appropriate number of eggs to deposit on it. If a host provides enough resources for only a few offspring, we expect a strong link between offspring number and fitness. An analogue to the Lack clutch size for these insects is the *single-host maximum,* which is the number of eggs that maximizes the fitness gained per host (Mangel 1987). Whereas the Lack clutch was originally defined as simply the number of eggs that produced the most fledglings, calculation of the single-host maximum typically integrates both the number and quality of surviving offspring from clutches of different sizes. Offspring quality can be indirectly estimated via body size, which is often positively correlated with longevity, fecundity, and mating success in insects. A related concept is the *larval competition curve,* which describes the combined fitness of surviving larvae as a function of the initial clutch size on a host (Smith and Lessells, 1985). If the per capita fitness of offspring declines linearly with each additional egg, then the total fitness gained per clutch rises in a decelerating fashion, and the larval competition curve resembles Lack's curve for altricial birds (figure 9.3).

Rate-Maximizing Behavior   The single-host maximum represents the number of eggs a female should lay if her fitness depended solely on the efficiency with which she exploits the current host. But each female must visit dozens or even hundreds of hosts in her lifetime, and she may be time-limited; that is, she faces a risk of dying (e.g., from senescence or natural enemies) before exhausting her supply of eggs. A female constrained by the amount of time available for oviposition achieves the highest fitness by maximizing her *rate* of fitness

gain (Charnov and Skinner 1984). According to the marginal-value theorem, females should leave a resource patch when this rate is higher elsewhere; time spent adding more eggs to a clutch reduces time available for searching and handling better hosts.

The threshold for departing a host in turn depends on the costs of searching. If hosts are plentiful (and an offspring's fitness is highest when it does not have to share a host), a female should deposit only one egg per host. When hosts are scarce, the amount of time that must be spent searching causes the optimal clutch size to rise and eventually shifts it toward the single-host maximum. Several empirical studies have found that herbivore and parasitoid clutch sizes increase with increasing search costs.

Estimates of the single-host maximum or the optimal clutch size can be derived even when offspring fitness does not decrease monotonically with offspring number. In some insects, per capita fitness initially rises with clutch size and declines only at the highest egg densities. This dome-shaped relationship between offspring number and fitness (sometimes called an *Allee effect*) arises in a variety of ways. For example, the presence of multiple feeding individuals may improve the quality of the resource for all, or a higher density of eggs or larvae may reduce the per capita risk to natural enemies. An Allee effect sets a lower limit to the optimal clutch size, even when search costs are negligible (Smith and Lessells 1985).

Clutch Size and Host Quality   Foraging insects encounter hosts that differ widely in their intrinsic quality for offspring. For herbivorous insects, plants present extensive intra- and interspecific variation in nutritional suitability and defensive chemistry, and host size can be an especially important determinant of quality for parasitoid wasps. Because host quality affects the relationship between the per capita fitness of offspring and clutch size, we might expect the optimal clutch size to increase with increasing host quality, at least when search costs are held constant (but see Roff 1992; Morris 1998). Variation in host quality can influence clutch size either by modifying average clutch sizes over evolutionary time scales or by promoting adaptive plasticity for the trait. In one herbivorous beetle, for example, mean clutch size increased linearly with the sizes of flower buds on five host plant species (Ekbom 1998).

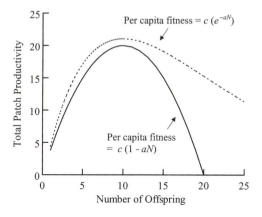

Figure 9.3 Total productivity of a resource patch used for oviposition by one or more female insects. Total productivity is the product of the number of eggs laid *(N)* and the per capita fitness of offspring. The solid line depicts expected total productivity when competition is severe and per capita fitness decreases linearly with N. The dotted line depicts total productivity when there is a negative exponential relationship between per capita fitness and N. The constant *a* determines the rate of the decline in per capita fitness, and *c* is a proportionality constant. (After Ives 1989 and Roff 1992.)

Bet hedging or "spreading the risk" has been invoked to explain insect responses to host quality in the same way that it has been applied to avian clutches in temporally varying environments. Unpredictable variation in host quality could be generated by induced host defenses or by density-independent variation in the risk of mortality from natural enemies. Because failure of entire clutches can be common among insects, bet hedging may help explain the frequent observation of clutches below the single-host maximum, especially among leaf-feeding insects that deposit only a few eggs even in lush patches of host plants.

Time versus Egg Limitation Early considerations of optimal clutch sizes in insects postulated that females should maximize the rate of fitness gain and assumed that females were mainly limited by the time available for oviposition. Several authors have questioned whether time spent laying eggs constitutes a significant fraction of the total time available for reproduction. It has been argued that egg limitation (i.e., the temporary or permanent depletion of mature eggs) is instead more likely to account for clutch sizes below the single-host maximum. Here, the most relevant variable is the fitness increment per egg (not per unit time), and the optimal clutch size is determined by the point at which the immediate fitness gain from adding an egg to a clutch is offset by the probability of running out of eggs. Egg depletion entails a cost because a female loses the opportunity to lay eggs on additional hosts where offspring fitness may be higher.

Both theoretical and empirical analyses of egg limitation have focused on parasitoid wasps. Rosenheim (1999) argued that egg limitation should be common among parasitoids as a consequence of both the trade-off between egg size and number and the trade-off between reproduction and somatic maintenance. In his model, the benefits of large eggs and improved somatic maintenance favor the production of fewer eggs, at least until egg limitation becomes severe enough to prevent further decreases in fecundity. Thus, some nontrivial fraction of individuals in any population should be egg-limited.

If we assume that the costs of reproduction incurred before each oviposition are unavoidable (the costs of egg maturation, the time required to locate a host, the increase in predation risk during search, etc.), then the primary cost that should affect female behavior is the cost of oviposition itself. As we have discussed, this cost is incurred because spending a given number of eggs or amount of time on a current host reduces future opportunities. Rosenheim (1999) placed egg and time limitation under a common currency of lost opportunity and then surveyed evidence for time or egg limitation among parasitoids. He concluded that females are more often egg-limited than time-limited, so that exhaustion of eggs is frequent enough to influence clutch size evolution.

Avoidance of Occupied Hosts For many insects, the most important determinant of host quality is whether a potential host already has received eggs or larvae. Oviposition by two or more females on the same resource obviously promotes the same intensity of larval competition as would occur if a single female laid an excessively large clutch. Females from a wide variety of insect species have evolved the ability to detect and respond to the presence of conspecific eggs or larvae on potential hosts (Messina 1998). Most studies have concentrated on how the presence of eggs alters the probability of host acceptance, but at least a few experiments have shown that females adaptively adjust clutch sizes on encountering occupied hosts.

At first glance, it might seem that a female should lay fewer eggs when there is a risk that other, later-arriving females may add eggs to the same resource patch (recall a similar argument for avian responses to potential brood parasitism). However, both Smith and Lessells (1985) and Ives (1989) have demonstrated that when multiple females exploit the same resource, the evolutionarily stable clutch size will depend strongly on the shape of the larval competition curve. A reduction in clutch size is expected when multiple females lay eggs in the same resource and there is a linear decline in the per capita fitness of larvae with an increasing number of larvae. In this case, exceeding the single-host maximum leads to a steep drop in the total productivity from a host (figure 9.3). Yet if increased crowding reduces the per capita fitness of larvae at a decelerating rate, then exceeding the single-host maximum causes only a gradual decline in total productivity (figure 9.3), and the optimal clutch size will be independent of the number of females that oviposit per patch. A literature survey uncovered wide variation among parasitoids in the shapes of larval competition curves (Ives 1989). We would therefore expect to find similar variation in egg-laying behavior.

Coevolution of Egg-Laying Behavior and Larval Competitiveness   We have emphasized the role of the larval competition curve in determining the single-host maximum. The shape of this curve does not solely depend on the number of offspring per unit of resource. Differences in larval competition curves (and hence differences in total patch productivity, figure 9.3) also depend on how larvae compete for resources. This character, which we call *larval competitiveness,* plays an important role in the evolution of female egg-laying decisions, including the number of offspring laid per host. Parasitoid wasps can be used to illustrate why it is necessary to consider the joint evolution of larval competitiveness and oviposition behavior.

Among parasitoids, the evolution of offspring number should depend in part on the probability of siblicide (Godfray 1987). As we noted earlier, each individual in a sibship is expected to seek a disproportionate share of limited resources. A female wasp has no contact with her offspring and cannot suppress such selfish behavior. Larvae of some parasitoids attack each other within a host, and each host yields a single survivor. Siblicide thus favors a clutch size of one and produces a parent-offspring conflict because the parent cannot deposit its own optimum clutch size. Godfray (1987) suggested that alleles for fighting behavior (and ultimate siblicide) would easily invade a population when clutch sizes are fairly small (two or three eggs). In contrast, alleles for nonsiblicidal behavior invade a population of fighters only under stringent conditions, one of which is that the per capita fitness of a larva is higher when it shares a host than when it develops alone (an Allee effect). Godfray (1987) concluded that clutch sizes should vary dichotomously, with so-called *solitary* species laying 1 egg per host and *gregarious* species depositing > 4 eggs per host.

Despite these predictions, subsequent studies have shown that small broods are common in some parasitoid taxa (Mayhew and van Alphen 1999). Moreover, mapping of larval behavior onto established phylogenies suggests that nonfighting, gregarious larvae have evolved many times from fighting, solitary larvae. Factors that would relax the assumptions of Godfray (1987) include high within-brood relatedness, Allee effects arising from the need to overcome host immune systems, and a high risk of injury to the eventual "winner" when larvae fight in a host. Fighting behavior may in fact be quite labile evolutionarily. Larvae of two closely related *Aphaereta* spp. possess sharp mandibles that can be used for siblicide, but only one species displays the behavior (Mayhew and van Alphen 1999). Whereas females that produce aggressive larvae lay one egg per host, those from the nonaggressive species lay multiple eggs. Data such as these actually support Godfray's (1987) original contention that offspring number and larval competitiveness should evolve in concert. Further evidence of a close relationship between the two traits is found in the seed beetle, *Callosobruchus maculatus,* which we discuss below.

Case Studies

To illustrate some of the concepts discussed in this chapter, we now turn to a group of beetles whose life cycle revolves around seeds (family Bruchidae). Seed beetles are excellent subjects for life history studies because they have short generation times, and for some species (those associated with human stores of legume seeds), the laboratory is a reasonable approximation of their natural environment. In the typical life cycle, the female lays her eggs either on a seed, inside a fruit, or on the surface of a fruit. Hatching larvae burrow directly into the seed, where they complete development. Adults emerge from an exit hole in the seed, and some species require neither food nor water during the adult stage.

Offspring Size

In central Arizona, the seed beetle *Stator limbatus* is abundant on the seeds of blue paloverde *(Cercidium floridum)* and catclaw acacia *(Acacia greggii).* Seeds of these two plants differ substantially in their suitability for the development of a beetle larva. Paloverde seeds are well defended against beetle larvae; more than 50% of the larvae usually die as they try to penetrate the seed coat. On acacia seeds, egg-to-adult survivorship of beetles is generally > 95%. If we examine the fitness consequences of variation in egg size, we find strong directional selection for females to lay large eggs on blue paloverde; larvae hatching from small eggs generally cannot penetrate the seed coat, but those from larger eggs can (figure 9.4A; selection intensity, i, ranges from 0.25 to 0.56). However, we find almost no mortality selection for large eggs on acacia (figure 9.4B; $i \approx 0$). Larvae from larger eggs do mature sooner on acacia, but directional selection toward large size remains weak relative to that on paloverde.

Because the intensity of selection differs between hosts, we expect the balance between selection for large eggs and selection for high fecundity to also differ between hosts, with the optimal egg size larger on paloverde than on acacia (Fox and Mousseau 1996). Interestingly, many populations of *S. limbatus* have access to both acacia and pa-

loverde. The result is disruptive selection within the beetle population, with selection sometimes favoring large eggs and other times favoring small eggs. This type of disruptive selection can maintain genetic variation in a population, cause population substructuring (a prerequisite to sympatric speciation), or promote phenotypic plasticity. In *S. limbatus*, females have evolved egg size plasticity; they adjust the size of eggs according to which host they have recently encountered (figure 9.4C). They are able to do this because they delay oviposition for at least 24 h after emergence, during which time they finish maturing eggs. Contact with host seeds during this egg maturation period allows a female to adjust her egg size in an adaptive way (larger eggs on paloverde, smaller eggs on acacia). As we

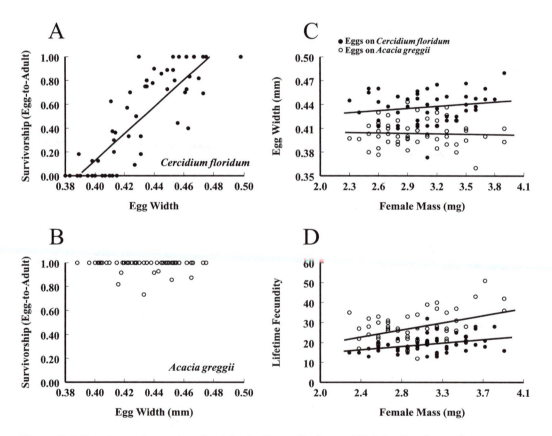

Figure 9.4 Egg size and egg size plasticity in *Stator limbatus*. (A) Selection favors large eggs on seeds of blue paloverde *(C. floridum)*. (B) There is no directional selection (through mortality) on egg size when offspring develop on seeds of catclaw acacia *(A. greggii)*. (C) Egg size is phenotypically plastic; females lay larger eggs on paloverde than on acacia. (D) Laying larger eggs comes at a cost, as females on paloverde have lower fecundity.

would expect, laying large eggs on paloverde reduces fecundity (figure 9.4D) as a consequence of the trade-off between egg size and egg number.

Within populations, paloverde trees vary substantially from one another in how resistant their seeds are to penetration by *S. limbatus* larvae; some seeds are almost completely impenetrable, and others permit almost all larvae to enter the seed. Selection on egg size also varies among individual paloverde trees; the amount of selection toward large eggs increases with increasing seed-coat resistance (Fox 2000). This result is consistent with theoretical models that predict especially strong selection toward large eggs in adverse or low-quality environments (review in Fox and Czesak 2000).

Offspring Number

If host seeds are small and larvae cannot move between seeds, we might expect seed beetle females to distribute their eggs in a way that reduces the severity of competition within seeds. Females of *Callosobruchus maculatus* typically lay only a single egg on each visit to a host seed, but the decision whether to accept a host that already bears eggs is analogous to choosing the optimal number of offspring per unit of resource. Females of *C. maculatus* avoid adding eggs to seeds that already bear eggs, and once all seeds are egg-laden, they preferentially oviposit on seeds with below-average egg loads (Messina 1991). The tendency to avoid occupied hosts varies considerably, however, both within and among populations.

Variation in egg-laying behavior among populations appears to reflect striking differences in the interactions of cooccurring larvae (Messina 1998,

and references therein). Females from an Asian population are unusually adept at distributing eggs to minimize larval competition, and they cease ovipositing once all seeds bear two or three eggs. Larvae from this population engage in a contest type of competition, so that small host seeds virtually never yield two emerging adults (table 9.1). In contrast, females of an African population are "sloppier" in their tendency to avoid occupied hosts, and they show no decline in realized fecundity even when seeds are scarce. African-strain larvae appear to "tolerate" each other, so that even small seeds that receive two larvae frequently yield two adults. Hybridization of the two populations indicated that this difference in larval competitive ability is inherited additively and cannot be explained simply by differences in body size.

A recent selection experiment (unpublished data) tested the hypothesis that intraspecific variation in larval competitiveness and adult egg-laying behavior ultimately depends on the typical sizes of host seeds. Game theory models have suggested that aggressive larvae (and interference competition) are superior in small host seeds, but passive behavior (and exploitation competition) is evolutionarily stable in a large host (Smith and Lessells 1985). The reason is that the cost of exploitation competition becomes relatively low as seed size increases. Replicate lines of the Asian strain either were maintained on their small, ancestral host (mung bean) or were transferred to a much larger, novel host (cowpea). After > 40 generations on cowpea, the competitiveness of Asian-strain larvae had lessened considerably; two adults frequently emerged from even small mung beans (table 9.1). Moreover, females from the cowpea-selected lines showed

Table 9.1 Differences in the competitiveness of larvae of *Callosobruchus maculatus*, as determined by the frequencies of mung bean seeds that received two equal-aged larvae and yielded zero, one, or two emerging adults.

	Percentage of seeds yielding			
	Zero adults	One adult	Two adults	No. of seeds
Beetle source				
African strain	0.0	45.7	54.3	105
Asian strain	3.9	96.1	0.0	102
Selection experiment				
Asian: kept on mung bean	12.5	85.4	2.1	120
Asian: switched to cowpea	6.7	28.3	65.0	48

"sloppier" oviposition behavior; that is, they were more likely to accept egg-laden cowpea seeds for further oviposition (unpublished data). One explanation for this result is that avoidance of occupied hosts entails a cost in terms of the time or energy available for oviposition, and that selection toward uniform egg laying is relaxed in the absence of strong competition between larvae. In any case, this selection experiment provided direct evidence that female egg-laying decisions and larval competitiveness tend to coevolve within insect populations.

Conclusions and Future Directions

We have emphasized two related concepts in our treatment of the evolution of offspring size and number. First, variation in these traits cannot be understood without a thorough consideration of the relevant trade-offs. Trade-offs generally arise because a finite number of resources must be partitioned among individual offspring, between reproductive and nonreproductive phases, or among separate bouts of reproduction. Unfortunately, determining the magnitude of potential trade-offs is not always straightforward. Some of the complexity is reflected in the long-standing debate over how to estimate the effect of current reproductive effort on subsequent parental fitness (Rose and Bradley 1998). In *Drosophila,* for example, laboratory selection experiments and nongenetic, environmental manipulations do not always produce the same conclusions with respect to the cost of reproduction. Among short-lived organisms, the most useful approach may be to combine environmental manipulations of offspring size or number with quantitative-genetic analyses in the same populations. Such a pluralistic approach can identify genetic trade-offs (via antagonistic pleiotropy) in the usual way and may also reveal the physiological (hormonal) mechanisms that underlie negative genetic correlations between fitness components.

Among organisms that are less amenable to genetic analyses, trade-offs will continue to be identified either by direct manipulation of phenotypes or by manipulation of their environments. Such manipulations are particularly likely to reveal trade-offs if the organism is a "capital" breeder (i.e., one that relies solely on stored energy for reproduction). Phenotypic manipulations will also have the best chance of uncovering trade-offs if multiple

traits (such as egg size, clutch size, and total reproductive investment) are modified simultaneously. Sinervo (1999) has shown how path analysis can be used to estimate pairwise correlations between related reproductive traits. Attempts to detect trade-offs in natural populations should be careful not to overlook potentially important costs, such as the production and incubation costs incurred by hole-nesting birds. Calculations of optimal clutch sizes have tended to ignore costs expressed when organisms prepare for breeding. These costs may be associated with mating behavior or with storage of energy and nutrients.

Coupled to the notion of trade-offs is the concept of natural selection as an optimization process; given certain trade-offs, there is an optimal size and number of offspring for an individual or population in a particular habitat. Not surprisingly, optimality models have played a prominent role in attempts to understand the evolution of progeny size and number. Such models have been criticized, however, because they incorporate unrealistic assumptions and ignore many taxon-specific biological details (although dynamic-optimization models have tried to increase realism by including physiological state variables; Mangel 1987). Nevertheless, these models can be used to explore how particular characteristics of a species or features of its environment should modify reproductive traits. For example, optimization models can identify the set of conditions under which a bird would increase its fitness more by parasitizing a conspecific than by adding an egg to its own nest (Lyon 1998).

Yet optimality models serve little purpose if they are not accompanied by empirical investigations of the role of natural selection, especially when we know that some variation in offspring size and number may be nonadaptive or may represent a correlated response to selection on other traits. With respect to offspring size, more field studies are needed to demonstrate that the relationship between progeny size and fitness indeed varies between local environments (Fox 2000). Why, for example, does selection appear to favor larger insect eggs at lower temperatures (Azevedo et al. 1996)? Similarly, the debate over whether insect egg-laying decisions reflect time or egg limitation has been fueled in part by a scarcity of relevant field data. Finally, we suggest that more empirical studies be aimed at determining how much variation in offspring size and number represents adaptive plasticity. Because the optimal clutch or litter size in any habitat depends on a large number of interacting variables, one might argue that selection should favor genotypes with wide reaction norms. Careful studies are needed to compare reaction norms among populations that may experience differing types and levels of environmental variation.

Notes

1. Nearly naked and usually blind when hatched; dependent on parents for food.

2. The mean of n numbers expressed as the nth root of their product.

3. A strategy that maximizes geometric mean fitness (and thus minimizes the variance in fitness) even if it reduces arithmetic mean fitness.

10

Senescence

MARC TATAR

At all taxonomic levels, there exists tremendous variation in life expectancy. A field mouse *Peromyscus* may live 1.2 years, while the African elephant may persist for 60 years, and even a mouse-sized bat such as *Corynorhinus rafinesquei* lives a healthy 20 years (Promislow 1991). Part of this variance is caused by differences in ecological risks, rodents being perhaps the most susceptible to predation, and to vagaries of climate and resources. Another portion is caused by differences in senescence, the intrinsic degeneration of function that produces progressive decrement in age-specific survival and fecundity. Senescence occurs in natural populations, where it affects life expectancy and reproduction as can be seen, for instance, from the progressive change in age-specific mortality and maternity of lion and baboon in East Africa (figure 10.1; Packer et al. 1998). The occurrence of senescence and of the widespread variation in longevity presents a paradox: How does the age-dependent deterioration of fitness components evolve under natural selection? The conceptual and empirical resolutions to this problem will be explored in this chapter. We shall see that the force of natural selection does not weigh equally on all ages and that there is therefore an increased chance for genes with late-age-deleterious effects to be expressed. Life histories are expected to be optimized to regulate intrinsic deterioration, and in this way, longevity evolves despite the maladaptive nature of senescence. From this framework, we will then consider

how the model is tested, both through studies of laboratory evolution and of natural variation, and through the physiological and molecular dissection of constraints underlying trade-offs between reproduction and longevity.

The Senescence Phenotype and Why It Evolves

Measures of Senescence

As humans are well aware from personal experience, performance and physical condition progressively deteriorate with adult age. And in us, as well as in many other species, mortality rates progressively increase with cohort age. Medawar (1955), followed by Williams (1957), stated the underlying assumption connecting these events: Senescent decline in function causes a progressive increase in mortality rate. Although mortality may increase episodically across some age classes, such as with increases in reproductive effort, we assume that the continuous increase of mortality across the range of adult ages represents our best estimate of senescence. To estimate senescence, or "demographic senescence," in this way requires that we construct life tables.

The cohort life table records number alive (N_t) in each age period, t to $t + 1$. From this vector, the

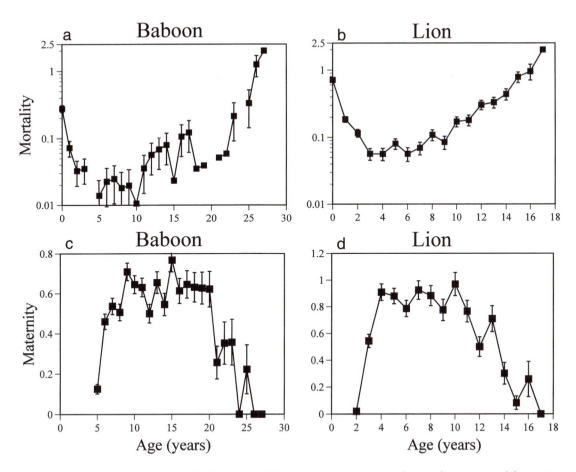

Figure 10.1 Vital statistics of female baboon and lion in Tanzania. Annual mortality estimated from (A) 201 yearling baboons and (B) 652 yearling lions based on censuses since 1972 (baboons of Gombe National Park) and 1967 (lions of Serengeti National Park and Ngoronoro Crater). Gross maternity of (C) baboon and (D) lion based on live births per number of females alive at midpoint of age class. (Redrawn from Packer et al. 1998.)

table represents the number of deaths (d_t), the cumulative proportion surviving (l_t), and age-specific mortality (q_t) or its complement, age-specific survival $(p_t = 1 - q_t)$. See Lee (1992) for useful background. Mortality rate, μ_t, is the conditional failure rate over a very short period t to $t + \delta t$; it is independent of census interval and, unlike q_t, ranges from 0 to infinity. When census intervals are small, μ_t can be estimated well by $-ln(p_t)$. Since mortality rates tend to increase exponentially with adult age, it is convenient to consider $ln(\mu_t)$. Furthermore, this transformation stabilizes the variance for μ_t across ages (Promislow and Tatar 1998). Although the distributions of l_t, d_t, and μ_t are functionally related, only mortality $(q_t$ or $\mu_t)$ estimates the age-

specific rate, our index for senescence. Indeed, patterns of l_t that superficially appear similar can be produced by cohorts where mortality does not increase in age (figure 10.2, case A) and those that do (Figure 10.2, case B). Mortality underlying this first set of hypothetical curves is age-independent and may represent extrinsic or environmental causes of death such as predation, disease, or physical disturbance. The mortality for the second set increases with age due to senescence and represents effects of intrinsic degeneration as well as interactions of functional decline with extrinsic risks originating in the environment. Both age-dependent and age-independent sources will simultaneously occur in natural populations, and then, we might

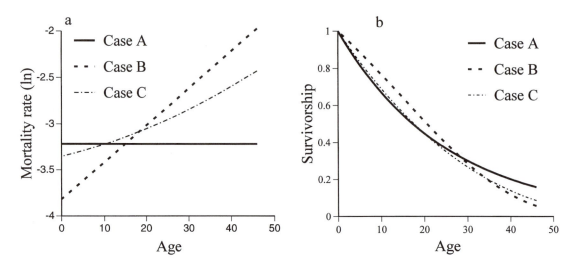

Figure 10.2 Mortality rate and survivorship compared. Each case assumes an underlying parameteric Gompertz-Makeham mortality model, $\mu_t = c + \lambda e^{\gamma t}$. In case A, mortality does not increase with age; $\gamma = 0$, $\lambda = 0.04$, $c = 0.0$. Case B represents demographic senescence, and mortality is progressively age-dependent; $\gamma = 0.04$, $\lambda = 0.22$, $c = 0.0$. In case C, mortality has both progressive age-dependent and age-independent components: $\gamma = 0.01$, $\lambda = 0.04$, $c = 0.025$. Case parameters illustrate that (a) qualitative differences in mortality rate $ln(\mu_t)$ can produce (b) ambiguous and similar trends of survivorship.

see a combined curve as illustrated by case C in figure 10.2. Since age-dependent change in mortality is used to represent underlying physiological deterioration, mortality models are used to parameterize and estimate this rate. These models may also help disentangle the age-dependent component from the age-independent aspects of extrinsic mortality.

Two common models that describe age-dependent mortality (or "hazard") are the Gompertz, $\mu_t = \lambda e^{\gamma t}$, and the Weibull, $\mu_t = \lambda^\beta \beta t^{\beta-1}$ (Lee 1992). In each, the rate of increase in mortality is regulated by the *shape* parameters γ or β, respectively. These index the rate of senescence since they dictate the rate of change in mortality; note that when $\gamma = 0$ or $\beta = 1$, mortality rates are age-independent. The *scale* parameter λ is a constant multiplier of the rate. This biologically represents the baseline susceptibility of individuals to changes in mortality risk—their "frailty"—and it contributes to the magnitude of the age-dependent mortality change. Extrinsic mortality can be added to either model so that $\mu_t = c + \lambda e^{\gamma t}$ (the "Gompertz-Makeham") or $\mu_t = c + \lambda^\beta \beta t^{\beta-1}$. Life tables constructed from very large samples reveal even more complex dynamics. The rate of change in mortality at the oldest ages may decelerate to become age-independent, or even

to become negative, whereby mortality rates decrease. The evolutionary significance of this trend hinges on whether it represents altered rates of individual aging or changes in cohort composition (Partridge and Mangel 1999). Given a life table of any size, we choose models and estimate parameters by maximum likelihood methods applied to the distribution of deaths, the probability density function $\hat{d}(t)$ (Pletcher 1999). Direct estimates of parameters from regression of mortality rate on age lead to bias as a result of sampling error at the youngest and oldest ages where either few individuals die or few remain at risk (Promislow et al. 1999).

Mortality isolates just one of two important demographic components of senescence. Physiological degeneration can also cause a progressive loss of reproductive capacity. Reproductive senescence, even in the absence of age-specific changes in mortality, can yield heavy fitness costs, which may especially represent the burden of aging on natural populations. In practice, however, it will be difficult to isolate degenerative causes of reproductive decline from adaptive and ecological alterations due to seasonality, nutrition, and host, prey, or oviposition availability. Unlike mortality, where any increase is likely to reflect some aspect of in-

mer of how extended longevity evolves. Inferences are based solely on the pattern of correlated selection responses, and a decade of careful work has revealed two challenges for these interpretations. First, selection on ability to reproduce at advanced age is often confounded with inadvertent selection on a correlated trait, such as development rate (for discussion, see Partridge et al. 1999). Furthermore, the environmental conditions of the selection and control regimes often entail subtle differences, for instance, in terms of food quality during larval development. Strains will evolve in response to these conditions but their phenotypes may not be assessed under the same environment; Gene-by-environment interactions will then obscure selection responses (Leroi et al. 1994).

Second, evolutionary responses are measured from the selection lines relative to their controls. The selected populations are taken to represent exceptional longevity, but an alternative view is that the control lines are abnormally short-lived. We are just beginning to appreciate that most laboratory strains used to initiate demographic selection programs are at risk of containing age-specific genetic artifacts that accumulated when the strains were domesticated (Promislow and Tatar 1998; Harshman and Hoffman 2000; Sgrò and Partridge 2000). Laboratory flies are typically propagated from bottle to bottle on 14-day cycles. Since development takes at least 8–10 days, adults can contribute progeny only before they are 5 or 6 days old. During domestication in bottles, selection will favor variants with high early reproduction while it ignores mutations with deleterious effects expressed at older ages. Under this regime, both pleiotropy and mutation accumulation will contribute to a shortening of longevity in the laboratory stain. Subsequently, when these lines are selected for late-age reproduction, we should not be surprised to see a rapid and large increase in longevity and in many other late-age fitness traits. Allele frequencies at pleiotropic loci can be restored toward ancestral equilibrium levels, or perhaps beyond, and the suite of solely late-acting deleterious mutations will be purged. The differences among these types of strains in longevity and in their correlated responses will therefore have multiple evolutionary origins, and the contribution of each cannot be easily disentangled.

Linda Partridge and colleagues (1999) provide a study with *D. melanogaster* that manages many of these concerns. The base strain was domesti-

cated to laboratory conditions in population cages rather than in bottles. Oviposition substrate in cages is continuously available and permits overlapping generations; during domestication, selection should act across most adult age classes. During the course of experimental evolution, lines were established with controlled larval density, and all adults were permitted to pupate and eclose to avoid selection on development rate. Artificial selection was on age of reproduction at either of two age classes: 1 week after final eclosion ("young") and 3–4 weeks after final eclosion ("old"). Five replicate lines were established for each treatment. After 31 "young" and 19 "old" generations, life history traits were measured under the environmental conditions of the selection regimes. There was no evidence for differences among base strain and "young" populations in longevity, fecundity, or development rate: Adaptation to bottles per se did not affect demography. Therefore, changes observed in "old" lines are attributable to demographic selection. In the "old" lines, mean longevity increased to 46.8 days relative to the life expectancy of 35.7 days in base and "young" controls. These mean differences are caused by a proportional decrease in mortality rate at most ages in the "old" lines (figure 10.3). Importantly, early fertility decreased in "old" lines by 30–40% over the first week of age relative to base and "young" but did not differ among these groups at later ages (figure 10.3). The correlated selection response unambiguously demonstrates a trade-off between early reproductive effort and survival to reproduce at late ages. Although such association cannot resolve genetic or proximal mechanisms, these results confirm that senescence evolves with life history constraints.

An elegant follow-up experiment with these lines provides the first definitive evidence for a mechanism that underlies laboratory evolution of *Drosophila* senescence. If differences in mortality among the "young" and "old" strains is directly caused by reproductive trade-offs, mortality patterns should become identical when adults are sterilized. On the other hand, if reproduction does not mechanistically produce line differences in mortality, then the differences are expected to persist among sterilized cohorts because, for instance, trade-offs with development or mutation accumulation are the underlying cause. Sgrò and Partridge (1999) made this test. Each strain of "young" and "old" flies was sterilized by one of two methods: gamma radiation or inheritance of the dominant, female

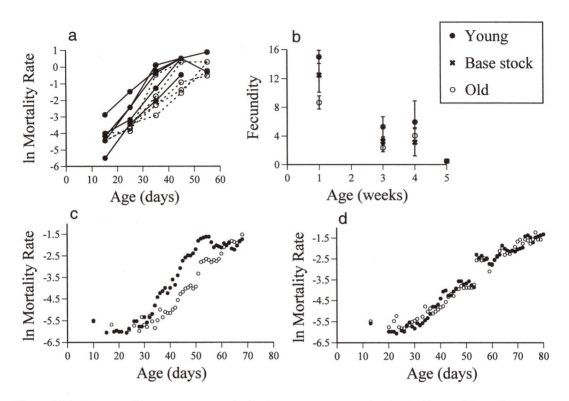

Figure 10.3 Demographic consequences of selection on age at reproduction in *Drosophila melanogaster*. (A) Mortality rate across 10-day intervals for replicate "young" control (solid lines and circles) and "old" selected (open lines and circles) strains after 31 "young" generations. (B) Fecundity as progeny per female per day measured at four times, with 95% confidence intervals; symbols as in (A); data include fecundity of females of the base strain (crosses) used to initiate "young" and "old" strains. (C) Mean of mortality rate for fertile females across replicate hybrid crosses of each "young" strain to base strain (closed circles), and each "old" strain to base strain (open circles). (D) Mean of mortality rate for sterile females across replicate hybrid crosses of each "young" strain to base strain males with ovo^D mutant, and of each "old" strain to base strain with ovo^D. (Panels A and B redrawn from Partridge et al. 1999. Panels C and D redrawn from Sgrò and Partridge 1999.)

sterile allele ovo^D. Mortality measured for the sterile flies revealed a complete lack of differences among the "young" and "old" selected strains, while nonsterilized controls produced the usual disparity (figure 10.3). Among these flies, selection on reproduction and the ensuing change in physiological trade-off with mortality explains the observed senescence evolution.

Comparative Studies of Senescence Evolution

Laboratory selection on life history elucidates evolution based on genetic variance and covariance within a population. Different approaches are required to understand the evolution of senescence among populations and among taxa. There is a long tradition of taxonomic comparison to search for correlates of lifespan, but it was not until 1991 that Daniel Promislow made a comparative test of predictions based on senescence evolutionary theory.

As we have seen, natural selection molds senescence indirectly, by favoring traits that involve trade-offs across age classes and by relaxing the fitness costs of genes expressed at late ages. Williams (1957) posited a set of testable deductions from this theory, among them that the onset of senescence should occur at the age of sexual maturation, and that rate of senescence should covary with the magnitude of extrinsic, age-independent mortality.

Promislow assembled life tables of mammals to test these hypotheses while controlling for taxonomic independence and covariates such as body size. Of particular interest was whether high age-independent (extrinsic) mortality associated with high rates of demographic senescence. The capacity to address this question depends on the quality of natural demographic data. Many life tables of field populations are cross-sectional; individuals of different age classes are observed for survival across a single interval, and often so when only a small number of individuals are at risk within each age class. Even for cohort life tables based on longitudinal follow-up, available sample sizes are usually small. These data constraints are important because point estimates of early adult mortality cannot be less than the inverse of the number at risk (Promislow et al. 1999). Yet these early ages are just where most deaths are extrinsic rather than senescent. Care must be taken to accurately estimate the rate across early adult ages. Promislow measured extrinsic mortality as the rate of change in survival at the age of first reproduction. The accuracy of this estimate is unknown, although the use of smoothing on cumulative survival very likely reduced point estimate bias. The rate of demographic senescence was characterized by the Gompertz mortality acceleration parameter γ estimated by regression from the period-specific survival after age of first reproduction. Contrary to the notion that senescence is a trait expressed only when wild animals are removed to protected environments, Promislow observed significant accelerations of age-specific mortality in 46 of 54 cases; senescence affects demographic fitness components in the field. However, among these species, there was no significant association between the estimate of extrinsic mortality and the rate of senescence.

Robert Ricklefs (1998) revisited this prediction with an expanded collection of life tables, adding a number of mammals and including a new class, birds. Extrinsic mortality, m_o, was estimated directly from a three-parameter Weibull mortality model $\mu_t = m_o + \alpha t^{\beta-1}$, with $\alpha t^{\beta-1}$ taken as the rate of senescence (note $\alpha = \lambda^\beta \beta$). This model was fit by nonlinear regression on age-specific mortality. Within and among the groups, Ricklefs noted a strong and significant association between extrinsic mortality and rate of senescence and concluded that patterns of aging may evolve in response to levels of ecologically imposed mortality. This re-

sult, however, is tentative. Bootstrapped life tables showed that reduced age class size increased the estimate of m_o, as we expect since regression on μ_t does not account for ages where no deaths are observed (Promislow et al. 1999). At the same time, since the analysis of taxonomic life tables used regression on mortality and took β as a constant among species, we should expect α to increase as the size of early age-classes decline. With these assumptions, the scale parameter is proportional to the age where mortality becomes apparently age-dependent—a value near the inverse of the number at risk. Therefore, both $\alpha t^{\beta-1}$ (the rate of senescence index) and m_o will correlate with early age class size and thus with each other. This reasoning illustrates the complexity of testing models of evolutionary senescence at taxonomic levels with demographic data. Ultimately, the taxonomic approach requires life tables constructed with finer time scales, larger age-class and cohort sizes, and refined estimation methods.

An approach related to taxonomic-wide comparison searches for ecological trends among locally adapted populations within a single species. The aim is to determine how patterns of demographic selection in the field correspond to intrinsic rates of senescence, which are measured in controlled, benign laboratory environments. This strategy is illustrated in an ongoing study of Trinidadian guppies by David Reznick (1997; see also Reznick, this volume). Across a series of streams, juvenile and early adult life history of guppies evolve rapidly in response to size-specific predation by the pike cichlid, *Crenicichla alta,* and the killifish *Rivulus harti* (Reznick et al. 1997). In lower reaches of streams, pike cichlid impose high rates of adult mortality, and experimental manipulation has demonstrated that this predation selects for increased early reproduction in guppies. In upper reaches of streams, the cichlid is absent, and a low level of predation is imposed by killifish. There, experimental trials have shown that guppies evolve a slower rate of development and an extended schedule of reproduction. Among fish of the lower reaches, selection for early reproductive effort and the reduced value of late adult life should favor more rapid rates of senescence in guppies. To test this hypothesis, Reznick is estimating the demographic rate of senescence under common laboratory conditions for second-generation offspring of fish derived from pools with and without the cichlid predator.

Relative to fish, it is somewhat simpler to generate laboratory life tables for field-derived insects and their offspring. In California, the lesser migratory grasshopper *Melanoplus* occurs in genetically distinct populations along the altitudinal gradient of the Sierra Nevada. In these sites, *Melanoplus* is univoltine, with a reproductive season beginning in September. Winterlike conditions, however, occur at earlier calendar dates at progressively higher elevations, and populations from high elevations are expected to experience a relatively truncated adult life history. Late-life reproduction is likely to be lost either by death in early winter storms or by the lack of suitable thermal conditions to mature and lay eggs. Under these conditions, high-elevation populations are predicted to evolve rapid rates of senescence as a result of selection favoring rapid nymph development or early reproductive investment, and through relaxed age-specific selection on deleterious traits expressible at later ages. Field demographic patterns have yet to be measured, but there is a clear gradient across populations for rates of senescence as measured in the laboratory (Tatar et al. 1997a). Females caught in the field

at five elevations from 130 to 1800 meters were maintained in the lab to produce eggs. Offspring were then reared under two temperature regimes, and life tables were generated from resulting adult males. An association was observed between population altitude and life expectancy under both sets of conditions; genotypes derived from high elevations lived 103–118 days, while those from lower elevations lived 122–149 days (mean longevity, cool and warm conditions, respectively).

To understand the evolution of senescence along this cline, an initial problem is to evaluate whether high rates of early reproduction are associated with rapid rates of senescence among the populations. Laboratory life tables with females are under construction for five populations ranging from 1111 to 2650 m. Field females of similar age when brought to the lab for egg collection presented intriguing preliminary observations. Remaining median longevity correlated negatively with elevation origin and ranged from 18 days (high elevations) to 59 days (low elevations) (figure 10.4A). Reproductive effort observed from the same females revealed that early age-specific fecundity was

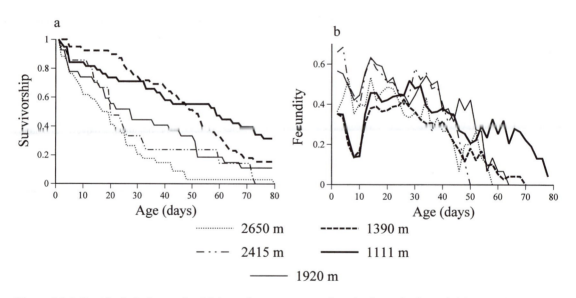

Figure 10.4 Residual vital rates for *Melanoplus sanguinipes* females brought from field sites of an elevation transect in the Sierra Nevada, California, to a common laboratory environment. Populations studied across 1111–2650 m are univoltine and nearly age-synchronized. Adult females were collected October 1998 and maintained individually on 12-h photoperiod at 28°C day, 24°C night with egg-laying medium and diet. Each cohort was initiated with between 21 and 40 individuals. Survivorship is based on daily census interval; fecundity is number of egg pods laid per alive female per 3-day interval. (Unpublished data, M. Tatar.)

greatest for females originating from high elevations (figure 10.4B). Among surviving females at late age, per capita maternity persisted the longest for those females of the low-elevation population (1111 m). Cohorts from each population reared under common environments are needed to confirm the genetic basis of this negative association between reproduction and survival, and to compare the schedule of fertility to that of age-specific mortality. Two interpretations can be suggested if a genetic correlation is confirmed. The pattern may be consistent with life history optimization where selection on reproduction constrains the potential longevity of adults. One the other hand, selection on early age classes at higher elevations may directly favor more intense early investment in reproduction without affects on longevity, but the diminution of selection on late age classes permits the accumulation of late-age-deleterious mutations. Both models will produce a negative correlation between reproduction and longevity across the cline.

A challenge for this study, as with guppies, concerns the novelty of laboratory conditions. Life histories are very sensitive to gene-by-environment interactions, and care must be taken to control for positive correlation among traits when genotypes segregate for predisposition to the laboratory regimes, although this bias makes the observation of any negative correlation even more impressive. Besides emulating natural conditions in the laboratory, it is important to assess senescence of natural genotypes under multiple lab environments, for instance, with different temperatures or levels of nutrients. Consistency in the qualitative rank of life table metrics helps to validate inferences made from lab cohorts to field populations.

Phenotypic Analysis of Trade-Offs

Experimental and comparative data show that longevity evolves under constraints set by trade-offs with other dimensions of fitness, in particular, reproduction. This fundamental concept is advanced if we can understand the mechanisms underlying trade-offs. To this end, manipulative studies of reproduction are useful. In many insects, longevity is increased when reproduction is experimentally curtailed (Bell and Koufopanou 1986). Often, this plasticity is attributed to the altered allocation of limiting nutrients from the metabolic demands of reproduction toward somatic maintenance. A second manipulative approach involves dietary re-

striction. In a number of organisms, a severe reduction in total caloric intake extends longevity but reduces or eliminates current reproduction. Again, the notion that limiting nutrients are competitively allocated can help explain this response—survival through lean times is ensured by allocating limited resources to the function of somatic persistence at the expense of current reproduction (Holliday 1989).

Manipulative studies, however, do not always produce a negative association between reproduction and longevity. This need not be a contradiction. Van Noordwijk and de Jong (1986) noted that when resource acquisition is taken into account, weak or even positive correlation may occur between reproduction and lifespan. An adequate level of acquisition reduces competition for resources among reproductive and somatic functions. Then, if individuals vary in their ability to acquire nutrients, lifespan and reproduction will positively covary. This notion is illustrated by the "Y model" (figure 10.5) (de Jong and van Noordwijk 1992; Tatar and Carey 1995). Somatic and reproductive functions are end points for resources derived from a common pool, the stem of the Y. Competitive allocation occurs as resources are moved from the pool and into one branch or the other. Resource acquisition occurs as input into the pool. Genetic variation may occur for both resource allocation and acquisition, but the expression of negative genetic correlation depends on the level of resource availability and the degree of variation for allocation relative to acquisition.

Recent advances in molecular biogerontology suggest a further set of mechanisms underlying demographic trade-offs observed in manipulative studies. Reproduction can directly compromise somatic function independent of nutrient allocation. Successful reproductive activity, for both males and females, requires elevated physiological activity, which must be supported by oxidative metabolism. Free radicals such as hydrogen peroxide and superoxide are produced by mitochondria during the respiratory production of ATP, and these reactive by-products may be central agents driving cellular and systemic aging (Martin et al. 1996). Reducing the level of reproduction, therefore, may reduce physiological metabolism and the production of oxidative radicals—longevity can increase. Countervailing the effects of metabolically driven damage are housekeeping mechanisms such as DNA-damage repair, molecular chaperones, and antioxidant systems, which, to an unknown extent, can

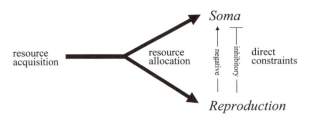

Figure 10.5 The "Y model" for trade-offs between somatic and reproductive function. (Modified from Tatar and Carey 1995 and de Jong and van Nordwijk 1992.)

keep in check the direct deteriorative effects of metabolic necessities (Lithgow and Kirkwood 1996). Recent work with transgenic *D. melanogaster* demonstrates that adult mortality rates are reduced by the overexpression of the molecular chaperone HSP70 (Tatar et al. 1997b) and of the antioxidant enzyme superoxide dismutase (Sun and Tower 1999). But as these agents can prolong survival, why are they not expressed at higher constitutive levels? At least for HSP70, there are costs associated with expression. *hsp70* is an induced gene which produces copious amounts of protein on stress, after which the chaperone is rapidly sequestered. In developing embryos and larvae, HSP70 expression protects against heat stress but interferes with normal development. In adults transgenic with additional copies of *hsp70*, very short heat shock induces a level of protein expression which extends longevity (figure 10.6A) but at the same time reduces viability of developing eggs within females (figure 10.6B).

Thus, housekeeping mechanisms can incur direct costs on reproduction, independent of nutrient allocation, and active levels of reproduction may involve suppression of systems that maintain cellar function. In the Y model, direct costs can occur between reproduction and somatic function in terms of both "negative" and "inhibitory" effects (figure 10.5). Reproduction has a direct negative cost, represented by the arrow, when its physiology or metabolism accelerates damage-inducing processes, for instance, when free radicals result from metabolism used to support oogenesis. Reproduction has inhibitory costs, represented by the terminal line, when mechanisms of somatic maintenance interfere with reproductive function. The cost arises because these functions are downregulated to permit optimal reproduction. For instance, heat shock proteins have the capacity to extend so-

matic survival but interfere with fertility; heat shock proteins are highly regulated both in the soma and in embryos during oogenesis. The occurrence of direct reproductive trade-offs requires us to reconsider interpretations of the classic manipulative studies. Both direct and allocation models of trade-offs predict the negative association between reproduction and extended longevity.

It is necessary, therefore, to devise new strategies in experimental design to determine the relative importance of these mechanisms as constraints on the evolution of lifespan. One approach is to manipulate both nutrient acquisition and allocation simultaneously. Under high acquisition, nutrients should be relatively unlimited, and the trade-off between reproduction and soma should be minimal. On the other hand, under low nutrients, competitive allocation should be more apparent because limited resources must be partitioned. These conditions were executed with the bean beetle *Callosobruchus maculatus* (Tatar and Carey 1995). *C. maculatus* is facultative aphagous: It completes its larval stage in beans, ecloses, and then can survive as a fully reproductive adult with or without food and water. It is easy to manipulate egg laying in these beetles by presenting or withholding oviposition substrate. Thus, it was possible to regulate both reproductive effort and nutrient acquisition in young adults to manipulate females to invest in eggs when resources were limiting or when they were abundant. Subsequent effects were measured as change in age-specific mortality at later ages, when both food and oviposition was equalized among groups (figure 10.7). In these trials, subsequent age-specific mortality was elevated among females that reproduced when nutrients were limiting compared to females that reproduced when nutrients were available. This interaction between resources and reproduction demonstrates

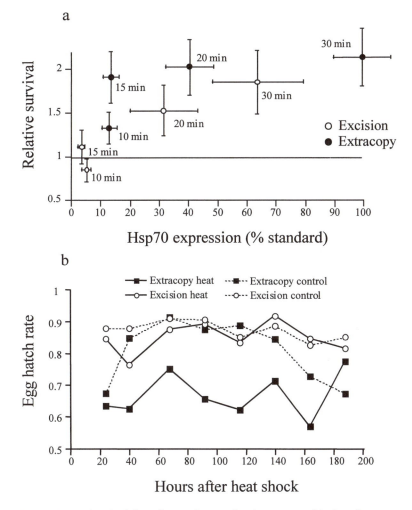

Figure 10.6 Survival benefits and reproductive costs of induced expression of heat shock protein-70 in adult female *Drosophila melanogaster*. Transgenic extracopy strain carries 12 additional *hsp70* genes; the excision strain carries the wild type copy number of *hsp70* genes and controls for *P*-element insert mutation present in the extracopy strain. (a) Induction of HSP70 protein by 37° heat treatment produces increased posttreatment survival. Short treatments are sufficient to induce gene expression (data not shown) and improved survival in extracopy flies, but not in excision flies. Longer heat treatments induce HSP70 protein and extended longevity in both genotypes. (Redrawn from Tatar et al. 1997b.) (B) Heat-treated and control females lay the same number of eggs (data not shown), but when mothers carry extra copies of *hsp70*, a 30-min heat treatment suppresses the hatch rate (Silbermann and Tatar 2000).

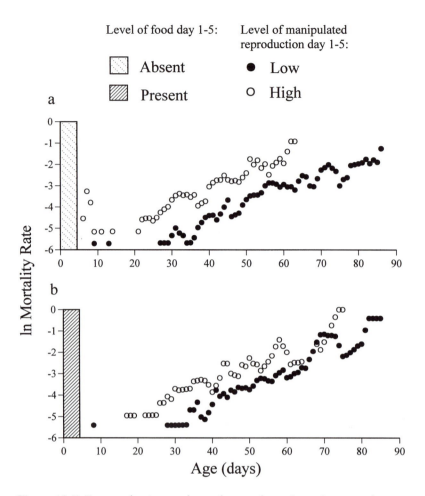

Figure 10.7 Egg production and supplemental nutrients (yeast and sugar water) at early ages interact to affect later age-specific mortality among facultatively aphagous, female seed beetles, *Callosobruchus maculatus*. (A) Without supplemental nutrients at ages 0–5 days, females manipulated to have high reproduction subsequently have high late-mortality rates when nutrients are provided and reproduction is curtailed (after age 5 days). (B) Young females provided with supplemental nutrients during the period of differential reproduction (age 0–5 days) subsequently present fewer differences in age-specific mortality when nutrients are provided and reproduction is curtailed. (Redrawn from Tatar and Carey 1995.)

that these life history trade-offs are mediated by competitive allocation of nutrients.

Direct costs, however, may still play a role, since even among fed females, the mortality rate was elevated for those under high relative to low reproduction. Females may incur some direct damage proportional to reproductive effort. Alternatively, adult feeding may not supplement all limiting nutrients so that competitive allocation still

occurs in the high nutrient treatments; some metabolites may be solely of larval origin, or the rate of acquisition in adults is physiologically limited.

Future Directions

We have a rich theory to aid us in understanding the evolution of senescence. Our empirical progress confirms the theory in its broadest sense. Senescence is the outcome of evolution, but the trait is not an adaptation. Multiple systems can and will deteriorate and lead to decreased function and ultimately death. The rate at which deterioration proceeds, however, is subject to selection, but only within constraints and compromises determined by the ecology and physiology of the organism. Taxonomic variation in lifespan is an end result, and individuals in natural settings conspicuously exhibit age-dependent loss of fitness—the consequences of senescence.

This summation resolves only the most general questions about aging and its evolution. Many problems are largely open and unsettled. For instance, mortality rate acceleration slows at the oldest ages in many organisms. This observation raises many questions: What is the age-specific activity of deleterious genes? Does the rate of senescence of an individual change with age or remain constant? How much variation exists among individuals for rates of senescence? We can note further observations and ask associated questions. For instance, why do individuals die at different ages even when cohorts are comprised of isogenic clones raised under extremely homogeneous environmental conditions? Why do modern humans exhibit marked postreproductive life expectancy? What are the relative contributions of antagonistic pleiotropy and mutation accumulation to the evolution of senescence? At what level of organization are trade-offs organized: at whole physiological systems or at specific loci? What role does antagonistic pleiotropy for life history traits play in the maintenance of polymorphism? Is senescence a single process driving disease and systemic degeneration, or is it a collective of independently degenerating systems? What are the mechanisms underlying costs of reproduction, and are the mechanisms that are plastic within species the same as those responsible for taxonomic differences?

To address these fundamental questions, future work will require refined strategies. Selection studies can be informative but must proceed from base populations with controlled genetic architecture. Natural populations can be profitably studied by combining field and laboratory demographic assays, but the uniqueness of the lab environment must be taken into account. Comparative approaches can reveal the broadest evolutionary history, but the demographic phenotypes must be based on better life tables amenable to rigorous analysis. These prescripts are short-term objectives. Future approaches are likely to involve explicit molecular and genomic analysis. Despite nearly a century of research, senescence and its evolution remain frontier problems of biological science.

11

Life Cycles

JAN A. PECHENIK

I have a Hardin cartoon on my office door. It shows a series of animals thinking about the meaning of life. In sequence, we see a lobe-finned fish, a salamander, a lizard, and a monkey, all thinking, "Eat, survive, reproduce; eat, survive, reproduce." Then comes man: "What's it all about?" he wonders.

Organisms live to reproduce. The ultimate selective pressure on any organism is to survive long enough and well enough to pass genetic material to a next generation that will also be successful in reproducing. In this sense, then, every morphological, physiological, biochemical, or behavioral adaptation contributes to reproductive success, making the field of life cycle evolution a very broad one indeed. Key components include mode of sexuality, age and size at first reproduction (Roff, this volume), number of reproductive episodes in a lifetime, offspring size (Messina and Fox, this volume), fecundity, the extent to which parents protect their offspring and how that protection is achieved, source of nutrition during development, survival to maturity, the consequences of shifts in any of these components, and the underlying mechanisms responsible for such shifts.

Many of these issues are dealt with in other chapters. Here I focus exclusively on animals, and on a particularly widespread sort of life cycle that includes at least two ecologically distinct free-living stages (figure 11.1). Such "complex life cycles" (Istock 1967) are especially common among amphibians and fishes (Hall and Wake 1999), and within most invertebrate groups, including insects (Gilbert and Frieden 1981), crustaceans, bivalves, gastropods, polychaete worms, echinoderms, bryozoans, and corals and other cnidarians (Thorson 1950). In such life cycles, the juvenile or adult stage is reached by metamorphosing from a preceding, free-living larval stage. In many species, metamorphosis[1] involves a veritable revolution in morphology, ecology, behavior, and physiology, sometimes taking place in as little as a few minutes or a few hours. In addition to the issues already mentioned, key components of such complex life cycles include the timing of metamorphosis (i.e., when it occurs), the size at which larvae metamorphose, and the consequences of metamorphosing at particular times or at particular sizes.

The potential advantages of including larval stages in the life history have been much discussed. They include minimizing competition with parents for both food and space; permitting development to take place in environments that are potentially safer and in which food is abundant; and, for species whose adults are sedentary or sessile, the opportunity for extensive dispersal (reviewed by Pechenik 1999). Key references on complex life cycles and their evolution include papers or books by Istock (1967), Thorson (1950), Jägersten (1972), Wilbur and Collins (1973), Strathmann (1985), Raff (1996), and Pechenik (1999) and compilations edited by McEdward (1995), Hayes (1997), and Hall and Wake (1999).

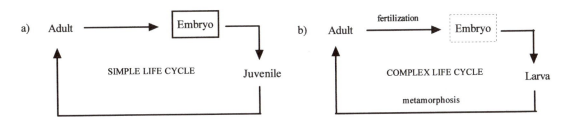

Figure 11.1 (A) Diagrammatic representation of a simple life cycle, with no distinctive free-living larval stage; the juvenile closely resembles the adult. The solid rectangle represents a protected stage; other stages are free-living. (B) Diagrammatic representation of a complex life cycle, which includes free-living larvae; larvae reach adulthood through a conspicuous metamorphosis. The dotted rectangle represents a stage that is protected (by brooding or encapsulation) in some species and free-living in others. Other stages are always free-living.

It is not possible to cover all aspects of life history evolution for all animals in a single chapter. Instead, I will first consider one major aspect of life history evolution for each major group (insects, amphibians, fishes, and marine invertebrates), partly as a way of introducing some of the major issues and partly to illustrate the range of approaches used to study life history evolution. Then, I will consider one major question that applies to all species with complex life cycles, but particularly to bottom-dwelling (benthic) marine invertebrates: the evolution and loss of larvae.

The Diversity of Animal Life Cycles

Life Cycle Evolution among Insects: The Proliferation of Species with Larvae

Not all insect species metamorphose. Indeed, among insects, metamorphosis is associated primarily with wing development: Bristletails and other species that do not develop wings and are not descended from winged ancestors exhibit no pronounced metamorphosis, only a gradual increase in size and an eventual arrival at reproductive maturity. Metamorphosis is most dramatic among holometabolous[2] species, which pass through a distinctive and largely inactive pupal stage of development; in such species, all of the transformations separating the larval morphology and physiology from those of the adult take place in the pupa. Wings, compound eyes, external reproductive parts, and tho-

racic walking legs develop from discrete infolded pockets of tissue (imaginal discs) that form during larval development (Rosmoser and Stoffolano 1998). Since metamorphosis is conspicuously associated with wing formation, the ancestral insects, which lacked wings, did not metamorphose. These early "apterygote" (without wings) insects first evolved some 420 million years ago. Approximately 100 million years later, only about 10% of fossilized species had holometabolous development, but the incidence of holometabolous development increased to over 60% of species over the next 100 million years. Over the most recent 200 million years, it has increased to about 90% of all described insect species. Clearly, metamorphosis has been favored during insect evolution. How can we account for this?

As originally stressed by Istock (1967), larvae are ecologically distinct from adults. They live in different habitats or microhabitats, have access to different resources, are exposed to different sorts of competitive interactions, and are vulnerable to different predators. Thus, to at least some extent, the proliferation of insect species with holometabolous development might reflect more favorable conditions in habitats occupied by larval stages. Among mayflies, to use an example offered by Istock (1967), individuals spend several years as larvae and only a few days as (nonfeeding) adults, suggesting that, in the past at least, selection strongly favored the larval stage of the life cycle.

The documented proliferation of insect species with pronounced metamorphosis may also reflect greater speciation rates for such species, perhaps associated with the proliferation and diversification of flowering plants that has taken place over the past 100 million years or so (Farrell 1998).

Life Cycle Evolution among Amphibians: Flexibility in the Timing of Metamorphosis

Like that of winged insects, amphibian metamorphosis involves extensive tissue remodeling. The transition from tadpole to juvenile involves resorption of the tail musculature and skeletal system; major reconstruction of the digestive tract; degeneration of the larval skin and alteration in skin chemical composition; growth of the hind and forelimbs; and degeneration of the gills and associated support structures. Physiological changes include a pronounced alteration of visual system biochemistry, replacement of larval hemoglobin with adult hemoglobin, and a shift from ammonia excretion to urea excretion. Typically, amphibian metamorphosis also involves a shift from aquatic to terrestrial habitats. In a number of species, all of these transformations can be brought about or accelerated by exposing advanced tadpoles to increasing concentrations of hormones produced by the thyroid gland (Gilbert and Frieden 1981).

The timing of metamorphosis has received considerable attention from ecologists over the past nearly 30 years, stimulated by the arguments of Wilbur and Collins (1973). Wilbur and Collins placed considerable importance on the trade-off between time of metamorphosis and size at metamorphosis, and they predicted that length of the larval period was not fixed but could be varied depending on environmental conditions. How might fitness be affected by the timing of metamorphosis? Metamorphosing too late might increase predation pressure on larvae or expose them to deteriorating environmental conditions. Metamorphosing too early, at a smaller size, might compromise the ability of juveniles to escape predators or might reduce adult fecundity. Metamorphosis might also be timed so that individuals leave the larval habitat before terrestrial predators become abundant or to coincide with an abundance of resources for the juvenile or adult stage. Much of the relevant research for amphibians has been summarized by Newman (1992). Although the research questions have been framed primarily for amphibians developing in pools that dry out as the summer progresses (vernal

pools), many of the issues addressed also apply to species developing in other habitats, and to both freshwater and marine invertebrates, too.

If ponds do not dry up, if food remains abundant, and if predators are rare, fitness should be enhanced if the larvae remain in the pond as long as possible and metamorphose at a large body size. Large juvenile size probably correlates with more effective escape from terrestrial predators, earlier maturation, and higher fecundity. On the other hand, if the pond is drying up or if predation on larvae is increasing or if food availability is declining, fitness should be increased by metamorphosing sooner, even at a smaller size. These expectations are supported by most experiments on a number of amphibian species (reviewed by Newman 1992). That is, larval development, and the timing of metamorphosis in particular, is a phenotypically plastic trait in many amphibian species.

Although the adaptive benefits associated with metamorphosing at particular times and at particular sizes seem clear in most of the experiments conducted to date, this does not mean that the responses have necessarily been selected for. For example, precocious metamorphosis as ponds are drying out could simply reflect an effect of elevated water temperature on larval differentiation rates (Smith-Gill and Berven 1979; Newman 1992). Similarly, the view that the larvae "decide" when to metamorphose (Wilbur and Collins 1973 and much of the subsequent literature) is probably misleading: There is no indication as yet that amphibian larvae become capable of metamorphosing and then either proceed with or postpone that process. As we will see, this is quite different from the situation for many marine invertebrates and some fish species. Finally, although much of the literature stimulated by Wilbur and Collins (1973) assumes that amphibian metamorphosis is somehow triggered by reaching a particular size and is thus related to rates of growth, there is no experimental basis for that assumption. Although there is still some uncertainty about the mechanism of amphibian metamorphosis, the process is clearly driven by hormonal activity; thus, the timing of metamorphosis probably has more to do with rates of differentiation than with rates of growth per se (Smith-Gill and Berven 1979). The mechanisms through which pond drying and other environmental factors affect the timing of metamorphosis are still being investigated (Hayes 1997).

Life Cycle Evolution among Fishes: The Capacity to Delay Metamorphosis

The morphological, physiological, and biochemical changes that accompany the metamorphosis of lampreys and bony fishes are well described in Gilbert and Frieden (1981). The emphasis of research in this area has long been either descriptive or focused on understanding the role of thyroid secretions in coordinating metamorphic changes. More recent studies, however, have considered ecological issues, particularly for coral reef species. In such species, larvae are passively dispersed by water currents from the parental reef. Later in development, they must recruit to another reef population before they can reach reproductive maturity. Once the larvae recruit, the juveniles remain in the same location through maturity. Note that these life histories differ from those of most insects and amphibians, in which dispersal is a primary responsibility of the adult stage.

For a number of years, it has been assumed that larvae of at least some fish species can select when and where they will metamorphose (reviewed by McCormick 1999). Implicit in this assumption are the ideas of "becoming competent" to metamorphose and delaying metamorphosis: The sensory and hormonal machinery of metamorphosis has developed in competent larvae, but metamorphosis does not occur until it is triggered by environmental cues associated with reefs. Until recently the evidence for these concepts has been circumstantial, and based on analysis of otolith growth bands (e.g., Sponaugle and Cowen 1997). Since otoliths often show daily deposition patterns as well as conspicuous marks at metamorphosis, it is possible to collect juveniles recruiting to reefs and determine their age at metamorphosis. Length-on-age regressions also then permit estimation of growth rates and size at metamorphosis. Otolith growth slows in some species late in larval development, and it has sometimes been assumed that this slowing indicates the onset of metamorphic competence (See Sponaugle and Cowen 1997). In at least some species, the estimated duration of larval life and size at metamorphosis show considerable variability from year to year or over the course of a single reproductive season, suggesting delayed metamorphosis in the absence of appropriate adult habitat (e.g., Sponaugle and Cowen 1997).

McCormick (1999) has now provided the first experimental evidence that these interpretations are probably justified. To do this, he collected larval manini *(Acanthurus triostegus)* from a reef in French Polynesia and confined some individuals in cages suspended above the reef and other individuals in cages anchored directly on the reef. All individuals confined on the reef surface metamorphosed within 24 h of collection from the plankton. However, most of the fish confined above the reef were much slower to complete the morphological changes of metamorphosis, and many still retained the full larval morphology by the end of the 5-day observation period. The experiment clearly demonstrates a capacity to delay metamorphosis: Larvae confined on the reef metamorphosed rapidly, showing that they were competent to metamorphose when they were collected, whereas larvae subsampled from the same population but confined in the water column generally did not metamorphose as quickly.

Clearly, then, larvae of at least some fish species differ considerably from those of insects and amphibians, in which competence and metamorphosis apparently go hand in hand. On the other hand, the temporal separation between becoming competent to metamorphose and actually metamorphosing recently demonstrated in fishes has a direct parallel among marine invertebrates, as discussed next.

Life Cycle Evolution in Marine Invertebrates: Stresses Experienced by Larvae Carried Over into Postmetamorphic Life

The phenomenon of delayed metamorphosis has been best studied for marine invertebrates. Numerous studies conducted over the past 80 years or so have shown that, for at least some species in most groups, larvae first become competent to metamorphose and then metamorphose only if they encounter certain external cues (reviewed by Thorson 1950; Pechenik 1990; Pechenik et al. 1998). If competent larvae fail to encounter these external cues, they remain active in the plankton, often continuing to feed and grow, sometimes for months (figure 11.2). The ability to delay metamorphosis has been demonstrated for the larvae of invertebrates as diverse as sponges, turbellarian and trematode flatworms, gastropod and bivalved molluscs, polychaete worms, crustaceans, bryozoans, and echino-

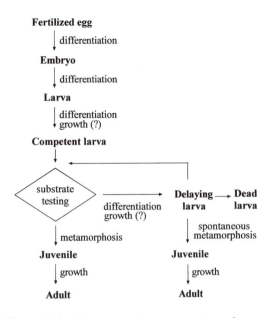

Figure 11.2 Diagrammatic representation of a marine invertebrate complex life cycle. Larvae typically become competent to metamorphose only after developing for a time in the plankton but do not necessarily metamorphose immediately on becoming competent to do so. If particular environmental cues are not encountered by competent larvae, larval life may be prolonged, sometimes for many months. Some fish species also show this ability to "delay metamorphosis."

derms (Thorson 1950; Pechenik 1990). The ability to postpone metamorphosis presumably increases the probability that larvae will eventually metamorphose into environments that are especially appropriate for juveniles and adults, thereby increasing the likelihood of survival and reproductive success (Thorson 1950; Pechenik 1990). Curiously, the larvae of most species that have been studied in the laboratory are not able to maintain the larval form indefinitely. In many species, competent larvae become less selective with time and eventually metamorphose despite our best efforts to prevent them from doing so (reviewed by Pechenik 1990). This observation led to the suggestion that fitness might decline as metamorphosis was delayed (Pechenik and Eyster 1989): Selection would not act to prolong larval life indefinitely if postponing metamorphosis beyond a certain point substantially re-

duced the ability of juveniles to compete for space, avoid predation, or tolerate physical stresses.

Such potential fitness costs have now been documented in a variety of species from a variety of groups. Prolonging larval life by only 72 hours decreased postsettlement or postattachment survival in the polychaetes *Capitella* sp. I and *Hydroides elegans;* postponing metamorphosis of barnacle cyprids *(Balanus amphitrite)* by 3 days decreased juvenile growth rates significantly; postponing metamorphosis by only 8–10 h decreased rates of colony growth by bryozoans in the genus *Bugula* (reviewed by Pechenik et al. 1998; Pechenik 1999). Decreased juvenile growth rates could affect many aspects of fitness, including vulnerability to predators, delayed maturation, and reduced ability to compete for space, particularly among sessile organisms like bryozoans and barnacles. Similar effects have also been reported for sponges and echinoids (reviewed in Pechenik 1990, 1999). Thus, to the extent that larvae delay their metamorphosis too long in the field, the presumed benefits long associated with prolonging larval life will not be fully realized.

Clearly, metamorphosis is not an entirely fresh start for many marine invertebrates, despite the great magnitude of ecological, morphological, physiological, and biochemical change that may accompany the transition. Similar effects have been documented for birds, humans, and amphibians (reviewed by Pechenik et al. 1998), so that the phenomenon is probably very widespread. To date there is little evidence of the extent to which postmetamorphic survival is affected by delayed metamorphosis in the field. When Wendt (1998) transplanted young colonies of *B. neritina* from the laboratory to the field, he found that differences in developmental rates persisted for the entire 2-week study period, resulting not only in slower rates of colony growth, but also in delayed maturation and reduced numbers of brood chambers per colony. Jarrett and Pechenik (1997) found that growth potential differed significantly among barnacles recruiting at different times during a single reproductive season. In general, growth potential was higher early in the season and reduced later in the season, possibly reflecting a higher incidence of delayed metamorphosis by larvae later in the season.

At least some other short-term stresses experienced by larvae can also carry over into life after metamorphosis, including nutritional stress (Pechenik et al. 1998) and salinity stress (Pechenik et al., in press).

Special Focus: The Evolution and Loss of Larval Stages from Complex Life Cycles

Within many animal groups we find some species that include one or more free-living larval stages in their development and some that do not. As discussed earlier, the number of insect species with holometabolous development has increased markedly over the past 300 million years or so, from about 10% to about 90% of described species. The evolutionary trend has clearly been toward proliferation of species with larvae. Among amphibians, on the other hand, the evolutionary tendency has apparently been toward the loss of larval stages. Larvae have been lost independently in anurans, salamanders, and caecilians; just among frogs and toads, larvae have probably been lost at least 10 separate times (Hayes 1997; Hall and Wake 1999). The morphological changes that have accompanied loss of larvae among amphibians, and the hormonal and molecular mechanisms associated with that loss have been studied for a number of species (reviewed by Hayes 1997; Hall and Wake 1999). However, the ecological consequences associated with the loss of larvae and the evolutionary forces that may have selected for such loss have not been widely considered. Ecological and evolutionary issues have been better explored, although certainly not resolved, for marine invertebrates.

As with amphibians, the evolutionary tendency among bottom-living (benthic) marine invertebrates seems to have been toward the loss of larvae (reviewed by Pechenik 1999). This has certainly been the case, for example, within groups of gastropods, ascidians, and asteroids (reviewed by Pechenik 1999). Yet, of those species for which reproductive patterns have been determined, approximately 70% still possess free-living larval stages (figure 11.3). Even when embryos develop initially within jelly masses, complex egg capsules, or internal brood chambers, the emerging organisms are often free-living larvae, either feeding or nonfeeding depending on species, rather than miniature adults (Pechenik 1979). For many marine invertebrate species, larvae are the primary dispersal agents in the life history and determine where adults will live. This is quite different from the situation among insects and amphibians, in which adults generally do the dispersing and select where the larvae will develop, but it is similar to the situation among coral reef fish.

Particularly for marine invertebrates that are stationary or sedentary as adults (e.g., clams, oysters, snails, barnacles, bryozoans, and tube-dwelling polychaete annelids), larval dispersal is generally viewed as advantageous (Dingle and Holyoak, this volume). It enables a species to quickly expand its range from one generation to the next, minimizes competition between adults and offspring of those adults, minimizes competition for food among siblings, potentially reduces the likelihood of habitat saturation, and minimizes the likelihood of inbreeding (reviewed in McEdward 1995; Pechenik 1999). It may also temporarily minimize predation on developmental stages, provided that predation intensity is lower in the plankton than it would be in or on the benthos (Pechenik 1979). Dispersive larvae are particularly advantageous when local conditions fluctuate radically and unpredictably in quality over short time periods (Jablonski and Lutz 1983; Holt and McPeek 1996). Over the longer term, extensive dispersal appears to reduce the likelihood of extinction, by spreading the species over a very wide geographical range and by facilitating the rapid recolonization of devastated areas if conditions later improve (Jablonski and Lutz 1983).

The apparent evolutionary trend toward loss of larvae from the life cycles of benthic marine invertebrates, and the apparent dominance of planktonic and mixed developments in the life cycles (figure 11.3) presents us with two intriguing questions: First, what are the disadvantages associated with larvae that would select for their elimination from life cycles? And second, if larval stages have been independently lost from many groups on many different occasions, how do we explain their apparent dominance in marine invertebrate life cycles today?

Potential Disadvantages of Having Larvae in the Life Cycle

Although many of the consequences of dispersal are beneficial, as discussed above, dispersal can also be disadvantageous (reviewed by Pechenik 1999). Indeed, there is no evidence that larvae have

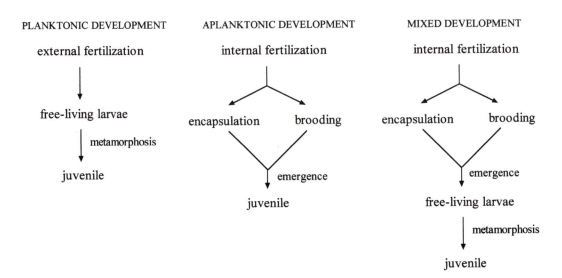

Figure 11.3 Summary of major reproductive patterns encountered among benthic marine inverte-brates. Mixed development is a combination of the other two patterns (Pechenik 1979). About 70% of species whose developments have been described fall into either the first or third categories and produce free-living larvae.

been selected for as dispersal agents; rather, dispersal may simply be a necessary by-product of having long-lived larvae in the life cycle (Pechenik 1980; Strathmann 1985). For one thing, when new recruits to local populations originate largely or entirely from distant sites, extensive dispersal can decrease overall fitness by preventing animals from becoming well adapted to local conditions (Hedgecock 1986; Storfer and Sih 1998).

But the most obvious disadvantage of larval dispersal is probably that offspring are generally denied the opportunity of recruiting to the location that supported the survival and reproductive success of their parents. Indeed, many may be swept by currents into unfavorable areas, never to return to habitats appropriate for their metamorphosis.

Many species exhibit two traits that may at least partially compensate for the likelihood of dispersal away from the favorable parental habitat. First, the larvae of many marine invertebrate species across the entire range of phyla do not metamorphose when they first become competent to do so. Rather competent larvae are triggered to metamorphose by contact with specific environmental cues, often chemical cues associated with particular prey species or with adults or juveniles of the same species (reviewed by Pawlik 1990; Pechenik

1990, 1999). In the absence of such cues, they continue to swim, and to feed if they are planktotrophic, periodically exploring substrates and swimming off again if they fail to encounter the appropriate cues for metamorphosis. Thus, as discussed earlier for larval reef fishes, marine invertebrate larvae can typically delay metamorphosis, for hours in some species and for many months in others (reviewed by Pechenik 1990). There seems to be no real parallel among insects and amphibians; again, in those organisms, larvae are not burdened with the responsibility of choosing where adults will mature and reproduce. Among marine invertebrates, the temporal disconnection between becoming competent and actually metamorphosing should increase the likelihood that dispersing larvae will eventually encounter (and metamorphase into) a habitat suitable for juvenile and adult existence.

One particularly good demonstration of the relation between selective metamorphosis and post-metamorphic survival was given by Olson (1983) for the colonial seasquirt (ascidian) *Didenmum molle,* a species that houses symbiotic algae. Larvae were released primarily near midday, were competent to metamorphose when released, and nearly always metamorphosed into shaded areas in shallow water, at a light intensity of about 100 $\mu E\ m^{-2}\ s^{-1}$. When individuals newly attached to the undersides of opaque panels at 2 m depth were exposed to full sunlight at the same depth, fewer than 3% of individuals survived for 4 days. Release from parents at midday obviously gives larvae of this species a good chance to assess local light regimes and metamorphose into areas that are not too brightly lit to bleach the symbiotic algae.

On the other hand, there is growing evidence that postmetamorphic growth rates and other potential fitness components can be compromised if metamorphosis is delayed too long, as reviewed in the previous section. To the extent that larval life is extended for long times in the field, the potential benefits of being able to delay metamorphosis may not be fully, or even partially, realized (reviewed by Pechenik et al. 1998). Metamorphosing in response to particular environmental cues may in itself present drawbacks (reviewed by Pechenik 1999). In particular, juveniles may be prevented from living in some favorable habitats by the narrow responsiveness of larvae to environmental cues. Another potential disadvantage, and one with no experimental support as yet, is the possibility that some predators may be able to mimic particular cues and stimulate larvae of other species to explore the substrate or metamorphose under false pretenses (Pechenik 1999).

The disadvantages of long-distance dispersal are best avoided by losing larvae from the life cycle, or by releasing larvae that can metamorphose very soon after entering the plankton.

Relative Susceptibility of Larvae to Predators and Environmental Stresses

Marine invertebrates that release free-living larvae produce tens of thousands or even millions of eggs each year; in the face of stable population sizes, this tells us that developmental mortality must be enormous (Thorson 1950). If most of that mortality occurs during planktonic life, as generally assumed, planktonic mortality could be an important force selecting for aplanktonic development. What proportion of developmental mortality occurs by planktonic predation? What proportion occurs during the time of substrate exploration? What proportion occurs between the time of metamorphosis and the time that juveniles are large enough to be readily sampled by ecologists? What proportion is simply due to permanent dispersal by water

currents away from favorable sites, or to developmental defects inherent in the larvae?

The sources of planktonic mortality have not been easy to determine (reviewed most recently in McEdward 1995 and by Pechenik 1999). Although direct observations have been made for several ascidian species living in clear waters and producing large, brightly colored larvae (e.g., Stoner 1990), most estimates of planktonic mortality have been understandably indirect and subject to alternate interpretation (reviewed in McEdward 1995). One key issue revolves around size-specific vulnerability to predators. It is commonly assumed that larger is safer, in which case larger size at emergence and rapid growth should reduce the magnitude of predation. However, many studies and models, primarily from the fisheries literature, indicate that larger larvae may in fact be more vulnerable to at least some planktonic predators due to greater visibility or greater predator-prey encounter frequency associated with faster swimming speed (reviewed by Pechenik 1999).

The plankton may actually be a safe place to develop, relative to the dangers imposed by the benthos. In the plankton larvae have a third dimension into which they may escape. The benthic world, however, is largely compressed into two dimensions; the time of substrate exploration may be the time of greatest risk from predators (reviewed by Pechenik 1979, 1999; Strathmann 1985; McEdward 1995).

Predation on newly metamorphosed juveniles is somewhat better documented, and extraordinarily high (reviewed by Gosselin and Qian 1997; Hunt and Scheibling 1997). Thus, there is no question that developmental mortality among species with larvae is enormous, but it is not at all clear that the dangers are primarily attributable to life in the plankton.

The question of whether encapsulation or brooding provides safety from predators is not well resolved. Clearly, predation pressure can select against free-living larvae only if survival is improved through encapsulation or brooding. Although the degree of safety associated with brooding has been little explored, there are many reports of predation on egg capsules and egg masses, which tend not to be chemically defended (reviewed by Pechenik 1979, 1999). Estimated mortality rates of encapsulated embryos often overlap those for planktonic larvae (reviewed by Strathmann 1985; Rumrill 1990; and in McEdward 1995). On the other hand, individuals emerging from benthic egg capsules and brood chambers tend to be about the same size or larger than the size at which temperate larvae metamorphose (Pechenik 1999). Aplanktonic development may thus confer a survival advantage, if larger individuals are less susceptible to benthic predation; however, there is little evidence of size advantage over the small range of sizes at which individuals first enter the benthos, regardless of whether they enter as metamorphosing larvae or as juveniles (reviewed by Pechenik 1999).

Perhaps larvae are more susceptible to bacterial and viral infection, thermal stress, salinity stress, UV irradiation, and other physical stresses. To date, however, there is little evidence that embryos are better protected from such stresses by being encapsulated or brooded (reviewed by Pechenik 1999); the functional properties of brood chambers, egg capsules, and egg masses merit further study. Indeed, encapsulation may actually increase exposure to such stresses, by imprisoning embryos in stressful situations that larvae could easily avoid through passive dispersal or active vertical migration (reviewed by Pechenik 1999).

In short, it is not yet clear whether free-living larvae make a species any more or less vulnerable to predators or to environmental stresses. Neither is there any compelling evidence of an energetic advantage associated with any particular reproductive mode; some data suggest that production of larvae entails a greater cost, while other data indicate that total energetic investment is comparable regardless of reproductive mode (reviewed by Pechenik 1999).

Explaining the Present Distribution of Larval and Aplanktonic Development

So why has selection favored the loss of larvae on so many occasions in so many different marine animal groups? And why are larvae so common in marine invertebrate life cycles?

The present distribution of larval and aplanktonic development cannot be explained by differential speciation rates. On the contrary, speciation rates have probably been greater for species with aplanktonic development, due to their generally smaller geographic ranges and generally lower dispersal capacity (reviewed by Pechenik 1999). Thus, differential speciation rates should favor the accumulation of species with aplanktonic development. Yet, such species are apparently in the minority. On the other hand, to the extent that species with aplanktonic development have smaller geographic ranges and less opportunity for extensive dispersal, such species should be more vulnerable to extinction than those with long-lived larvae. Thus, differential extinction could have increased the proportion of species with planktonic development, despite the lower speciation rates of such species. For the most recent, end-Cretaceous extinction, species with aplanktonic development were apparently neither more nor less vulnerable to extinction than species with planktotrophic larvae (Jablonski 1995; Smith and Jeffery 1998). The precise role played by macroevolutionary forces in shaping present patterns of reproduction is difficult to assess since the various forces act in different directions, some favoring the accumulation of species without larvae and some favoring the preferential elimination of such species.

It is also difficult to explain the low percentage of species with aplanktonic development based on developmental constraints, physiological constraints, or behavioral and anatomical constraints (Raff 1996; Pechenik 1999). Within mixed life histories, for example, adults release free-swimming larvae from egg capsules, gelatinous egg masses, or brood chambers—free-living, dispersive larvae persist in the mixed life cycle even though their elimination should be easily achieved, by increasing egg size, for example, or providing embryos with nurse eggs or other forms of extraembryonic nutrition (Pechenik 1979, 1999).

The prevalence of marine invertebrate life histories with planktonic larvae might reflect the difficulty of losing larvae from life cycles more than selection for maintaining them. In each generation, for species with planktonic larvae, individuals newly recruiting into a given adult population are largely or entirely derived from distant populations, and individuals in those distant populations may be experiencing very different selective pressures. Thus, gene flow, mediated by dispersive larvae bringing with them the genes for good dispersal capability, may inhibit the transition to aplanktonic development despite selective pressures favoring such a transition (reviewed by Pechenik 1999). If so, we should be paying increased attention to the peculiar conditions under which aplanktonic development can evolve. Larvae might be lost, for example, only when they are retained in certain areas by particular hydrographic conditions, and when those same conditions prevent input of larvae from outside populations for long periods of time.

More speculatively, larvae could conceivably be lost despite extensive dispersal if certain life history traits correlate with strongly heritable differences in tolerance of environmental stress (Pechenik 1999). For example, suppose that larger eggs are associated both with shorter larval life and with greater tolerance of low salinity. Under such a situation, estuarine salinity gradients could select simultaneously for reduced dispersal from upstream parental populations and greater tolerance of low salinity stress in those upstream populations. Even better, the best dispersers would be selected against in upstream populations by the low salinities before they reach reproductive maturity, increasing the isolation of the upstream population. To date, this potential mechanism for the evolution of aplanktonic development from species with planktonic development is entirely hypothetical.

Implications of Human Activities for the Future Evolution of Reproductive Patterns

Clearly we are not yet able to fully explain the present distribution of marine species with planktonic and aplanktonic developments. Life histories need to be determined for more species, and we need an improved understanding of mortality sources during development and the degree to which embryos are protected from particular mortality sources by brooding or encapsulation. We also need a better understanding of the evolutionary relationships among species with different reproductive patterns, as well as more detailed consideration of the conditions under which larvae can be lost from life cycles despite the potential for high gene flow from distant populations. These issues are of more than passing interest. Although their resolution may help explain the present distributions of life history patterns among benthic marine invertebrates, they may also indicate the extent to which humans are now selecting for or against species with particular life history characteristics (Pechenik 1999).

If free-living larvae are more heavily exposed to UV and chemical pollutants and are more sensitive to those stresses, and if brooding and encapsulation confer substantial protection from those stresses, then human activities may be selecting against species with larval development. Moreover, transi-

tions from planktonic to aplanktonic development could be favored if extensive habitat degradation and small-scale extinctions effectively isolate subpopulations from the potentially homogenizing effects of dispersal.

On the other hand, since species with planktonic development often have greater geographic ranges, they should be less vulnerable to extinction. Thus, human activities may be selecting against species with aplanktonic development to the extent that ranges are smaller and opportunities for recolonization following local extinctions are reduced for such species; dispersal provides a potential escape from local habitat deterioration and a mechanism for rapid recolonization if local conditions improve.

Similar issues can be considered for life history evolution among insect and amphibian species. Whether human activities are indeed influencing life history evolution and the direction that such influences are taking may be interesting and productive lines of future ecological and evolutionary research.

Case Study: Determining the Direction and Frequency of Shifts in Life History Pattern

I am using this case study in part to draw attention to one more aspect of the problem: the role of phylogenetic reconstruction in raising and resolving issues regarding life cycle evolution. In particular, I wish to make the point that the growing use of molecular techniques by marine ecologists and evolutionary biologists is both increasing some of our uncertainty about the details of life history evolution and offering some hope of eventually resolving some of those uncertainties.

Implicit in most discussions of life cycle evolution is the assumption that the direction of evolutionary change is known. That is, we like to assume that we know, for any given group of related species, what the ancestral reproductive mode was. Earlier, I noted that for marine invertebrates, larvae are generally considered ancestral, and the absence of free-living larvae is generally thought to be a derived condition reflecting loss of larvae from the life history. If so, we need to explain both the disadvantages of larvae that could cause their selective elimination from life cycles and their apparent dominance today despite those selective disadvantages, as discussed in the previous section. However, some workers have argued that larvae represent the *derived* condition and that the ancestral mode involved brooding of offspring, at least within some marine invertebrate groups (reviewed by Pechenik 1999). If so, we need mostly to consider the selective *advantages* of larvae, and why some species have not lost larvae from the life cycle, and how they have been able to succeed without dispersive larval stages. If the direction of evolutionary change has been toward the evolution of larvae in some groups or at some times, and toward their loss in other groups or at other times, then we need to understand how selection could simultaneously favor transitions in different directions within some groups or in different but related groups.

Hart et al. (1997) sought to explore the pattern of life history evolution for a group of seastar species in the family Asterinidae, a family that incorporates a considerable diversity of reproductive pattern: some species develop as free-living feeding larvae, some develop as free-living but nonfeeding larvae, others develop completely within benthic egg masses, and still others are brooded within the body of the hermaphroditic parent. In the latter two cases, offspring emerge as miniatures of the adult; there is no free-living larval stage. What was the ancestral reproductive mode for this family? And how many times were free-living larvae gained or lost within the group? The first step in their analysis was to determine the phylogenetic relationship among species within the family.

One dozen species were included in the study, representing two genera: *Asterina* and *Patiriella*. Tissue samples were taken from one or two representatives of each species, the DNA was extracted, and a portion (about 2400 bp) of the mitochondrial genome was amplified by PCR (polymerase chain reaction). Some of the PCR products were then sequenced, and sequences from the different species were then aligned so that the degree of difference and similarity among species could be determined (Raff 1996). Homologous sequences from a distantly related seastar species *(Pisaster ochraceus)* were used to root the phylogenetic tree. Figure 11.4 shows the single phylogenetic hypothesis that was generated by all three algorithms used (maximum parsimony, neighbor joining, and maximum likelihood).

Assuming that feeding, dispersive larvae were ancestral in the group, the results imply that feed-

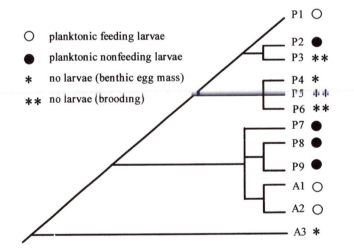

Figure 11.4 Life history evolution within the asteroid family Asterinidae, inferred from the phylogenetic analysis by Hart et al. (1997). A represents species in the genus *Asterina*, while P represents species in the genus *Patiriella*. P3 was incorrectly listed as *Asterina p. pacifica* in the original paper; it was officially transferred into the genus *Patiriella* in 1993 (M. Byrne, personal communication). Reproductive patterns for each species are indicated to the right of each species name. If the ancestral species had free-living, feeding larvae, then larvae were lost independently at least three times (once on the way to A3; once on the way to P4, P5, and P6; and once on the way to P3).

ing larvae were lost independently at least four times, and that larvae of any sort, whether feeding or nonfeeding, were lost independently at least three times (figure 11.4). This would imply substantial selection against larvae, as discussed earlier.

But how secure is the assumption that the ancestral condition is represented by feeding larvae, and what would be the impact of altering that assumption? In fact, the ancestral reproductive pattern remained unresolved by the analysis, and the authors present several scenarios based on different assumptions—for example, that the ancestral reproductive pattern included nonfeeding larvae, or benthic nonfeeding development without larvae. If the common ancestor did *not* release larvae, then the analysis summarized in figure 11.4 suggests that free-living larvae must have been *gained* at least three times. That would imply substantial selection *against* brooded or encapsulated development, and *for* dispersive larvae! Clearly, knowledge of ancestral characteristics is crucial to understanding life history evolution; without that, we can't even know what we are trying to explain. Similarly, it becomes important to understand whether shifts in life history pattern are readily reversible or not, or whether they must always occur in a particular ordered sequence, for example, from larvae to encapsulated development to brooding, or from feeding larvae to nonfeeding larvae to encapsulation. Indeed, one of the biggest problems in determining the ancestral reproductive mode for these animals is that we don't really know whether evolutionary shifts in one direction (from feeding to nonfeeding larvae,

for example) are less likely than shifts in the other direction (e.g., from nonfeeding to feeding larvae).

In addition, there is obviously a lot riding on the degree of confidence we can have in the phylogenetic reconstructions themselves. Knowing the direction of evolutionary change clearly requires that phylogenetic relationships be firmly understood. Unfortunately, it is not easy to evaluate the accuracy of phylogenetic reconstructions (Raff 1996; Pechenik 2000). One potentially troublesome outcome in the present results is that the two genera included in this survey do not appear to be monophyletic. Under most circumstances that would call the analysis itself into question. In this case, morphological considerations apparently also indicate that the family requires a thorough taxonomic overhaul, but it is too soon to tell where that will lead. Obviously, if the proposed phylogeny is incorrect, any conclusions about the evolutionary trajectory of life histories within the group will also be incorrect.

The molecular techniques involved in extracting these sorts of data are so sophisticated, and the algorithms used to work with those data are so complex, that one can easily forget that the cladograms generated are working hypotheses. They are not, as yet, final answers, but works in progress—just like most other research, in fact.

Notes

1. A change in form during development.
2. Often referred to as *complete metamorphosis.*

12

Sex and Gender

TURK RHEN
DAVID CREWS

In this chapter, sex will refer to the central process of meiosis and syngamy[1] in eukaryotic organisms. Although some form of sexuality characterizes the life cycle of many eukaryotic organisms
(i.e., virtually all fungi, plants, and animals), not
all eukaryotes are sexual (e.g., many protists)
(Margulis 1970, 1996; Bell 1982). Certain asexual
protists, for example, only undergo mitosis and
never alternate between haploid and diploid stages
by way of meiosis and syngamy. Consequently,
one of the most fundamental questions in biology
is: Why do certain organisms go through meiosis
and syngamy while others do not? Despite the apparent simplicity of this query, evolutionary biologists have not provided an entirely satisfactory explanation for the evolution of sex. Much of the
difficulty arises because there appears to be no single answer. Moreover, sex is often confused with
other associated phenomenon. For instance, one
completely subordinate, but intimately related, occurrence is the evolution of gender in organisms
that go through meiosis and syngamy.

In his essay on the evolution of sex, Ghiselin
(1988) aptly wrote, "Gender means the differentiation into males, females, and such alternatives as
hermaphrodites. It also includes the differences between sperm and eggs. Such differences are important because they create the circumstances that
make sex a puzzle" (p. 9). Yet he dismisses this
subject in the next sentence: "Otherwise we are not

much concerned about gender either." Here we clarify the relationship between the evolution of sex and
the evolution of gender. This is a critical concept to
comprehend because gender differences are nearly
universal in sexual organisms. We also discuss some
of the major hypotheses proposed to explain why
sex exists and recent empirical work that sheds light
on the factors that may favor meiosis and syngamy,
regardless of gender differences. In the remainder of
the chapter, we present a more thorough analysis of
the evolution of gender, including a discussion of
what the fundamental gender difference is and why
there are so many different mechanisms that produce more derived gender differences.

The latter question has been hard to address because gender differences have usually been framed
in terms of such contrasting alternatives as simultaneous versus sequential hermaphroditism,[2] hermaphroditism versus gonochorism,[3] and environmental sex determination versus genotypic sex
determination. Sex allocation theory (Orzack, this
volume) indicates that these phenomena are simply
life history adaptations that maximize individual
fitness through male and/or female function in
light of various constraints. To illustrate how natural and sexual selection act to mold gender differences within specific genetic, developmental, and
ecological constraints, we review some of our own
work on the evolution of temperature-dependent
sex determination in reptiles.

The Evolution of Eukaryotes, Mitosis, and Sex

Although the exact historical events that led to the evolution of eukaryotes are not clear, Margulis (1970, 1996) has proposed that eukaryotes arose as the result of a series of endosymbiotic relationships among various archaebacterial and eubacterial cells. In this scenario, an initial symbiosis or symbioses produced the first eukaryotic (i.e., nucleated) cells. These primordial eukaryotes then acquired other symbionts, which gave rise to organelles like mitochondria and chloroplasts. The serial endosymbiotic theory is supported by phylogenetic evidence that mitochondrial DNA and chloroplast DNA are related to the genomes of various prokaryotes (e.g., Williamson 1993; Turmel et al. 1999). In other words, eukaryotes are composite organisms made up of heterospecific cells and their genomes. The concomitant evolution of a regular pattern of inheritance of these genomes (i.e., mitosis and sex) represented a major evolutionary innovation in the history of life.

In addition, Margulis proposed that the evolution of mitosis involved a symbiotic association between early eukaryotes and motile prokaryotes that contained an actinlike protein. This symbiotic relationship was presumably the progenitor to the eukaryotic flagellum and its basal body. The protoflagellum, in turn, differentiated via gradual evolutionary steps into the mitotic apparatus that we recognize today (see chapter 8 in Margulis 1970 for a series of hypothetical transitions based on putative intermediate forms of mitosis in extant protists). Whatever the evolutionary history of the mitotic and meiotic apparatus, it is generally accepted that mitosis evolved first as a means to efficiently distribute chromosomes to daughter cells and that meiosis and syngamy followed.

Although there are a number of potential costs to sex that make its evolutionary origin and subsequent maintenance paradoxical (see Ghiselin 1988; Shields 1988), the only inescapable consequence of meiosis and syngamy is the alternation of ploidy. In unicellular and small multicellular eukaryotes, this process bears a significant and direct "cellular-mechanical cost" in terms of additional time required for meiosis, gametic union, and nuclear fusion when compared to mitosis (Lewis 1983). It was estimated, for example, that the production of two daughter cells via sex takes at least twice (and up to 10 times) as long as the production of two daughter cells via mitosis. Accordingly, and in the absence of other mitigating factors, asexual forms should easily supplant sexual forms when they are placed in direct competition with each other.

In an elegant study of sex in yeast, Birdsell and Wills (1996) eliminated this "cellular-mechanical cost of sex," as well as most of the other potential costs of sex. They found that one round of meiosis and syngamy bestowed a significant competitive advantage in all of their replicate competitions between sexual and asexual strains; strains that had gone through one sexual bout replaced asexual strains in as few as 100 generations. Birdsell and Wills were able to ascribe the payoff of sex to a number of factors (i.e., overdominance at the mating type locus, meiotic recombination). However, it is unclear from this study whether the demonstrated benefits of sex would be enough to outweigh the intrinsic "cellular-mechanical cost of sex" or any of its various other costs. Nevertheless, this experiment provides a classic model for the careful dissection and evaluation of the theoretical costs and benefits of meiosis and syngamy. In a general sense, this work also simulates the presumed circumstances during the origin of sex: the first sexual organisms were almost certainly simple eukaryotes in which reproduction was normally via mitosis and the costs of sex were relatively low.

Another frequent, though not inescapable, consequence of sex is mixis, or the rearrangement of genetic material into new combinations. Mixis occurs during meiosis through intrachromosomal recombination and/or independent assortment of different chromosomes. Effective mixis, however, ultimately depends on syngamy between gametes that were produced by different individuals (i.e., outcrossing). This union of gametes from genetically unrelated individuals has been simultaneously touted as the "two-fold genetic cost of sex" and as the adaptive explanation for the evolution of sex. Organisms that reproduce by outcrossing contribute only half of their offspring's genes. It has been argued on these grounds that, to obtain an equal genetic representation in the next generation, a sexually reproducing organism would need to contribute twice as many surviving progeny as would an asexually reproducing organism. In short, sexual reproduction is an inefficient way to pass on one's genes. On the other hand, outcrossing organisms produce offspring with the maximal level of genetic diversity compared to asexual organisms.

Various hypotheses posit that such diversity provides a selective advantage in spatially and/or temporally variable environments (Bell 1982; Ghiselin 1988; Shields 1988). While none of these hypotheses have been directly tested, certain results from the yeast experiment discussed above support the thesis that mixis can increase adaptation to a given environment. Birdsell and Wills (1996) found that meiosis and syngamy (i.e., mixis) produced an advantage in competitions between "sexual" strains that were genetically identical to "asexual" strains prior to sex. Since the strains of yeast used in the relevant competition were heterozygous diploids that went through only one round of meiosis and were self-fertilized, there was mixis without the "two-fold genetic cost of sex" (i.e., genetic dilution due to outcrossing; see Waser and Williams, this volume). Thus, it was inferred that the improved competitive ability of the "sexual" strains was due to the production of at least a few genetic recombinants that were better adapted than the parental genotype to the stable culture conditions of that particular experiment.

By analogy, these findings suggest that mixis may provide a competitive advantage to sexual organisms that are exposed to spatially and/or temporally heterogeneous environments. Despite these tantalizing empirical results, much more work needs to be done to carefully tease apart all the interacting factors that incur costs and/or provide benefits to sexual reproduction. For example, mixis not only creates but also breaks up the best genotypes. In the example just discussed, another round of meiosis and syngamy may very well have destroyed the successful recombinant genotypes. Therefore, it is still uncertain whether the benefits of mixis can counterbalance the "two-fold genetic cost of sex" or the various other costs of sex, for that matter.

In addition to the creation of novel genetic combinations, sex is thought to play a role in the elimination of deleterious mutations. Such mutations occur in all organisms whether they are sexual or asexual. However, it has been posited that these mutations accumulate in asexual organisms via Muller's ratchet. Picture an asexual population that is experiencing a given rate of deleterious mutations and is being purged of those mutations by selection. In an infinite population, a theoretical equilibrium is reached between mutation and selection with variation in the number of mutations per individual (see figure 12.1). A similar equilibrium distribution occurs in a finite population, but there is a chance that those individuals with the fewest mutations will not reproduce. When this occurs, the frequency distribution of deleterious mutations in the population advances one notch in the ratchet (figure 12.1). There is then little chance to reestablish the group with the fewest mutations because the probability of a back mutation of a harmful allele is assumed to be very low. Successive turns of the ratchet result in the accumulation of deleterious mutations in asexual organisms and their eventual extinction. In contrast, harmful mutations can be eliminated (or their numbers reduced) in sexual organisms because recombination can repeatedly produce individuals with fewer mutations than found in the parental generation.

In short, evolutionary biologists now contemplate a synthetic perspective in which the prevention of Muller's ratchet and the generation of new genetic combinations (i.e., mixis) work together to

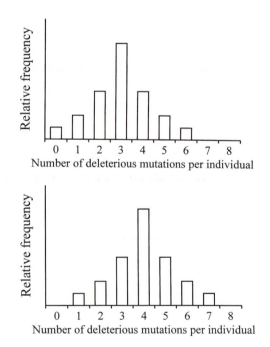

Figure 12.1 Frequency distribution of individuals with variable numbers of deleterious mutations. The distribution shifts to the right one notch when all individuals in the group with the fewest mutations fail to reproduce (an event that would be expected to happen fairly frequently in finite populations).

favor the evolution of sex. However, recent work on Bdelloid rotifers indicates that this ancient and diverse group of invertebrates has gone without sex for tens of millions of years (Welch and Meselson 2000). This observation suggests that meiosis and syngamy and its attendant effects on mixis and/or mutation elimination are not essential for long-term evolutionary success.

The Evolution of Gender Differences

Until now we have not mentioned gender differences in our discussion of the evolution of sex. This is because gender differences are theoretically unnecessary for meiosis and syngamy to occur. In fact, there is apparently no gender differentiation of any sort in at least a few eukaryotes, but they nonetheless go through the processes of meiosis and syngamy (Hurst and Hamilton 1992). These curious creatures support the notion that the evolution of gender is entirely subordinate to and unnecessary for the evolution of sex. They also raise the question of why gender differences (i.e., sperm and egg, males and females, various types of hermaphrodites) ever evolved in the first place. Those alternative traits listed in the introduction are certainly gender-related, but they are also more derived characters. What, then, is the fundamental gender difference?

Hurst and Hamilton (1992) proposed a useful definition for male and female based on the inheritance of cytoplasmic genes: Organelles are inherited via "female" gametes during syngamy. In this delineation, virtually all eukaryotes either have no genders (both parents contribute organelles) or just two separate genders (uniparental contribution of organelles). These authors also proposed that this decisive gender difference evolved to minimize conflict between cytoplasmic genes from different parents within the newly formed zygote. Whatever the selective forces were that favored this asymmetric inheritance of organelles, this definition is of great heuristic value. For example, complementary mating types (which number in the thousands in some fungi) that once may have been considered the primordial manifestations of gender simply become mechanisms that ensure outcrossing, much like incompatibility alleles in higher plants. Therefore the only "mating types" of direct relevance to our discussion of gender evolution are those that govern syngamy between male (without organelles) and female (containing organelles) gametes.

Proteins that mediate the interaction between gametes are diverse and appear to be evolving rapidly. In animals, sperm-egg interactions are complex, with species differences occurring at various stages in the fertilization process (Vacquier 1995). For example, there are species differences in the proteins that mediate chemoattraction of the sperm to the egg, binding of the sperm to the egg envelope, exocytosis of the acrosomal contents (which create a whole in the egg envelope), and fusion of the sperm and egg membranes to restore diploidy. Analogous processes undoubtedly occur at different stages during pollination and fertilization in plants. In some angiosperms, pollination even depends on an exclusive one-to-one, interspecific interaction between insect pollinator and plant. Once pollen is deposited on the stigma, there are also species-specific interactions that influence growth of the pollen tube through the style toward the ovule and egg, release of sperm from the pollen tube in the event that it reaches the ovule, and fusion of the sperm and egg within the ovule. Thus, in addition to the obvious differences in size and mobility between male and female gametes, there are many subtle ways in which gametes have differentiated at the molecular level to govern syngamy.

While it has often been stated that all other gender differences are derived from an initial difference between male and female gametes, a bewildering array of gender-related traits hindered an all-inclusive theory of gender evolution. Nevertheless, Charnov (1982) began such a synthesis with his seminal work on the theory of sex allocation. The theory quite simply predicts that selection acts to maximize the product of the fitness gain through male function, or W_M (i.e., sperm, pollen, sons), times the fitness gain through female function, or W_F (i.e., eggs, seeds, daughters) (see Orzack, this volume). This theorem holds true in the choice to reproduce as a gonochorist (i.e., as a male or female but not both) versus reproducing as a hermaphrodite (i.e., through both male and female function). Likewise, the theory successfully deals with sex ratio evolution in gonochorists and allocation to male versus female function in hermaphrodites (i.e., simultaneous versus sequential hermaphroditism). While gender is labile in numerous organisms, many others are unable to alter their reproductive contribution through male versus female function.

Some species, for instance, have major genotypic factors that determine gonadal phenotype and that dictate an inflexible sex ratio of 50% males and 50% females (i.e., the X-Y sex determining mechanism of mammals or the Z-W mechanism of birds). Despite this constraint on sex ratio evolution, other mechanisms have evolved to produce gender differences. For example, it is thought that genetic differences between males and females evolve when alleles with antagonistic effects on the two genders are linked to a new sex-determining gene. This pattern of selection favors suppressed recombination between the loci which, in turn, allows genetic differentiation of new sex chromosomes (Rice 1996). Close examination of nucleotide sequences of gene homologs on the X and Y chromosomes in humans supports this scenario for the evolution of mammalian sex chromosomes (Lahn and Page 1999). While an initial event (i.e., mutation) conferred the primary sex determining role on SRY in early mammals, subsequent evolutionary events, presumably Y chromosome inversions, suppressed recombination between the nascent sex chromosomes. Once recombination was suppressed, alleles on the X and Y chromosomes were then able to fully differentiate into gene homologs with distinct functions. Some Y-linked genes have roles in male-specific processes like spermatogenesis, and certain X-linked genes function during the female-specific process of X chromosome inactivation. In contrast, alleles at other loci on the Y chromosome degenerated into pseudogenes, and their counterparts on the X chromosome were left as intact, functional genes.

Notwithstanding the evolution of genotypic sex determination and sex chromosomes, many genes that produce gender differences are autosomal. At its most fundamental level, sex-limited expression of autosomal genes involves the activation or repression of different genes in males and females. Sex-limited expression of such loci can result from interactions with genes located on the sex chromosomes, as in *Drosophila melanogaster,* or hormone-dependent mechanisms, as found in many vertebrates. For example, androgenic steroids, like testosterone, are produced in different amounts in male and female vertebrates. Androgens at the high levels typically found in males then act via androgen-specific receptors to initiate (or inhibit) gene transcription. A recent theoretical model for the evolution of sex limitation at such loci indicates that, much like models for sex allocation, se-

lection acts to maximize the product of the fitness gain through males times the fitness gain through females (Rhen 2000). These models suggest that various mechanisms of gender differentiation evolve according to this unifying principle.

To illustrate this theorem in more detail, let us examine some examples of optimal life history strategies for hermaphrodites. The following model for sequential hermaphroditism has been dubbed size-dependent sex reversal. Consider a hermaphroditic fish in which fitness through male function does not vary with body size but fitness through female function increases with size (i.e., because larger individuals can produce more eggs). In this simple case, sequential hermaphroditism is favored over simultaneous hermaphroditism because an individual can maximize its lifetime fitness by reproducing first as a male (i.e., at a small body size) and then switching to female function later in life (i.e., at a larger body size) rather than reproducing as both genders throughout its life (figure 12.2). In other words, selection acts to maximize $W_M \times W_F$ in light of the developmental constraint that the fish increases in size during its lifetime. Other hermaphroditic fish appear to maximize their fitness by reversing this sequence of gender change. In some populations of the blue-headed wrasse, for example, ecological conditions permit the largest male to dominate smaller males and monopolize matings with all resident females. If the male dies or is removed, the next largest individual (a female) changes gender to become the sole reproducing male. This example can be referred to as behavior-dependent sex reversal because it is the conduct of the dominant fish that suppresses sex change in the other resident fish. Selection again acts on the pattern of sex allocation to maximize the product of $W_M \times W_F$, but this time it does so under both developmental (i.e., growth) and ecological (i.e., mating system) constraints.

In contrast to hermaphrodites, gonochorists are developmentally committed to reproduce as only one gender. Nevertheless, they can still maximize gender-specific fitness by altering their sex ratio (the proportions of males and females in a population) to prevailing ecological conditions (Charnov 1982). Sex ratios can vary in an adaptive manner in spatially structured populations. Under these conditions, strong local mate competition is thought to favor female-biased sex ratios because only a few males are necessary to fertilize all the females within a local population. The most convincing ev-

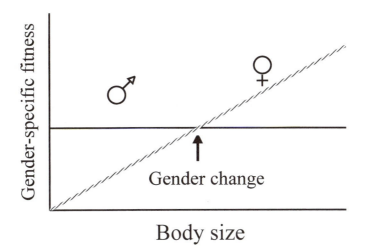

Figure 12.2 Gender-specific fitness as a function of body size in a hypothetical sex-changing fish. The optimal size for gender reversal is indicated by the arrow.

idence for this model comes from haplodiploid hymenoptera, in which females can control the sex of their offspring by allowing or not allowing fertilization of their eggs (males are haploid and females are diploid).

Sex ratios also vary in an adaptive manner when sex determination occurs after fertilization and is under the control of the embryo. Competition for host resources or, more precisely, variation in nutritional content of hosts favors one form of environmental sex determination in parasitic nematodes in the family Mermithidae (Charnov 1982). In this group, sex ratios are female-biased on large hosts or at low rates of infection when food is relatively abundant. Individual worms under these environmental conditions are able to grow to a larger size, which presumably benefits females more than males. In contrast, sex ratios are male-biased on small hosts or at higher rates of infection when food is more restricted. Feeding the host, while holding the number of parasites constant, alters the sex ratio of the parasites (increases the proportion of females produced). Overall, these results strongly support the idea that nutritional content of the host, rather than some other factor like local mate competition, influences sex determination in these parasites.

Abiotic factors can also influence sex determination. Conover and Heins (1987), for instance, clearly demonstrated adaptive variation in environ-

mental versus genotypic sex determination in the Atlantic silverside, *Menidia menidia*. In southern populations of this species, water temperature acts as a reliable cue that indicates the length of the growing season and therefore adult body size. Temperature also has a strong effect on sex ratios, with female-biased ratios produced below approximately 18°C and male-biased ratios above this temperature. This phenomenon is called *temperature-dependent sex determination,* or TSD. Since larger size differentially benefits females (larger females produce more eggs than smaller females), it pays for larvae to develop as females when water temperatures are relatively cool early in the season. Males, whose fitness is not size-dependent, develop later, when temperatures are warmer. In contrast, sex ratios in northern populations are insensitive to the effect of temperature (gender is determined by genetic factors) because the breeding and growing seasons are too short for substantial differences in growth to occur between offspring produced early and late in the season. In other words, there is an ecological constraint on the evolution of TSD in circumstances where fitness does not vary with environmental temperature.

In sum, various mechanisms have evolved to produce gender differences. These include genetic differences between males and females (i.e., sex chromosomes), autosomal loci that are expressed in a sex-limited manner, gender change in hermaphrodites as a function of body size, and sex determination that is influenced by environmental factors. As a general rule, these gender-related traits evolve to maximize the product of $W_M \times W_F$ in the context of specific genetic, developmental, and ecological constraints. To expound on these concepts in more detail, we will discuss how sex ratio appears to vary in an adaptive manner in two reptiles with TSD.

Case Studies: Temperature-Dependent Sex Determination in Reptiles

Temperature during embryonic development determines gonadal sex in all crocodilians, many turtles, and some lizards (figure 12.3). Although the adaptive significance of TSD in reptiles has been debated, there is growing evidence that temperature may have gender-specific fitness effects in this group, much as it does in the Atlantic silverside. Some important discoveries have been made in the common snapping turtle, *Chelydra serpentina*, and

the leopard gecko, *Eublepharis macularius*. In these species, there is convincing evidence that certain gender differences result from direct incubation temperature effects on sex determination as well as its indirect effects on other traits.

In the northern population of snapping turtles studied by Rhen and Lang (1998), low temperatures ranging from 20 to 22.5°C produce a high proportion of males (60–90% males, respectively), only males are produced between 23 and 27°C, mixed sex ratios with increasing proportions of females at increasing temperatures are produced between 27 and 29.5°C, and all females are produced above 29.5°C. Importantly, there is evidence for genetic variation for temperature effects on sex ratio in this population, which suggests that the pattern of TSD may be relatively free to evolve. In support of this inference, there is substantial geographic variation in the sex-ratio reaction norm (Ewert et al. 1994). For example, low temperatures produce fewer males (< 10% males) in southern populations. The upper transition from male- to female-biased sex ratios is also shifted, but to a lower temperature range. Consequently, there is a

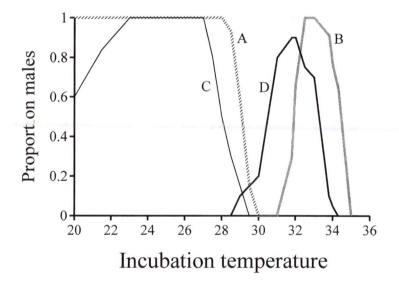

Incubation temperature

Figure 12.3 Representative effects of embryonic incubation temperature on sex ratio in various reptiles. (Redrawn from Rhen and Lang 1998) and Viets et al. 1993.) (A) Painted turtle. (B) American alligator. (C) Common snapping turtle. (D) Leopard gecko. Many reptiles with TSD have rather abrupt transitions (i.e., approximately 1°C) from temperatures that produce only males to temperatures that produce only females (or vice versa).

narrowing of the temperature ranges that produce males in southern populations. These data, although not as dramatic as the shift from TSD to genotypic sex determination in the Atlantic silverside, strongly suggest that the pattern of TSD has evolved and still has the potential to evolve in the snapping turtle.

Assuming that sex ratio evolves in this species, one can ask whether it does so in an adaptive manner. To address this question, Rhen and Lang (1995, 1999a,b) examined the effect of incubation temperature on a number of traits that could plausibly have differential fitness effects on males and females. Foremost among these was the effect of incubation temperature on posthatching growth. Snapping-turtle males reach sexual maturity earlier and at a larger size than females, which indicates that males grow faster than females. Consequently, if growth influences fitness in a gender-dependent manner, sex allocation theory predicts that there should be covariance for growth and for sex ratio as a function of incubation temperature. Yet any effort to associate a growth advantage with a particular incubation temperature would be confounded by potential sex effects on growth.

To resolve this dilemma, Rhen and Lang (1994) altered the gender (i.e., gonadal phenotype) of snapping-turtle embryos via hormonal manipulations at three representative incubation temperatures. The experimental manipulations produced females at two male-producing temperatures (i.e., 24 and 26.5°C) and males at a temperature that usually produces mostly females (i.e., 29°C). Thus, the normally confounded effects of incubation temperature and gonadal sex were separated. In short, embryonic incubation temperature had a strong effect on subsequent growth of hatchlings, whereas gonadal sex did not: Temperatures that normally yield males produced faster growth than the temperature that produces mostly females (Rhen and Lang 1995). These results imply that the gender differences in growth observed in nature are due to the matching of gonadal phenotype to particular incubation conditions during embryonic development (figure 12.4).

Further experiments were designed to determine how embryonic temperature regulates subsequent growth. Those studies revealed that behavioral thermoregulation of juvenile turtles was also influenced by incubation temperature (Rhen and Lang 1999a). Turtles from the low and intermediate incubation temperatures were found in the warm section of a thermal gradient more than was expected by chance, while turtles from the high incubation temperature were evenly distributed between the warm and cool sections of the gradient. Since that study also showed that warm ambient temperatures enhanced growth (some turtles were held in constant warm or cool environments that precluded thermoregulation), a portion of the incubation temperature effects on growth may be mediated indirectly via its effect on thermoregulation. However, the effect of incubation temperature on growth remained significant even in those turtles held at constant ambient temperatures. This finding suggests that part of the incubation temperature effect on growth was due to differences in growth physiology that were not mediated by behavioral thermoregulation. Similar results in other studies of the snapping turtle support the notion that embryonic temperature has both direct and indirect (i.e., through temperature choice) effects on posthatching growth (see discussion in Rhen and Lang 1999a).

A final study demonstrated incubation temperature and gender effects on total body mass and energy reserves in snapping turtles shortly after hatching (Rhen and Lang 1999b). Since hatchling snapping turtles struggle, often for a considerable period of time, to emerge from their subterranean nest and then must traverse large distances to reach water, all without access to food, incubation temperature and gender effects on the initial levels of energy reserves or how those reserves are utilized prior to feeding may have implications for survival. More important, perhaps, incubation temperature effects on posthatching growth could influence male and female fitness differently. In some populations of snapping turtles, female fecundity increases with body size so that larger females generally lay more eggs than smaller females. Bigger male snapping turtles also presumably have a fitness advantage because males have been observed fighting during the breeding season: It is a virtual axiom in behavioral ecology that larger individuals win aggressive encounters. Thus, if larger and dominant males are able to monopolize or obtain a greater share of matings than smaller males, the relative gain in fitness with body size may be greater for males than for females. In sum, embryonic temperature has numerous phenotypic effects on snapping turtles that may be related to fitness.

Relative to snapping turtles, which take 4 to 5 years to mature in populations with the fastest

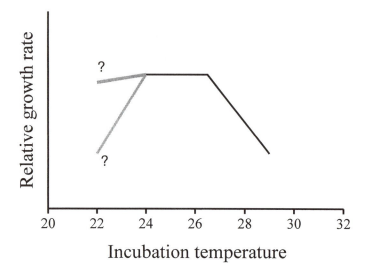

Figure 12.4 Composite effect of embryonic incubation temperature on growth rate after hatching in the common snapping turtle. (Solid line redrawn from Rhen and Lang 1995, 1999a.) Dashed line represents possible patterns of growth at low temperatures that may vary geographically with sex ratio.

growth, leopard geckos mature at a young (~45–50 weeks) age. Thus, it is feasible to directly determine if incubation temperature has any phenotypic effects that persist into adulthood and whether such phenotypes influence fitness differently in males and females. Female leopard geckos are normally produced across the entire range of viable incubation temperatures, while males are also produced across a fairly broad range of temperatures, although in varying proportions. An incubation temperature of 26°C produces all females, 30°C produces a female-biased sex ratio (~20% males), 32.5°C produces a male-biased sex ratio (~70% males), and 34°C again produces a female-biased sex ratio (5% males) (Viets et al. 1993). This pattern of TSD facilitates the investigation of incubation temperature effects in both sexes without experimental manipulations of sex ratio.

Incubation temperature has been shown to influence various traits in adult female and male leopard geckos (Gutzke and Crews 1988; Flores et al. 1994; reviewed in Crews et al. 1998). For example, females from the intermediate temperature that produces a male-biased sex ratio are less attractive (i.e., elicit less male courtship behavior) than females from temperatures that produce female-biased sex ratios. Conversely, those unattractive females are more aggressive toward males than their attractive counterparts from other temperatures. Hormone levels also vary in accord with embryonic temperature, so that less attractive, more aggressive females have higher androgen levels than more attractive, less aggressive females (Gutzke and Crews 1988). Moreover, females from the male-biased temperature are larger than females from other temperatures (Tousignant and Crews 1995). Thus, incubation temperature has effects on female behavior, endocrine physiology, and body size that are correlated with its effects on sex ratio.

It was originally hypothesized that the masculinized females from male-biased temperatures have compromised reproductive success (Gutzke and Crews 1988). Further study, however, indicates that temperature does not have any major influence on female fitness. For instance, there are no incubation temperature effects on fecundity, age at sexual maturity, or other measures of reproductive success (Tousignant and Crews 1995; Tousignant et al. 1995). Likewise, recent analyses of a much larger set of cumulative data on hundreds of animals from our colony at the University of Texas indicates that important measures of female fitness are not influenced by incubation temperature (J. Sakata, T. Rhen, and D. Crews, unpublished data). Fecundity does not vary with temperature because females have a determinate clutch size of just two eggs (females occasionally lay one egg in a clutch). Moreover, females from the entire range of temperatures produce approximately the same number of eggs during their lifetime. A slight incubation temperature effect on the fertility of eggs shows that, if anything, females from the high temperature lay a higher proportion of infertile eggs than females from the three other temperatures. This finding suggests that female attractivity and aggressiveness are not indicators of actual mating success: Relatively unattractive and aggressive females from the male-biased temperature still copulate with males and produce as many fertile eggs as females from temperatures that produce mostly or all females. Finally, the viability of embryos does not vary with maternal incubation temperature.

Hence, if TSD has any adaptive significance in the leopard gecko, temperature effects on fitness must be present in males. In the following discussion, we focus on males from the lower female-biased temperature (i.e., 30°C) and males from the male-biased temperature (i.e., 32.5°C). These incubation temperatures produce sufficient numbers of males for robust statistical comparisons of intra- and intersexual behavioral interactions.

In general, intermale aggression is extreme in the leopard gecko. In an encounter with a conspecific of either gender, male leopard geckos will often raise their body off the ground, standing on all four limbs in a high-posture display. The next step in a typical interaction depends on whether the conspecific is perceived as a male or a female (high postures may convey the signaler's gender because females do not display this behavior nearly as much as males do). Consequently, when two males are placed together they usually exhibit the high-posture display in unison. The males then slowly approach and lick each other. Pheromonal cues then elicit almost instantaneous aggression that entails reciprocal episodes of biting. Fights regularly include rapid body rolls that are reminiscent of crocodilians tearing flesh from their prey. Such agonistic encounters, if not stopped, can lead to severe skin lacerations, loss of limbs, or even the loss of a tail. Considering that combat is so costly and that high-posture displays might serve as a way to evaluate ones' opponent without fighting, the effects of relative body size on agonistic encounters were investigated (J. Sakata and D. Crews, unpublished data).

Most staged interactions between males resulted in fights regardless of the animals' relative body size. Males with a 10% weight advantage always won contests; the loser fled and rapidly waved his tail in defeat. Clearly, relative body size plays a critical role in determining dominance relationships in male leopard geckos. And since males from a male-biased temperature are on average larger than males from a female-biased temperature (Tousignant and Crews 1995), we hypothesize that it may be advantageous for males to develop at the temperature that produces more males. This scenario is plausible if large size and dominance enable individual males to control or gain access to multiple females. For example, male leopard geckos could establish territories that encompass the home ranges of multiple females. In fact, males from a male-biased temperature scent-mark significantly more than males from a female-biased temperature (Rhen and Crews 1999), a finding suggesting that male leopard geckos may be territorial. However, males from the female-biased temperature are more sexually active than males from the male-biased temperature (Rhen and Crews 1999). Thus, an interesting possibility is that males from different temperatures have different reproductive tactics, some males adopting an aggressive, territorial strategy (i.e., those males from the male-biased temperature) and others a satellite strategy (i.e., those males from the female-biased temperature). Although polygyny arising from territoriality is typical of many animals and is plausible for the leopard gecko, this explanation for the adaptive significance of TSD in the leopard gecko is speculative. There is currently nothing known about the mating system of leopard geckos in nature.

Future Directions

Persuasive evidence indicates that some gender differences in the snapping turtle and the leopard gecko result from the influence of incubation temperature directly on gonadal sex and indirectly on the individual's subsequent physiology, growth, and behavior. Although it is plausible that the observed correlations between sex ratio and such traits are adaptive, it remains to be clearly demonstrated that temperature-induced phenotypes actually have differential fitness effects on males and females. Future studies of the phenomenon of TSD in reptiles should focus on this critical link between phenotype and lifetime fitness. Studies of TSD in reptiles, including those described above, have been conducted in the laboratory under constant incubation temperatures, but temperatures in nature fluctuate on multiple spatial and temporal scales. Consequently, another critical point that needs to be addressed is whether temperature effects are observed under natural conditions and whether such effects are related to the outcomes that have already been detected in the laboratory. To date, a few studies have shown that the process of TSD is operable in the field and that it is correlated with laboratory results, but none have shown that temperature influences the development of traits other than gonadal sex in nature.

In conclusion, various developmental mechanisms have evolved to produce gender differences. Sex allocation in hermaphrodites and mechanisms of sex determination in gonochorists evolve to maximize the product of fitness gain through male function (gametes that do not contain organelles) times fitness gain through female function (gametes that contain organelles). Interestingly, this principle also governs the evolution of sex-limited

expression of autosomal loci. A major question is whether this principle also applies to the evolution of other gender differences like gamete dimorphism, sex chromosomes, and genomic imprinting (*imprinting* refers to alleles at a given locus that are expressed differently when inherited via the mother vs. the father). Perhaps this principle can be generalized. In any event, the fact that gender differences are theoretically uneccessary for the evoluton of meiosis and syngamy but are nearly ubiquitous in sexual organisms is an evolutionary puzzle that needs explanation. Finally, additional empirical studies on the adaptive significance of sex are clearly needed.

Acknowledgments This work was partially supported by an Individual National Research Service Award MH#11369 from the National Institute of Mental Health and a National Science Foundation Dissertation Improvement Grant #IBN-9623546 to TR and Grant MH#57874 from the National Institute of Mental Health to DC.

Notes

1. The fusion of two gametes in fertilization.
2. Possessing both male and female reproductive organs.
3. Separation of the sexes into separate individuals; in plants, this is referred to as dioecy; see Sakai and Westneat (this volume).

13

Sex Ratios and Sex Allocation

STEVEN HECHT ORZACK

Understanding the allocation of energy is the goal of the evolutionary analysis of sex allocation. Whether one is concerned with the relative sizes of male and female flower parts in plants like those discussed by Campbell (1998), the ratio of males and females in insects like those discussed by Orzack et al. (1991), or the relative sizes of male and female reproductive organs in hermaphroditic fish like those discussed by Leonard (1993), one is concerned with how energy allocated toward reproduction is apportioned into one sex as opposed to the other (or more in the case of some kinds of organisms). Here, the sexes are entities that at regular or irregular intervals produce gametes, some of which come together to produce zygotes. The abstract nature of this description underscores the degree to which there are common evolutionary aspects to all of these problems, despite the fact that the biological details involved are so diverse.

One of the most influential and important agendas for evolutionary studies of sex allocation was laid out by Charnov (1982). He described the underlying evolutionary similarities between phenomena as diverse as sex change in shrimp and sex ratio in vertebrates like us. Even more important, he promoted sex allocation as a central evolutionary problem by describing how seemingly unrelated allocation problems could all be analyzed with a kind of mathematical approach elaborated by Shaw and Mohler (1953). I consider in turn four important examples of this approach.

The Theory of Sex Allocation: Sex Ratio

Shaw and Mohler's goal was to understand the evolution of the proportions of males and females. This problem of the sex ratio was most famously addressed by Darwin in his 1871 book, *The Descent of Man, and Selection in Relation to Sex,* as well as by others in the subsequent decades. The most influential analysis is that of Fisher (1930); however, Carl Düsing, who worked in the 1880s, can rightly be regarded as the progenitor of modern sex ratio theory (see Edwards 1998).

Shaw and Mohler showed that the fitness, W, of a rare female producing $\hat{m}$ males and $\hat{f}$ females in a population in which other females produce m males and f females is proportional to

$$\frac{\hat{m}}{m} + \frac{\hat{f}}{f} \tag{13.1}$$

This "Shaw-Mohler" expression is a fundamental result; it relates the traits of a mutant individual to those of the resident population and thereby allows one to make inferences about evolutionary dynamics and equilibria. If the mutant female is the same as all others, the value of this expression is 2. Hence, one can infer the fate of a mutant simply by determining whether the value of equation 13.1 is less than or greater than 2. So, for example, the frequency of rare females producing 50% sons and 50% daughters will increase in a population in which the resident females produce, say, 53% sons and 47% daughters since the value of the Shaw-Mohler expression is 2.007 > 2. The inherent symmetry of the expression indicates that the same type of rare female will increase in a population with females that produce 47% sons and 53% daughters. The implication is that a female who produces equal numbers of males and females will increase in frequency and, as a result, will eliminate all other types of females. In other words, an even (or 1:1) sex ratio will evolve in the population. In addition, one can use this Shaw-Mohler expression to show that an even sex ratio is stable against invasion of a rare mutant producing an uneven sex ratio. In qualitative terms, the basic dynamic underlying evolution of the sex ratio in this instance is that individuals with a tendency to produce more of the sex that is rare at a given time leave more descendants (who inherit their ancestors' tendency) because their rarity increases their average mating success. This increase in frequency acts to reduce the disparity in frequency of the two sexes; this process stops when the sex ratio is even. This result does not mean that all families have a 1:1 sex ratio, even on average. Instead, it simply means that an equilibrium is reached when equal numbers of males and females are produced by the population as a whole. (If one makes the additional assumption that the population is finite in size, one can show that individual allocation ratios closer to 1:1 can have higher fitnesses when the population is perturbed from the 1:1 equilibrium; the strength of this selective force can be weak, however.) It is important to keep this distinction between population-level and individual-level predictions in mind when assessing the explanatory power of this evolutionary argument.

Beyond the basic assumptions that there are two sexes and that alleles at genetic loci affecting the sex ratio obey Mendel's laws, two assumptions of note underlying this analysis are that the population is panmictic[1] and that a given unit of energy allocated toward the production of sons results in an equivalent decrease in energy allocated toward the production of daughters (and vice versa). Violation of any of these assumptions will change the predictions of the model. In some instances, the change is modest; in others, it is dramatic (see Bull and Charnov 1988 for examples and discussion). For example, if the energy required to produce, say, a son is less than that to produce a daughter, the sex ratio will evolve to be *uneven* and biased toward males to an extent where equal total energy investments in the two sexes are present in the population. Typical data for vertebrates imply that such a sex ratio bias could be 5–10%. However, in many species the supposition is that producing a male and producing a female require equivalent amounts of energy; therefore, one predicts an even sex ratio.

The Theory of Sex Allocation: Sex Change (Sequential Hermaphroditism)

Consider the timing of sex change, such as occurs in species in which a given individual may reproduce as a female at one period during life and as a male during another period. In some cases, an individual may even change more than once. Some of the most dramatic examples of sex change involve fish; the variety of organisms undergoing sex change is surveyed by Policansky (1982).

Formal analysis has concerned organisms that change sex once in their lifetimes (Warner 1975; Leigh et al. 1976). Leigh et al. showed that the fitness, W, of a rare individual changing from, say, male to female at time $\hat{t}$ in a population in which all other individuals change at time t is proportional to

$$\frac{m(\hat{t})}{m(t)} + \frac{f(\hat{t})}{f(t)} \qquad (13.2)$$

where $m(\ )$ and $f(\ )$ denote the contribution to fitness gained through the production of sperm and the production of eggs, respectively. As in the case of sex ratio, the critical value for this Shaw-Mohler expression is 2; it determines whether a rare mutant will increase in frequency or not. Since the values of $m(\ )$ and $f(\ )$ are a function of age, specific predictions depend on assessing how the sexes differ in mortality or growth rates or in how fertility changes with age. If, for example, female fertility increases at a greater rate with age than does male fertility, sex change from male to female is favored. Equation 13.2 can even encompass cases in which, say, a mutant undergoing sex change arises in a population in which individuals do not change sex (see Charnov 1982, pp. 133–134).

The Theory of Sex Allocation: Simultaneous Hermaphroditism

Consider simultaneous hermaphroditism, such as occurs in those plant species in which an individual can have separate male and female flowers or have a "perfect" flower, that is, one with male and female structures. The evolutionary issue is investment in pollen production via the allocation of energy to male structures such as stamens versus investment in ovule production via allocation of energy to female structures such as pistils.

Charnov et al. (1976) showed that the fitness, W, of a rare individual that gains an amount of fitness $\hat{m}$ from pollen production and a fitness $\hat{f}$ from ovule production in a population in which each other female gains m and f is proportional to

$$\frac{\hat{m}}{m} + \frac{\hat{f}}{f} \tag{13.3}$$

As in the other sex allocation problems, the critical value for this expression is 2. Here, one can infer whether, say, a hermaphroditic population is stable against invasion by mutants of either pure sex. This depends on how change in energy investment in male and female structures affects male and female fitnesses. An important determinant of stability for a hermaphroditic population is whether a given increase of energy investment in, say, pollen production results in a smaller increase in male fitness when the absolute level of investment is higher as opposed to when it is lower. This pattern of diminishing returns implies that individuals who are purely male cannot invade the population. A pattern of diminishing returns for investment in ovule production implies that individuals who are purely female also cannot invade. If hermaphroditism is evolutionarily stable in this way, the relative proportions of energy investment in male and female function in the hermaphroditic individuals depends on the similarity between the curves relating performance to investment in the two sexes. If the curves are identical, equal investment in the two sexes has the highest fitness (see Lloyd 1984 and Campbell 1998 for further details).

The Theory of Sex Allocation: Reproduction in a Social Context

Consider the sex ratio produced by a group of individuals or colony of, say, ants in which some individuals reproduce while others do not, there is joint care of the offspring, and colonies have a temporal integrity greater than the lifetime of individuals. Some species with these traits have spectacular biology, especially as colonies may include millions of individuals. Such "eusocial" organisms are found mainly among the Hymenoptera and the Isoptera (termites).

The sex ratio produced by a eusocial colony could be controlled by the one or a few individuals who actually produce offspring and/or by the more numerous offspring who raise their younger siblings until they become functional adults. In the former case, an individual might produce more of one sex than the other (as in many Hymenoptera; see below for further details), while in the latter case, an individual might nurture or eliminate more of one sex than the other in order to alter the sex ratio of the offspring present at the time when they emerge (see Crozier and Pamilo 1996, p. 44). In either case, females are usually regarded as the controlling agents of the colony's sex ratio (as in the following discussion), although one can conceive of males as being involved as well. A female producer of offspring is often called a queen, while the nurturer of offspring is often called a worker. In the Hymenoptera, workers are female, while in the Isoptera they are male and female.

Crozier and Pamilo (1996) showed that the fitness, W, of a female producing $\hat{m}$ males and $\hat{f}$ females in a population in which other females produce m males and f females is proportional to

$$g_m v_m \frac{\hat{m}}{m} + g_f v_f \frac{\hat{f}}{f} \tag{13.4}$$

Here, g_m and g_f denote the relatedness of the female to the male and female offspring, respectively. In effect, these quantities measure the degree to which a female and a given offspring share genetic material. v_m and v_f denote the reproductive value of males and females, respectively. For a given sex, this quantity measures the probability that an allele found in the population at some distant future time is derived from that sex.

When a single queen reproduces in the colony, the implied evolutionary context of equation 13.4 is the population of single-queen colonies, whereas when workers reproduce, the evolutionary context is the population of workers.

Values of the relatedness and reproductive value coefficients depend on the sexual system. In some eusocial organisms, such as the Isoptera, males and females are diploid. As a result, save for cases in which chromosomal translocations link autosomes and a sex chromosome, when the queen is mated by a single male this diplodiploidy implies that $g_m = 0.5$ and $g_f = 0.5$ for the queen, and $g_m = 0.5$ and $g_f = 0.5$ for a worker. In this case, $v_f = v_m$ because each offspring gets half of its genome from each parent. In other eusocial organisms, such as the Hymenoptera, in which males are haploid and females are diploid (see below), if a queen is mated by a single male this haplodiploidy implies that $g_m = 1.0$ and $g_f = 0.5$ for the queen, while $g_m = 0.5$ and $g_f = 0.75$ for a worker. In this case, $v_f = 2v_m$.

As in the other sex allocation problems, the critical value for equation 13.4 is 2. As in the simple case of sex ratio I first discussed, one can infer that a population containing diplodiploid or haplodiploid queens producing, say, an even sex ratio cannot be invaded by a queen producing any other sex ratio. The same is true of a population of diplodiploid workers. However, the asymmetries in relatedness imply that a population containing haplodiploid workers producing three females for every one male cannot be invaded by a worker producing any other sex ratio. Conversely, a worker producing this sex ratio can invade a population consisting of workers producing any other sex ratio. The distinction between the optimal queen and worker sex ratios in haplodiploids implies that intracolony conflicts should occur; whether they have and how they may have been resolved evolutionarily are the basis for considerable controversy.

Regardless of how colony sex ratios are controlled, when females abstain from reproduction and instead allocate energy toward the production of their mother's subsequent offspring, it is an example of so-called kin selection. The potential evolutionary equivalence of expending energy on the production of one's own offspring or on those of a relative is the essential basis for this kind of selection. Crozier and Pamilo (1996) provide an excellent introduction to the voluminous literature on this and other phenomena related to the evolution of eusocial insects.

The Theory of Sex Allocation: An Overview

Each of the analyses underlying these Shaw-Mohler expressions has assumptions that are particular to each evolutionary question. One important feature common to all is that the evolutionary fate of a particular trait—say, a given proportion of females or a given time to switch between male and female reproduction—is not fixed. Instead, the fitness of the trait depends on the frequencies of all of the traits in the population. This frequency dependence is not an assumption, as it is an inherent aspect of a trait relating to the production of offspring of different mating types (e.g., males and females). In addition, in all of these analyses, one has determined what trait is stable against invasion of rare alternative traits. Therefore, the traits predicted at equilibrium by these analyses are local optima. This aspect of their biology can be demonstrated formally with a calculation of the relative fitnesses of resident and alternative traits at the equilibrium. These analyses are examples of the evolutionarily stable strategy (ESS) approach (although not all of the predictions described here are ESSs in the strict sense; see Hines 1987 for an excellent introduction to this approach to modeling and for further details about the nature of equilibria).

What Does the Theory of Sex Allocation Tell Us?

The theoretical analyses of sex allocation problems I outlined above are important simply because they make readily apparent the common aspects of the diverse biological problems being analyzed. This is no small accomplishment. Nonetheless, to understand their full importance it is also essential to carefully assess what power they have to explain facts about nature.

Consider a specific example, the sex ratio. We know that even sex ratios will evolve as predicted in experimental populations (e.g., Conover and Van Voorhees 1990; Basolo 1994). What about sex ratios in natural populations? It has often been claimed that most sex ratios are the result of the within-population evolutionary dynamic described above. Historically, almost all of these claims have related to even or near-even sex ratios in species like our own. Yet, there is potential ambiguity in this inference for two reasons. One is that the model described above predicts only the allocation ratio for the population as a whole, and how it should be applied to data relating to individuals is a matter of debate. The second reason is that many of the species with such sex ratios also have two sex chromosomes, whose 1:1 segregation during meiosis results in an even number of male and female offspring on average. To this extent, an even sex ratio might simply be at most a by-product of natural selection for "honest" meiosis (cf. Herre et al. 1987). Only in the last two decades or so has research made this claim less tenable (see below).

What about the evolutionary claims I've outlined that concern sex change and simultaneous hermaphroditism? What do they explain about nature? These phenomena have received less attention than sex ratios; what is clear in specific instances is that the direction of sex change is consistent with that predicted by observed demographic differences; that is, the direction of sex change is toward the sex with greater reproductive success when it is older (e.g., Policansky 1981). Similarly, specific instances of hermaphroditism are consistent with a pattern of diminishing returns with respect to energy investment; however, most investment ratios in hermaphroditic plants are significantly distinct from 1:1 in selfing and in outcrossed species. One particular difficulty in the context of testing explanations for the evolution of hermaphroditism is knowing how to meaningfully distinguish between the apportionment of energy into one sex as opposed to the other since they reside in the same individual.

What is generally left unexplained in regard to sex change and simultaneous hermaphroditism is their taxonomic distribution. So, for example, there are many species of fish without sex change in which it "should" occur because the sexes differ in how reproductive success changes with age. It is unclear whether such cases mainly reflect a mistaken assessment of realized patterns of energy investment (implying that the evolutionary analysis is correct, but that the prediction is mistaken due to faulty data) or reflect a mistaken analysis (one whose structure omits an important aspect of the relevant biology). On the other hand, many species of flowering plants are hermaphroditic (with perfect flowers), although energy investment does not obviously exhibit a pattern of diminishing returns. This may be due to an insufficient amount of data, as distinguishing between a pattern of diminishing returns for energy investment (see above) and, say, a linear pattern may require a substantial sample size. Both of these examples underscore the fact that the quality of data is often a real but underappreciated issue in the evolutionary analysis of sex allocation.

The theory of sex allocation for eusocial insects also has successes and failures. Sex ratios for species of ants in which a single queen is present are broadly consistent with the female bias predicted by worker control of the colony sex ratio (see Crozier and Pamilo 1996, pp. 202–204). This supports the claim that the absence of reproduction by female workers has evolved because the workers are more closely related to the sisters they raise than to their mothers. On the other hand, sex ratios in termite colonies appear to be not generally consistent with the even sex ratio predicted by theory. More generally, we lack an explanation for the evolution of eusociality in termites. Females are no more closely related to one another than they are to their mother; to this extent, it is unclear why they abstain from reproduction. It may well be that selection at the level of the colony is involved.

The Shaw-Mohler expressions that I've outlined are manifestations of a theoretical approach to understanding sex allocation whose conceptual core has remained unchanged for the last several decades, although important elaboration and refinement continue to occur. This constancy stands in marked contrast to the change in our empirical understanding of sex allocation. I illustrate this change by discussing three extraordinary phenomena in the biology of sex ratios. The dynamic nature of the empirical study of sex allocation is underscored by the fact that all were unknown or little known to most evolutionary biologists prior to the mid-1960s.

Extraordinary Sex Ratios of the First Kind

Integral to the history of the study of sex ratio evolution has been the claim that sex ratios are even or near-even in most species. It is important to understand the genesis of this claim. Many investigators in the 19th century reported sex ratios in humans and other species. As one can guess, few reports, if any, contained any statistical analysis to test the claim. Darwin's 1871 analysis is the most famous. What is essential to note is that many of the sex ratios he reported *are* biased to an astatistical eye (1871, Part 1, pp. 300–315). Indeed, his overall claim about the data is a claim for bias; he wrote (1871, Part 1, p. 265), "As far as a judgement can be formed, we may conclude from the facts given in the supplement, that the males of some few mammals, of many birds, of some fish and insects, considerably exceed in number the females." So how is it that even or near-even sex ratios came to be regarded as a general fact about nature that was in need of explanation? Of course, some of these and similar examples of bias would evaporate after being subjected to statistical scrutiny. In addition, some well-studied species, such as *Drosophila melanogaster, Homo sapiens,* and some domesticated plants and animals, were found to have even or near-even sex ratios and little or no genetic variation for this trait. In his influential 1930 argument about the evolution of the sex ratio and why energy investment in the two sexes should be even, Fisher refers to "organisms of all kinds" (p. 158), strongly implying that he regarded his argument as applying generally. These findings and claims generated a claim that sex ratio biases and sex ratio adjustment or manipulation rarely occurred in nature, if at all. In his influential book, Williams (1966, p. 151) would express as well as sustain this perspective that most sex ratios are even or near-even, by writing, "In all well-studied animals of obligate sexuality, such as man, the fruit fly, and farm animals, a sex ratio close to one is apparent at most stages of development in most populations."

Yet, it is clear that such claims never reflected all available data. After all, numerous reports of biased sex ratios appeared as early as the beginning of the 19th century. Many but not all involve "well-studied" species of Hymenoptera with "obligate sexuality" (see below). The point is to provide a cautionary lesson about the power of general claims about nature; they can describe *and* circumscribe the facts we seek to explain. At the very least, the empirical basis for a general claim about nature needs to be carefully enumerated so that the validity of such a claim can be meaningfully judged.

The plastic character of our expectations about nature is underscored by recent numerous reports of biased sex ratios and of facultative sex ratio adjustment in vertebrates. These are extraordinary sex ratios of the first kind. Consider the report by Nager et al. (1999); they experimentally changed the nutritional condition of female lesser blackbacked gulls *(Larus fuscus)* and could thereby cause them to produce more of one sex. More females were produced by poor-condition mothers

and more males by good-condition ones. There are many other reports of sex ratio biases associated with differences in nutritional or social status in species that, like this one, have sex chromosomes (see Hardy 1997 for examples and discussion). It would not be much of an overstatement to say that reports of such biased or "manipulated" sex ratios predominate in the recent literature. Of course, publication bias is a concern in this context; after all, unbiased, unmanipulated sex ratios are not regarded as newsworthy. Nonetheless, in contrast to the view expressed by Williams and many others, it is no longer reasonable to regard most sex ratios in most populations as "close to one" or to regard sex chromosomes as necessarily constraining a species to have such a sex ratio.

What remains unclear in regard to most of these putative examples of facultative sex ratio adjustment are the mechanisms underlying the adjustment. Nonetheless, plausible mechanisms exist, although there are few cases in which data clearly support one mechanism (see Krackow 1995 for discussion). To some extent, the effect of this lack seems to cause claims of facultative adjustment to be viewed skeptically by many and, if accepted, to be regarded as anomalous; the best antidote to this kind of intellectual inertia is to remember that Darwin knew nothing about proximate mechanisms when he asked (1871, Part 1, p. 316), "Can the sexes be equalized through natural selection?"

The primary evolutionary explanation of many of these examples relates to how the sexes differ in regard to mating success as their nutritional or social status changes. In particular, Trivers and Willard (1973) postulated that mothers in poor condition should produce a female-biased sex ratio, while females in good condition should produce a male-biased sex ratio. Their underlying assumption was that competition among males for mates is more intense than it is among females; to this extent, a poor-quality male may not end up mating at all, while a poor-quality female may still be mated. This explanation appears to be correct in specific instances, although in others there appears to be no association between female condition and direction of bias. In addition, as should be no surprise, there are many species with biased sex ratios for which we have no information on competition for mates. The consequence of these conflicting results is that the explanatory power of the Trivers and Willard hypothesis is unclear. However, their claim stands as an important conceptual touchstone and organizer in much current work. The following kind of extraordinary sex ratio provides an example of this influence.

Extraordinary Sex Ratios of the Second Kind

The second extraordinary phenomenon is temperature-dependent sex ratio determination. Although this occurs in many disparate taxa, perhaps the most dramatic examples occur in the Reptilia (for further details see Janzen and Paukstis 1991; Rhen and Crews, this volume). For most species studied thus far, mixed sex broods are produced over a small range of intermediate temperatures, unisexual broods of "opposite" kinds being produced on opposite sides of this transition zone. A number of explanations have been proposed; most investigators, being influenced by Trivers and Willard's hypothesis, posit differences in the way the fitness of each sex is affected by temperature, the idea being that, say, a female might benefit more than a male from emerging at a cold temperature because her resulting increased body size at sexual maturity affords her a greater increase in fecundity than a similar increase in body size of a male would increase his fitness. Such explanations have been partially confirmed in a number of species, but there is no case in which there are reasonably complete data, especially in regard to the putative difference between the sexes in regard to the effects of temperature on fitness. We lack a general evolutionary understanding of this phenomenon (see Shine 1999 for further details).

Extraordinary Sex Ratios of the Third Kind

The last extraordinary phenomenon is the female bias in the sex ratios of many species of insects, especially among the Hymenoptera. There are many species in which males constitute a few percent or less of individuals sampled. Such sex ratios have been noted and commented on since the 19th century, if not earlier.

What could explain such an extreme bias? In the Hymenoptera, most of which are haplodiploid (see above), a common proximate explanation is that females control the sex ratio by the release of stored sperm as a haploid egg is about to be oviposited; unfertilized eggs develop into males (apparently because of being stretched during oviposi-

tion), whereas fertilized eggs develop into females. While this mechanism very likely underlies most female-biased sex ratios in Hymenoptera, unmated females in some species are known to produce such sex ratios as well (see Orzack 2002 for further details). For this reason, controlled observations are needed for assessment of the proximate basis for the sex ratio of any species under investigation.

For species in which each individual can reproduce on its own and in which there is no necessary social behavior involved in reproduction or rearing of offspring (as discussed above), the most famous evolutionary explanation is that of W. D. Hamilton. In 1967, he defined (p. 481) a "biofacies of extreme inbreeding and arrhenotoky [haplodiploid determination of offspring sex]" in which a finite, usually small fixed number of females (often assumed to be 10 or fewer) oviposit together, leaving offspring that mate among themselves. Males remain, while mated females depart and distribute themselves randomly into new groups of ovipositors, and the cycle is repeated. This local mate competition (LMC) model predicts that the optimal sex ratio will be female-biased to a degree dependent on the number of females ovipositing together. The presumption is that a female adjusts her sex ratio so that her daughters' and sons's matings in the local mating group will maximize the number of their (mother's) genes transmitted into the next generation. Since a male can inseminate more than one female, it is evolutionarily advantageous to produce more females than males when all mating is local and only mated females emigrate to find new resources and reproduce.

Since the appearance of Hamilton's paper, the theory of LMC has been extensively refined; so, for example, we now know the effects on sex ratio optima of variable foundress number, of emigration of males from the natal patch, of relatedness of females ovipositing together, and of different kinds of genetic determinants of sex ratio behaviors (see Orzack 2002 for references to these theoretical analyses). It is also clear that population structure (of which LMC is one kind) is necessary but not sufficient for the evolution of a female-biased sex ratio; the sexes must also differ ecologically (as when females have a greater tendency than males to emigrate from their mating site). As a whole, this body of theory is one of the most important in evolutionary biology, especially as there has been extensive analysis of how changes in model assumptions result in changes in model predictions. One of the most important insights to be gained from this work is that sex ratio evolution with LMC involves a balance between group selection favoring a biased sex ratio and individual selection within groups favoring an even sex ratio.

Is Hamilton's LMC hypothesis a correct explanation for most female-biased sex ratios? Unfortunately, we lack sufficient data to answer this question. Some comparative observations lend some support to this hypothesis. So, for example, in some cases, species that may have less subdivided populations have less female-biased sex ratios than those of species that may be more subdivided. This result is important, but not compelling, mainly because we lack any rigorous ecological or genetic assessment of population subdivision for most species. A second kind of support for the hypothesis is that some species, such as the fig-pollinating wasps discussed by Herre et al. (1997), have an ecology that strongly implies that LMC occurs and that the females of these species produce female-biased sex ratios on average. This result supports Hamilton's claim, although it does not necessarily rule out other causes for the observed sex ratios; some recent work suggests that the behavioral microcosm Hamilton described is not present in some fig wasp species with female-biased sex ratios (so, for example, mating may not be local in that males may leave their natal fig in order to mate elsewhere; see Orzack 2002 for references). Nonetheless, the work of Herre and his colleagues is an important support for Hamilton's claim that biased sex ratios do evolve in response to population subdivision. What is at issue is whether their results are typical of what one would find in other species with biased sex ratios.

The most intricate experimental studies of sex ratio traits involve the parasitic wasp, *Nasonia vitripennis,* a parasite of filth flies. Limited ecological and genetic analyses imply that population may often be subdivided and that females may disperse more widely than males. To this extent, this species may possess something like the "biofacies of extreme inbreeding" postulated by Hamilton. Do females produce the female-biased sex ratios one expects in such a circumstance? This question has been extensively investigated, most intensively by Orzack et al. (1991), who used 12 isofemale strains sampled from localities in Sweden in order to study the "conditional sex ratio," the set of sex ratios produced by an individual female when ovipositing with varying numbers of other females over the

course of her life. Identification of the "target" female's offspring was made possible by the use of genetic markers. This is one of a few sex ratio studies, and the only study of this behavior in particular in which such identification is possible; all other analyses, such as of the fig wasps mentioned above, have focused on the sex ratio of the group of offspring produced by a set of females (see Orzack 2002 for references). While such observations are important, they do not relate directly to the predictions of the model. Some of the results of these experiments are shown in figure 13.1, which depicts the relationship between the average sex ratio produced by target females of an isofemale strain and the size of the group in which the female oviposited. What is striking is the variety of patterns. Strain MS1 has the expected pattern to the extent the average sex ratio increases monotonically as group size increases. Strain MS37 exhibits no apparent relationship between sex ratio and group size, whereas strain MS56 exhibits a relationship consistent with females knowing when they are alone but not being able to accurately assess group size. Finally, MS71 exhibits a nonmonotonic relationship between sex ratio and group size. Visual comparison of these patterns serves a heuristic purpose; one might infer that there has been natural selection to assess group size at least in an average sense, but variation occurs because the strength of selection for this ability has been quite weak. However, this inference is not deeply compelling, especially since one can fool oneself into thinking that a diversity of patterns exists, when, in fact, there is only one underlying pattern that has been distorted by experimental and sampling errors. To this extent, a more satisfactory way to assess the relationship between sex ratio and group size is via statistical analysis. The specific predictions tested are those of Frank (1985) and Herre (1985), who analyzed an ESS model of this behavior. In particular, they showed that

$$r_N^* = \frac{1+F}{1+2F} \frac{N-1}{2N} \tag{13.5}$$

where r_N^* is the optimal proportion of males produced by a female in a group of females of total size N, and F is the inbreeding coefficient of the population. This equation is based on particular but plausible assumptions about the control of sex ratio by the female, about the relative information that females in the group possess, about mating

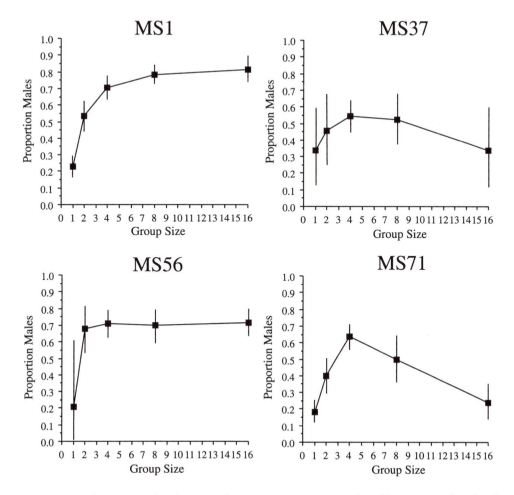

Figure 13.1 The relationship between the average sex ratio produced by a target female of an isofemale strain and the size of the group in which the female oviposited. Strains MS1, MS37, MS56, and MS71 are derived from a population in Sweden; 95% confidence intervals are shown. See Orzack et al. (1991) for further details.

among offspring within groups, and about the settling of females into groups.

A statistical assessment of the accuracy of model predictions is shown in figure 13.2, which depicts the results of a hierarchical logistic regression analysis. Such an analysis is well suited to this purpose, both because with logistic regression one can account for the two-state nature (male, female) of the trait and because its hierarchical aspect allows one to assess whether there is a significant overall relationship between model predictions and data and whether there is also strain specificity in this relationship. Adding the ESS sex ratios as an overall predictor (βx) in the model indicates that it

successfully predicts the overall trend in the data (e.g., $C + \alpha_i$ vs. $C + \alpha_i + \beta x$, $\chi^2 = 581.05$, 1 d.f., $P \leq .01$). Allowing for strain specificity when adding the ESS sex ratios as a predictor ($(\beta + \gamma_i)x$) in the model indicates that the strains differ in their fit to model predictions (e.g., $C + \alpha_i + \beta x$ vs. $C + \alpha_i + (\beta + \gamma_i)x$, $\chi^2 = 57.60$, 1 df, $P \leq .05$). Examination of the strain-specific regression coefficients $(\beta + \gamma_i)$ and their standard errors indicates that some are significant, while others are not (not shown). The results of this analysis are consistent with the claim that natural selection has caused the average sex ratios to move toward the optimal ones; the strength of this selective force appears to be weak,

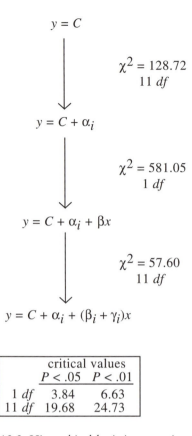

$$y = C$$

$$\chi^2 = 128.72$$
$$11\ df$$

$$y = C + \alpha_i$$

$$\chi^2 = 581.05$$
$$1\ df$$

$$y = C + \alpha_i + \beta x$$

$$\chi^2 = 57.60$$
$$11\ df$$

$$y = C + \alpha_i + (\beta_i + \gamma_i)x$$

	critical values	
	$P < .05$	$P < .01$
1 df	3.84	6.63
11 df	19.68	24.73

Figure 13.2 Hierarchical logistic regression analysis of the relationship between predicted *(x)* and observed *(y)* conditional sex ratios. The inbreeding coefficient of the population is assumed to be 0.50. C and α_i denote overall and strain-specific y intercepts. β and γ_i denote overall and strain-specific regression coefficients. See Orzack et al. (1991) for further details.

at least in the recent past. Another possible explanation for the heterogeneity of behaviors is that "genetics has gotten in the way"; that is, the nature of the underlying loci is such that it is impossible for the optimal traits to be expressed or to be true-breeding. Whatever the explanation, these analyses have yielded important and unique insights into the evolution of conditional sex ratios in this species.

In no sense does the mixed nature of the support for Hamilton's claim detract from its heuristic importance. Indeed, it is hard to overestimate the importance of this paper. It established population subdivision as an important topic in the study of sex ratio evolution and laid the groundwork for analysis of a broad spectrum of other important topics in this area, including the effects of resource quality (such as Trivers and Willard's work). Although all three kinds of extraordinary sex ratios that have come to prominence since the 1960s have markedly changed the study of sex allocation, the one garnering the most attention has been the female-biased sex ratios in many species of Hymenoptera, and the most influential claim about this phenomenon has been Hamilton's.

How Do We Know about Sex Allocation?

One aspect to consider is how we come to determine the facts about sex allocation that are taken to be in need of evolutionary explanation. One "fact" of central conceptual importance that is worth considering in some detail is the idealization that there is *a* sex allocation for a given population. Yet, we know that populations wax and wane and males and females are born and die. What this means is that the notion that there is, say, *a* sex ratio is always false. Instead, the sex ratio measured at one time and place will differ from the sex ratio measured at another time and place (cf. Clausen 1939, p. 1). In many instances, the falsehood associated with the assumption that there is a sex ratio that is stable in a spatial and temporal sense may be evolutionarily unimportant, but this is not a priori true. The point is that all empirical claims about sex ratios are claims with implicit spatial and temporal limits. Yet, this fact has often been obscured, the reason being that theory has greatly influenced empirical expectations; theorists often allude to *the* sex ratio of this or that hypothetical population. There are positive and negative consequences of this tendency. The positive consequence is that the progress in our theoretical understanding of sex allocation is striking. The negative consequence is that the true nature of many sex ratio data has tended to become obscured. Although any field of science may have a conceptual framework informed more by theory than by data, I believe that the future promise of sex allocation studies will be best realized by greater integration of theory and data, along with a clearer understanding of the extent, validity, and nature of the phenomenon being explained. Only in this way can we meaningfully distinguish between what is explained and what remains a mystery.

Future Directions—Empirical and Theoretical Issues to Be Explored

The study of sex allocation has exemplified the adaptationist approach to evolutionary explanation. What this means is that evolutionary explanations of sex allocation traits invoke only natural selection as an important causal force; explanations that assign an important causal role to developmental or phylogenetic constraints or nonselective forces such as genetic drift have generally been ignored (cf. Orzack and Sober 1994). The evolutionary insights I've discussed above are testimony to the power of this approach. However, it is worth recalling that explanations invoking the primacy of natural selection do not have an inherent precedence, even in the context of sex allocation. Although sex allocation traits involve numbers of individuals, the currency of fitness, such traits can certainly evolve via nonselective means. For example, selection for an optimal sex ratio may be very weak in some instances since the fitness surface surrounding the sex ratio can be almost flat (see Orzack et al. 1991 for illustrations); this implies that the frequencies of different sex ratio traits could be affected by genetic drift.

Nonetheless, it is plausible that one need invoke only natural selection when providing an evolutionary explanation for a sex allocation trait. What does this mean in practice? There must be some substantive biological framework for any evolutionary analysis. Such a framework involves assuming that some part of the biology is, in effect, held fixed while another part is free to evolve. An example would be a model of energy allocation in which the total amount of energy allocated is fixed but the allocation toward male and female function can evolve. There is no necessary conflict between this kind of constraint and the possible evolution of an optimal allocation ratio.

It is exactly this central aspect of present sex allocation studies, that natural selection plays the only important causal role in a trait's evolution, that makes this field such an important one in regard to providing a strong test of the important claims that phylogenetic inertia plays a substantial role in trait evolution.

Just as the empirical study of sex allocation has changed dramatically in the last three decades, so, too, has the historical study of adaptation changed dramatically during that time. The rise of cladistics and of claims that the testing of adaptive hypotheses with comparative data must account for the phylogenetic relationships of the species involved have markedly changed the way in which adaptive hypotheses are tested (see Harvey and Pagel 1991 for further details). The basic claim underlying this change is that traits of related species may not count as independent evidence for or against a particular adaptive hypothesis. A descendant species may have simply inherited the trait from an ancestral species; in this way, phylogenetic "inertia" is said to explain the presence of the trait.

These developments and the empirical study of sex allocation have largely transpired independently of one another. Accordingly, many current sex allocation analyses involve the comparison of predictions and data from only one species or population. More important, when many species have been studied, their phylogenetic relationships have usually been ignored. To this extent, these analyses have at least implicitly (and sometimes explicitly) assumed that phylogenetic inertia is not involved in the trait's evolution.

The tendency toward such an explanation is not a matter of parochialism. If natural selection *is* as important to the evolution of sex allocation traits as has been traditionally assumed, inertia will matter little in the context of understanding the distribution of traits across taxa. Assessing whether this is true should be one focus of the study of sex allocation. One way to do this is to compare the results of comparative tests of adaptive hypotheses in which putative phylogenetic dependencies have been accounted for with the results of tests in which such dependencies have been ignored. Whatever the results of such comparisons, it is clear that they constitute an important test case whose insights "cut both ways" and should benefit both the study of sex allocation and the historical study of adaptation.

Just as we require such investigations of long-term evolutionary dynamics, so, too, is there a need to investigate shorter-term evolutionary dynamics. Most empirical analyses of sex allocation traits rely on the assumption that populations are at evolutionary equilibrium. The use of ESS models has reinforced the tendency to make this assumption. They have proved a boon in terms of advancing our conceptual understanding, but the larger question as to whether evolutionary equilibrium is usually an appropriate assumption to make still remains. More particularly, the question as to whether it is appropriate to focus on near-monomorphic equilibria remains to be answered.

The answer to these questions will come only with future work that involves assessment of the temporal dynamics of sex allocation traits and of the quantification of within- and between-population trait variation (such as the work of Orzack et al. 1991 described above). Even limited amounts of data—say, time series involving just 10–20 years of data—would be an important advance in terms of assessing the validity of assuming that most sex allocation traits are at or near evolutionary equilibrium. What should be paramount in this general context is a focus on what "is" as opposed to what "must be." ESS models and optimality models more generally have been attacked by those who plausibly claim that populations "must be" genetically polymorphic, that genetic constraints "must get in the way" of adaptation, that temporal variability "must" stymie the evolutionary attainment of optima, and so forth. On the other hand, proponents of such models have made plausible claims about how the world "must be" that are counter to every one of these claims. This debate about adaptationism is a central one in evolutionary biology. It is also a debate that can be resolved; what is at the base of such a resolution is an ensemble of tests, each using a specific data set to assess the qualitative and quantitative accuracy of an optimality model. Central to this assessment is an investigation of the nature and extent of trait variation. In this sense, the quantification of temporal and spatial variation in sex allocation traits has significance that extends to the field of evolutionary biology as a whole (Orzack and Sober 1994 provide more detail about the debate over

adaptationism, how it can be resolved, and how tests of specific optimality models are involved.)

The degree to which evolutionary equilibria are meaningful motivates future theoretical work as well. By definition, ESS analyses concern trait stasis, since they define a trait resistant to evolutionary change. To this extent, they provide limited inferences about evolutionary transitions, especially on the long-term time scales. We need more theoretical work to elucidate the dynamics of these near-monomorphic systems and of truly polymorphic systems as well (cf. Eshel and Feldman 2001). This should include theoretical work to elucidate how the evolution of allocation patterns may affect the long-term fate of the population. For example, at present, we know little empirically or theoretically about how the evolution of sex allocation may affect extinction dynamics of populations. Yet, such an interaction may be important in terms of explaining larger-scale patterns in the evolution of sex allocation across taxa.

The Future of Sex Allocation Studies

I hope I have clearly described the advances in the study of sex allocation and *why* they are important. Our future challenge is to make good on the achievement of such studies by using them as motivation for new explorations of the biology of sex allocation. Surprises await us.

Acknowledgments   I would like to thank the editors of this volume for providing the difficult challenge of writing this chapter and for their forbearance, support, and comments. E. Leigh provided useful comments as well. Work on this paper was partially supported by NSF grants DEB 9407965 and SES 9906997.

Note

1. Characterized by random mating among all members of a population.

14

Ecological Specialization and Generalization

DOUGLAS J. FUTUYMA

Anyone who is even slightly acquainted with plants or animals knows that different species inhabit different parts of the world, live in different habitats, and, in the case of animals, eat some imaginable kinds of food and not others. As with many other familiar facts, it may not occur to us to ask why the geographic and ecological ranges of species are limited, until we realize that species vary drastically in their geographic, ecological, and physiological amplitudes.

Examples

The bracken fern *(Pteridium aquilinum)* is broadly distributed in temperate climates of every continent (except Antarctica), whereas the curly-grass fern *(Schizaea pusilla)* is limited to parts of eastern Canada and central New Jersey in the United States. The black-billed magpie *(Pica pica)* is a familiar bird from western Europe through eastern Asia and from Alaska to the Great Plains of North America, but the very similar yellow-billed magpie *(Pica nuttalli)* is restricted to central California. What accounts for the much narrower distribution of one than the other species?

Related species often differ in the variety of habitats they occupy. The thistle *Cirsium canescens* is restricted to well-drained sandhills in the American prairie, whereas *Cirsium arvense* is a European species that has become a rampant weed in North America, growing in many types of soil. The endangered Kirtland's warbler *(Dendroica kirtlandii)* nests only in stands of jack pine of a certain age, while its relatives, such as the yellow warbler *(Dendroica aestiva)*, nest in many types of vegetation and have far broader geographic ranges as well. (Species with narrow and broad habitat associations are referred to as *stenotopic* and *eurytopic*, respectively.)

Stenotopic species or populations frequently have a narrower tolerance of certain physical variables than do others. Most plants and animals from warm tropical environments cannot survive freezing temperatures, and Antarctic notothenioid fishes cannot tolerate temperatures above 6°C. In contrast, species that inhabit environments where the temperature varies widely often have broad temperature tolerance. In many such species, individuals are capable of biochemical and physiological alterations that *acclimate* them to pronounced changes in temperature.

The diet of an animal may be broad *(polyphagous)*, as is true of humans, raccoons, and insects, such as migratory locusts, that eat a great variety of plants. *Oligophagous* species, in contrast, have narrow diets. The Colorado potato beetle *(Leptinotarsa decemlineata)* will not eat anything except certain species of *Solanum* (including potato), and many other insects are restricted to a single species of host plant. Parasites and symbiotic mutualists are often extremely host-specific. *Plasmodium fal-*

ciparum, a sporozoan that causes malaria, must spend part of its life cycle in human red blood cells; the mite *Macrocheles rettenmeyeri* lives attached to the feet of the army ant *Eciton dulcius;* almost every one of the approximately 700 species of figs *(Ficus)* is pollinated by a single species of small wasp (Agaonidae) that develops only in the flowers of that species of fig.

Specialization, Generalization, and Ecological Niches

Compared to a generalist, a specialist species has a narrower tolerance of one or more abiotic variables (such as temperature) or uses a narrower spectrum of resources (such as food or nesting sites). Due to geographic variation, a species as a whole may have wider tolerance or span than any one local population. A species may be more specialized than another in some respects but not others; for example, the Kirtland's warbler has much the same diet of insects as other warblers, even though it is more exacting in its nest site. Thus, the terms *specialized* and *generalized* should be applied not to species, but to particular ecological attributes of species: particular axes of their ecological niches (see the section on ecological niches below).

Implications

Ecological specialization and generalization have profound implications for both evolution and ecology. (Useful reviews of parts of this subject include Futuyma and Moreno 1988; Jaenike 1990; Hoffmann and Parsons 1991.) The nature of coevolution among interacting species depends on how specific their interactions are (Thompson 1994). Broad or narrow tolerances affect the geographical distributions of species and thus influence geographical patterns of diversity. Potentially competing species are more likely to coexist if they specialize on different resources, and indeed, the biodiversity of the world stems largely from the astonishing variety of organisms' specializations. But at the same time, specialization can make a species vulnerable to extinction when environments change, resulting in a loss of biodiversity.

Questions

The chief question we will address is what factors may account for the evolution of broader or nar-rower tolerance or resource use. A related question is why the relative numbers of specialized or generalized species differ among clades.

This topic encompasses physiological tolerance, morphology, and behavior and their effects on a species' ecological and geographical distribution. It would be surprising if a single, simple hypothesis were to explain so many characteristics, in all organisms, so several hypotheses will claim our attention, certain of which are perhaps more applicable to some features and to some organisms than to others.

History

Paleontology and Morphology

From the late 19th century through the period of the "modern synthesis" of evolutionary theory in the 1930s and 1940s, specialization was discussed mostly by paleontologists, who often used the term to refer to morphological departure from an ancestral ground plan. Many, such as Edward Drinker Cope and Henry Fairfield Osborn, held that organisms tend to evolve from generalized to more specialized states, and that specialized forms have had a higher rate of extinction, because they were incapable of adapting to environmental changes. Hence, Cope proposed, in his "law of the unspecialized," that generalists have survived longer and have been the progenitors of new clades with new adaptations. Some paleontologists, rejecting Charles Darwin's theory of natural selection, believed that an intrinsic evolutionary momentum caused some species to become "overspecialized" and virtually predestined for extinction. During the modern synthesis, however, biologists such as Ivan Schmalhausen argued that specialization evolves by natural selection, because it increases efficiency and provides a competitive advantage. George Gaylord Simpson, Ernst Mayr, and others argued that specialized organisms have been the ancestors of some major adaptive radiations. For example, legless, burrowing lizards seem less versatile than their limbed relatives, yet one lineage of burrowing lizards gave rise to snakes, which occupy an astonishing diversity of ecological niches.

Physiology and Ecology

In modern usage, *specialization* and *generalization* refer to relative breadth of physiological tolerances or resource use. The field of ecological physiology, which took shape in the 1940s, includes studies of the mechanisms of acclimation and other features that enable some organisms to tolerate a wider environmental range than others. Ecological physiology has given rise to evolutionary physiology (Feder et al. 2000), which includes phylogenetic and genetic studies of physiological tolerances and capacities.

Modern interest in specialization and generalization has grown mostly from population and community ecology. In the 1950s, zoologists such as David Lack began to show differences in resource use (especially food) between closely related species (e.g., Darwin's finches in the Galápagos Islands) and proposed that the species had evolved differences to reduce interspecific competition. In 1957, G. Evelyn Hutchinson proposed a definition of ecological niche, according to which, theoretically, resource-limited species can coexist only if their ecological niches do not entirely overlap. In the 1960s, Hutchinson's student Robert H. MacArthur provided influential theoretical and empirical studies of the niche relationships of competing species, and at the same time, the mathematical biologist Richard Levins developed a theory of the conditions under which broad versus narrow niches might evolve. Levins and MacArthur jointly developed a theory of the evolution of resource use among multiple species, in which the necessity for different specializations of coexisting species was prominent. Subsequently, the evolution of niche breadth became a topic to which both population ecologists and population geneticists made further contributions.

Concepts and Hypotheses

Ecological Niches

G. E. Hutchinson provided a formal definition of the *fundamental niche* of a population (or species): It is the set of environmental conditions and resources within which the population can sustain a positive growth rate. The niche is multidimensional. A species of clam, for instance, may be able to survive and reproduce within a certain range of temperature and within a certain range of salinity, and to be capable of feeding on a certain range of food particles. These three variables are three axes of the niche of the species. Mathematically, we can

add other dimensions to this niche (copper concentration, bottom type, etc.), although we cannot envision a more-than-three-dimensional space. The *niche breadth* (or width) of a population refers to the amplitude of its tolerance along one or more dimensions. Much ecological literature discusses niche width, as well as niche overlap of competing species or genotypes, with respect to a single dimension, such as food (e.g., the sizes of insects or seeds consumed by lizards or birds).

The *realized niche* of a species, the set of environments that it actually occupies, may be smaller than the fundamental niche, either because the species has not colonized places where those conditions prevail or because it has been excluded from such places by competing species or predators. For instance, competition imposed by the barnacle *Balanus balanoides* restricts another barnacle, *Chthamalus stellatus,* to the higher levels of the intertidal zone; *Chthamalus* thrives lower in the intertidal zone if *Balanus* is experimentally removed (Connell 1961).

We are concerned with the evolution of the fundamental niche. Genotypes may differ in their location (e.g., best temperature for growth) or their breadth along one or more niche axes. Therefore, a population's niche can evolve. There are, of course, limits to such evolution, for the characteristics bequeathed to an organism by its phylogenetic history may make the evolution of some conceivable adaptations so improbable as to be effectively impossible. It is hard to imagine a jellyfish expanding its habitat to include land, for lack of any supporting structures and many other prerequisites.

What We Need to Explain

Often, related species collectively span a range of conditions or resources, but each species occupies only part of this span. For example, bark beetles in the genus *Dendroctonus* feed on conifers in North America and Europe. From a phylogeny based on mitochondrial DNA sequences (figure 14.1), Kelley and Farrell (1998) inferred that *Dendroctonus* ancestrally fed on *Pinus* (pine) and that one lineage gave rise to species that switched to *Picea* (spruce), *Larix* (larch), or *Pseudotsuga* (douglas-fir). Most species are relatively generalized, feeding on 5–12 species of *Pinus*. Six species, designated as specialists by Kelley and Farrell, feed on, at most, three congeneric species of plants. All of these specialists are at separate tips of the phylogeny, a fact implying that they evolved independently from more

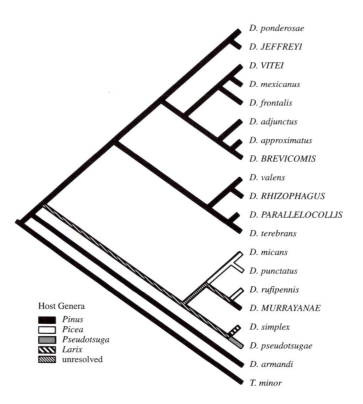

D. ponderosae
D. JEFFREYI
D. VITEI
D. mexicanus
D. frontalis
D. adjunctus
D. approximatus
D. BREVICOMIS
D. valens
D. RHIZOPHAGUS
D. PARALLELOCOLLIS
D. terebrans
D. micans
D. punctatus
D. rufipennis
D. MURRAYANAE
D. simplex
D. pseudotsugae
D. armandi
T. minor

Host Genera
Pinus
Picea
Pseudotsuga
Larix
unresolved

Figure 14.1 A phylogeny of the bark beetle genus *Dendroctonus*, based on DNA sequence of a mitochondrial gene. The most parsimonious reconstruction of the history of host associations is indicated by crosshatching or coloring of the branches. The names of the host-specialized species are in uppercase letters. The specialized habit seems to have arisen independently in each species and relatively recently, since no clade of more than one species is specialized. This pattern also suggests that specialists have had higher rates of extinction than generalists. (After Kelley and Farrell 1998.)

generalized ancestors. Most of the specialists occur in regions where the host plants of their close generalist relatives grow; nevertheless, they do not feed on them.

If we were to find that these insects could not develop on Australian conifers of the genus *Araucaria*, we would not be surprised, because they have not experienced selection for adaptation to this plant. But why do no species of *Dendroctonus* feed on firs (*Abies*), which are abundant conifers within these beetles' range? Perhaps selection has never favored adaptation to *Abies* (an untestable hypothesis), or perhaps, the genetic variation required for adaptation to *Abies* is lacking in *Dendroctonus* populations. (This possibility might be explored by screening for relevant genetic varia-

tion, as Futuyma et al. 1995 have done with other host-specific species of beetles.) Next, we note that no species of *Dendroctonus* feeds on more than one genus of conifer. The immediate common ancestor of *D. simplex* and *D. pseudotsugae,* which feed, respectively, on *Larix* and *Pseudotsuga,* presumably fed on one or the other of these plants, or on both. Whichever it fed on, one or both of the descendant species lost the habit of feeding on a plant to which its ancestor had presumably been adapted.

Three patterns in the evolution of diet have been revealed by phylogenetic studies. In some instances (although not in *Dendroctonus*), a specialist lineage gives rise to generalist species. In other cases, generalists give rise to specialists (as in *Den-*

droctonus). In yet other cases, a lineage specialized for one resource gives rise to a descendant specialized for a different resource (as is likely for the *Dendroctonus* species that feed on *Larix* and *Pseudotsuga*). The chief puzzle in the evolution of niche width is posed by the latter two cases: Why does a lineage become specialized, losing a past adaptive capacity in the process? And why, if one species can adapt to a resource or environment, have closely related species, with presumably much the same genetic potential, not done so? The advantages of generalization seem self-evident: In a variable environment, a generalist genotype can use alternative resources or tolerate variable conditions. The loss of such flexibility, or the failure to evolve it in a manifestly variable world, requires explanation.

Major Hypotheses

Hypotheses for the evolution of specialization are often framed as models in which there exist two "habitats," 1 and 2, which may represent different temperatures, soil types, plant species on which an animal feeds, or the like. Results of such models generally apply also to evolution in multiple "habitats." A population may consist of a specialist genotype, restricted to one of the habitats, or a single generalist genotype that can occupy both; or the population could be polymorphic for genotypes that specialize on each habitat.

Two major classes of hypotheses have been advanced to account for why a specialist genotype may replace a generalist: (1) The generalist has a selective disadvantage, because of a *trade-off* or *cost*; that is, the advantage of a genotype in one context is accompanied by a disadvantage in another context. (2) The generalist is selectively *neutral* and so is not maintained by selection. That is, the ability of the population to use one of the two habitats is eroded by mutation and genetic drift. The trade-off hypothesis is the traditional one and has several elaborations and ramifications.

Trade-Off Hypotheses

Levins (1968) first elaborated a theory of niche breadth based on trade-offs. Assume that genotypes differ in a phenotypic character that affects fitness. It might be an enzyme, the activity of which declines as temperature departs from an optimum, or it might be the size of a bird's beak, which presumably handles one kind of food best,

such as seeds of a certain size, and is less effective the more a food item deviates from this optimum. Variant enzymes or beak sizes have different optima, so a plot of the genotypes' fitness values in two different habitats would show a generally negative correlation between fitness in habitat 1 and in habitat 2 (figure 14.2). Some genotypes might have broader temperature tolerance than others, or broader seed-handling facility, but Levins assumed that for each habitat (i.e., temperature or seed size), there exists a possible specialized genotype that has higher fitness than the generalist, perhaps due to its greater efficiency. For example, if g is a generalized genotype, a is a specialist in habitat 1, and b is a specialist in habitat 2, the fitnesses *(W)* of these genetically different phenotypes might be

In habitat 1: $W_a > W_g > W_b$
In habitat 2: $W_b > W_g > W_a$

Levins proposed that the outcome of evolution would depend on the frequencies (c_1, c_2) of habitats 1 and 2, on just how inferior the generalist is to the specialists in their optimal habitats, and on whether the members of the population experience the two habitats as *fine-grained* or *coarse-grained* variation. If the habitat is fine-grained, most individuals experience both habitats during their lifetime, in roughly the same frequencies (c_1,c_2) as they occur in. (For example, an antelope may eat two plant species as a matter of course.) If the environment is coarse-grained, an individual experiences only one of the two habitats during its lifetime. (For instance, a leaf-mining caterpillar completes its development within a single leaf of one of the two plant species that the antelope may browse—showing that the grain of the environment is determined by the biology of the organism that experiences it.)

Levins assumed that at equilibrium, the population would attain the genetic composition that maximizes mean fitness. Using a graphical method (figure 14.3A) to solve for the equilibrium composition, he concluded that (1) a single specialist should prevail if one habitat type were more common and if the specialist's fitness in that habitat greatly exceeds that of the generalist (figure 14.3B); (2) a generalist should prevail if it has broad tolerance, that is, fitness not much less than that of the specialists in each habitat (figure 14.3C); and (3) the population should be polymorphic for the two specialists if their fitnesses greatly exceed the gen-

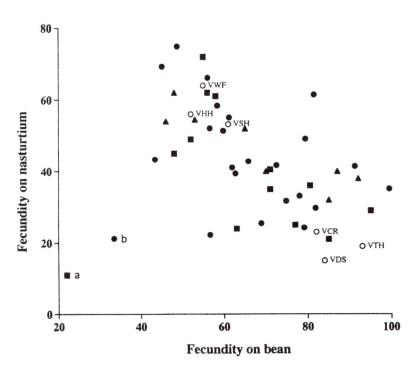

Figure 14.2 A pattern indicative of a negative genetic correlation in fitness in two environments, providing evidence for a trade-off. Each point indicates the fecundity of a cloned genotype of the aphid *Aphis fabae* in two environments, bean and nasturtium. The genotypes labelled *a* and *b* have low "general vigor" in both environments; if there were many such genotypes, they could mask the trade-off that is evident in this figure. The genotypes represented by solid points in this figure were not collected directly from a natural population; they are F_2 genotypes from crosses between three field-collected clones that had high fitness on nasturtium and three that had high fitness on bean. These are indicated by the groups of three open, labeled circles. (From Mackenzie 1996.)

eralist's in their respective habitats and if the environment is coarse-grained (figure 14.3D).

Population geneticists have used explicit genetic models to explore this topic. Howard Levene (1953) provided the first such model, of two alleles at a single locus. Survivors from two habitats mate at random, their offspring are distributed at random between the habitats, and the survival of the offspring determines their relative fitnesses:

	A_1A_1	A_1A_2	A_2A_2
In habitat 1:	W_1	1	V_1
In habitat 2:	W_2	1	V_2

Levene showed that a stable "niche polymorphism" will result if the harmonic mean[1] fitness

of both A_1A_1 ($1/[c_1/W_1 + c_2/W_2]$) and of A_2A_2 ($1/[c_1/V_1 + c_2/V_2]$) is less than 1 (the fitness of the heterozygote). (Note that a sexual population cannot consist entirely of heterozygotes—the simplest example of how genetic factors can prevent a population from attaining an optimal adaptive composition.) Otherwise, the advantage of one homozygote in its habitat so outweighs its disadvantage in the other that this homozygous specialist becomes fixed. In fact, polymorphism is stable only within a narrow range of fitnesses and habitat frequencies.

This model is based on *antagonistic pleiotropy*, in which a gene affects two (or more) characteristics in opposite ways (see Tatar, this volume). In this case, the two characters are fitness in habitat 1 and in habitat 2. Antagonistic pleiotropy is a ge-

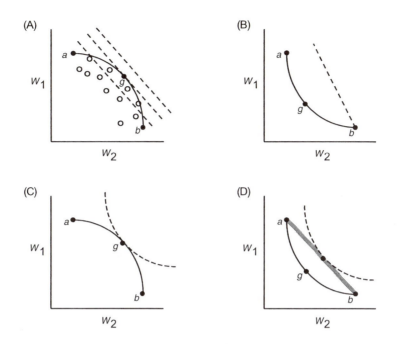

Figure 14.3 The Levins model of optimal composition of a population in a variable environment. (A) In this graph, as in figure 14.2, the fitness of each genotype (solid or open circle) in habitat 1 (W_1) is plotted against its fitness in habitat 2 (W_2). This set of points is the "fitness set." Genotypes a and b ("specialists") are the fittest genotypes in habitats 1 and 2, respectively. Genotypes such as g, with roughly equal fitness in both habitats, are "generalists." The fittest genotypes lie along the outer border (curved line) of the fitness set; interior genotypes (open circles) are less fit and need not be considered. The fitness set is convex in this case, because generalists (g) are only slightly less fit than the specialists in each habitat. The broken lines represent the "adaptive function" $A = c_1w_1 + c_2w_2$ (plotted as $w_1 = [A - c_2w_2]/c_1$), where c_1 and c_2 are the frequencies of habitats 1 and 2, w_1 and w_2 are the mean fitness values in habitats 1 and 2 of a population composed of one or more specified genotypes, and A is the overall fitness of the population. The slope of the adaptive function depends on c_1 and c_2. This adaptive function describes fitness in a fine-grained environment, in which most individuals experience both habitats. The point on the fitness set to which the adaptive function is tangent maximizes A and therefore is the "optimal" population composition. In this case, a generalist genotype (g) is optimal. If the environment consisted only of habitat 1, the adaptive function would be horizontal (with slope 0) and the specialist a would be the optimal genotype. (B) The fitness set is concave if generalists have much lower fitness in each habitat than the specialists a and b. The adaptive function for a fine-grained environment is tangent to the fitness set at either point a or point b, depending on its slope, so a specialist genotype is optimal. (C) In a coarse-grained environment, in which individuals experience one or the other habitat, an adaptive function $A = w_1^{c_1}w_2^{c_2}$ may be defined (broken curve). Unless c_1 or c_2 equals 1.0, the adaptive function is tangent to the fitness set at a point representing a somewhat generalized genotype such as g. (D) The gray line represents the set of populations polymorphic for two specialists, a and b; points along the line closer to a are populations in which the frequency of a is higher. If the fitness set is concave and the environmental variation is coarse-grained, the optimal population (point of tangency) is polymorphic for the two specialist genotypes. Such a composition is stable if the genotypes reproduce asexually, but if genotypes a, g, and b are sexual genotypes A_1A_1, A_1A_2, and A_2A_2 at a single locus, the polymorphism is unstable, and one of the specialist genotypes will be fixed. Thus, genetic processes may prevent a population from attaining its adaptive optimum.

profitable type of prey (that yielding the highest income of energy per unit time), and to take less profitable items only if the most profitable prey is not abundant enough to satisfy an individual's requirements (Krebs and Davies 1987). This suggests that a broader diet or habitat use should evolve in resource-limited species than in species that are limited by other factors.

Costs of *information processing* have been suggested as an advantage of feeding specialization in animals (Bernays and Wcislo 1994). An animal may find its optimal prey more effectively if it pays "selective attention" to stimuli from that prey and "screens out" stimuli from others.

Reaction Norms and Plasticity

Costs that exist at a lower level of organization can often be reduced by the evolution of compensatory features at a higher level within the organism. For example, the activity of a particular enzyme may decline as temperature departs from some optimum. However, a series of isozymes, variant enzymes encoded by members of a gene family, may differ in temperature optima, so transcription of different genes may provide the organism with broader temperature tolerance. Changes of this kind occur during physiological acclimation.

Physiological acclimation is an example of *phenotypic plasticity*, the expression by a genotype of different phenotypes under different environmental conditions (Pigliucci, this volume). Other examples include short-term changes in morphology such as seasonal changes in the plumage of birds, developmental switches that produce lifelong differences in morphology such as habitat-dependent differences in growth form in many plants, and the expression of many behavioral traits, which may be affected by an individual's history of experience.

Why does phenotypic plasticity not always increase in evolution, leading to broader niches? One possible answer is that environmental cues are sometimes unreliable (which could induce development of the wrong phenotypic state). Another is that the ability to change phenotype might itself entail a cost.

Effects of Demography on Niche Evolution

The smaller a population is, the stronger selection must be to overcome the effect of genetic drift (i.e.,

random fluctuations in allele frequency). For example, if N is the effective size of a population (see Nunney, this volume) and s is the coefficient of selection for an advantageous allele that has frequency q, the probability of fixation of the allele is $P = 1 - e^{-4Nsq} / 1 - e^{-4Ns}$. Therefore, advantageous alleles are less likely to be fixed in small than in large populations, and disadvantageous alleles are less likely to be eliminated.

Suppose a species occupies a "main" habitat (1) in which the rate of population growth (fitness) is high and a suboptimal habitat (2) in which it is lower or even negative. Overall, gene flow between subpopulations is from the main into the suboptimal habitat. Assume that the main habitat supports a larger population, and consider a mutation that has a "+, −" effect, increasing fitness in habitat 1 and decreasing it in habitat 2. It is more likely to be fixed than a mutation with equal but opposite effects (−, +), because selection is more effective in the main habitat (Holt and Gaines 1992). If the effects of the mutation are +, 0 (increasing fitness in habitat 1 without altering fitness in habitat 2), it clearly is likely to be fixed. Fixation of either a +, − or +, 0 mutation will increase the difference in mean fitness between the subpopulations in the two habitats.

Consequently, the growing disparity in population growth rates may increase the asymmetry in gene flow between habitats, further lowering the fitness of the population in habitat 2. (Therefore, a population may more readily adapt to the suboptimal habitat if it is geographically isolated. This may be the principal source of species with novel ecological specializations.) Moreover, as the disparity in mean fitness between the habitats increases, it becomes more advantageous to use the superior habitat. Therefore, selection may favor specialized habitat preference, further restricting the population to that habitat and enhancing selection for adaptation to that habitat. This theory predicts that ecological niches should evolve slowly. Indeed, species in a clade often retain similar ecological characteristics for many millions of years, as is true of the host-plant affiliations in many groups of insects (Futuyma and Mitter 1996).

Neutral Evolution of Specialization

We now consider the chief alternative hypothesis to explanations of specialization based on trade-offs: random fixation of neutral or slightly deleterious habitat-specific mutations (Kawecki et al. 1997).

If a population is isolated in a new habitat, mutations that degrade its adaptations to the ancestral habitat are selectively neutral and may be fixed by genetic drift. Thus, such characters may be lost, and the altered population may be confined to the new habitat. Now suppose a population occupies both habitats. As we have seen, "+, 0" mutations that enhance fitness in the larger population in habitat 1 are more likely to be fixed than "0, +" mutations with the opposite effect, because selection is more effective in the larger population. But even some slightly deleterious mutations (0, −) that lower fitness in habitat 2 will be fixed by genetic drift. The accumulation of many such mutations can reduce the population's fitness in habitat 2 and may even lead to extinction of populations in that habitat. Thus, specialization would evolve due to genetic drift. Selection would contribute, insofar as it would favor alleles for a specialized behavioral preference for the increasingly superior habitat 1. If some slightly deleterious mutations are fixed that are expressed independently in each habitat, it would be advantageous for individuals to mate and lay eggs in the habitat in which they are more fit. It is conceivable that the evolution of such habitat fidelity could lead to the sympatric origin of two specialized species.

Case Studies

Two of the subject areas in which the evolution of niche breadth has been studied are thermal tolerance and diet specialization. The latter has been analyzed especially in herbivorous insects.

Some studies have challenged the presumption that trade-offs account for specialized tolerance or diet and show that trade-offs must be demonstrated; they cannot be assumed a priori. For example, in experimental populations of *Drosophila* selected for activity at different temperatures, tolerance to acute heat shock was correlated with the temperature of the population's selection regime, but tolerance of cold shock was not (Gilchrist et al. 1997).

Albert Bennett and Richard Lenski and their colleagues have examined patterns of pleiotropy for temperature-dependent population growth in experimental populations of *Escherichia coli* (Mongold et al. 1999 and included references). Their "ancestral" population, derived from a single hap-

loid cell, was maintained for 2000 generations (about 300 days) at 37°C. Replicate populations derived from the ancestral population were propagated for another 2000 generations in several temperature regimes, including 32°, 37°, and 42°. (The upper critical temperature, above which the ancestral population could not grow, is about 43°.) At intervals, Bennett et al. estimated the relative fitness of an experimental population by competing it with a genetically marked sample from the ancestral population, which had been kept in a frozen but viable state.

All the populations improved in relative fitness at temperatures near their "selection temperature." For example, the populations maintained at 32° became fitter at temperatures near 32°. Surprisingly, almost no populations showed a decline of fitness, relative to the ancestor, when tested at other temperatures. For example, when measured at 42°, the fitness of populations adapted to 32° was no lower than the ancestor's fitness. Thus, the effects of advantageous mutations were remarkably temperature-specific, providing little evidence of trade-offs. The only exceptions were genotypes that improved fitness at extreme temperatures. For example, several mutations that conferred an ability to survive at 44° (above the ancestor's thermal maximum) reduced fitness at 41° or below.

Comparisons of related species often show that each performs better in its typical environment than in its relatives' environment. In the leaf beetle genus *Ophraella*, for example, the sister species O. *slobodkini* and O. *notulata* feed, respectively, on *Ambrosia artemisiifolia* (ragweed) and *Iva frutescens* (marsh elder). Keese (1998) grew larvae of both species on both plants, in both the laboratory and natural habitats, and measured survival to adulthood. For O. *slobodkini*, the mean ratio of survival on *Ambrosia* relative to *Iva* was 2.13 in the laboratory and 3.08 in the field, whereas for O. *notulata*, these ratios were 0.81 and 0.46. Even though these species means present a pattern expected of a trade-off, however, they do not provide evidence that specialization evolved due to antagonistic pleiotropy. Any of the hypotheses we have considered would give rise to the observed pattern. For this reason, researchers have sought evidence of antagonistic pleiotropy *within* populations, where the genetic dynamics we have discussed occur.

Among studies of the diet of herbivorous insects, perhaps the clearest evidence of a trade-off was found in the polyphagous aphid *Aphis fabae*, which has a succession of rapidly developing, viviparous, parthenogenetic generations during the summer, followed by a sexual generation in the fall. The parthenogenesis enabled Mackenzie (1996) to rear each of many genotypes on several different plants, and to estimate fitness by the lifetime production of offspring per female. In eastern England, the aphid's main host is bean (Fabaceae), but small populations are found on dock (Polygonaceae) and a garden plant, South American nasturtium (Tropaeolaceae). Among numerous clones collected from bean, fecundity on bean was not correlated with fecundity on dock, but it was negatively correlated with fecundity on nasturtium. A similar correlation was found in F_2 progeny of crosses between bean- and nasturtium-adapted genotypes (figure 14.2). This finding suggests that the correlation is due to antagonistic pleiotropy, because recombination among genes with host-specific effects might be expected to have generated some offspring well adapted to both hosts.

Most other studies of herbivorous insects have provided equivocal evidence on trade-offs. For example, Futuyma and Philippi (1987) studied parthenogenetic genotypes of a polyphagous moth, the fall cankerworm (Alsophila pometaria). Each genotype was reared on four of the species' normal host plants, belonging to two families (Fagaceae [oaks and chestnut] and Aceraceae [maple]). The investigators measured larval weight at a specific age and found that it varied significantly among genotypes and among hosts. In both field- and laboratory-reared larvae, the genetic correlations were positive; that is, the "best" genotypes on one plant tended to be "best" on other plants as well. Similar patterns, providing no hint of trade-offs, have been found in studies of several other insects (Jaenike 1990; Fox 1993).

As we noted earlier, a trade-off may be masked by the existence of genotypes that display low "general vigor," perhaps owing to mutations that reduce fitness in both environments (e.g., points a and b in figure 14.2). Statistical methods might reveal obscured trade-offs in such cases; for example, Futuyma and Philippi's data on Alsophila revealed a negative correlation between genotypes' fitness on Fagaceae and Aceraceae when they were analyzed by principal components analysis (Jaenike 1990). Other authors have cautioned that such methods may yield misleading results and have proposed that correlations among genotypes do not yield as reliable evidence of trade-offs as selec-

tion experiments do (Fry 1993). For instance, in experiments by Gould (1979) and Fry (1990), spider mite (Tetranychus urticae) populations became adapted to marginal hosts (cucumber or tomato) that initially caused high mortality. Subpopulations were then reared on a favorable host (bean) for a number of generations. The ability of these "reversion" lines to survive on the marginal hosts declined, a finding suggesting that genotypes with high fitness on these plants were disadvantageous in the bean environment and were reduced in frequency.

Several recent studies have provided support for the hypothesis that trade-offs arise from neural limitations on information processing. The proposition is that a specialized forager might distinguish suitable from unsuitable plants more effectively if it needs to make few distinctions ("accept if plant is A, reject if not-A") than many ("accept if A or B or C, reject otherwise"). Bernays and Funk (1999) contrasted a population of the aphid Uroleucon ambrosiae from the eastern United States, where it feeds almost solely on Ambrosia trifida, with a western population that feeds on both this plant and several others. They found that even in small cages, eastern aphids find A. trifida faster, and then begin feeding more rapidly, than western aphids. In a similar vein, Janz and Nylin (1997) contrasted three polyphagous and two monophagous species of butterflies, all of which lay eggs on nettle. Egg-laying females of the monophagous species discriminated against poor-quality, senescent nettle plants in favor of young plants, whereas the generalist species did not make such a distinction even though their larvae, like those of the specialist species, grow more slowly on senescent foliage. A parallel difference was found between a specialized and a generalized population of one species of butterfly. Rapid, efficient decision making may not only result in using better resources but also enhance the rate of resource acquisition and reduce the risk of predation while the animal's attention is engaged.

Future Directions

The answers to many questions about the evolution of generalization and specialization remain uncertain or entirely lacking. Important problems for future research include the following.

1. How ubiquitous are trade-offs, and how can they best be detected? The assumption that a trade-off exists seems not always to hold true, so further documentation, with respect to a variety of physiological and ecological traits, is needed. Moreover, methods of testing for trade-offs need attention. How can trade-offs, if they exist, be detected if they are masked by variation in "general vigor"? Should tests be based on the fitness effects of single genes or characters, on fitness components of genotypes in different environments, on genetic correlations, on selection experiments, or on population means? Can trade-offs that operate in a population's natural environment be measured in a laboratory environment?

2. Do trade-offs evolve? If so, when might we expect to find them, and when not? Selection of modifier genes has been shown to ameliorate trade-offs in both natural and laboratory populations. Rausher (1988) suggested that as a population adapts to a novel habitat, selection of genes that independently improve fitness might reduce trade-offs that existed initially. Conversely, Joshi and Thompson (1995) expect trade-offs to be most pronounced in populations that have adapted to both habitats and are closer to genetic equilibrium. They argue that the only loci that remain polymorphic, generating a genetic correlation, are those at which different alleles enhance fitness in each habitat.

3. What are the mechanistic bases of trade-offs? Evolutionary ecologists postulate trade-offs as ad hoc explanations of observations, but whether or not the trade-offs might be expected on mechanistic molecular or physiological grounds is

often uncertain. Should we really expect phenotypic or behavioral plasticity to be costly, and why? If costs in fitness exist but are not appreciable in terms of energy, wherein do they lie? In this subject lies opportunity for syntheses between evolutionary biology and physiology or functional morphology.

4. How is the evolution of behavioral specialization (of resource or habitat use) in animals related to the evolution of morphological and physiological adaptations? Does a behavioral shift generally precede and engender selection for morphological and physiological adaptations? Are the behaviors employed in resource or habitat use genetically determined, and are they genetically independent of the morphological and physiological adaptations?

5. The theory that greater efficiency of natural selection in large than small populations results in evolutionary conservatism of niches has not been explored empirically. The related hypothesis that gene flow from central to marginal populations limits the geographic range has been tested hardly at all. Do marginal populations expand their range by evolving reproductive isolation?

6. Does specialization ever evolve due to accumulation of deleterious mutations with habitat-specific effects?

7. What can phylogenetic patterns tell us about the relative rates of evolutionary transition between generalists and specialists?

Note

1. The reciprocal of the arithmetic mean of the reciprocals of a set of numbers.

PART III

BEHAVIOR

15

Mating Systems

ANN K. SAKAI
DAVID F. WESTNEAT

The study of mating is one of the most active areas in evolutionary ecology. What fuels this research is curiosity about a stunning diversity of ways in which zygotes are formed. Many plants and some animals can reproduce without combining gametes. Many other plants combine gametes but do so within the same individual (selfing). Still other plants and animals require a gamete from another individual to stimulate reproduction but do not incorporate the genetic material contained in that gamete in the offspring. Finally, many organisms combine gametes produced from different individuals in sexual reproduction, but the ways in which these individuals get together to reproduce are also amazingly diverse and have major implications for how selection acts in these populations.

Why are there so many different ways to reproduce? Answering this question is a major challenge for evolutionary ecologists. Our approach begins with how a variety of ecological factors affect selection on reproductive traits. Because many reproductive traits show genetic variation, diversity in selective pressures can lead to a diversity of evolutionary changes. Thus, understanding the evolutionary ecology of mating systems can help to interpret the significance of this variation and can provide new insight into related phenomena. For example, costs of female reproduction associated with development of offspring greatly impact other aspects of the life history, and males are often limited by mates (Savalli, this volume). Factors such

as levels of selfing, inbreeding depression, and allocation of resources play a part in mating systems of both plants and animals (Waser and Williams, this volume), and sex allocation theory has been used in both plants and animals to explore the evolution of hermaphroditism and unisexuality (Campbell 2000; Orzack, this volume).

This chapter explores some of the major forces affecting mating systems. Our treatments of plants and animals differ in emphasis, but our goal is to use the perspective of evolutionary ecology to define more fully the similarities, differences, and diversity in plant and animal mating systems, and to highlight potentially interesting yet currently unanswered questions.

Concepts

Diversity in patterns of zygote production arises in part from ecological factors influencing two issues: selection on the evolution of sexual reproduction itself and differentiation of the sexes. Sexual reproduction occurs with production of gametes through meiosis and subsequent fertilization and formation of the diploid zygote. It results in genetic recombination that may have both costs and benefits. Recombination of genes with another individual may be costly in two ways. First and foremost is the "twofold cost" of sex (Williams 1975): An asexually reproducing individual transmits all of its genes to offspring, but a sexual female transmits only half the genes in offspring from sexual reproduction. Second, sex recombines different gene variants in new ways, potentially breaking up groups of genes that work well together in that environment (coadapted gene complexes). In a constant environment, this cost may be prohibitive. In variable and uncertain conditions, however, sexual recombination provides a greater variance in genotypes that increases the chance that some offspring will have better genotypes for changing or novel environments. Other theories suggest that sexual reproduction may be advantageous because of environmental changes caused by coevolving parasites, or as a means to purge deleterious mutations (see related chapters in Hurst and Peck 1996). Most animals are obligately sexual; many plants can switch between sexual and asexual reproduction.

Once sex has evolved, it is usually characterized in both flowering plants and animals by anisogamy, in which gametes are dimorphic in size. Disruptive selection for polymorphism in gamete size arises because of a trade-off between production of fewer, larger gametes that increase zygote survival via extra nutrients, and production of smaller, more numerous gametes that increase the number of possible zygotes. By convention, the sex producing the smaller, mobile gametes (sperm) is the male, and the sex producing the less mobile, larger gametes with food reserves (ova or eggs) is the female (but see Rhen and Crews, this volume, for an alternative definition). In both plants and animals, similar evolutionary processes may act on the two sexes. Male gametes must reach female gametes and successfully compete with each other for fertilization of the eggs, females may differentially limit access to eggs, and fitness of the progeny may depend on available resources as well as parental care. Once established, the divergent forces that fostered anisogamy may continue to have major effects on mating. Changing environmental conditions can alter selective forces, which in turn can elaborate other differences between the sexes, contribute to conflicting selection pressures on each sex, and thereby help explain diversity in mating.

Reproduction in Plants and Animals

Studies of the reproductive biology of plants and of animals have proceeded independently for many years, largely because of an emphasis on more descriptive aspects of animal mating behaviors and plant pollination systems. As a consequence, differences in approaches and terminology have sometimes masked similarities in underlying evolutionary processes. For example, dioecy and gonochorism describe populations with unisexual male and female individuals in plants and animals, respectively. Even the term *mating system* has different meanings. Emlen and Oring (1977) described avian mating systems based on "the ecological and behavioral potential to monopolize mates, and by the means through which such monopolization takes place" (p. 217), including features such as the number of mates, how mates are obtained, characteristics of pair bonds, and patterns of parental care. More recently, Reynolds (1996) used the term *breeding system* to describe the relationship of mating behavior and parental care by both sexes, with less emphasis on ecology and greater emphasis on the role of females and individual interactions. In plant biology, the many mechanisms that control who mates with whom (differences in flowering time, spatial separation of the sexes, incompatibility systems, etc.) are described as aspects of *breeding systems;* the term *mating system* is usually more narrowly defined to describe the amount of selfing (fertilization of the ovule by pollen from the same plant). This greater emphasis on selfing occurs because most flowering plants are hermaphroditic with both male and female function and thus have more potential for selfing than animals. In most animal systems, unisexuality of individuals is taken as a given, and questions focus more on sex ratios, sex allocation, and sexual selection (Orzack, this volume; Savalli, this volume).

In a typical vertebrate, meiosis occurs in the testes and ovaries to form haploid gametes, and fertilization results in a diploid embryo. In flowering plants, meiosis occurs in the stamens and pistil of the flowers, and additional mitotic divisions follow meiosis. As a result of these mitotic divisions, two to three identical haploid nuclei (including the sperm nucleus) occur within each pollen grain, and typically, eight identical copies of the egg nucleus are contained within the embryo sac. After transfer of the pollen to the stigmatic surface of the pistil (pollination), the pollen tube grows through the style of the pistil until it reaches the ovary and the egg. The diploid embryo is formed by fertilization of the haploid egg by the haploid sperm nucleus. Unlike animals, most flowering plants undergo double fertilization. In most plants, the second fertilization occurs when two of the haploid nuclei identical in genotype to the egg are joined by a haploid sperm nucleus, fusing to form the triploid endosperm. The triploid endosperm (or in some cases the cotyledons [seed leaves]) provide nutrition for the developing seedling. As a result, the diploid embryo has equal genetic contributions from the mother and father, while the triploid endosperm has twice the genetic contribution of the mother relative to the father. The typical flowering plant disperses a seed with a protective seed coat of maternal genotype, often a triploid endosperm, and the diploid plant embryo. These different levels of relatedness have been used to interpret patterns of seed abortion in plants as a part of female choice in sexual selection (Willson 1994).

Flowering plants also differ from most animals by their lack of mobility and modular construction and the fact that most flowering plants are hermaphroditic (Barrett and Harder 1996). Mobility in animals means mating behavior often includes movement and advertisement by one or both sexes. Because most adult plants are not mobile, mating opportunities and the transfer of male gametes are mediated through abiotic or biotic pollination vectors, such as wind, water, insects, birds, and mammals. Historically, studies of the reproductive biology of plants centered on pollination biology, with descriptions of flowers and rewards characteristic of particular pollinators (reviewed in Bertin (1989)). In contrast, Waser et al. (1996) pointed out the importance of considering more generalized pollination systems. The amazing diversity of flowers has largely been attributed to attracting pollinators to maximize female fitness through seed production. The effect of floral morphology and displays on male fitness can now be considered because of better molecular and statistical techniques to measure male fitness. Some animals with limited adult dispersal, such as mollusks and parasites, have reproductive biology very similar to that of plants, including self-fertilization.

Because of their modular nature, many plants can produce new individuals asexually as branches produce roots or through more elaborate means. In addition, many plants have multiple flowers and thus multiple sites of sexual reproduction. Individual flowers may be hermaphroditic (perfect), male (staminate), or female (pistillate). As a result, sex expression in plants can be quite complex, and flowers on the same plant may vary in sex expression by position or through time. Variation in sex expression can also occur among plants within populations. The majority of flowering plants are hermaphroditic, with both male flower parts (stamens with anthers and pollen) and female flower parts (pistils with ovaries and eggs forming fruits) within the same flower (hermaphroditic flower). An individual plant may be hermaphroditic because it has hermaphroditic flowers or because it has both unisexual male and unisexual female flowers (see Sakai and Weller 1999 for terminology). With hermaphroditism, there is the potential for selfing (ovules fertilized by pollen from the same plant). The amount of selfing depends on many traits, including timing of maturity of the receptive stigma and pollen grains (dichogamy), floral morphology and the relative position of the stigma and anthers in the flower (e.g., herkogamy), chemical interactions of the stigma and pollen grains (e.g., self-incompatibility), and pollinator behavior (Waser and Williams, this volume).

A first step in understanding the evolutionary ecology of mating and breeding systems is an understanding of the variation that occurs in these systems. Below, we give a brief review of plant breeding systems and animal mating systems and provide a few hints about forces that may be shaping them.

Classification of Plant Breeding Systems Many traits used to characterize animal mating systems incorporate behaviors that are not applicable to plants (e.g., male parental care, defense of territories or mates, establishment of pair bonds) or traits that have until recently been difficult to measure (e.g., paternity, number of mates). Sexual selection theory (Savalli, this volume) and Bateman's principle (stressing the importance of male-male competition for mates and female choice of mates because of limiting resources) are most fully developed in the animal literature but have also been applied to plants (Willson 1994). In plants, females have relatively little control over the source of pollen, and often the only mobile stage when males compete with other males is during pollen dispersal and pollination. As a result, male-male competition in plants has focused on prepollination aspects of floral displays for pollinator attraction, as well as pollen tube competition in the style of the pistil. Female choice in plants has focused on postpollination pollen tube growth, as well as on postfertilization studies of selective abortion of developing seeds and fruit (Willson 1994). More recently, evolutionary approaches to plant reproductive biology have incorporated general theories of sex allocation to explain the diversity of plant breeding systems.

Over three-quarters of flowering plant species are hermaphroditic, and a significant number of these exhibit some selfing. Selfing also occurs in a few animals (many snails and earthworms). Variation in plant breeding systems has traditionally been described by the degree of separation of the sexes on plants, both in space (herkogamy, dioecy) and through time (dichogamy, sex switching). As a result of this variation, plants also vary in the relative amount of inbreeding that occurs. Genetic systems include apomixis (formation of seeds without fertilization), self-fertilization of hermaphroditic plants, mixed mating systems (with both selfing and outcrossing), and outcrossing promoted by a variety of characters (e.g., dichogamy, herkogamy, dioecy, self-incompatibility).

Selfing and outcrossing Several hypotheses have been proposed to explain the advantages of selfing (Waser, this volume). In the "reproductive assurance" model, selfing is favored because selfing guarantees that there will be sufficient pollen to ensure seed set. In the "automatic selection" hypothesis, selfers are favored because they may have a significant transmission advantage, with up to 50% greater genetic contribution to the next generation through maternal and paternal contribution in selfed seeds plus paternal contribution in outcrossed seeds. In contrast, outcrossers contribute only as the maternal parent in their own seeds and as a paternal parent in outcrossed seeds.

Selfing may occur through autogamy (selfing within the same flower) or through pollen flow between flowers on the same plant (geitonogamy). Geitonogamy may be common in wind-pollinated plants as pollen is blown within and among inflorescences on a plant. It may also be promoted by large floral displays that encourage insects to forage among flowers on the same plant. Selfing plants are characterized by a number of features reflecting modified allocation patterns, including lower pollen:ovule ratios, smaller flowers, and reduction in attractants such as nectar. In many selfing species, floral morphology promotes selfing, with stamens and stigmas maturing synchronously and with anthers adjacent to stigmas. Some species—for example, some violets *(Viola)*, and touch-me-nots *(Impatiens)*—have mixed mating systems

with two types of flowers. The more familiar showy flowers (chasmogamous flowers) are capable of outcrossing with insect pollinators; smaller inconspicuous flowers never open and only self (cleistogamous flowers).

Not all hermaphroditic plants self, and the selfing advantage is reduced under several circumstances. A major factor favoring outcrossing is the expression of inbreeding depression, or the reduction in fitness of selfed progeny (w_s) relative to outcrossed progeny (w_o). Inbreeding depression is expressed as $d = 1 - (w_s/w_o)$. If inbreeding depression is high or outcrossers produce more than twice as many progeny as selfers, then the selfing advantage disappears (Waser and Williams, this volume). Models of plant mating system evolution by Lande and Schemske (1985) suggesting that completely selfing or completely outcrossing populations should evolve stimulated a flurry of empirical studies. As a result, a number of mixed mating systems with intermediate levels of selfing have been described, and alternative models predicting stable mixed mating systems have been proposed (Waser and Williams, this volume). As statistical and molecular methods to determine paternity and outcrossing improve, the relationship between outcrossing rates and inbreeding depression has started to be examined at both the population and family level. Other factors may also reduce the automatic selection advantage and lead to mixed mating systems, including situations where selfing reduces the amount of pollen available for outcrossing (pollen discounting) and the timing of self-pollination relative to cross-pollination (Waser and Williams, this volume).

Apomixis Apomixis occurs when there is reproduction without fertilization. In the broadest definition, apomixis includes vegetative propagation with establishment of separate genetically identical individuals through vegetative growth and fragmentation of roots, runners, and so on. This process is common in plants because of their modular nature with many undifferentiated growing tips. Many unicellular plants and invertebrates with relatively immobile adult stages (e.g., hydras, corals) also can reproduce vegetatively through budding. In agriculture, vegetative propagation has been used extensively in propagation (e.g., new plants from the "eyes" of old potatoes). Several noxious invasive plants (e.g., water hyacinth, *Eichhornia*) spread rapidly because of vegetative propagation.

Apomixis more narrowly defined includes only agamospermy. Agamospermy is widespread in flowering plants, although it is particularly common in the rose family (Rosaceae; e.g., blackberries, *Rubus*) and in the sunflower family (Asteraceae; e.g., dandelions, *Taraxacum officinale*). In agamospermy, seeds are formed, but the normal processes of meiosis and fertilization are bypassed. Zygotes are identical in genotype to the seed parent and are derived from either maternal tissue or an unreduced ($2N$) egg. In some cases, pseudogamy occurs, where pollination is necessary to form the triploid endosperm, but the zygote is not the result of fertilization (Briggs and Walters 1997). A few other organisms (e.g., some *Cnemidophorus* lizards, *Daphnia* and other cladocerans) also can produce eggs by apomixis, and many apomictic organisms are capable of both apomixis and sexual reproduction. Apomixis in plants and animals has been associated with higher elevations and latitudes in harsher physical environments. This pattern may result from reduced selection for sexual reproduction because of fewer biotic interactions with parasites, competitors, and predators, more limited mating opportunities, or because apomixis is often correlated with polyploidy, and polyploidy confers greater fitness in harsher environments (Bierzychudek 1987).

Dichogamy Flowers may have male and female function at different times (dichogamy), either because stigmas mature before the release of the pollen (protogyny), or because pollen matures before stigma receptivity (protandry). Dichogamy has traditionally been viewed as a mechanism that promotes outcrossing. The effectiveness of dichogamy in promoting outcrossing varies with the degree of temporal separation of sexes within the flower, inflorescence, and the plant. In some cases, individual flowers may be dichogamous, but because flowers open sequentially, selfing can still occur through geitonogamy. Dichogamy may also be favored because it reduces the amount of self-pollen on the stigma (stigma clogging) that may reduce outcrossing.

Gender changes Some plants and animals (e.g., several fish species) are capable of changing sex expression and functional gender through time. Sex switching may be more common in plants than in animals because sex lability may confer a selective advantage to stationary organisms that cannot exercise habitat choice. Sex-switching organisms have been popular, not only because of their unusual natural history, but also to test models of sex allocation. Small plants of jack-in-the-pulpit (*Arisaema triphyllum,* an herbaceous perennial of the spring flora) are male and switch to female function with larger plant size, presumably increasing fitness through fruit production. These changes in sex expression with plant size are remarkably consistent with the size advantage model for sequential *hermaphroditism*, with sex allocation patterns that optimize an individual's total fitness (Charnov 1982).

Heterostyly Heterostylous breeding systems have plants with hermaphroditic flowers of two (distyly) or three (tristyly) different style and stamen lengths (reviewed in Barrett 1992). For example, the hoary puccoon (*Lithospermum caroliniense,* Boraginaceae) has yellow tubular flowers that are distylous. Pin flowers have long styles and short stamens, and thrum flowers have short styles and long stamens (figure 15.1). This reciprocal positioning favors outcrossing with transfer of pollen between the pin and thrum flowers by pollinators. Pin × thrum and thrum × pin crosses produce many more seeds than "illegitimate" pin × pin or thrum × thrum crosses. Heterostyly has arisen independently several times and occurs in about 25 plant families. Heterostylous species include the shamrock genus (*Oxalis,* figure 15.1), *Primula*, and purple loosestrife (*Lythrum salicaria),* a tristylous species first studied by Darwin. Genes controlling morphological differences are linked to genes controlling sporophytic self-incompatibility (see below). This combination of heterostyly and sporophytic self-incompatibility may have evolved as a highly effective outcrossing mechanism.

Self-incompatibility Self-incompatibility is a genetically based physiological system where fertile hermaphroditic flowering plants are prevented from producing offspring with pollen possessing the same mating type, resulting in a highly effective outcrossing mechanism. Inhibition of pollination occurs because the pollen does not adhere to the stigma, or because the pollen fails to germinate or grow down the style. Based on recent molecular work (Franklin et al. 1995), self-compatibility is the ancestral condition in flowering plants. Self-incompatibility has evolved independently in over 71 families, in both monocotyledons and dicotyledons (Weller et al. 1995a). Although self-incompatibility is a prezygotic isolating mechanism and inbreeding depression is a postzygotic mechanism,

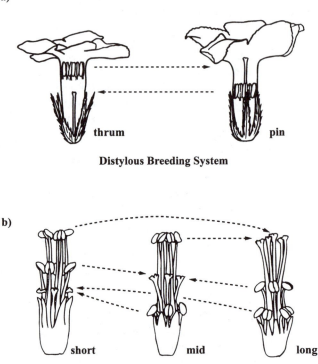

Figure 15.1 Heterostyly. (A) *Lithospermum caroliniense* (Boraginaceae) has a distylous breeding system, and plants have either pin or thrum flowers. Pollen is transferred between anthers and stigmas at similar positions in the flower. (B) Some species of *Oxalis* (shown here without petals; modified from Weller 1976) have a tristylous breeding system. Pollen flow between flowers is indicated by arrows.

it is often difficult to distinguish between them except through studies of pollen tube growth and embryo development. Lack of seed set alone may result from either early-acting inbreeding depression or self-incompatibility.

Several types of self-incompatibility have been well documented (Briggs and Walters 1997; Richards 1997). Gametophytic self-incompatibility occurs in several species, including tobacco *(Nicotiana)*. Compatibility is determined by interactions of the S alleles, found in the diploid maternal stigma and style, and the haploid genotype of the pollen. If the S allele present in the haploid pollen is also present in the maternal genotype, an incompatibility reaction occurs. For example, if the genotype of the stigma is S_1S_2, an incompatibility reac-

tion will occur if the pollen is either S_1 or S_2. S_3 pollen tubes grow normally and result in fertilization. With gametophytic self-incompatibility, pollen tube development is arrested on the stigma (e.g., grasses) or as the pollen tube grows down the style, often forming a callose plug and preventing fertilization. The presence of callose plugs is indicative of a self-incompatibility reaction.

In sporophytic self-incompatibility, the incompatibility reaction is controlled by the interaction of the maternal genotype and the diploid paternal genotype of the anther (not the haploid pollen genotype). When the pollen is formed, the anthers produce proteins that are present in the pollen wall and cause the incompatibility reaction. In sporophytic self-incompatibility, the incompatibility reaction is very rapid, and pollen tube development is arrested at or near the stigmatic surface, along with formation of callose plugs. In the simplest situation with no dominance, the S alleles act independently, and an incompatibility reaction occurs if any of the S alleles are the same in the maternal and paternal parents. For example, pollen from an S_1S_3 parent is incompatible with an S_1S_2 female parent (regardless of the haploid genotype of the pollen) but compatible with an S_2S_4 female parent. Sporophytic self-incompatibility alleles may also show dominance, resulting in far more complex incompatibility reactions.

In both gametophytic and sporophytic self-incompatibility, the ability of plants to successfully mate is frequency-dependent because alleles in low frequency will have fewer self-incompatibility reactions with other individuals. New mutations of S alleles occur initially in low frequency and will be favored, and as a consequence, there are dozens of S alleles in many species. Given this frequency-dependent selection, it remains a mystery why distylous systems of self-incompatibility should have only two alleles.

Dioecy Sex expression in flowering plants can vary at the level of the flower, the individual plant, or the population, resulting in a bewildering diversity of systems and terminology. Individual flowers may be hermaphroditic (perfect), male (staminate), or female (pistillate), and individual plants may have one or more floral forms in any combination on a plant. Populations may be monomorphic with all plants similar to each other, or dimorphic with plants differing in floral types. For example, in monoecious breeding systems, separate male and female flowers occur on each plant, but the population is monomorphic because all plants are similar. Dimorphic breeding systems include gynodioecy (populations of plants with female flowers and plants with hermaphroditic flowers), subdioecy (populations of females, males, and a few hermaphrodites), and dioecy (populations with only males and females). Dioecy has arisen independently many times and occurs in about 6% of flowering plant species. These different patterns of spatial separation have different effects on outcrossing, with selfing through geitonogamy possible through breeding systems such as monoecy, but excluded by dioecy. Even in strictly dioecious species, however, biparental inbreeding (inbreeding among close relatives) may occur if dispersal distance of pollen and seeds is limited.

Selective factors promoting the evolution of dioecy can be grouped into those stressing the classic view of dioecy as a mechanism to avoid inbreeding depression, and those stressing factors related to allocation of resources to male and female function (Sakai and Weller 1999). Dioecy, with obligate outcrossing of unisexual plants, may have evolved as a mechanism to promote outcrossing and avoid inbreeding depression. One pathway to dioecy is through gynodioecy and involves initial introduction of male sterility (females) into a population of hermaphrodites. Several studies suggest that high levels of inbreeding depression coupled with high selfing rates have favored the persistence of females in initially hermaphroditic populations (Sakai and Weller 1999).

A second hypothesis that explains dioecy uses sex allocation theory to relate factors that may affect fitness through male and female function (Charnov 1982). Sex allocation models assume a trade-off between male and female function in hermaphroditic plants (Orzack, this volume). Dioecy is predicted to evolve in ecological conditions where there are disproportionate (accelerating) fitness gains associated with specialization in one sex. These fitness gain curves may be affected by many factors, including availability of resources, pollinator attraction and other factors affecting pollen export and import, and seed dispersal. For example, dioecy has been associated with fleshy fruits, and fitness gains have been proposed to occur because larger fruits crops may disproportionately attract more specialized seed dispersers that carry seeds to more favorable habitats (accelerating female fitness gains). Dioecy also has been associated with small flowers pollinated by small generalist insects. Charlesworth (1993) argued that this association occurs because unspecialized pollinators cannot discriminate against unisexual flowers that do not have a reward, permitting evolution of dioecy. It has been difficult to empirically test resource allocation theories because of difficulties in understanding the correct currency for comparison, measuring fitness (particularly male reproductive success) in the field, and measuring trade-offs in allocation patterns (Campbell 2000).

Classification of Animal Mating Systems   Some animals reproduce without fertilization (pathenogenesis occurs in some invertebrates, fish, amphibians, and lizards), and some are hermaphroditic as in plants. But the vast majority of animals must reproduce sexually, a male combining gametes with a female. Because of widespread anisogamy, most studies of animal mating focus on differences between males and females in selective pressures that lead to divergence in physiology; behavior, including parental care and degree of monopolization of mates; and responses to ecological conditions, particularly the distribution of resources in nature.

Classification of animal mating systems has been based primarily on how mates are acquired, the number of mates and their monopolization, and characteristics of pair bonds and patterns of parental care (Emlen and Oring 1977; Davies 1991). Promiscuity is the most widespread of animal mating systems and is characterized by short-term associations between males and females that generally cease after eggs are laid. The selective pressures producing dimorphism in gamete size continue to act in such systems, yet with a variety of interesting twists. Males, with their numerous, small, and mobile gametes, typically attempt to fertilize as many females as possible. Ecological factors, however, sometimes constrain which males can achieve fertilization (Emlen and Oring 1977). These factors favor a wide range of alternative mating behaviors, in which a subset of the population behaves very differently in attempts to acquire mates. Alternative tactics include forced copulation instead of providing nutritive courtship offerings, quick and inconspicuous spawning instead of elaborate courtship, and searching instead of defending.

Recently, researchers have focused on variation in female behavior as a critical aspect of many animal mating systems, especially in promiscuous ones. Sometimes, females attempt to copulate with more than one partner. A wide array of ecological factors might affect this behavior. Multiple mating could maximize the benefits of sex by producing highly variable broods of offspring. Alternatively, females might copulate with any male to ensure fertilization and then obtain genes for valuable traits for the offspring by copulating with particular males. Yet another possibility is that females acquire valuable nutrients from males during courtship and copulation. Below, we describe one case study in which parental care is a major benefit of multiple mating to females. Evolutionary ecologists are only starting to explore the diverse ways ecology might influence variation in female behavior.

The behavior of females, in turn, has a powerful effect on the evolution of male behavior. If females copulate with two or more males, fertilization success for any one male will be reduced. Traits that bias fertilization in a male's favor will be under selection, and a wide variety of such traits have been described in every taxon of animals (e.g., Birkhead and Møller 1998). Males of many taxa defend females during the period before she lays eggs, sometimes fighting off other males or preventing the female from moving about and encountering other males. Males of some insects have intromittent organs that can remove the sperm of previous males, and in insects and mammals, males produce substances in their ejaculates that block or inhibit insemination by other males.

Such male adaptations to control fertilization appear to influence female success. There is growing evidence that females have evolved mechanisms to prevent males from controlling fertilization completely. Mechanisms for avoiding mate guarding or for sequestering sperm and preventing its removal may be common in insects. These results support the view that mating is the outcome of multiple bouts of evolution in which the sexes are under differing selective pressures.

Male and female animals in a variety of taxa, particularly in birds, associate for periods of time far longer than necessary to attract a mate and fertilize eggs. A diverse array of different groupings of males and females is possible. In fish, mammals, and some birds, multiple females can be associated with one male. Such polygynous systems can arise in a variety of ways that represent a continuum of levels of prolonged social association. Access to females may be controlled either directly by defense of a group of females (female defense polygyny) or indirectly by defense of clumped resources that attract more than one female (resource defense polygyny). Female defense polygyny is rare in birds but common in many ungulates where females and young stay in small herds. Males may defend either the herd itself or the preferred habitat of the group. Whether or not polygyny occurs depends on the value of male care to either male or female fitness. Many mammals may be polygynous because male care has little impact on offspring survival. Polygyny may be rare in birds because male care is valuable to females, favoring avoidance of polygyny. It is likely to occur when a female can obtain greater reproductive success by mating with a male on better territory with more than one female than with a monogamous male on poor territory, despite the loss of male care. Exactly how female fitness is influenced by polygyny is potentially complex (Searcy and Yasukawa 1989).

In male dominance polygyny, males gather during the breeding season, and females select mates based primarily on male status in the group. These groups enhance male-male competition and lead to greater variance among males in reproductive success. This mating system is often associated with unpredictable resources or resources that are difficult to defend or are highly dispersed. One extreme type of male dominance polygyny is lek polygyny, where males cluster to attract females at a display site (lek), and females visit to mate. Females then leave to rear the young alone, and so social associations beyond copulation are minimal (leks thus are more of a promiscuous system than a polygynous one). Leks are often found in organisms with long breeding seasons, including species of insects, mouth-breeding fish, bullfrogs, and some mammals and birds. Ruffs *(Philomachus pugnax)* are European wading birds; males form leks where ranking is determined by contests. In leks up to about five males, only the highest-ranking male mates. As the number of males on the lek increases, other males are able to sneak matings (Widemo and Owens 1995). In scramble competition polygyny, reproductive success is highest for those males that are best at finding females rather than at male-male encounters. This mating system is often associated with an extremely short receptive period for females. In the wood frog, all females in a population are receptive for only one night a year, and male frogs spend their time finding as many receptive females as possible, rather than defending territory.

In monogamous mating systems, one male pairs with one female for a prolonged period. Conditions that tend to promote monogamy include young that require much parental care and conditions where parents can share in parental care. Although less common in mammals, monogamy is the most common mating system in birds. As an example, American robins *(Turdus migratorius)* are monogamous, and both parents prepare the nest, incubate the eggs, and feed the young. Monogamy is apparently favored because the survival rate of the young and the reproductive success of the parents are highly dependent on the rate of delivery of food to the young.

Polyandry occurs when the female pairs with more than one male per breeding season. This pattern is much less common but has been documented in some bird species and several fish species. Control of males may occur directly because of interactions among females, or indirectly by control of resources. In polyandrous fish, fertilization is often external, and males often do all the incubation and brood care. This care, rather than the number of eggs, may be the most limiting factor in the number of successful progeny for a female. In extreme cases, behavioral roles are reversed; females compete for males, and females are often bigger and more brightly colored than males. In the mating system of the jacana *(Jacana spinosa)*, a large tropical marshland bird in Central and South America, females compete for territories that attract males, which provide parental care to a clutch.

Regardless of social mating system, promiscuity still occurs. In many monogamous birds, males and females frequently copulate outside established pairings, leading to sometimes very high frequencies of extrapair fertilizations. Considerable variation in these frequencies, and in which sex initiates the matings, exists among birds. Such events are likely to be influenced by sexual conflict of various sorts. Explaining this variation is a continuing challenge for evolutionary ecologists.

Case Studies

In this section, we highlight two case studies, one from flowering plants and one from birds, that illustrate how general theories of mating-system evolution can be studied empirically. Both cases use a comparative approach combining ecological and evolutionary perspectives to examine selective forces resulting in a diversity of mating systems.

Evolution of Breeding Systems in Schiedea

The Hawaiian Islands, with their great isolation and limited colonization, have led to many insights into mating-system evolution, including the classic work of Carson and colleagues on the Hawaiian pictured-wing *Drosophila* (Wagner and Funk 1995). The native Hawaiian flora also has remarkable examples of breeding-system evolution. In the carnation family (Caryophyllaceae), the single lineage of the Hawaiian Alsinoideae (with the endemic genera *Schiedea* and *Alsinidendron*) has a diversity of breeding systems, including selfing, gynodioecy, and dioecy. Ecological studies of pollination vectors as well as micro- and macroevolutionary approaches (phylogenetic analysis, examination of the importance of resource allocation and inbreeding depression) have been used to study breeding-system evolution in this group.

With both molecular and morphological traits used to construct a phylogeny, there are four major clades, two with only hermaphroditic species (both highly selfing and outcrossing species) and two that include dimorphic species (with gynodioecy, subdioecy, and dioecy). Dimorphism probably arose at least twice within this lineage, and the most likely common ancestor was an outcrossing, biotically pollinated, hermaphroditic herb (figure 15.2; Weller et al. 1995b). Evolution of breeding systems has been tightly linked to changes in habitat and pollination system. Most hermaphroditic species grow in mesic or wet habitats and are biotically pollinated, with distinctive fragrances and nectar in both genera (Weller et al. 1998). In contrast, all dimorphic species are found in dry, windy habitats and are wind-pollinated. Adaptations to wind pollination include more exposed, highly condensed inflorescences with better aerodynamic properties for pollen dispersal and receipt; smaller, more numerous pollen grains easily dispersed by wind; and higher pollen:ovule ratios characteristic of wind-pollinated plants where pollen is dispersed more indiscriminately than with many animal vectors.

What selective factors could cause such shifts in breeding systems? Clues can be found through comparative studies of inbreeding depression (d), selfing rates (s), and relative resource allocation patterns of females and hermaphrodites ($k = 0$ when females and hermaphrodites have equivalent seed production; $k = 1$ when females produce twice as many seeds as hermaphrodites). Under the simplest models, females will be favored in hermaphroditic populations (leading to gynodioecy) when there are sufficiently high levels of inbreeding depression, selfing rates, and seed production by females ($k + sd > 0.5$; Charlesworth and Charlesworth 1978).

S. membranacea is characteristic of hermaphroditic species, occurring in mesic-wet habitat. Most offspring are outcrossed (selfing rates are low, $s = 0.13–0.38$), and thus, the high levels of inbreeding depression found ($d = 0.70$) are not expressed in these progeny (Culley et al. 1999). As a result, hermaphroditism is a stable breeding system. In contrast, gynodioecious species of *Schiedea* have high

levels of inbreeding depression as well as moderate-high selfing rates, a combination favoring females. *S. salicaria,* a gynodioecious shrub growing on the dry slopes of the West Maui Mountains, has a low frequency of females (13%), and females and hermaphrodites are not well differentiated. *S. salicaria* shows no differences between females and hermaphrodites in resource allocation ($k = 0$, as measured by seed production). Hermaphrodites have high levels of inbreeding depression in the greenhouse ($d = 0.82$) and selfing rates are moderate (0.44–0.51), a combination near values that favor females. In contrast, *S. adamantis* is a wind-pollinated gynodioecious shrub with a high frequency of females (39%) found on the dry slopes of Diamond Head Crater overlooking Waikiki. Females are strongly favored because of a combination of relatively high selfing rates (0.68) and inbreeding depression (0.60). In addition, *S. adamantis* shows differences in resource allocation, females producing over twice as many seeds as hermaphrodites ($k = 1.3$; Sakai et al. 1997). The high levels of inbreeding depression in both of these gynodioecious species suggest that inbreeding depression may be an important factor in selection for dimorphic breeding systems. Differences in resource allocation between females and hermaphrodites may be a consequence rather than a cause of the presence of females in the population.

Hermaphroditic species found in dry, windy habitats more characteristic of dimorphic species may reveal the most about how gynodioecy evolves. *S. menziesii* is a dry-site hermaphroditic species growing on the windy, steep, rugged cliffs of the West Maui Mountains. Inbreeding depression levels are extremely high ($d = 0.61–0.87$), but selfing rates are variable from year to year ($s = 0.43–0.70$), and plants do not appear to be well adapted to either wind or insect pollination. Although females have not been found in the field, female progeny are found in field-collected seeds grown in the greenhouse. *S. menziesii* may be a hermaphroditic species on the brink of evolving gynodioecy. A shift to wind pollination may be critical for females to avoid pollen limitation and to increase in numbers.

Selfing has evolved at least twice within this lineage (*Alsinidendron* species, *S. diffusa* and others; Weller et al. 1995b). Despite high selfing rates, *Alsinidendron* species often produce copious amounts of nectar (up to 9.8 μl/24 h in *A. obovatum),* a finding suggesting that these pollination shifts may

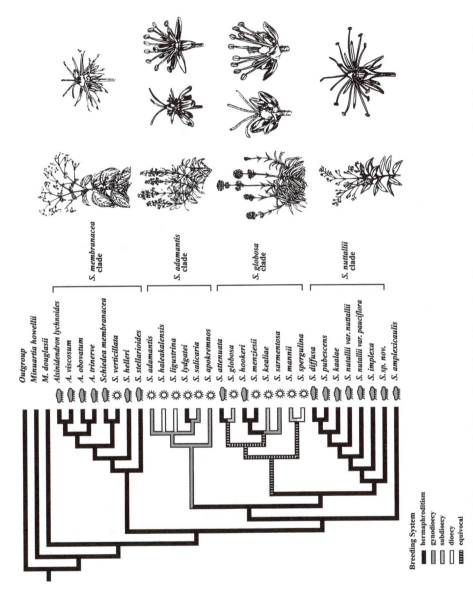

Figure 15.2 Phylogeny of the Hawaiian Alsinoideae. Breeding systems are mapped on the phylogeny. Habitat is indicated as dry (sun) or mesic/wet (rain clouds). Clades are named after the illustrated species. Smaller leaves are typical of dry habitats; more condensed inflorescences are associated with wind pollination and dimorphic species. A female flower (left) and hermaphroditic flower (right) are shown for gynodioecious *S. adamantis*; a female flower (left) and male flower (right) are shown for subdioecious *S. globosa*; hermaphroditic flowers are shown for *S. membranacea* and *S. nuttallii*. (Modified from Sakai et al. 1997.)

Breeding System
- ■ hermaphroditism
- ▨ gynodioecy
- ☐ subdioecy
- ☐ dioecy
- ▥ equivocal

have occurred quite recently. *S. diffusa* often grows as single vines with pendant inflorescences in very wet rainforests. Selfing may be favored if reproduction is limited by mates. In *A. trinerve*, flowers hang down and barely open at maturity. These morphological changes may be under selection because they keep the pollen dry and also promote selfing by putting stamens and stigmas in close proximity. Relatively low pollen:ovule ratios in both these species are consistent with high selfing rates. *A. viscosum* may be a highly selfing species that has previously purged deleterious alleles; inbreeding depression is extremely low ($d = 0.01$–0.21; Weller, Sakai, and Thai, unpublished).

These studies suggest that changes in breeding system have resulted from shifts in habitat and pollination vectors. In areas of low population density and small population size, particularly in wetter habitats where floral morphology promotes selfing, inbreeding may have effectively purged deleterious alleles, resulting in the evolution of highly selfing species (e.g., *Alsinidendron*). In larger populations, levels of inbreeding depression were high, but selfing rates were apparently low. With colonization of dry, windy areas, pollinator availability may have been limiting, and as a result, selfing rates increased substantially; expression of strong inbreeding depression may have led to selection for outcrossing (e.g., *S. menziesii* and *S. salicaria*). Changes in adaptations to wind pollination or conditions affecting selfing rates as well as genetic drift may greatly influence whether females are lost or become established in these populations. In other species, high inbreeding depression, continued high selfing rates, and evolution of wind pollination apparently favored mutations for male sterility (females) and gynodioecious breeding systems. Once females were present in the population, hermaphrodites were under strong selection for greater male function (e.g., *S. adamantis*). Continued selection and differentiation may have led to subdioecious and ultimately dioecious breeding systems (e.g., subdioecious *S. globosa*, dioecious *S. spergulina*).

Mating Systems of Dunnocks

Detailed ecological and evolutionary studies of mating systems exist in several animal taxa, including scorpionflies, dungflies, and coral reef wrasses. However, several studies of birds have driven theory on mating systems by combining data on ecological factors, male and female mating tactics, and the influence of parental care. Davies's (1992) studies of the variable mating system of the dunnock *(Prunella modularis)* perhaps most ideally illustrate the power of evolutionary ecology to uncover fascinating yet subtle details of an organism's mating system. These studies also illustrate the interplay between theoretical analyses and empiricism that is at the heart of an approach using evolutionary ecology.

Dunnocks are small, nondescript, sparrowlike birds inhabiting hedgerows and gardens in Europe. In some places, males and females appear like most other birds; they pair monogamously and cooperate to raise dependent young to independence. In other locations, a male may be paired to two females, and in still others, a female can be paired with two males. Occasionally, two or more males breed with two or more females (polygynandry).

Emlen and Oring (1977) suggested that mating systems were an outcome of resources affecting female distribution that in turn affects the ability of males to monopolize more than one mate. The variation in dunnock mating arrangements arises out of differing ways in which male and female territories overlap. When females defend large territories, no one male can monopolize them, and breeding is polyandrous. When females defend small territories, a single male can sometimes monopolize two females, and so breeding is polygynous. Variation in female territory size is linked to habitat; female territories are larger when vegetation is sparse. Foraging is most successful in dense vegetation, so that that food supply may affect the size of female territories. A food supplement experiment confirmed this; females on supplemented territories defended smaller areas. Male territory size did not change, a finding indicating that male territoriality was focused on acquiring mates. In supplemented areas, because of the smaller territory they defended, females were more likely to be paired with a polygynous male than in unsupplemented areas.

As mentioned above, a dominant theme in studies of animal mating is that the two sexes often are under conflicting selection pressures. The dunnock system clearly illustrates this theme. The number of fledglings a female produces is highly dependent on the social situation. A female with two males as mates produces the most fledglings, monogamous females are intermediate, and females that share a mate with another have the lowest success. The same is true for males; males that have two mates fledge the most young, and those that share a mate with another have the fewest (Davies 1992). This creates conflict between the sexes, as the situation that is best for females is the worst for males and vice versa. In the case of the dunnocks, the pairing pattern and hence the pattern of conflict emerges from the distribution of resources in the habitat.

Conflict over pairing is not the only dynamic within the dunnock system. Polyandrous trios produce some of the most intriguing male-female interactions yet uncovered. In these territories, one male (the "alpha") is often dominant to another (the "beta"). Alpha males typically stay closer to the female and exclude the beta from the area when the female is fertilizable. The female, however, often moves toward the beta male and sometimes manages to approach him without the alpha noticing. When this happens, the female solicits a copulation, and the beta attempts to inseminate her (Davies 1992). Nests of females in which beta insemination has happened are often composed of chicks with mixed paternity. Theory suggests that sometimes this affects how much care a male provides to those offspring. Indeed, when dunnock chicks need food, it is only the broods of those females with whom the beta has copulated that receive food from both males. This additional help from the beta male is the major reason why female fledging success is highest when she has two mates. The dunnocks remain one of the clearest cases of paternity influencing paternal behavior that is known today.

Theory about the effects of paternity on parental behavior also suggests that paternity is more likely to have effects if there are strong cues of paternity available to males. Davies's (1992) experimental studies of dunnocks also provide convincing evidence of the mechanisms by which males judge the level of paternity in their broods. Males were removed from their territories for 3 days either during the period when the female normally was receptive or during incubation. These latter males all fed young normally. All other males fed chicks only if they were present at least some of the time (and so had access to the female) during egg laying. When males were removed a day after a model egg was placed in the nest, their subsequent parenting was normal even though they had had no access to the female during her actual egg laying. Hence, males use the appearance of an egg, and their mating access to the female with respect to egg appearance, as cues for deciding to care for

offspring. Such cues are fairly good rules of thumb for actual paternity.

The dunnock case is a superb example of the power of using evolutionary ecology to study mating. Resource distribution, as demonstrated by manipulations, affects male and female spacing, which in turn influences pairing patterns. Those pairing patterns influence fitness and induce potential sexual conflict on males and females. Within particular social units, divergent selection on males and females has resulted in a suite of complex behaviors, such as mate guarding, female soliciting tactics, and male parental behavior, all of which are linked behaviorally and ecologically. This approach is being applied to many other animal systems, and many interesting twists on these themes have been uncovered.

Future Directions

Much of our recent understanding of plant and animal breeding systems has resulted from evolutionary approaches examining the fitness consequences of ecological factors. The advent of better molecular and statistical methods to infer paternity has permitted better measures of male fitness in both plants and animals and has also allowed better tests of sex allocation theory. Results measuring male and female fitness gain curves in plants show that shifts in allocation may be heritable, but support for the assumption of trade-offs is not strong, and more studies are clearly needed (Campbell 2000). Studies of suites of physiological traits associated with these morphological and ecological changes are needed to fully understand the trade-offs and mechanisms involved in evolution of breeding systems (Geber et al. 1999). With increasingly sophisticated molecular approaches, it may be possible to identify the actual genes important in evolution of behavior and other traits related to mating systems.

Similarly, the application of phylogenetic techniques to mapping behavioral and reproductive traits in a historical and comparative framework has resulted in significant insights into the evolution of breeding systems and mating systems (e.g., Harvey and Nee 1997; Weller and Sakai 1999). More microevolutionary approaches involving quantitative genetics may suggest the underlying genetic basis and genetic potential for changes in sex allocation. For example, positive genetic correlations between male and female traits will make it more difficult to evolve dioecy than if there are negative genetic correlations.

Studies of mating systems have become increasingly important in conservation and restoration biology, where understanding the interrelationships of the genetic consequences of small population size, genetic structure of populations, and mating and breeding systems is of immediate concern. Knowledge of the breeding and mating systems of both invasive species and species of invaded communities may well be important in management decisions. Lack of critical pollinators, behavior of alien pollinators, and competition for pollination by alien species may all affect the genetic structure of both alien and native species. Mating systems of only a small fraction of organisms are known. Studies of mating systems of a greater diversity of insects, algae, bryophytes, ferns, fungi, and other organisms will provide many special opportunities to examine mating systems and breeding-system theories for biologists with an understanding of the natural history, ecology, and evolution of these organisms.

Acknowledgments   We thank Stephen Weller for helpful discussions and Theresa Culley for help with figures. This work was supported by grants from the National Science Foundation (NSF DEB 9815878 to AKS and NSF IBN 9816989 to DFW) and the University of Kentucky (DFW).

Note

This chapter is written in remembrance of Warren H. Wagner and with special thanks to Richard Alexander.

16

Sexual Selection

UDO M. SAVALLI

When Charles Darwin wrote *On the Origin of Species* (1859), he issued a challenge to potential critics: "If it could be proved that any part of the structure of any one species had been formed for the exclusive good of another species, it would annihilate my theory, for such could not be produced through natural selection" (p. 229). Darwin went even further by identifying several traits that seemed especially problematical: the sterile worker castes of social insects and extravagant ornaments that appear to benefit potential predators more than their bearers. It is this latter problem and Darwin's solution to it—sexual selection—that are the focus of this chapter.

I begin by providing a very brief historical overview of sexual selection, focusing on the initial controversies and its resurgence in the 1970s. I then provide an overview of the conditions that lead to sexual selection and the kinds of traits that are favored by it. Sexual selection usually involves evolutionary changes in both males and females. Thus, I first address the evolution of extravagant male traits (it is typically males that exhibit such traits). Since female choice is one of the mechanisms that can lead to the evolution of extravagant male traits, I also address the evolution of female preferences. Finally, I will identify those areas of the field that are the most controversial, unresolved, and promising for future research.

This review is, of necessity, brief and selective. I have tried to cite recent reviews rather than the extensive primary literature to provide an easy entry point into the literature. Readers wishing for more detailed treatment would do well by starting with Andersson's (1994) excellent book.

Definitions and Historical Overview

Darwin introduced the concept of sexual selection in *On the Origin of Species* (1859) and greatly elaborated the idea in *The Descent of Man and Selection in Relation to Sex* (1871). Darwin defined sexual selection as depending "not on a struggle for existence, but on a struggle between males for possession of the females" (p. 136). In modern parlance, sexual selection can be defined as a special case of natural selection in which selection is acting on variation among members of one sex in the ability to obtain mates, matings, or fertilizations. Although sexual selection is a subset of natural selection, it is a convenient and common practice—which I will follow here—to limit the term *natural selection* to nonsexual mechanisms of selection (such as selection on survival and fecundity).

Darwin identified two modes by which sexual selection could operate: contest competition (involving fights or contests) among members of one sex for access to mates (or for resources that will attract mates)—known as *intrasexual selection* or, because it usually involves males, simply *male-male*

207

207

competition—and competition to attract members of the other sex, referred to as *mate choice, intersexual selection,* or (the now rarely used) *epigamic selection*. The common use of either *intrasexual* or *male-male competition* referring to contest competition alone has been criticized, since both male-male competition and mate choice involve competition among males for mates (analogous to the ecological concepts of interference competition in the former and exploitative competition in the latter) (Andersson 1994). This dichotomy also obscures the fact that competition among males can take other forms besides combat, such as being the first to locate and reach females, endurance rivalry, and sperm competition. The distinction is nonetheless useful because male-male contest competition and female choice are the primary mechanisms likely to lead to the evolution of the kinds of extravagant traits most studied by students of sexual selection. Furthermore, if *male-male (or intrasexual) competition* is defined as including all modes of sexual selection, the term simply becomes synonymous with sexual selection and is thus unnecessary. Herein I will use *male-male competition* in the traditional sense of contest competition and distinct from the effects of mate choice and other forms of competition, and I thus avoid the more problematic *inter-* and *intrasexual selection* altogether (which, in any case, sound far more similar than any antonyms should). This approach does leave a terminological void, however, since we have no term for female choice from the males' perspective. I propose using the term *attractiveness competition* for competition among males deriving from differential attractiveness to females. Thus, attractiveness competition will lead to the evolution of traits that make males more attractive to females than their rivals. Attractiveness competition can lead to other forms of competition, such as contest competition, if males must compete for access to resources that will make them attractive.

The term *mate choice* (or, more commonly, *female choice*) is also open to multiple interpretations. Mate choice has been divided into active versus passive choice and into direct versus indirect choice. Active mate choice involves females sampling several males and then rejecting some, while passive choice reflects a tendency for females to orient toward or seek out the most conspicuous signal, perhaps simply because that signal is detected sooner than others (Andersson 1994). From a male's perspective, both active and passive choice can select for more conspicuous male signals. The difference between active and passive choice is more relevant to understanding the evolution of female choice, since active choice implies an evolved behavior, while passive choice could come about as a result of simple biases in the female's sensory systems that evolved for reasons other than mate choice (see below). Another distinction is between direct and indirect choice (Wiley and Poston 1996). Direct choice involves discriminating among the attributes of males and includes both active and passive choice as defined above. Indirect choice, on the other hand, includes any other behaviors that will lead to nonrandom mating, for example, behaviors that lead to increased male-male competition, such as forming same-sex aggregations, always mating in specific areas, or signaling sexual receptivity. Females may also exhibit a generalized reluctance to mate without any overt discrimination, which would favor males that can overcome such reluctance (such as through forced copulations). Many advantages that might accrue from direct mate choice may also be gotten from indirect choice; behaviors leading to indirect mate choice may have evolved in part because females derived benefits from mating with particular males (such as those vigorous enough to successfully defend a group of females). Here, I will use the more traditional definition of *mate choice* as direct choice that includes both active and passive mate choice but excludes indirect choice.

Darwin (1871) argued that male-male combat would lead to the evolution of male weaponry and other traits that enhanced fighting ability, such as large mandibles, horns, or tusks. The evolution of showy traits, such as elaborate male plumage or song, Darwin attributed to female choice. The idea that male-male contests could lead to the evolution of male weaponry was readily accepted, but the possibility of females choosing among males was met with considerable resistance. Many otherwise staunch supporters of Darwin, most notably Alfred Russel Wallace and Julian Huxley, criticized the idea of female choice, in part because it seemed to require that nonhuman females have an arbitrary "aesthetic sense" (the history of this debate is thoroughly detailed in Cronin 1991). Thus, Wallace and Huxley indirectly identified the biggest weakness of Darwin's sexual selection hypothesis: Darwin did not provide an explanation for why female choice should evolve.

Fisher (1915) was the first to suggest mechanisms for the evolution of female choice. Although best known for what we now refer to as the runaway or Fisherian model (below), Fisher also proposed an indicator model, in which the degree of expression of a male ornament indicated the male's quality (Fisher 1915). However, sexual selection did not attract much attention until after the republication of Fisher's book, *The Genetical Theory of Sexual Selection,* in 1958 and the rise in interest in evolutionary questions and an emphasis on individual selection in the 1960s (Cronin 1991). Perhaps most influential in renewing interest in sexual selection was Zahavi's (1975) proposal of an alternative to runaway selection: the handicap principle. Zahavi's hypothesis—which suggested that elaborate traits were deliberate handicaps, the very costliness of which indicated the bearer's fitness—stirred up considerable controversy as various theoretical models (discussed below) were developed initially suggesting such a mechanism was not plausible. The result was a great deal of interest in sexual selection by both theoreticians and empiricists that continues to this day.

Concepts

The Origin of Differences between the Sexes

Sexual selection can occur in any sexually reproducing organisms, even if monoecious. The most striking examples, however, will be found in organisms with separate sexes, where differences between males and females can evolve. The origin of gender is reviewed elsewhere (Rhen and Crews, this volume); here, I discuss its consequences.

In dioecious,[1] anisogamous[2] organisms, the sexes differ in their initial reproductive investment (Trivers 1972). Females, by definition, produce large eggs that are individually costly to produce. Males, on the other hand, produce plentiful, individually cheap sperm. As a result, a female can often obtain sufficient sperm to fertilize all of her eggs with just a single mating. Assuming no other reproductive investment beyond gamete production, female reproductive success is largely limited by the resources available to produce eggs. Males, on the other hand, can increase their reproductive success more or less linearly with each additional female they fertilize. This fundamental difference in the reproductive investment of the sexes leads to the evolution of different mating strategies of the sexes and, ultimately, sexual selection: Females, gaining little by seeking multiple matings, need mate only enough to obtain sufficient sperm, while males should seek as many matings as possible. As a consequence, females will be sexually receptive less often than will males, the result being a skew in the operational sex ratio (the ratio of sexually receptive males to sexually receptive females). This skew leads to competition among males for access to a limited supply of females, and thus, to greater variance in male reproductive success. The differences in parental investment resulting from anisogamy lead to sexual selection acting primarily on males rather than females and thus the most common pattern of sexual selection: choosy females and competitive males. The intensity of sexual selection also depends on the degree to which resources or females are monopolizable. This will depend on the distribution of females and is expected to be most intense in those species where females (or resources attractive to females) are highly clustered in space and time, and thus defendable (Sakai and Westneat, this volume).

Investment in gametes is not the only form of parental investment (Trivers 1972; Clutton-Brock 1991). In some species, males invest as much as females or more than females in reproduction by providing nutrients to females (Savalli and Fox 1998), by providing parental care, or even by gestating the young (Clutton-Brock 1991). In such cases, the sex roles and reproductive tactics may be similar between males and females or even reversed: Both sexes may be choosy or exhibit mate competition (Sakai and Westneat, this volume). Thus, even though I will, for simplicity, refer to male-male competition and female choice in subsequent discussions, it should be understood that these concepts can also be applied to female-female competition and male choice in role-reversed species. Recent theory suggests that differences in male and female strategies are not as absolute as is suggested by most models of sexual selection. Instead, both sexes should potentially choose their mates (Johnstone et al. 1996). In particular, females may vary considerably in fecundity, so that male choice of the most fecund females is favored. Female choice is more common than male choice because the cost of being choosy, in terms of lost reproductive opportunities, is generally much higher for males than for females.

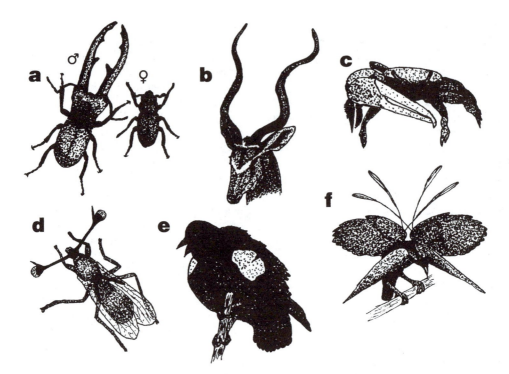

Figure 16.1 Examples of secondary sexual characters resulting from sexual selection. The most common kinds of differences include traits involved in contest competition (a–c) and traits used as signals (c–i). Traits used in contests include large body size and weaponry, such as large size and mandibles of male stag beetles *(Cyclommatus imperator)* (a); the horns of male greater kudu *(Tragelaphus strepsiceros)* (b); or the enlarged claw of male fiddler crabs *(Uca* sp.) (c), which is also used as a signal. Signals may be used in male-male contests or to attract females; they may be visual (c–f), auditory (g), chemical (h) or tactile (i); examples include the eye stalks of stalk-eyed flies *(Cyrtodiopsis whitei)* (d); the red epaulets (shoulders) of male red-winged blackbirds *(Agelaius phoeniceus)* (e); the elaborately ornamented plumage of a male standardwing bird-of-paradise *(Semioptera wallacei)* (f).

The fundamental difference in the size and motility of gametes produced by the two sexes means there must be differences in the gamete-producing organs (testes and ovaries) as well as in the organs responsible for gamete delivery (e.g., ovipositors, penises). Such primary sexual characteristics are not of concern here since they arise from natural selection rather than sexual selection. (Some modifications of primary sexual characteristics have functions other than gamete delivery and may have evolved via sexual selection; see figure 16.1I and Fairbairn and Reeve, this volume, for examples.) Of interest here are sexual differences in traits that do not have an immediate reproductive function. Such secondary sexual characteristics are very diverse and often quite striking (figure 16.1), despite the fact that males and females typically have iden-

tical ecological roles. Secondary sexual characters are usually attributed to sexual selection.

Mechanisms of Sexual Selection

Most kinds of sexual differences are thought to have arisen through one or more mechanisms of competition among males for access to females (table 16.1). As in ecological competition, males may compete for females by diverse mechanisms that include contests, endurance rivalry, scramble competition, attractiveness competition (arising from mate choice), and sperm competition (Andersson 1994). Contest competition involves males fighting for access to females or resources that are attractive to females. It favors the evolution of traits that enhance fighting ability, such as weapons or large

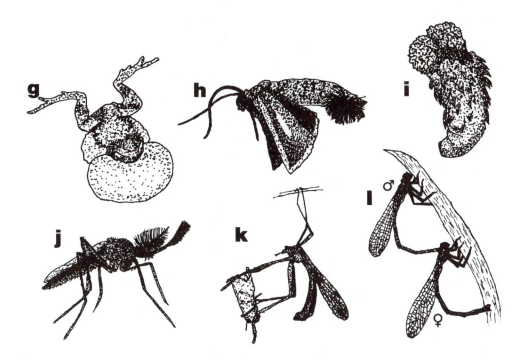

Figure 16.1 (*continued*)
The vocalizations of a male túngara frog *(Physalaemus pustulosus),* shown inflating his vocal sac while calling (g); the enlarged and everted pheromone-producing glands of a male looper moth (*Plusia* sp.) (h); and the complex structure of the hemipene of a snake (*Rhadinea* sp.), which may function as a tactile signal (i). Other kinds of sexually selected traits include enhanced sensory structures used to locate females, such as the elaborate antennae of male mosquitos (j); providing resources, such as the nuptial gifts provided by male hangingflies *(Hylobittacus apicalis)* (k); and mate-guarding behavior to prevent sperm competition, as in these damselflies (l), in which the male continues to grasp the female as she lays eggs.

size, as well as traits that can signal a male's fighting ability. It can also lead to the evolution of traits (such as small size, inconspicuousness, and sneaky or even female-mimicking behavior) that allow males to avoid direct contests by pursuing alternate reproductive tactics (Gross 1996). Scramble competition arises when males gain a mating advantage by locating females before other males do and favors well-developed locomotor and sensory abilities. Since such traits may also be favored by natural selection—such as to locate food or escape predators—differences between the sexes may not be as pronounced, and the role of sexual selection may be more difficult to ascertain. Endurance rivalry favors males that can remain active or at a display location (e.g., a lek) for the longest time (like scramble competition, it may not lead to any obvious sexual differences). Attractiveness competition involves males competing in their ability to attract and stimulate females. It favors the evolution of traits that are attractive to females, such as behavioral and morphological displays or the ability to sequester resources (nutrients, nest or oviposition sites, or territories) that attract females.

Until recently, most research, both theoretical and empirical, has focused on the premating aspects of sexual selection. However, sexual selection does not end with mating. For example, if females mate with multiple males, there may be competition between the males' sperm for the ability to access the eggs for fertilization, resulting in various adaptations to enhance fertilization success. These adaptations include mechanisms that remove or displace a rival's sperm (such as copious ejaculates and scrapers) and traits (such as mate-guarding behavior and sperm plugs) that prevent rivals from

Table 16.1 Principle modes of sexual selection (after Andersson 1994; Andersson and Iwasa 1996).

Mechanism	Characteristics	Kinds of traits favored
Contest competition		
Combat	Engagement in fights or other physical contests	Weapons, strength, large size (figure 16.1A–C)
Strength signals	Signals of strength or fighting ability used to assess each other and avoid direct combat	Various visual, auditory, tactile or chemical signals (figure 16.1B–E)
Alternative tactics	Poor competitors avoid direct contests by using alternative behaviors	Sneaky behaviors; inconspicuous or femalelike signals
Endurance rivalry	Mating success dependent on duration of tenure at breeding or display grounds	High vigor, nutrient reserves, and other traits that allow male to remain reproductively active
Scramble competition	Competition to be the first to gain access to females	
Via mobility	Rapid or more efficient locomotion	Well-developed locomotor traits; small size may allow for more efficient locomotion
Via searching	Detection of females before others	Well-developed sensory systems that are used to detect females (e.g., figure 16.1J)
Via protandry	Emerging or becoming reproductively active before others	Rapid development (may result in small size)
Attractiveness competition	Results from female choice: Males compete to be most attractive to females	Various visual, auditory, tactile, or chemical signals; nuptial gifts, sequestering nutrients (figure 16.1A–I, K)
Coercion	Males coerce or force copulations	Large size and strength; coercive behaviors
Infanticide	Males kill dependent offspring so females become sexually receptive sooner	Infanticidal behavior
Sperm competition		
Prevent copulations	Prevent other males from subsequently mating with female	Mate guarding behavior (figure 16.1L), sperm plugs
Displace rival's sperm	Remove or displace a previous male's sperm	Production of copious sperm; scrapers and other ways to displace rival's sperm

mating with females in the first place (Andersson 1994; Birkhead and Møller 1998). Females can continue to exhibit mate choice after insemination by controlling which sperm get used to fertilize the eggs (Eberhard 1996). Sexual selection can even occur after fertilization if females invest differentially in the offspring fertilized by different males.

The Evolution of Male Secondary Sexual Traits

The least controversial aspect of sexual selection is the evolution of male weapons. The potential use of such weapons in male-male contests is obvious and frequently observed. There is evidence from a variety of taxa, from insects to mammals, showing that males with larger, more robust weaponry are more successful at winning contests, although body size (which typically exhibits a positive correlation with weapon size) is rarely controlled experimentally (Andersson 1994). In addition to use in fights among males, weapons can also be used as a signal of fighting ability to intimidate rivals without a fight. Male weapons could also function in predator defense or as ornaments that are attractive to females, but there is little evidence to support either of these hypotheses.

Large body size is also likely to incur a competitive advantage and thus may be favored by contest competition or endurance rivalry. Sexual size dimorphism is common and widespread, but there is considerable variation in the nature and degree of dimorphism (Andersson 1994). The issue of body size is complex, however, as body size impacts all aspects of life history that must be considered when investigating patterns of sexual dimorphism. Sexual selection appears to be the most probable explanation for larger male size that is common in birds, mammals, and those insects equipped with male weaponry. There is less consensus on the cause of the more widespread pattern of females that are larger than males. The most common explanation is that large female size is favored by fecundity selection (larger females produce more or larger eggs or offspring). Fecundity selection may be sufficiently strong to favor larger females even when there is sexual selection for large male size. Sexual selection is sometimes also considered: Small males may be favored by female choice (either directly or through preferences for more agile aerial courtship displays) or scramble competition (small males may be more mobile or agile or may develop more rapidly). It should be noted, however, that numerous other hypotheses have been proposed to account for sexual size dimorphism; detailed treatment is beyond the scope of this chapter (see Fairbairn and Reeve, this volume, for an example of several forms of selection on male and female body size).

In many taxa, males have distinctive and often conspicuous visual or auditory signals. As with weapons, male signals may function to mediate male-male contests by signaling a male's dominance status or fighting ability (Zahavi 1977; Rohwer 1982). Some behavioral displays, such as the roaring and parallel walks of red deer or face-to-face displays of stalk-eyed flies, obviously allow males to assess each other. In other cases, the signal may simply indicate a male's presence on his territory. Although less obvious, ornamental traits, such as bright colors or long tails, can also function to signal dominance or intimidate rivals. Such status signaling has been most thoroughly studied in wintering flocks of birds (and thus outside the context of sexual selection) (e.g., Rohwer 1982) but could as easily apply to status signaling among males during the breeding season, as is the case with red-winged blackbird *(Agelaius phoeniceus)* epaulets (Searcy and Yasukawa 1995) and some widowbird tails (see case study, below). Cheating is prevented

either by the use of costly traits (weaker males would not be able to invest as much in the trait; Zahavi 1977) or through occasional escalated fights that test the signal bearer (Rohwer 1982).

Although conspicuous sexual ornaments could arise as signals used in male-male contests, more often they are attributed to female choice (often without direct evidence). If females exhibit a preference for males with more elaborate signals, the males exhibiting more attractive signals will obtain more matings or fertilizations than males with less preferred traits. It is clear that a strong female preference, especially in polygynous or promiscuous mating systems, can lead to high variance in male mating success and intense selection for favored traits. Female choice can also lead to the evolution of male ornaments in monogamous species if preferred males pair with more fecund females, pair earlier in the breeding season (which affects reproductive success), or obtain more extrapair matings (Darwin 1871; Møller 1994). Female preferences have been demonstrated by both correlative and experimental studies (reviewed in Andersson 1994; see case studies for examples). Correlative studies are often simpler technically but suffer the drawback that one cannot be certain of causation. Thus, a correlation between the expression of a male ornament and mating success is not sufficient evidence for female choice of that ornament: Females may be selecting males on the basis of some unmeasured trait that happens to correlate with the ornament. A more rigorous approach is to experimentally manipulate the ornament in question or to provide artificial signals (such as playbacks of vocalizations).

Which mode of sexual selection predominates can vary among closely related species (e.g., widowbird case study, below) or even among populations. For example, in sand gobies, *Pomatoschistus minutus,* male-male contest competition predominates in a population where nest sites are scarce, with little indication of female choice, while in a population with an abundance of nest sites, females preferentially mate with large males (Forsgren et al. 1996). Of 232 sexual selection studies (including all taxa, traits, and methodologies) surveyed by Andersson (1994), 167 (72%) reported evidence for female choice, although only 91 (39%) were under circumstances in which male-male competition was not prevented. In contrast, 58 studies (25%) reported sexual selection by male-male contests, while smaller numbers reported male mate choice or other kinds of male-male competition. In many cases, the sexually selected traits have dual functions, serving as both signals in male-male contests and signals to attract females. Although it is tempting to conclude that female choice is a more common mechanism of sexual selection than male-male contest competition, this pattern is likely to be biased, as there has been much more interest in female choice and because some forms of male-male competition are much more difficult to detect. Individually testing many species is impractical because of the time needed to conduct experiments, practical considerations that can make it difficult to detect one or the other form of sexual selection, and the presence of biases in the selection of study organisms and traits.

An alternative to experimental approaches is comparative approaches, such as that carried out by Irwin (1994) on icterid blackbirds. In these birds, plumage dimorphism is greater in polygynous species than in monogamous species. Irwin found that when male and female plumage brightness was mapped onto a phylogeny, female plumage color was more evolutionarily labile, and that changes in sexual dimorphism were due more often to changes in female plumage than to changes in male plumage. This finding suggests that plumage evolution may be determined more by selection on female social signaling (such as for territory defense in monogamous species) than by sexual selection on males. If female plumage does function in territory defense, it is likely that male plumage has a similar function, although Irwin's study did not test this hypothesis directly.

The Evolution of Female Choice: Direct Benefits

If male traits evolve in response to female choice, this begs the question: Why do female preferences evolve? Females can obtain a variety of benefits, both direct and indirect, by mating with certain males (table 16.2). Note that female choice evolves primarily through natural selection (via increased fecundity, offspring quality, or survivorship gained from mating with certain males) rather than through sexual selection (female-female competition for access to mates), although it typically leads to sexual selection on males.

Females often choose mates on the basis of resources controlled by males. Such choice can take

a variety of forms, such as females trading copulations for access to honeycombs defended by male orange-rumped honeyguides *(Indicator xanthonotus)*, females selecting territories defended by males, or females choosing to mate with males (or mating longer with males) that can provide nuptial gifts (such as prey items, nutritive spermatophores, or even the males' bodies). The role of nuptial gifts and other nutritive contributions has been well studied in invertebrates, where this form of male paternal investment is common (e.g., Savalli and Fox 1998; Sakaluk 2000). In vertebrates, on the other hand, female choice of resources (often in the form of defended resources such as territories) has not received as much empirical support as might be expected. One reason may be that it is often difficult for biologists (though the same may be true for the animal in question) to determine what attributes define a quality territory. Even such well-established hypotheses as the polygyny threshold model (which has as a fundamental assumption that females are choosing resources) have had only limited empirical success. Studies that have experimentally manipulated resources are uncommon but generally support the idea of female choice of resources (e.g., Searcy and Yasukawa 1995).

In addition to material resources, females may choose males that have demonstrated an ability to provide parental care. Unless females have had previous experience with particular males, they will need to rely on an indicator of potential male parental quality. Potential indicators include courtship feeding, displays of skill or agility (especially important for some predatory species), or the ability to sequester carotenoids. Additional forms of choice are listed in table 16.2, but these have been little studied.

The Evolution of Female Choice: Indirect Benefits

Much of sexual selection theory since the 1960s has focused on the possibility that females choose males on the basis of male ornaments or other traits that do not produce an immediate fitness benefit. Hypotheses for the evolution female choice fall into two broad groups: runaway or Fisherian sexual selection and good-genes models (reviewed in Andersson 1994; Andersson and Iwasa 1996; Ryan 1997).

Table 16.2 Potential benefits of choosing mates (after Andersson 1994).

Direct benefits
    Obtain access to resources (such as territories or nest/oviposition sites)
    Obtain food or other nutrients captured or synthesized by males
    Obtain quality paternal care
    Obtain sufficient sperm to fertilize eggs
    Mate with the correct species
    Avoid harassment
    Reduce risk of predation
    Avoid direct transmission of parasites
Indirect benefits
    Obtain good genes for offspring
    Runaway selection: Offspring inherit favored trait and preference

According to the runaway sexual selection hypothesis, an initial female preference for males with particular traits, such as long tails, would lead to a mating advantage for long-tailed males. Females that exhibited a stronger than average preference for long tails would be more likely to mate with long-tailed males, the result being a genetic correlation in the offspring between the alleles for long tails and the preference for long tails. Since the females' sons would exhibit the preferred long tails, they would have greater than average mating success. Because these sons also carry alleles for preferring long tails, the alleles for preferring longer tails will also spread through the population. In effect, the preference for long tails is self-reinforcing.

Good-genes models, also known as handicap or viability indicator models, postulate that the degree of expression of male ornaments reflects the overall viability or quality of the male. If viability is heritable, mating with the brightest or most ornamented males would yield offspring that inherit their father's superior genotypes. Male indicator traits are typically costly in that they reduce male survivorship (hence their characterization as "handicaps"). Their costliness ensures that they are honest indicators: Only the most fit males can invest fully in the display. Resistance to diseases and parasites may be an important component of male fitness. A coevolutionary arms race with parasites could maintain genetic variation in fitness, thereby preventing one potential problem with this mechanism: the loss of genetic variation in fitness while under selection. Other potential genetic benefits include an optimal level of outbreeding, genetic

compatibility, and increased heterozygosity of offspring.

Both the good-genes and runaway hypotheses have been evaluated extensively by means of genetic models, as well as some ESS models. These include both few-locus models (usually two or three loci in either haploid or diploid systems) and quantitative genetic models. The models are very general and simple and therefore have little quantitative predictive power and make too many assumptions (typically several dozen) to be effectively tested (Andersson 1994). Recent models do, however, suggest that both mechanisms are possible, and that typically, both processes can work simultaneously, enhancing each other's effects. Indeed, to test the good-genes hypotheses in the absence of a runaway effect it is necessary to design models that assume a monogamous mating system with equal reproductive success for all surviving males. Recent sexual selection models have been developed to include multiple traits and multiple preferences (Andersson and Iwasa 1996).

Empirical tests of these hypotheses have not kept pace with theoretical developments. Most tests of good-genes hypotheses demonstrate a relationship between male viability or fitness and the expression of the male trait. For example, the role of parasites has garnered a great deal of attention, in part because parasite levels can be measured quite easily and thus provide a quantifiable indicator of male quality. A variety of studies have demonstrated correlations between male parasite loads and degree of development of a sexually selected trait (e.g., Møller et al. 1998). Another potentially useful indicator of male viability is fluctuating asymmetry, the small, random deviations from perfect symmetry of bilaterally symmetric traits that result from developmental instability. Individuals of superior condition or viability should exhibit fewer developmental problems and thus be more symmetrical. Fluctuating asymmetries have been used to assess male quality and in some cases may be used by females to assess potential mates (Møller and Swaddle 1997).

Although evidence of correlations between male ornaments and some indicator of male fitness is predicted by and consistent with the good-genes models, it is not sufficient evidence to demonstrate that females are obtaining a good-genes benefit. A more robust test is to demonstrate that females mating with preferred, more viable males produce offspring of greater viability. Such experiments need to be carefully designed to exclude potential maternal and paternal effects, however. Few studies have adequately tested this prediction, and the results have been mixed (Ryan 1997). On average, female choice of males accounts for only a small proportion of the variance in offspring fitness (1.5% across studies, including possible maternal and paternal effects) (Møller and Alatalo 1999). Unfortunately, no easily testable predictions have been developed to test the runaway hypothesis; as a result, it is sometimes considered the null model when there is no support for direct benefits or good-genes benefits to female choice.

A final possibility that must be considered is that female preferences did not evolve for obtaining higher-quality mates but evolved instead through natural selection on females' sensory systems for unrelated functions (such as predator detection or locating food). Male traits then evolved to take advantage of preexisting biases within the female sensory system (Ryan et al. 1990), referred to as *sensory exploitation*. Even traits such as the presentation of nuptial gifts may have evolved via sensory exploitation (Sakaluk 2000). The possibility of sensory exploitation has been identified by a comparative approach in a few taxa (one of which is discussed in the *Physalaemus* case study below). If the phylogenetic distributions of the preference and the male trait suggest an earlier origin of the preference than the male trait, then males in at least some lineages must have been able to take advantage of a preexisting bias (see the *Physalaemus* case study and Figure 16.3, below). However, such data must be interpreted with caution since the

conclusions are highly dependent on the validity of the phylogeny used.

Case Studies

The Evolution of Long Tails in Widowbirds

Perhaps the best known of all sexual selection studies are Malte Andersson's (1982) experiments with the long-tailed widowbird, *Euplectes macrourus*. The males of this species are black with red shoulders and have enormously long tails (up to 1 m) during the breeding season. Females and nonbreeding males are streaky brown with short tails. Andersson manipulated tail length by cutting off part of the tail feathers of some males and gluing these onto others. He also had sham (tails cut and reglued to the same male) and unmanipulated controls. Males with elongated tails had more females subsequently nesting in their territory than did males with shortened tails (figure 16.2A). On the other hand, all males retained their territories, regardless of manipulation, indicating no effect of tail length on male-male competition. It should also be noted that there was no correlation between natural tail length and pairing success, suggesting that female preferences may often be too subtle to easily detect with only the natural range of variation, and that either experimental manipulation or extraordinarily large samples sizes will be necessary before we can claim the absence of female choice with any confidence.

Andersson did not follow up on his initial experiments, but other widowbirds have been studied by other researchers. Widowbirds exhibit considerable variation in tail length (see sketches in figure 16.2), even though all species are polygynous and ecologically similar, being predominantly granivorous birds of grasslands and marshes. In most species, males defend territories (usually quite large) and build the initial nest frame but provide little or no parental care. Multiple females may nest in a male's territory (up to nine in both long-tailed and yellow-shouldered widowbirds). One species, Jackson's widowbird, *E. jacksoni*, is unusual in that males display on a lek rather than defending all-purpose territories.

As in long-tailed widowbirds, female Jackson's widowbirds preferentially visit males with normal

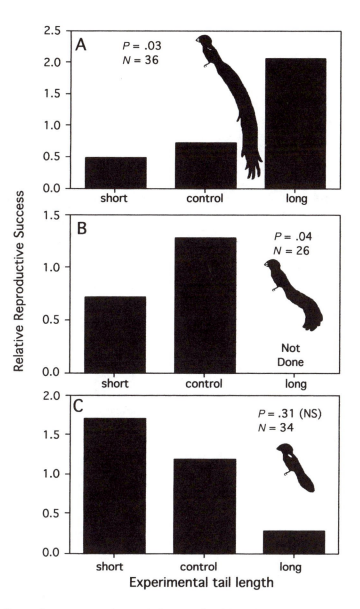

Figure 16.2 The effects of experimental manipulation of tail length on measures of male attractiveness in three species of widowbird *(Euplectes)*. (A) Long-tailed widowbird, *E. progne*. (After Andersson 1982.) (B) Jackson's widowbird, *E. jacksoni*. (After Andersson 1992.) (C) yellow-shouldered widowbird, *E. macrourus*. (After Savalli 1994b.) For comparison among species, attractiveness is presented as treatment mean/experimentwide mean. Attractiveness was measured as the number of females nesting in a male's territory for *E. progne* and *E. macrourus* or the number of females visiting a male's display site for *E. jacksoni*. Unmanipulated and sham controls were pooled as they did not differ in any study (there were no sham controls for *E. jacksoni*). The small silhouettes of widowbirds are drawn to the same body size to illustrate differences in relative tail length. *P* values represent the results of the original analyses, which differ among studies and do not necessarily reflect the way the data are presented here. Sample sizes are experimentwide (i.e., all treatments).

tails over those with experimentally shortened tails (tails were not elongated) (S. Andersson 1992; figure 16.2B). In my study of the yellow-shouldered widowbird, *E. macrourus,* on the other hand, manipulations of tail length had no effect on female preferences (the nonsignificant preferences were opposite what was expected; figure 16.2C) but did have a large effect on male-male competition: Males with shortened tails were more likely to lose their territory than males with normal (sham control or unmanipulated) or lengthened tails (Savalli 1994b). This species has the shortest tail of the widowbirds that have been studied, although the male's tail is still twice as long in the breeding season as in the nonbreeding season or in females.

Malte Andersson's study did not examine any variables other than tail length, while both Steffan Andersson's and my studies examined additional morphological traits as well as aspects of the male's territory. In yellow-shouldered widowbirds, the only variable that consistently correlated with male pairing success was the number of nest frames (cock's nests) that a male built (Savalli 1994a). This finding suggests that in this species, females may be selecting males on the basis of the nests they provide, a behavior which may be primitive in the genus, as males in most other weaverbirds are also responsible for building the nests. In the lekking Jackson's widowbird, a central display tuft seems to be necessary to attract females (Andersson 1991): Preferences for display courts with grass tufts may be homologous with preferences for nests. In addition to examining the role of sexual selection, I also considered nonsexual selection alternatives to the evolution of long tails—neither the species recognition nor the unprofitable prey hypotheses were supported by field and comparative data, a finding further strengthening the case for the role of sexual selection in the evolution of widowbird tails (Savalli 1995).

Long tails may have evolved initially as a male-male signal, possibly to make displaying males conspicuous to neighboring males across large territories. In Jackson's and long-tailed widowbirds, the resources provided by males have become less important, relative to the presumed ancestral condition, the result being a switch to female choice based on plumage characteristics. Since tail length may have been used to signal male quality to other males, it would make a good indicator for females as well.

Despite evidence that sexual selection favors long tails in three species of widowbirds, it is still not clear why there is so much geographic and interspecific variation in tail length. One possibility is that it is due to different modes of sexual selection: In the species in which male-male competition is more important, the tail is relatively short (10 cm), while in species in which female choice is important, the tails are longer (> 25 cm). Currently, there is neither an empirical nor a theoretical basis to indicate whether female choice should lead to more extravagant traits than male-male competition. Another possibility is that the cost of having long tails varies between species. One possible cost is that long tails make flight difficult in rain (long-tailed widowbirds are incapable of flying in the rain). This possibility is supported by comparative data I collected that show that species living in drier climates tend to have longer tails than those in wetter climates.

The Evolution of Female Preference for Complex Calls in Physalaemus *Frogs*

Male túngara frogs, *Physalaemus pustulosus,* produce complex (for a frog) vocalizations that include a drawn-out whine and zero to six shorter broad-frequency chucks (Ryan 1985; figure 16.3A). Extensive studies by Ryan (1985), using observational data as well as playback of natural and artificial calls, demonstrated that females prefer whines with chucks over whines alone (although they will mate with males producing only whines) and prefer chucks with lower fundamental frequencies. Since, as in most vocal species, large males can produce lower sounds than small males, large males have greater mating success. Producing these vocalizations is costly for males. Calling uses 4.5 times as much energy as resting and also increases the risk of predation, especially from frog-eating bats, *Trachops cirrhosus.* These bats are preferentially attracted to calls with chucks, presumably because they are easier to localize than the whines alone. Consequently, males vary the number of chucks they produce, depending on the risk of predation (producing few chucks if alone and more if in a large chorus) (Ryan 1985 and references therein).

Male túngara frogs do not defend territories (instead, they are attracted to chorusing males),

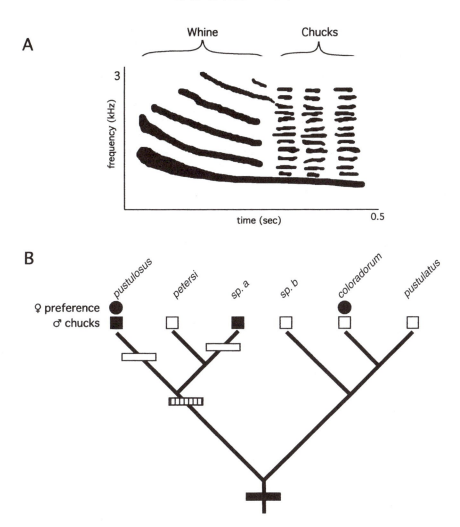

Figure 16.3 (A) Sonogram of a typical túngara frog, *Physalaemus pustulosus*, vocalization, showing the whine and three chucks. (After Ryan 1985.) (B) A cladogram illustrating the relationships between the six species in the *P. pustulosus* species group (after Ryan 1997) and the presence (solid figures) or absence (open figures) of male chucks (squares) and female preference for calls with chucks (circles). Solid bar: the probable origin of the female preference for chucks. The chuck component of the call evolved either once (hatched bar) with a subsequent loss or evolved twice independently (open bars).

and male vocalizations do not appear to influence male-male competition (Ryan 1985). Furthermore, ovipositioning takes place away from the males' calling sites, where they pair with females. Thus, there is no evidence and little opportunity for either male-male competition or for female choice of resources. Nor can species recognition account for the evolution of the chuck component, since only the whine is necessary to attract females. This system is one of the clearest demonstrations of female preference for a costly trait and has also proven to be very suitable for physiological and comparative studies of that preference.

Ryan et al. (1990) studied the neurophysiology of signal reception in female *Physalaemus* frogs in detail. The whine component of the call is processed primarily by the amphibian papilla region of the inner ear (which is most sensitive to low frequencies). The chuck call, however, consists of a broader range of frequencies and stimulates both the amphibian and basilar papilla (which is sensitive to higher frequencies). Thus, stimulation of the amphibian papilla is necessary to stimulate a female response, but females will respond more strongly if the basilar papilla is also stimulated. The female preference for lower-frequency chucks comes about because the basillar papilla produces greater neural stimulation at lower frequencies.

The *Physalaemus pustulosus* species group consists of six allopatric neotropical frogs that all have a whine component to their vocalizations (Ryan et al. 1990; Ryan 1997). However, the vocalizations of these species differ in other respects, some having only the whine and others having additional components, such as the chucks, before or after the whine. *P. coloradum,* for instance, has only a whine, but females nonetheless prefer *P. coloradum* calls with artificial chucks added over conspecific whines alone. This similar behavioral response of the two species can be explained physiologically: The basilar papilla of *P. coloradum* is tuned to the same frequencies as the basilar papilla of *P. pustulosus* (Ryan et al. 1990). Thus, the most parsimonious interpretation is that the female auditory system, including the basilar papilla, and thus female preferences, did not coevolve with the male trait but originated in a common ancestor of the *P. pustulosus* species group (figure 16.3). The male chuck component subsequently evolved in some of these species to exploit this bias in the female auditory system. The discongruity between the origin of female preferences and male traits has been demonstrated in only a handful of species (Ryan 1997); its importance as a mechanism for the evolution of female preferences remains uncertain.

Future Directions

Despite considerable research in sexual selection in recent decades, many questions still remain to be satisfactorily answered. The most basic of these is the relative importance of attractiveness versus contest signaling for the evolution of male ornaments. Comparative studies may prove to be fruitful in addressing this area. The second major question in sexual selection, for which there are even fewer data, concerns the relative importance of runaway

selection, good-genes selection, and sensory exploitation in the evolution of female preferences. Also of interest is whether the strength of selection varies with the mechanism of sexual selection involved. To date, few empirical data have been brought to bear on this question. Current models are just being developed to account for some of the complexity of sexual selection, such as multiple secondary sexual traits (e.g., the black plumage, colored shoulders, head ruff, songs, and larger size of male widowbirds, in addition to their long tails) and the considerable divergence often seen among populations and closely related species.

Mate Choice Copying

Recent experiments and observations, primarily with birds and fish, suggest that females may regularly copy the mate choice of others (e.g., Kirkpatrick and Dugatkin 1994). The theoretical implications and empirical basis need to be explored further. For instance, under what circumstances should females copy others rather than make their own decisions? What are the costs (such as making inappropriate choices) and benefits (such as reduced search costs) of copying? Is there heritable variation in the tendency to copy? What is the effect of mate choice copying on the evolution of male traits? Although it seems intuitive that copying should increase the overall similarity of female preferences, thereby increasing variance in male mating success, mate choice copying could, under some circumstances, have the opposite effect—increasing variance in female preference, thus decreasing variance in male mating success (Kirkpatrick and Dugatkin 1994). Which outcome prevails depends in part on how consistent females are in their mating preferences in the absence of copying. Probably most critical for understanding copying, and also applicable to many mate choice models, is the cost of mate choice. Unfortunately, there are few data available for estimating such costs.

Multiple Mating and Postcopulatory Sexual Selection

Although females often obtain sufficient sperm to fertilize all of their eggs with just a single mating, females often mate multiple times with the same or different males. This is true even for species that are socially monogamous: There is a growing body of data, using biochemical or genetic markers, showing that extrapair copulations are common in many birds. At least 16 hypotheses have been proposed to account for multiple mating, both with the same male and with different males. Benefits may be direct (such as fertilization assurance, access to nuptial gifts, or avoidance of male harassment) or indirect (such as promoting sperm competition or allowing for sperm choice) (reviewed in Birkhead and Møller 1998). Unfortunately, empirical work has not kept pace with theoretical developments. Many of the hypotheses have not been adequately tested; we do not have a clear understanding of what accounts for interspecific variation in multiple mating and extrapair copulations.

One of the key consequences of multiple mating by females is that sexual selection may continue past mating. Sperm competition has been recognized for several decades now, and a variety of mechanisms have been described (Birkhead and Møller 1998). Only recently has the possibility of postmating female choice been emphasized (Eberhard 1996). Much work remains in this area, from documentation of the methods used by females to choose among sperm (or offspring) to distinguishing between female choice, sperm competition, and male manipulation.

Notes

1. Separation of the sexes into separate individuals; in animals, this is often referred to as *gonochorism*.

2. The situation where male and female gametes differ in size.

17

Cooperation and Altruism

DAVID SLOAN WILSON

People have always been fascinated by cooperation and altruism in animals, in part to shed light on our own propensity or reluctance to help others. Darwin's theory added a certain urgency to the subject because the principle of "nature red in tooth and claw" superficially seems to deny the possibility of altruism and cooperation altogether. Some evolutionary biologists have accepted and even reveled in this vision of nature, giving rise to statements such as "the economy of nature is competitive from beginning to end . . . scratch an 'altruist' and watch a hypocrite bleed" (Ghiselin 1974 p. 247). Others have gone so far in the opposite direction as to proclaim the entire earth a unit that cooperatively regulates its own atmosphere (Lovelock 1979).

The truth is somewhere between these two extremes; cooperation and altruism can evolve but only if special conditions are met. As might be expected from the polarized views outlined above, achieving this middle ground has been a difficult process. Science is often portrayed as a heroic march to the truth, but in this case, it is more like the Three Stooges trying to move a piano. I don't mean to underestimate the progress that been made—the piano has been moved—but we need to appreciate the twists, turns, and reversals in addition to the final location.

Historical Overview

To see why cooperation and altruism pose a problem for evolutionary theory, consider the evolution of a nonsocial adaptation, such as cryptic coloration. Imagine a population of moths that vary in the degree to which they match their background. Every generation, the most conspicuous moths are detected and eaten by predators while the most cryptic moths survive and reproduce. If offspring resemble their parents, then the average moth will become more cryptic with every generation. Anyone who has beheld a moth that looks *exactly* like a leaf, right down to the veins and simulated herbivore damage, cannot fail to be impressed by the power of natural selection to evolve breathtaking adaptations at the individual level.

Now consider the same process for a social adaptation, such as members of a group warning each other about approaching predators. Imagine a flock of birds that vary in their tendency to scan the horizon for predators and to utter a call when one is spotted. It is not obvious that the most vigilant individuals will survive and reproduce better than the least vigilant. If scanning the horizon detracts from feeding, the most vigilant birds will gather less food for themselves than their more oblivious neighbors. If uttering a cry attracts the attention of the predator, then sentinels place themselves at risk by warning others. It is entirely possible that birds who do not scan the horizon, and who remain silent when they see a predator, survive and reproduce better than their vigilant neighbors.

These two examples show that the evolutionary concept of adaptation does not always conform to the intuitive concept, especially at the group level. It is easy to imagine a bird flock as an adaptive unit. We would expect members of the flock to adopt the creed "All for one and one for all." We might expect sentries to be posted at all times to detect predators at the earliest possible moment and to relay the information to feeding members of the flock. Unfortunately, individuals who possess these behaviors do not necessarily survive and reproduce better than individuals who enjoy the benefits and do not share the costs. Since Darwin's theory relies entirely on differential survival and reproduction, it appears unable to explain groups as adaptive units. This can be called the fundamental problem of social life. Groups function best when their members produce benefits for each other, but it is difficult to translate this kind of social organization into the currency of biological fitness.

Darwin was aware of the fundamental problem of social life and proposed a solution. Suppose there is not just one flock of birds but many flocks. Furthermore, suppose that the flocks vary in their proportion of vigilant callers. Even if a vigilant caller does not have a fitness advantage within its own flock, groups of vigilant callers will be manifestly more successful than groups whose members do not look out for each other. In short, group-level adaptations can be favored by a process of natural selection among groups, just as individual-level adaptations are favored by a process of natural selection among individuals within each group (see Nunney, this volume). This became known as the *theory of group selection.*

Darwin's solution to the fundamental problem of social life is elegant, but it was rejected by most evolutionary biologists during the 1960s (Williams 1966). The critics accepted Darwin's basic reasoning but argued that group selection was too weak as an evolutionary force to make a difference. Virtually all adaptations in nature must be explained in terms of natural selection within single groups. This became known as the *theory of individual selection.* Of course, nature isn't entirely brutish, and

animals do seem to be nice to each other at least some of the time. Two theories arose to explain such apparent niceness without invoking group selection. *Inclusive fitness theory,* also called *kin selection theory* (Hamilton 1964), explains how niceness can evolve among genetic relatives. The *theory of reciprocal altruism* (Trivers 1971) explains how niceness can evolve among nonrelatives as long as favors are returned in an I-scratch-your-back-and-you-scratch-mine arrangement. Both theories portrayed cooperation and altruism as forms of self-interest consistent with individual selection theory and in contrast to a more genuine form of cooperation and altruism that requires group selection. In a somewhat separate development, even individual-level adaptations were portrayed as a form of genetic selfishness in what became known as *selfish gene theory* (Dawkins 1976). Finally, a branch of economics developed to study cooperation in humans was modified to form *evolutionary game theory,* which broadly overlaps with the theory of reciprocal altruism (Maynard Smith 1982; Axelrod and Hamilton 1981).

The reader may feel overwhelmed by all of these theories. However, most of them are not alternative theories in the classical scientific sense, meaning that some can be rejected and others accepted on the basis of empirical tests. Inclusive fitness theory and evolutionary game theory will never be rejected in favor of selfish gene theory. Instead, these theories are seen as different ways of studying the same evolutionary process. The gene's-eye view offers one useful perspective, the individual's-eye view another, and the two perspectives are mutually consistent. The only exception to this rule is group selection theory. It invokes a process—natural selection at the group level—that can be shown by empirical test to be insignificant in nature. Thus, group selection deserves to be rejected in the classical scientific sense, while the other theories happily coexist with each other.

At least, that was the common wisdom for most of the second half of the 20th century, still reflected in many college textbooks today. It turns out that the rejection of group selection theory was premature and was always based more on theoretical considerations than empirical tests. All of the theories outlined above assume that social interactions take place in groups that are small relative to the total population. When these groups are carefully identified, it can be shown that cooperation and altruism are selectively disadvantageous within groups and evolve only because of group-level selection. In other words, Darwin's elegant solution to the fundamental problem of social life emerges as the general solution. All challengers have the same solution embedded in their own structures. The general theory is often called *multilevel selection theory* to emphasize that natural selection can and does take place on a hierarchy of levels, from genes within single individuals (Hurst et al. 1996) to entire ecosystems (Swenson et al. 2000).

Working with private correspondence in addition to public records, Schwartz (2000) shows how Hamilton first developed inclusive fitness theory as an alternative to group selection, only to later see it as a kind of group selection. Sober and Wilson (1998) recount the fascinating history of this subject in more detail. Of course, from the conceptual standpoint, the history is not fascinating but confusing and even maddening. If scientific progress is like the Three Stooges moving a piano, students should be able to learn the final position rather than the tortuous path taken to get there. Having provided a historical overview, I will now attempt to explain how altruism and cooperation evolve from first principles, with minimal reference to history.

One final comment is in order before continuing. It is common to distinguish between evolutionary definitions of cooperation/altruism, which are based on survival and reproduction, and psychological definitions, which are based on motives. How or even whether animals think or feel as they perform their actions is irrelevant as far as the evolutionary definitions are concerned. Even plants can be altruistic, for example, if they evolve to refrain from overshadowing their neighbors. The psychological concept of altruism is analyzed from an evolutionary perspective in Sober and Wilson (1998) and will not be further considered here.

Coordination and Distribution

Groups of individuals can often achieve more together by coordinating their activities than they can alone or in less coordinated groups. The advantages of coordination include combined action, division of labor, parallel processing, and a host of other factors that are familiar to us in our own efforts to achieve more in groups than we can alone. Of course, some activities are best performed by single individuals or by an optimal group size, and additional individuals only get in the way. It is important to appreciate the potential advantages of living in groups without being an uncritical "groupie."

To explain the evolution of coordinated action, we must show how the genes responsible for coordinated action evolve in competition with alternative genes that code for other behaviors. Obviously, there must be advantages of coordination, but this is only a necessary and not a sufficient condition. In addition, we must know how the costs and benefits of coordinated action are distributed among members of the group. If the genes responsible for coordinated action bear most of the costs while the alternative genes receive most of the benefits, then coordinated action may not evolve despite its benefits. Just as people are familiar with the advantages of coordination, they are all too familiar with the problems of distribution, which go by names such as *selfishness* (in the everyday sense of the word), *social dilemma, the free-rider problem,* and *the tragedy of the commons* (Hardin 1968).

The distinction between altruism and cooperation is made primarily on the basis of distribution. An altruistic behavior benefits others at the expense of the self and therefore poses a severe distribution problem. A cooperative behavior benefits others along with the self and therefore poses less of a distribution problem. However, this distinction is not nearly as clear-cut in the evolutionary literature as it might seem. A common model of altruism is to let c = the cost to the altruist and b = the benefit to the recipient of the altruism. This seems clear enough but consider what happens when we pair altruists and selfish individuals with each other. An altruist paired with another altruist gets a net fitness gain of $b - c$; its own cost plus the benefit from its partner. An altruist paired with a selfish partner gets a loss of $-c$, its selfish partner gets a benefit of b, and a selfish individual gets nothing when paired with its own kind. These numbers qualify as a prisoner's dilemma in evolutionary game theory, which is the paradigmatic model of *cooperation,* not altruism! This transformation occurs because altruism is defined on the basis of the altruist's effect on its own fitness ($-c$), regardless of whom it is paired with, while cooperation in game theory is defined on the basis of the interaction between the so-called cooperator and its partner. In this fashion, even very costly altruistic acts can be made to appear to be self-interested acts of cooperation by pairing the altruist with another altruist. To avoid this kind of confusion, we need to focus on the problem of distribution rather than making a rigid distinction between cooperation and altruism.

Coordination and distribution together provide the subject matter for an evolutionary theory of cooperation/altruism. It is important to appreciate that both are interesting and complicated subjects in their own right and even more interesting and complicated when they interact with each other. Historically, interest in distribution has tended to overshadow interest in coordination. The challenge has been to show how altruistic behaviors that benefit others at the expense of the self can evolve by natural selection. E. O. Wilson (1975, p. 3) even called altruism "the central theoretical problem of sociobiology." Since coordination without sacrifice easily evolves, it seems that little more needs to be said about it.

Individual Organisms: The Ultimate Cooperative Groups

To see why coordination remains interesting in the absence of distribution problems, consider individual organisms. Selfish gene theory envisions individuals as groups of genes that are social actors in their own right. Against this background, imagine a mutant moth gene that improves the cryptic coloration of the moth. This gene must *coordinate* its effect with other genes that code for other aspects of cryptic coloration. The gene is not altruistic, benefiting other genes in the moth at the expense of itself. Neither is the gene exploiting the other genes in the moth for its own benefit. It simply benefits all the genes in the moth, including itself, to an equal degree. Natural selection is not taking place within the group of genes formerly known as the moth, but groups with the mutant gene collectively survive better than groups without the mutant gene. In short, the standard process of individual selection (all we have done is describe it from a novel perspective) is about the evolution of coordination in the absence of distribution problems— but that does not make it uninteresting! We are fascinated by the evolution of beneficial genes and how they coordinate with each other to produce the wonderful and intricate adaptations that allow organisms to survive and reproduce in their environments. By analogy, if groups of individuals can evolve into coordinated units, unfettered by problems of distribution, it would be bizarre to call these group-level adaptations uninteresting just because they easily evolve (Wilson 1997a)!

Modeling Coordination and Distribution

Now that we have both coordination and distribution firmly in mind, we can proceed to model them theoretically. As we just saw, genes interact primarily with other genes in the same individual, which are a very small part of the total gene pool. The gene pool is a population of individuals in addition to a population of genes. Similarly, individuals usually interact in social groups that are a small part of the total population. The gene pool is a population of groups in addition to a population of individuals and a population of genes. For example, if we are interested in how individuals fight over territories, the individuals contesting a given territory are obviously interacting with each other and obviously comprise a small subset of the total population. The disease organisms in a single host interact with each other, in part through their effect on their host. The number of bacterial and viral organisms in a single host often exceeds the number of people on the planet earth, yet they are still only a small subset of all the disease organisms in all hosts. A plant might interact only with its immediate neighbors, which are a very small subset of the total plant population. In this case, the interacting individuals do not form a discrete group. Unlike disease organisms in hosts, each individual stands at the center of its own group, and the effect of individuals on each other declines continuously with distance rather than encountering a sharp boundary. Nevertheless, even in this case, the social interactions are localized and take place among a small subset of the total population. Social interactions sometimes encompass an entire population, either because the population is very small (e.g., a bird species on an island) or because the effects of the behavior are very widespread (e.g., global warming caused by human activity). As we shall see, behaviors that encompass entire populations present grave problems for the evolution of cooperation/altruism.

The fact that social interactions have a *population structure* (Nunney, this volume), a term that loosely refers to large populations broken up into smaller groups, is critical for the evolution of cooperative and altruistic behaviors. Seemingly small differences in the population structure, such as whether a disease organism is transmitted through the air or through water, can have a large effect on the degree of cooperation/altruism that evolves (Ewald 1994). To understand these complications, it is helpful to begin by understanding the evolutionary consequences of behaviors within the localized groups of individuals that are actually interacting with each other. To fix ideas, imagine a single group of N individuals composed of two types, A (for altruistic) and S (for selfish), in proportions p and $(1 - p)$, respectively. All individuals have a baseline fitness of X. In addition, each A type performs a behavior that increases the fitness of everyone in its group (including itself) by an amount b, at some expense, c, to itself (throughout this chapter, I will be changing the specific definitions of b and c to fit particular cooperative/altruistic behaviors). A types therefore have a fitness of $W_A = X - c + pNb$ and S-types have a fitness of $W_S = X + pNb$. If we treat this single group as an evolving population in its own right, does evolution favor the benefit-producing A type or the deadbeat S type? The answer is that the deadbeats win as far as evolution within the single group is concerned when $W_S > W_A$, which works out to $c > 0$. The value of b is irrelevant because the benefits affect everyone in the group equally. Evolution is about *differences* in fitness. Benefiting everyone in the group is like buying a lottery ticket for everyone in the lottery. The values of N and p are irrelevant for the same reasons. Even if the A types manage to benefit the group at no cost to themselves ($c = 0$) the best they can do is break even as far as their relative fitness is concerned $(W_A = W_S)$. A types can be favored by evolution within the group only when their effect on themselves is positive ($c < 0$), apart from their effect on the group of which they are a part. There is no way to explain the evolution of benefits to the whole group on the basis of the evolutionary forces taking place within the group (Williams 1966). This is the fundamental problem of social life, stated in mathematical form.

Luckily, evolution within a single group is only part of the story of evolution in structured populations. Table 17.1 shows a simple structured population in which a population is divided into two groups that start with the same number of individuals but differ in their proportions of the two types. The A types have the lowest fitness within each group, as we have already seen. However, the group with more A types grows larger than the other group because the A types, after all, are con-

Table 17.1 The evolution of cooperation/altruism in a structured population.

Total population before selection $N = 200, P = .50$	
Group 1	Group 2
$n_1 = 100$	$n_2 = 100$
$p_1 = 0.3$	$p_2 = 0.7$
$W_A = 3.8$	$W_A = 7.8$
$W_S = 4.0$	$W_S = 8.0$
$n'_1 = 394$	$n'_2 = 786$
$p'_1 = 0.29$	$p'_2 = 0.69$

Total population after selection $N' = 1180, P' = .56$

Note: A population of $N = 200$ individuals with altruists *(A)* at a frequency of $P = .50$ is divided into two groups that are initially equal in size ($n_1 = n_2 = 100$) but differ in their proportion of altruists ($p_1 = 0.3$, $p_2 = 0.7$). Both types have a baseline fitness (measured as number of offspring) of 1 in the absence of altruism. Altruists confer a benefit of $b = 0.1$ on everyone in their group (including themselves) at a cost of $c = 0.2$. Selfish individuals enjoy the benefits of altruism without paying the cost. The altruists decline in frequency within each group (compare p'_1 with p_1 and p'_2 with p_2), but the group with more altruists produces more offspring than the group with fewer altruists (compare n'_2 with n'_1). When both opposing levels of selection are considered, the frequency of altruists in the total population has increased from $P = .50$ to $P' = .56$.

ferring benefits on their whole group. The S types outreproduce A types within groups, but groups with more A types outreproduce groups with more S types. This is Darwin's solution to the fundamental problem of social life, stated in mathematical form. Fitness differences at both levels must be included if we are to calculate evolutionary change in the total population.

To proceed further, we must know how the groups interact with each other. If they are permanently isolated, then the evolution of selfishness will run its course within each group and there was no warrant for thinking of them as part of the same structured population. If they are spatially isolated groups that trade a fraction of dispersers every generation, we would need to calculate the new values of N and p for a number of generations to see what evolves. If the groups compete by direct warfare, we would need to follow a different procedure. These details depend on the particular organism and behavior we are studying and can have a large effect on what evolves. One common model assumes that the individuals simply disperse from their groups, like butterflies flying away from the plants on which they grew as larvae, to lay eggs on new plants that become the groups (for the larvae) of the next generation. This allows one to simply combine the individuals from both groups to determine the average fitness of A and S in the total population. It turns out that A is *more* fit than S overall in table 17.1, even though it was *less* fit than A within each group (see Sober and Wilson 1998, pp. 23–25, for a fuller discussion of this paradox). The effect of A types on their whole group makes a difference that is great enough to counteract their disadvantage within groups.

It is clear from this example that the evolution of altruism/cooperation depends critically on variation among groups. If there was no initial variation among the two groups ($p_1 = p_2 = 0.5$), there would be no group selection, and the outcome of this model would be the same as for a single group. Conversely, if the two types were completely segregated into different groups ($p_1 = 0$, $p_2 = 1$), there would be no individual selection, and A types would evolve whenever they benefit their group as a whole ($bN - c > 0$). In general, variation can be partitioned into within-group and between-group components, which strongly influence the balance between levels of selection.

I have described the evolution of cooperation/altruism as a process of multilevel selection, but all other theories fit easily into the same framework, even though they were originally proposed as alternatives to group selection. In addition, multilevel selection theory identifies new possibilities for the evolution of cooperation/altruism that were not imagined by its apparent rivals. In the rest of the chapter I will show how multilevel selection theory both *includes* and *goes beyond* previous formulations. In addition, I will provide some real-world examples of cooperation/altruism to put some empirical flesh on the theoretical bones.

Old Facts and New Possibilities

Factors That Influence Variation among Groups

We have seen that the partitioning of variation within and among groups is critical for the evolution of cooperation/altruism. One important factor that influences this partitioning is genealogical relatedness. Suppose we want to predict the behaviors that evolve among a category of genealogical relatives, such as members of the same clone. Implicitly, we are assuming that a large population is broken up into groups of individuals that are genetically identical to each other. In this case there is no variation within groups, and individuals should evolve to be maximally altruistic toward each other. Other categories of genealogical relatives such as "full siblings," "cousins," and so on assume different kinds of groupings with different partitionings of variation. The higher the degree of relatedness, the more variation among groups (and less within groups). The coefficient of relatedness *(r)*, which is the centerpiece of inclusive fitness theory and is often defined as the probability of sharing genes identical by descent, can also be understood as an index of variation among groups in a structured population, *regardless* of whether the genes are identical by descent (Hamilton 1975).

Although genealogical relatedness is *one* mechanism for increasing variation among groups, it is by no means the *only* mechanism. Another possibility is for altruists and selfish types to sort on the basis of their social interactions. After all, if individuals can learn about each other's behavioral dispositions and can freely choose their social partners, altruists should band together and force the deadbeats to interact with each other by default (Wilson and Dugatkin 1997). In a recent study of American college students (Sheldon et al. 2000), incoming freshmen were given a questionnaire measuring their degree of helpfulness. Four months later, they retook the same test and were asked to give copies to three friends that they had made in college. These groups of friends ($N = 4$) then played a social dilemma game in which individuals could earn points for their group or for themselves at the expense of their group. Within each group, less helpful members (as measured by the questionnaire) earned more points than more helpful members, demonstrating the advantage of selfishness within groups. However, more helpful groups (measured as the average score of the members) earned more points than less helpful groups, demonstrating the advantage of helpfulness between groups. The partitioning of variation in test scores within and among groups was nonrandom and corresponded to a coefficient of relatedness of r = 0.18. Thus, after only 4 months of social interactions, these groups of genealogically unrelated students had achieved a degree of sorting approaching that of groups of first cousins ($r = 0.25$). It would be interesting to know if sorting on the basis of behavioral interactions takes place in nonhuman species.

In the spirit of simplicity, most evolutionary models assume that behaviors are coded directly by genes (i.e., one allele that codes for altruism and another that codes for selfishness at a single locus). Unfortunately, this assumption has the effect of linking behavioral variation within and among groups directly with genetic variation. For example, a group must be genetically uniform to be behaviorally uniform. In reality, we know that genes usually code for complex psychological mechanisms, even in so-called simple creatures such as invertebrates, which interact with the environment to produce behavior. As a result, the *behavioral* partitioning of variation within and among groups no longer corresponds directly to genetic variation. It becomes possible for a group to be behaviorally uniform even though it is genetically diverse, or for groups to become behaviorally different even though they are genetically the same. This more realistic portrayal of the gene-behavior relationship has implications for multilevel selection theory that are only beginning to be explored. For example, consider a gene that codes for the rule "copy the most common behavior in your group" (Boyd and Richerson 1985; Wilson and Kniffin 1999). This gene does not code directly for a behavior; rather, it provides a rule for acquiring a behavior from the surrounding environment. Individuals who follow this rule will quickly become behaviorally uniform within single groups, while groups will differ depending on which behavior was originally in the majority. In other words, all of the behavioral variation will exist at the between-group level, a population structure that maximally favors the evolution of cooperation/altruism. This example is admittedly oversimplified, but it is a step in the right direction. Human behavioral variation within and among groups is influenced far more by social norms, imitation, and other complex psychological and cultural processes than by genetic variation per se. The same can probably be said for many nonhuman species. It is humbling to think how much evolutionary models of cooperation/altruism have ignored by assuming a direct connection between genes and behaviors.

Factors That Influence Selection within Groups

Extreme variation among groups allows coordination to evolve even in the presence of distribution problems. Another way for coordination to evolve is by reducing the distribution problem within groups. A good example is the human social convention of drawing straws. This practice is a solution to a distribution problem, in which providing benefits for the group requires a high cost on the part of a single individual. If the behavior is performed voluntarily in the absence of additional incentives it would count as strongly altruistic and would require extreme variation among groups to evolve. Drawing straws solves the distribution problem by randomizing the probability of becoming the altruist. In a group of size N that draws straws, all individuals have an expected payoff of $[b(N-1)-c]/N$, where b is the benefit to those who draw the long straw and c is the cost to the one who draws the short straw. All members of the group have the same payoff in a probabilistic sense, while groups that draw straws will outperform groups that don't draw straws (and don't perform the behavior) whenever the behavior increases the fitness of the group as a whole [$b(N-1)-c] > 0$—a degree of "altruism" that would evolve only among genetically identical individuals in a kin selection model! Of course, the practice of drawing straws is a miniature social contract that must be enforced. The person who draws the short straw cannot request a rematch and is obliged to perform her duty. If she doesn't, she will probably experience the moral outrage that is so characteristic of our species. To explain the evolution of a social convention such as drawing straws, we must explain the evolution of the underlying behaviors, but such explanations turn out to be plausible and do not require extreme self-sacrifice (see Sober and Wilson 1998, chapter 4, for a review).

Lottery processes similar to drawing straws may be more common in nonhuman species than we currently realize. Two possible examples are specialized foraging in a leaf-cutting ant *Acromyrmex versicolor* (Rissing et al. 1989) and stalk development in the slime mold *Dictyostelium discoideum* (Matapurkar and Watve 1997). The ant colonies are initiated by several genetically unrelated queens, but only one engages in the risky task of foraging above ground for the leaf material that will be shared by all members of the group. The slime mold is an amoeba whose life cycle includes both a solitary stage for foraging and a group stage for dispersal. When the local environment deteriorates, the individual amoebae stream together to form a collective body that migrates to a suitable spot and develops into a tower of nonreproductive stalk cells supporting a ball of spores. The stalk cells facilitate the dispersal of the spore cells at the cost of their own reproduction—the ultimate distribution problem. For both the ants and the slime mold, biologists have tended to assume that members of a group must be genealogically related for such high-cost altruism to evolve. However, it is equally possible that the groups are genetically diverse and that social interactions within the group randomize the probability of becoming the "altruist," making extreme variation among groups unnecessary. Future research is required to decide among these possibilities.

Social Groups All the Way Down

Evolution is often envisioned as a long series of mutational steps, from the origin of life, to bacteria, to the multicellular organisms of today. It is almost certain that another evolutionary process is at work, in which social groups become so functionally integrated that they become a new organism. The individuals of today are actually the social groups of past ages, whose members led a more autonomous existence. The history of life on earth has been punctuated by a number of these coalescing events; social insect colonies from single insects, multicellular organisms from single cells, eukaryotes (nucleated cells) from prokaryotes (bacterial cells), chromosomes from isolated genes, and so on, all the way down to the origin of life as social groups of interacting molecules (Maynard Smith and Szathmary 1995; Michod 1999). Each transition has required a solution to the fundamental problem of social life. For example, imagine a protocell in which the genes exist as independent units. Some genes work for the good of the cell while others use the resources of the cell to replicate themselves. This is the age-old problem of altruism and selfishness, frame-shifted downward to the level of genes in cells. A chromosome can be interpreted as a solution to the problem of cheating, by linking all the genes into a single structure that replicates as a unit.

Viewing organisms as groups has had a profound effect on multilevel selection theory. Never again can it be said that higher-level selection is always weaker than lower-level selection; you and I are shining contradictions of that statement! Theories of cooperation/altruism have vastly expanded their domain to include the sciences of genetics and development. The laws of genetics and develop-

ment, which previously referred merely to general patterns, are being found to bear an eerie resemblance to the other meaning of the word *law*: a social contract enforced by punishment. Organisms are being found to be not entirely integrated, with rogue elements that continue to benefit themselves at the expense of their group (Hurst et al. 1996). Finally, it is inconceivable that higher-level selection stops at the level currently known as the individual organism. Selection at the level of social groups is likely to be an important, if not a dominating, evolutionary force in thousands of species. It is even possible that human social groups, like social insect colonies, can legitimately be compared to single organisms in many important respects (Sober and Wilson 1998).

Absence of Cooperation/Altruism Where It Was Previously Thought to Exist

At the beginning of this chapter, I stated that evolutionary theory is achieving a middle ground in which cooperation/altruism can evolve but requires special conditions. The purpose of multilevel selection theory is not to show that altruism/cooperation is everywhere but to identify the special conditions. Inclusive fitness theory is summarized by a famous inequality called *Hamilton's rule, $b/c > 1/r$*, where c is the cost to the altruist, b is the benefit to a single recipient, and r is the coefficient of relatedness. According to Hamilton's rule, r is the *only* piece of information required to predict the degree of altruism that evolves between social partners.

Limited dispersal causes relatives to become neighbors. Thus, altruism has been widely assumed to be most common in species with limited dispersal, such as plants whose seeds and pollen are carried only a small distance from the parent plant. However, there is more to the evolution of altruism/cooperation than relatedness. Recall from table 17.1 that the advantages of selfishness are local (within groups), while the advantages of altruism take place at a larger spatial scale (between groups). Altruism evolves in this example because a period of local interactions alternates with a period of dispersal, allowing the altruistic groups to export their productivity throughout the larger population. In populations characterized entirely by limited dispersal, relatives do indeed become neighbors, but the abundant progeny of altruistic neighbors remain in the same vicinity and are not exported to other regions of the population. At the same time, selfish individuals who border on or invade the interiors of an altruistic patch prosper at the expense of their neighbors and ultimately take over the patch. If the landscape is completely saturated with individuals, limited dispersal does not favor the evolution of altruism, despite the fact that one's neighbor tends to be one's relative (Queller 1992a). The situation changes, however, when disturbance events create gaps that are filled in by dispersal from around the edges (Mitteldorf and Wilson 2000). Altruistic patches spread into the gaps faster than selfish patches, only to succumb to selfishness after the gaps have been filled. It is fascinating to watch graphic representations of these computer simulations, which look like a game of tag in which altruistic patches "chase" gaps created by disturbance events and, in turn, are "chased" by selfish patches. A degree of altruism can evolve in these simulations, although less than would be predicted on the basis of Hamilton's rule. These are cases in which multilevel selection theory reveals conditions that are *more* stringent for the evolution of cooperation/altruism than previously thought.

Cognitive Coordination

Cooperation/altruism is usually studied in the context of physical activities such as hunting, predator detection, and food gathering. However, cognitive activities such as memory and decision making are also demanding tasks that can benefit from the coordinated action of groups. The concept of a "group mind" may appear to be science fiction, but it has been well documented in social insect colonies and may exist elsewhere.

As one example, a honeybee colony must make decisions about which flower patches to visit and which to ignore over an area of several square miles; whether to gather nectar, pollen, or water; the allocation of workers to foraging versus colony maintenance; and so on. T. Seeley and his colleagues have worked out in impressive detail how these decisions are actually made (Seeley 1995). In one experiment, a colony in which every bee was individually marked was taken deep into a forest where virtually no natural resources were available. The colony was then provided with artificial nectar sources whose quality could be experimentally manipulated. When the quality of one source was lowered below the quality of other sources, workers ceased to visit the inferior source—proof

that the colony can perceive changes in its environment and respond adaptively. Yet, individual workers visited only one patch and therefore had no frame of comparison. Instead, individuals contributed one link to a chain of events that allowed the comparison to be made at the colony level. Bees returning from the inferior source danced less and themselves were less likely to revisit. With fewer bees returning from the inferior source, bees from better sources were able to unload their nectar faster, which they used as a cue to dance more. Newly recruited bees were therefore directed to the best patches.

Mechanisms of coordination go beyond the famous symbolic bee dance that allows bees to communicate the location of resources to each other. In fact, many aspects of coordination are remarkable for their lack of sophistication as far as individual behavior is concerned. The individuals respond to environmental cues and each other in a simple fashion, but the interactions have emergent properties that result in complex and adaptive behaviors at the colony level. For example, the colony acts as if it is hungry when its honey supplies are low, sending more workers to collect nectar, yet no individual bee is physically hungry. Instead, the state of the colony is communicated by the amount of time that returning foragers must wait to regurgitate their load of nectar to other workers who carry it to empty cells. When resources are scarce and many cells are empty, returning foragers can immediately unload their nectar, which serves as the cue for increasing foraging effort. Even the physical architecture of the colony, such as the location and dimensions of the dance floor, honeycomb, and brood chambers, has been shown to contribute to group coordination.

There is a tendency to assume that these wonderful examples of group-level cognition should be restricted to the social insects. After all, members of a social insect colony are highly related to each other and tend to reproduce through a single queen. However, this reasoning confuses the problem of coordination with the problem of distribution. It is true that highly costly altruistic behaviors might be restricted to species that achieve a high degree of variation among groups, via genetic relatedness or another mechanism, but forms of coordination that can be achieved without self-sacrifice should be far more widely distributed in nature. Prins (1996) has convincingly shown that African buffalo herds *(Syncerus caffer)* pool information in a beelike fashion to decide where to forage. Less convincing but still promising evidence exists for groups as diverse as primate troops, bird flocks, fish schools, and human societies.

Practical Implications

Disease organisms provide some of the best examples of multilevel selection and also show how the theory can be used for practical purposes. Cooperation/altruism in disease organisms can take many forms, depending on the details of the life cycle. In the trematode parasite *Dicrocoelium dendriticum*, an intermediate stage of the life cycle resides in ants. One parasite burrows into the brain of the ant and changes its behavior, causing it to walk to the tips of grass blades, where it is more likely to be eaten by a cow or sheep, which is the host for the adult stage of the parasite. However, the brain worm, as it is called, dies in the process. It sacrifices its life so that the other members of its group can complete their life cycle (see Sober and Wilson 1998 for a more detailed account of how the brain worm evolves by multilevel selection).

In other cases, the death of the host spells death for the group of parasites or disease organisms within. From the perspective of the whole group, the host should be kept alive and circulating among its own kind so that future hosts can be infected. However, as we have seen, natural selection within single groups is insensitive to the welfare of the whole group. Parasite and disease strains that maximize their reproduction within a host will outcompete their more prudent competitors, regardless of the consequences for the group. The process of within-group selection for more virulent strains has been documented a number of times, most recently for malaria (Mackinnon and Read 1999; see also Frank 1996). Only competition between groups favoring more prudent strains can counter this trend. The balance between levels of selection depends on a number of factors, some of which can be manipulated to favor the evolution of "altruistic" mild strains over "selfish" deadly strains.

A natural multilevel selection experiment occurred in 1991, when a virulent strain of cholera appeared in Brazil and spread throughout South and Central America as far north as the state of Texas. Cholera is an intestinal disease that spreads by direct contact. In countries with poor sanitary conditions, it spreads easily from person to person. In this case, it is adaptive for cholera to reproduce as fast as possible, even to the point of killing present hosts, to reach as many future hosts as possible. In other words, the kind of explosive reproduction that leads to virulence is favored by *both* within-group *and* between-group selection. In countries with good sanitary conditions, the vast majority of disease progeny that leave a given host perish without reaching new hosts. In this case, it is adaptive at the group level to keep the host alive and circulating. Evolution in disease organisms is sufficiently rapid that within a few years, this single cholera strain had become locally adapted to the sanitary conditions of each country, remaining deadly where conditions were poor but becoming mild where conditions were good. The important point is that good sanitary conditions not only decreased the number of disease organisms but also changed the basic character of the disease. Indeed, in areas with the best sanitary conditions, the once-deadly strain of cholera became so mild that its effects were subclinical and did not even require a visit to the doctor (Ewald 2000)!

The Future Study of Cooperation and Altruism

The rejection of group selection during the 1960s was treated as a major event, allowing all of nature to be explained in individualistic terms. This paradigm led to many advances in our understanding of cooperation and altruism but ultimately became limiting. Multilevel selection theory provides a far more satisfying framework that allows cooperation and altruism to be explained at face value, as adaptations that allow groups to function as adaptive units. As much as we have learned about cooperation and altruism during the second half of the 20th century, we stand to learn much more in the future.

18

Foraging Behavior

DONALD L. KRAMER

Foraging is the set of processes by which organisms acquire energy and nutrients, whether the food is directly consumed (feeding), stored for later consumption (hoarding), or given to other individuals (provisioning). Foraging behavior plays an important role in evolutionary biology, not only because it is a major determinant of the survival, growth, and reproductive success of foragers but also because of its impact on predator avoidance, pollination, and dispersal adaptations of potential food organisms. From a contemporary perspective, it is surprising how generally the fundamental role of behavior was neglected in early-20th-century studies of evolution and ecology. Following the development of quantitative techniques and field-oriented approaches by European ethologists, however, interest in foraging, along with other aspects of behavior grew rapidly. Most of this research has sought to describe, explain, and predict foraging behavior quantitatively. The development of an a priori predictive approach using optimality theory, in particular, has revealed a richness and complexity in the patterns of foraging that could not have been imagined only a few decades ago. My goal in this chapter is to provide a brief overview of the main issues in foraging behavior and the logical basis of current approaches. I wish to highlight the successes and potential value of these approaches, while recognizing the gaps and challenges for future research.

Historical Context

Contemporary studies of foraging by evolutionary ecologists are based on the synthesis of two research traditions, both emerging during the 1960s. The ethological approach to behavior is illustrated by the research of K. von Frisch and his associates on honeybee foraging and N. Tinbergen and his group on searching behavior of birds. The ethologists' recognition of behavior as an evolved phenotype, their emphasis on its ecological context, and their careful quantitative and experimental fieldwork set the stage for behavioral ecology (Curio 1976). They classified the behavioral components of foraging, an important contribution to much of the ecological work that followed, and identified a number of widespread characteristics such as localized search following the discovery of a prey ("area-restricted search") and enhanced detection following experience of a particular prey type ("search image").

The theoretical approach to population ecology was foreshadowed by the Russian V. S. Ivlev. His earlier research and conceptual framework for the ecological determinants of foraging rate and food selection became widely available with the publication of a book in English in 1961. At about the same time, C. S. Holling, interested in the role of predators in the regulation of prey populations, produced an influential series of papers based on the idea that individual components of foraging behavior could be combined into a model that would predict foraging rates. His papers included both a theoretical framework and experimental studies showing how the relationship between predation rate and prey density (the functional response) would arise from components of predation and influence the persistence of prey populations. A very different theoretical approach was proposed by population ecologists with a more explicitly evolutionary framework (Schoener 1987). In a series of papers starting in 1966, J. M. Emlen, R. H. MacArthur, E. R. Pianka, T. W. Schoener, E. L. Charnov, and others began to develop models predicting the rate of energy gain arising from alternative behavioral rules in different foraging environments. Initially, they focused on food selection, often from the perspective of diet overlap and community ecology. They argued that the diet yielding the highest rate of gain should be the one that occurs in nature because natural selection is an optimizing process. To many students of behavior, this "optimality" approach seemed to demand an unlikely level of sophistication in animals. Most psychologists of the time regarded animals as very simple learning machines, and ethologists were examining the "release" of supposedly fixed sequences of social behavior using crude dummies. Thus, the success of early experimental tests of optimality models was particularly striking, stimulating a rapid increase in theoretical and empirical studies, as this approach quickly dominated the study of foraging.

The 1970s and 1980s witnessed continued growth of the field as behavioral ecologists were emboldened to ask ever more subtle questions and to develop new theoretical tools. In response to the challenges of some articulate criticism from outside the field and new questions within the field, the logical structure and assumptions of optimality models were examined more closely (Stephens and Krebs 1986). The most important development during this period was the incorporation of frequency dependence into the study of foraging using game theory (Giraldeau and Caraco 2000). Harper (1982) successfully applied a large-scale model of habitat selection called the *ideal free distribution* (Fretwell 1972) to local-scale competitive foraging of ducks in a park pond. Barnard and Sibly (1981) recognized the inherent frequency dependence of some individuals' exploitation of the foraging effort of others (kleptoparasitism). These developments stimulated many researchers to recognize that games among foragers were likely to be widespread as well as theoretically and empirically tractable.

At the start of the 21st century, the literature on foraging is growing rapidly, and its concepts are now incorporated into much of ecology and evolution. This is especially true in studies of spatial distribution, predation risk, pollination, and seed dispersal. Foraging theory plays an important though still limited role in fundamental and applied studies of population dynamics and community structure (Fryxell and Lundberg 1997). There are signs of a new appreciation of the importance of understanding the mechanisms underlying foraging behavior. However, the potential for a strong predictive approach to this key ecological interaction is far from realized, and many important questions remain to challenge researchers.

Concepts

Basic Elements of the Foraging Process

Foraging Cycles and Their Components   Foraging is a cyclical activity in which a series of behavioral acts leads to the final consumption of each unit of food. To facilitate the development of general theory, the behavior comprising a foraging sequence is divided into functional categories called *components* (table 18.1). For animals that feed on discrete items, whether mobile or not, the "prey cycle" is the basic unit of foraging; this includes search, assessment, pursuit, and handling. When food items are aggregated, multiple prey cycles occur within a patch cycle comprising patch search and/or travel, patch assessment, and patch exploitation. When foragers return to a fixed location to consume or hoard their prey or to provision other individuals and carry multiple prey per trip, prey and patch cycles can be nested within a central place cycle consisting of travel, loading, and unloading components. Multiple prey, patch, and central place cycles are often nested within a meal or foraging bout cycle, which in some species may include travel to and from a foraging site as well as an obligate period of nonforaging while food is digested.

Although useful in the establishment of a general theoretical and empirical framework for foraging, the division of a continuous sequence into

Table 18.1 Components of the foraging process.

1. **The prey cycle**—acquisition of individual food items.
   1.1. **Search**—leads the forager to come into sensory contact with potential prey and terminates when a prey is detected; for cryptic prey, may be divided into phases in which prey are **encountered** (potentially detectable) and **not-encountered** (out of detection distance); may be **active** (involving movement by the forager) or **sit-and-wait** (forager not moving during search).
   1.2. **Assessment**—leads the forager to pursue or abandon detected prey; may also occur during pursuit and handling.
   1.3. **Pursuit**—leads the forager to come into physical contact with detected prey (capture); may include **ambush** (forager not moving during pursuit), **stalking** (approach, often slow, that is difficult for prey to detect), and **overt attack**.
   1.4. **Handling**—leads to consumption of captured prey; may include **food preparation** (e.g., killing, removing shell or spines) and **ingestion** (e.g., grasping, masticating, swallowing).

2. **The patch cycle**—foraging on aggregations of prey.
   2.1. **Search**—leads the forager to detect a patch whose location was previously unknown; when movement is between patches of known location, **interpatch travel** is a more appropriate term.
   2.2. **Assessment**—leads the forager to begin to exploit or to abandon a patch.
   2.3. **Exploitation**—series of prey cycles (sometimes without additional prey search) that leads to consumption of some or all prey in patch.

3. **The central place trip cycle**—foraging that involves movement between a foraging site and a fixed location to which the forager returns with prey.
   3.1. **Outward trip**—movement from the central place to the foraging site.
   3.2. **Loading**—one or more prey or patch cycles leading to accumulation of a prey load.
   3.3. **Return trip**—movement from the foraging site to the central place carrying prey.
   3.4. **Unloading**—deposition of the prey load in the central place (may be replaced by **handling** when prey are consumed rather than stored or provisioned to others at the central place).

4. **The meal/foraging bout cycle**—foraging that occurs in more-or-less discrete periods separated by bouts of other activities.
   4.1. **Travel**—movement to a foraging area from a location at which other activities take place.
   4.2. **Feeding**—sum of activities in prey, patch, and central place cycles.
   4.3. **Processing**—digestion and assimilation of food; although some digestion occurs during feeding and other activities, processing is relevant as a separate category when food consumption is very rapid relative to digestion resulting in a required pause between bouts of feeding; this phase is sometimes called **handling** in ecological (but not behavioral) analyses.
   4.4. **Other activities**—not foraging; may overlap with processing.

separate components is somewhat arbitrary. For example, in some sit-and-wait predators, handling one prey overlaps with search for the next, and wolves may assess the vulnerability of a potential prey by initiating a testing pursuit distinguishable from the all-out effort to capture a prey. Categories may be subdivided, combined, or deleted according to the organism, food type, and question being asked. For example, assessment is often included as part of search, while pursuit can be usefully divided into stalk and attack components.

Measures of Foraging Success   Ideally, evolutionary studies of foraging behavior should use measures of foraging success that are correlated as closely as possible with fitness. However, the nutrients and energy obtained by foraging are often allocated simultaneously to survival, growth, and reproduction, making it impossible to examine a single major component of fitness. Furthermore, foraging is so flexible and most studies of it are so brief that it is rarely possible to identify different foraging phenotypes, much less to compare growth or reproductive success among them. (For an exception see Altmann 1998.) Thus, comparisons of foraging behavior are based on estimates of the success in gaining food and the costs of doing so, although the quantitative relationships of these measures to fitness are usually unknown. Any foraging cycle may be terminated before a prey or patch is consumed, as a result of prey escape, abandonment decisions, or nonforaging interruptions. The forager pays costs in time, energy, and sometimes increased risk of mortality as it engages in each component. The benefits in the form of energy and nutrients come only at the end of successful cycles when the food is consumed.

Net rate of energy gain is frequently considered the ideal measure of foraging success. Maximizing this rate provides the most energy for fitness-related activities and permits the animal to minimize its foraging time to allow for other important activities. The estimation of net rate of energy gain requires behavioral measures of foraging time and prey consumption, as well as bioenergetic estimates of the value of the prey and the costs of foraging. Energy costs of foraging are typically obtained from physiological and biomechanical estimates of costs of locomotion under steady-state conditions in the laboratory. When the energy costs of different components of foraging are similar and when costs are very small relative to rates of gain, costs are sometimes ignored and the gross rate of gain is used as a measure of foraging success. On the other hand, net rate of energy gain fails to account for the value of specific nutrients and nonenergy costs such as predation risk and the expenditure of other resources. Thus, appropriate measures of foraging success vary with the species and situation. See Ydenberg et al. (1992) for further discussion of these issues.

Foraging Decisions   A key aspect of foraging behavior is its flexibility. Often, an animal has the option of continuing what it is doing, switching to an alternative form (or "mode") of the same component, or switching to another component altogether. For example, when stalking a prey, a lion may at any moment continue the stalk, switch to direct attack, or begin to search for an alternative prey. Furthermore, an animal can switch to alternative aspects of foraging, such as information gathering or aggressive defense of feeding areas, or it can cease foraging altogether. Behavioral ecologists refer to the performance of one of these qualitatively or quantitatively different activities as *decisions*, a term that is meant to reflect the availability of alternatives rather than to define a particular process by which one of the alternatives is selected. *Decision rules* are the relationships between foraging decisions and environmental conditions, such as food density, or organismal states, such as the

individual's fat level. They are therefore similar to the concept of *norms of reaction* as applied in studies of phenotypic plasticity (see Pigliucci, this volume), but with the important difference that changes in phenotype (decisions) occur on a smaller time scale. Predicting decision rules is a major goal of evolutionary studies of foraging. Some of the principal foraging decisions are listed in table 18.2.

The degree of flexibility in decision making is potentially highly variable. For example, the decision whether to consume a particular type of potential prey could be fixed for an entire species, could vary among populations exposed to different densities of that prey or alternative prey, or could vary within an individual, according to prey abundance and the individual's current handling skills, physical condition, or nutritional needs and so on. The finer the adjustment of foraging decisions to local conditions, the greater the need for information about those conditions. Such information might come from directly relevant experience (for example, the individual's recent foraging success). Decisions might be based alternatively on simpler rules of thumb that relate to easily measurable environmental characteristics (such as light level or temperature), to the expected abundance of prey, and to the effectiveness or abundance of predators and competitors. The less flexible a decision rule is or the more removed a rule of thumb is from the relevant characteristics, the more one would expect to find discrepancies between observed foraging decisions and the decision offering the highest success. This would be especially true in the case of evolutionarily novel situations. For example, a forager using size as a cue to prey quality might make the wrong choice when presented with a novel food type that is small but nutrient-rich. Conversely, highly flexible decision rules based on an individual's experience of the effects of alternatives should lead to novel forms of foraging behavior and to evolutionary innovation (Lefebvre 2000).

Foraging Constraints and Trade-Offs   Some determinants of foraging success are not under the direct control of the organism. In the framework of foraging theory, these are referred to as *constraints*. Constraints are often regarded as being either extrinsic or intrinsic to the organism. The primary extrinsic constraints are the distribution, abundance, and defensive adaptations of potential food items. The animal can select or ignore different food types or forage in different areas but cannot

Table 18.2 Some foraging decisions studied by behavioral ecologists.

1. **Time budget decisions**
   When to start a foraging bout (e.g., relation to time of day, local conditions, internal state)
   When to stop a foraging bout
   When to initiate and terminate controlled interruptions of a foraging bout (e.g., vigilance, grooming)
2. **Spatial distribution decisions**
   Which specific site to search and the sequence in which to visit multiple sites
   When to switch to another site
   How close to other foragers to search (e.g., foraging group size, local density)
3. **Movement decisions**
   Locomotor mode (e.g., fly versus walk)
   Speed and gait of movement
   Duration, timing, and location of pauses during movement (intermittent locomotion)
   Timing and direction of turns and intervals between them
   Specific route
4. **Selectivity decisions (choice)**
   Microhabitat choice (e.g., substrate types, proximity to other foragers)
   Diet choice
   Patch choice
   Behavioral sequence choice: In which order to perform different activities involved in assessment, handling, and patch exploitation
5. **Persistence decisions**
   Whether to continue assessment, pursuit, handling, patch exploitation, or loading versus returning to search
6. **Food allocation decisions**
   Whether to consume or hoard a particular item or to provision others
   Where to hoard
   Which individual to provision
7. **Defense decisions**
   Whether to defend
   What specific area to defend and not defend
   When to patrol and display
   Which intruders to respond to and in which order
   Whether to threaten or attack
   Attack and display decisions (mode, speed, duration)
   (*Note:* Intruders will have a parallel set of decisions with regard to defenders.)
8. **Information acquisition decisions**
   Whether to sample other prey and sites or other foragers
   When to sample
   Which sites to sample and in which order
   How long to sample a particular site

directly control what is actually available. Other extrinsic constraints include the distribution and abundance of other foragers and predators, as well as relevant aspects of the physical structure of the foraging environment (e.g., vegetation density). Intrinsic constraints determine how a particular decision rule, under a particular set of extrinsic constraints, translates into foraging success. Intrinsic constraints include the limited availability of time and energy, bioenergetic limitations (e.g., moving faster often requires a higher rate of energy expenditure per unit distance), sensory capacities (e.g., ability to discriminate colors, to detect immobile prey), and central nervous system capabilities (e.g., ability to remember and integrate recent foraging success, capacity to attend to more than one activity or prey type at the same time). There is also an interaction between extrinsic and intrinsic constraints. Potential food items are a characteristic of the organism, including its ability to recognize, capture, consume, and digest them, and these characteristics can change with experience.

Foraging decisions can affect different aspects of the same foraging component as well as affecting other foraging components and nonforaging components of fitness. For example, a forager switching from sit-and-wait to active search (a movement speed decision) at a given abundance of a particular cryptic prey type (extrinsic constraint) may increase the rate with which prey are encountered but decrease the proportion of those prey that are detected as a result of the forager's sensory physiology (intrinsic constraint). This increase in encounter and decrease in detection rate may either increase or decrease the time cost of search per detected prey. In addition, moving may increase both the rate of energy expenditure as a result of muscle physiology (intrinsic constraint) and the probability of being detected by a predator during search (extrinsic constraints of predator abundance and sensory capabilities). Situations in which decisions that increase one component of fitness decrease other components are called *trade-offs,* and they are ubiquitous (see Roff, this volume). Understanding foraging behavior depends on recognizing the major trade-offs resulting from foraging decisions.

Optimal Foraging

Logical Framework   If natural selection has favored the evolution of decision rules that maximize foraging success, or of flexible behavioral systems that can learn which decisions maximize foraging success, the foraging patterns observed in animals should be those that give the greatest foraging success. When we know what alternative decisions are possible and understand the most important trade-offs arising from them under a given set of constraints, we should be able to predict which decision will yield the greatest foraging success under a given set of conditions. In essence, then, the assumption that foraging is a well-designed system allows us to predict its properties using an optimality approach, as in other areas of evolutionary ecology (see Roff, this volume). We use theory to explore the expected properties of foraging systems. Often referred to as an *economic approach,* optimality analyses have been the key to the development of the predictive study of foraging. Differing views concerning how often the basic assumptions underlying the optimality approach are met, or how important deviations from these assumptions are, have led to very different perspectives on the value of this approach. In the resulting controversies, protagonists often ignore its limitations and antagonists ignore its power.

An optimal foraging model is a numerical or graphical hypothesis designed to predict the rela-

tionship between foraging decisions and a set of constraints consisting of the environmental conditions and the state of the organism. A model requires the a priori selection of a measure of foraging success to be maximized (the "currency") and a set of mathematical relationships between the decision and the currency for each condition or state. These relationships are a set of constraints that specify the effect of the decision on the costs and benefits of different components of the foraging cycle and the way in which these costs and benefits are combined to determine the currency. Foraging models can predict qualitative trends in decisions with changes in conditions or state or the quantitative value of the decision for specific conditions.

Optimality models are risky enterprises, and there are sound theoretical reasons why they may fail, even if the calculations are correct. As with any adaptive trait, the optimal decision rule may lag behind a fluctuating environment, may not occur at most locations in a spatially variable environment, and may be constrained by the underlying genetics. A model may fail if important contributions to the currency are ignored or if the wrong currency is used. Certain food items, competitors, or mortality risks during foraging may be too recent to have had an effect on the evolution of foraging decisions, and others of historical importance may be absent. Some environmental circumstances may have been too rare or have had too little impact on fitness to lead to the evolution of adaptive responses. Finally, cues relevant to important variables may not be available to the animals' sensory systems.

Pyke (1984), Stephens and Krebs (1986), Mangel and Clark (1988), and Houston and McNamara (1999) and references therein provide detailed discussions of optimal foraging models and their assumptions. The last two references in particular summarize recent advances in the use of dynamic optimization models to integrate the effects of current and future decisions and predict optimal decision trajectories.

Approaches to Testing   The primary goal of testing models is not to determine whether or not the model is "correct" (as a simplification of nature, the model is bound to be wrong at some level), but to determine how well it predicts behavior. When a model predicts well, the challenge is to find the range of species and situations in which it continues to predict. When a model predicts poorly, examining the assumptions used in constructing the model may not only improve the model but also generate new discoveries about nature.

Optimal foraging models provide some of the most favorable situations for testing precise, quantitative predictions in the entire field of evolutionary ecology. Foraging is a common activity, often observed and measured with relative ease, and the foraging environment can often be manipulated experimentally. In many areas of evolutionary ecology, only a limited range of decisions is available to be examined, so the consequences of many alternatives are difficult to determine. (If phenotypes are optimal, the strongly suboptimal traits required to demonstrate this should not exist.) In foraging, by contrast, it is often possible to generate foraging situations that permit estimation of the consequences of a full range of decisions. Nevertheless, testing is a more demanding task than is sometimes acknowledged, with its own set of assumptions, and tests can fail (or succeed) for the wrong reasons.

The simplest test of an optimal foraging model is to examine whether the decision rule changes qualitatively with environment or state as predicted by the model. While useful in a preliminary way, such tests are ultimately unsatisfying. It is much more useful to be able to predict the quantitative value of the decision and to be able to measure the discrepancy between prediction and observation. Indeed, this is the only way to determine whether the qualitative prediction is actually valid in a particular case.

All tests make assumptions about the flexibility of the decision rule and the information available to the animal. In some experimental tests, the decision rule predicted by a general model would not be optimal for an animal that actually had perfect knowledge of the test situation and experimental protocols. In other cases, the protocols are appropriate, but the animal may not be aware of the situation. Early tests of foraging models made considerable progress despite ignoring the behavioral mechanisms by which animals gather and use information. However, it is becoming increasingly clear that an integration of the economic approach with these mechanisms will improve both the predictive power of the models and the strength of the tests.

Tests of optimal foraging models typically use observational and experimental methods. Observational tests have a high level of external validity; the data gathered are directly relevant to the ecology of the study population. However, it may be difficult to find a sufficient range of variation in the environmental conditions to test the model, or the variation that occurs may be confounded with other important changes in the environment or the state of the animals. Furthermore, determining the effect of alternative decisions on the currency requires care that individuals with different decisions do not differ in other important ways. Experimental tests can achieve a wider range of environmental conditions or organismal states while controlling for confounding variables, and they facilitate the determination of decision/currency relationships. However, they typically require manipulations involving unnatural food types, food densities, or foraging environments. Poor predictive power of the models is sometimes related to a lack of evolutionary history relevant to the foraging situation under examination.

Comparative tests require evolutionary differentiation of foraging behavior in relation to different environmental conditions. Most tests of optimal foraging models, however, involve short-term responses to variation in environmental conditions, implicitly assuming highly flexible decisions. While comparative study of the evolution of such flexible response systems would be of great interest, I am not aware of any such investigations.

Test Results The most common result of published tests of optimal foraging models is partial support of the predictions. Models often correctly predict qualitative trends but not the quantitative values, or the observations support some but not all predictions. This suggests a "half full and half empty glass," in which proponents point to the success (e.g., Stephens and Krebs 1986; Schoener 1987) and critics emphasize the failures (e.g., Gray 1987). In an evaluation of 125 tests, Stephens and Krebs (1986) concluded that predictions were at least partially validated in 71% of tests and clearly not supported in 13% (the rest were ambiguous). Even tests with novel foods and artificial environments have had considerable success. For example, Carlson (1983) obtained a very close fit to the quantitative prediction of a simultaneous choice model when wild red-backed shrikes collecting food for their offspring were offered a choice between mealworm pieces (an unfamiliar food) of different sizes and with different handling times produced by threading the pieces onto wires attached to a shrub that was moved to different locations (an unfamiliar foraging site and handling behavior). Even pigeons and rats pressing keys in a psychologist's operant chamber can come quite close to the predictions of models developed for foraging in the field.

It is likely that the published literature is biased by the difficulty of publishing negative results and a tendency for researchers who obtain negative results to continue searching for additional relevant variables or invalid assumptions. Nevertheless, it is clear that the optimality approach has predictive power. Evaluation of this power would be improved if it could be quantified on a common scale. A promising approach comes from the mathematical technique of dimensional analysis (Stephens and Dunbar 1991), which transforms the mathematical relationships of a model into a reduced set of unit-free variables. This unit-free statement of the decision rule permits different studies to be compared within the same framework and reveals the relative magnitude of discrepancies between observation and prediction. Using this approach, a preliminary analysis of patch exploitation studies by Kramer and Giraldeau (unpublished) confirmed the overall positive relationship between proportional patch use and search or travel between patches when scaled to patch "half-life."

Foraging Games

An important part of the foraging environment for many species is the presence of other foragers, of the same and sometimes other species, which can increase or decrease foraging success (table 18.3). Such effects are most evident when animals forage in groups, but even "solitary" foragers can affect each other, for example, by reducing prey abundance, alerting prey to the presence of predators, or revealing new food sources by their foraging activity. Many processes involving interactions with other foragers show density dependence because their occurrence, magnitude, and sometimes direction depends on the number of other individuals present. Density dependence is often negative, in that some currency of foraging declines with the number of other individuals, but it may also be positive (also known as an *Allee effect*) over at least part of the density range. Sometimes, social effects may be affected by the proportion of individuals making different alternative decisions. This

frequency dependence is an important consequence of social foraging, often in combination with density dependence.

Optimality theory can be used to predict foraging decisions in relation to the abundance or tactics of other foragers as with other environmental conditions. However, a major difficulty in testing such predictions is that the other foragers may also change their behavior in relation to the change in their foraging environment. Thus, to predict foraging behavior in a social context, the results of this series of mutual interactions must be predicted. Game theory provides the mathematical tools for this prediction. See Dugatkin and Reeve (1998) and Roff (this volume) for summaries of game theory in evolutionary ecology. Giraldeau and Caraco (2000) provide a detailed recent review along with new applications of game theory to social foraging.

The basic approach of game theory is to find, from a given set of decision rules, the rule or mix of rules among the interacting individuals such that no individual using an alternative rule would have greater success than the individuals using the established rule or rules. From the perspective of evolutionary genetics, this situation prevents genes for alternative rules from invading the population and is therefore called an *evolutionarily stable strategy* (ESS). *Evolutionarily stable state* is a more appropriate term to account for situations in which no single rule is unbeatable but a mix of rules is. The "strategies" in most models tend to be rather simple fixed responses or simple switches between two responses based on a single environmental factor. However, as with applications of optimality theory, in foraging game situations that have actually been studied, the decisions are typically flexible individual responses.

Game theory is so important because the outcome of games is often strikingly different from simple optimal solutions. The selection of foraging sites offers a simple illustration. Consider a choice of potential foraging sites that differ only in their food availability and consequently in the foraging success they offer. For a single forager, the optimal solution is to forage at the site with the highest food availability, and the model does not change the prediction if several individuals are considered. If foraging rate is negatively density-dependent, however, and there are many foragers using the same set of sites, the foraging rate in the patch with the highest food availability may drop below that of other unoccupied patches if all go to the site

Table 18.3 Social influences on foraging.

1. **Changes in food availability.**
   - **Exploitation**—removal of food from the foraging area.
   - **Passive interference** (also called **prey depression**)—reducing foraging rates by making prey less available, for example, by inducing its antipredator defenses.
   + **Facilitation**—making prey more available, for example, by confusing them or making them more visible while fleeing another forager.
   + **Risk reduction**—lowering the variance of foraging success by sharing food discoveries among individuals.
2. **Changes in costs or benefits of search, pursuit, and handling.**
   + **Cooperative hunting**—improved pursuit and handling success or decreased pursuit and handling time by groups.
   + **Increased rate of discovery** of shareable prey or patches by foragers in groups.
   - **Scramble competition**—animals foraging in groups may have search areas that overlap with those of other foragers and therefore require more search time to discover the same number of prey; they may have to increase their rate of movement or change other foraging tactics to avoid having other foragers discover or capture the prey first.
3. **Kleptoparasitism**—exploiting the search, pursuit, and handling of others.
   + For the kleptoparasite, more potential victims decrease search, pursuit, or handling time by allowing exploitation of prey or patches in which another individual has already invested.
   –/+ For the kleptoparasite, more kleptoparasites may change food availability and change costs and benefits of foraging as in sections 1 and 2.
   - For the victim, more kleptoparasites result in increased effort in search, pursuit, or handling as a result of losing prey or all or part of patches, or additional effort to avoid or defend against kleptoparasites, or reduced success as a result of choosing prey less vulnerable to kleptoparasitism.
   + For the victim, more victims can improve defense or dilute the impact of kleptoparasites.
4. **Food defense** (also called **active interference** or **interference competition**)—use of aggressive behavior to reduce the foraging of other individuals on particular prey (**food guarding**) or specific locations (**territoriality**).
   + For the defender, increased prey availability as a result of reduced exploitation and passive interference by other individuals.
   + For the defender, effectiveness of defense may be increased by cooperative defense.
   - For the defender, costs of defense increase with the number of potential intruders.
   + For intruders, foraging in groups can increase access to defended areas or prey by overcoming the defense.
   - For intruders, decreased access to particular prey or foraging sites as a result of effective defense by other individuals.
   - For intruders, increased effort or risk of injury to obtain resources by intruding in locations defended by others.
5. **Changes in foraging time availability and risk of attack from predators or conspecifics.**
   + Grouping can reduce predation risk during foraging; this may permit less time to be spent on vigilance during the foraging period or the use of sites that would be too dangerous for a solitary individual.
   - Foraging in groups may attract more predators, increasing predation risk or reducing the areas that are safe to use.
   +/– When attacks from conspecifics are a threat, groups may either increase or decrease the risk, depending on the situation.
6. **Foraging information.**
   + Other individuals can be used to obtain information about beneficial foraging locations, food types, and foraging techniques.
   - Information scrounging may reduce the number of accurately informed individuals and provide wrong information.

Note: Minus and plus signs indicate how each process will affect the foraging success of a focal individual as the number of other foragers increases.

with highest food availability. The game theory solution, known as the *ideal free distribution,* is for individuals to distribute themselves among sites so that the combination of forager density and food availability results in similar foraging rates in all sites, a very different result from the prediction of simple optimality. This is an example of frequency dependence in that the best site for any one forager depends on the distribution of the foragers among the other sites. The ideal free distribution is clearly too simple to apply in many real-world situations, but it has been repeatedly observed in controlled laboratory studies, where foragers often show an impressive capacity to adjust quickly to deviations in foraging success. Furthermore, it has formed the basis for a series of more complex models that have strongly affected how behavioral ecologists think of foraging distributions and habitat selection.

Group size presents a similar problem. For certain foraging tasks such as the hunting of large prey by carnivores, it is possible to calculate an optimal group size that yields the highest rate of foraging success per individual. However, group formation constitutes a game: depending on how groups are formed and how access to groups is controlled, groups may form that are much larger than optimal size with foraging success much lower than the optimum. This is representative of a widespread finding in game theory models that individual success will be lower than that predicted by simple optimality.

A fundamental characteristic of social foraging is that the effort of individual foragers creates an opportunity for parasitism of that effort by other foragers. Kleptoparasitism, however, is not a viable foraging strategy if all foragers are attempting to do it and none are looking for their own food. This is another case where frequency dependence is clearly involved in the success of alternative forager decisions, such as searching for food patches independently or watching other foragers and attempting to join patches that they discover. "Producer-scrounger" models predict the occurrence and frequency of these alternative foraging decisions.

Game theory is also relevant to interactions between predators and prey and between plants and their pollinators and seed dispersal agents. A review of such coevolutionary interactions is beyond the scope of this chapter (see section "Interspecific Interactions," this volume). However, recent models of forager distribution are beginning to take into account distribution games between predators and prey as well as among foragers (e.g., the chapter by Sih in Dugatkin and Reeve (1998)).

Case Study

The Load Size Decisions of Chipmunks

Theory   For central place foragers that carry more than one prey per trip, one foraging decision is when to stop loading and return to the central place. This decision determines the load size carried and is part of the foraging process for seed-hoarding rodents, for parental birds provisioning their young, and for nectar- and pollen-collecting bees and wasps. In this section, I will illustrate the interplay between foraging theory and experimental tests using the load size decision of eastern chipmunks, *Tamias striatus*. These small (100-g), terrestrial, forest-dwelling sciurid rodents feed primarily on the seeds of deciduous trees, though a wide range of plant and animal material is sometimes consumed. Food is carried to a burrow for both short-term consumption and long-term storage. In the winter, animals hibernate without accumulating significant fat reserves, waking periodically to feed on stored food. Except for females with young, each animal occupies its own burrow, which is vigorously defended against conspecifics. Although animals chase other individuals near food sources and in the vicinity of their burrows, home ranges overlap extensively. Chipmunks exposed to human presence and provisioning become very tame. At our study site, in the public portion of a university nature reserve, many animals can be observed from a few meters or less without evident disturbance.

Our interest in load size decisions arose from one of a set of models of central place foraging published by Orians and Pearson (1979). This model is conceptually derived from an influential model of patch use known as the *marginal value theorem* (Charnov 1976). It predicts that, if loading rate decreases as the size of the load increases, the optimal load size should increase with distance to the central place and with food density. The derivation of this prediction is shown in a step-by-step analysis in figure 18.1, but an elegant graphical solution (figure 18.2) improves the intuitive understanding of the predictions. A decline in loading rate with size of the load already accumulated seems likely because holding prey could reduce the success or rate of search, pursuit, and handling of additional prey, depending on the morphology of the forager and the type of prey. The trade-off here is between the gain per trip, which is maximized by taking the largest load possible, and the number of trips per unit of foraging time, which is maximized by taking the smallest load. The rate of food delivery to the central place is the product of these components and is maximized at an intermediate load size (figure 18.1C). Note that if the declining loading-rate constraint is removed so that load size in-

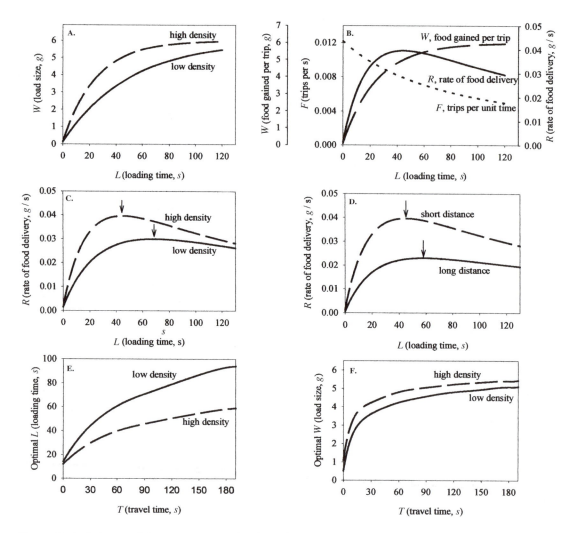

Figure 18.1 Orians and Pearson's (1979) load-size/loading-time decision model for a central place forager. The model assumes a currency of gross rate of food delivery to the central place with constraints as follows: (1) The food density and distance from the central place are fixed by the environment. (2) Loading rate decreases with decreased food density and with the amount of load already acquired; therefore, the total load increases with loading time, but at a decreasing rate, and approaches the maximum load more slowly when food is scarce. Panel A illustrates loading functions for patches of high [$W = 6(1 - exp\ (-0.04L - 0.02))$] and low [$W = 6(1 - exp\ (-0.02L - 0.02))$] density. (3) Round-trip travel time includes the outward and return trip plus unloading time and is a fixed cost for any particular distance; it has a positive value at zero distance because of unloading time and increases linearly with distance. (4) Gross rate of food delivery is the product of food gained per trip (load size) and trips per unit time (1/(loading + travel time)). Panel B shows these values for a high-food-density patch 20 m from the burrow, assuming that $T = 50 + 1.6D$, where T is travel time (s) and D is distance (m). The animal is free to cease loading at any time, and the decision rules to be determined relate loading time to travel time and loading time to food density. For any travel time and food density, food delivery rate can be calculated for all possible loading times. The loading time that provides the maximum delivery rate is predicted to be the one used by the animal. Panel C compares gain curves for high- and low-density patches 20 m from the burrow; panel D compares gain curves for high-density patches at 20 and 80 m. Optimal load sizes are indicated by arrows. The predicted decision rules in panel E relate optimal loading time to travel time for high- and low-density patches. In panel F, these are converted to optimal load sizes, using the loading functions. These show that optimal load sizes are larger in high-density patches and increase in a curvilinear fashion with travel time.

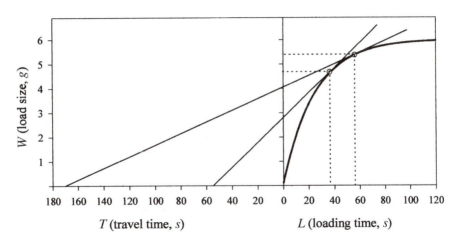

Figure 18.2 A graphical formulation of the optimal solution to the load size decision model based on Orians and Pearson (1979) and using the values for the high-density patch presented in figure 18.1. Load size in relation to loading time increases to the right and travel time increases to the left of the origin. For any travel time, the optimal loading time and load size are indicated by the tangent to the loading curve. This is because the slope of the loading curve equals the rate of food delivery (load divided by loading time + travel time). Any point along the curve represents a possible decision, but the tangent represents the greatest slope and hence the highest possible food delivery rate. This is illustrated for a short and a long travel time. Dotted lines drawn from the points of tangency to the axes indicate the optimal loading time and load size.

creases linearly, the optimal load is always the maximum (Kramer and Nowell 1980).

Tests   We chose to make a quantitative, observational test of the predictive power of the decision rule relating load size to travel time, under field conditions (Giraldeau and Kramer 1982). If Orians and Pearson's model is correct and if chipmunks show a curvilinear loading function, these animals could have been selected to take a species-characteristic, fixed load size corresponding to the average distance and seed density at which they forage. Our design assumed, however, that individuals can adjust their load size to recent experience of distance and loading characteristics, including evolutionarily novel food types and foraging situations. To determine the average loading curve for our study population, we recorded the weight of seeds taken in 738 uninterrupted loads and 113 loads in which the experimenter interrupted the chipmunk early in the loading process. We used a fixed quantity of sunflower seeds (10 g) on a bare tray (23 × 24 cm) placed at different distances (0.2–135 m) from the burrows of 21 different adult animals. By

using data from some animals at long distances, we were able to estimate the consequences of taking large loads, even if the animals only took small loads when near their burrows. In addition, we determined the round-trip travel time and distance to the burrow for the same trials.

The data showed clearly that load size increased at a decreasing rate with loading time (figure 18.3A) and that travel time increased linearly with distance. The optimal load size for each travel time was calculated with the equation $T = (f(P^*)/f'(P^*)) - P^*$ where $f(P^*)$ and $f'(P^*)$ are the loading function and its derivative evaluated at P^*, the optimal loading time. (See Giraldeau and Kramer 1982 for details.) The result was a prediction for a strong increase in load size with increases in travel time over the range of travel times observed (figure 18.3B).

Data from the 738 uninterrupted loads showed that chipmunks did take larger loads when they were farther from their burrows, a finding providing qualitative support for the predictive capacity of the model (figure 18.3B). However, load sizes were considerably smaller than predicted. In addi-

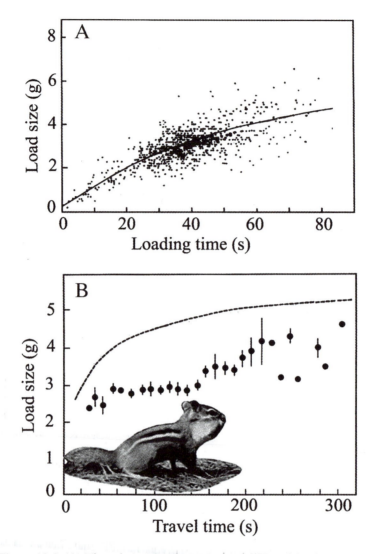

Figure 18.3 (A) The relationship between load size and loading time for eastern chipmunks hoarding sunflower seeds at different distances from their burrows. Each point represents a different load. The line shows the best fitting exponential equation, $W = 6.0 (1 - exp (-0.018 - 0.023)$, $r^2 = 0.56$; other curvilinear functions gave a similar fit. (B) The relationship between travel time and load size. The solid circles show observed mean load sizes ± 1 SD, and the dashed line shows the load size predicted to maximize the rate of food delivery. (Modified from Giraldeau and Kramer 1982, figures 1 and 3.)

tion, instead of a relatively sharp increase in load that gradually leveled out size with increasing distance, chipmunks showed little change in load size over the first 50 m, with a sharper increase at greater distances. In a subsequent study, using a uniform spatial array that allowed us to determine the loading time for each seed, we also found curvilinearity of the loading function and an increase of load size with distance (Giraldeau et al. 1994). However, this study also found highly significant individual differences, as well as effects of distance and the presence of competitors, on the loading functions. Travel time was also affected by the presence of competitors. Unfortunately, this prevents the generalization of loading curves to predict what animals would have obtained if they had taken larger loads than the ones observed. Nevertheless, the data clearly showed that the chipmunks would have had lower rates of food delivery if they had used loads smaller than those observed, and an extrapolation strongly suggested that they would have done better by taking larger loads. In other words, chipmunks in this different test situation were also very likely to be taking smaller loads than predicted by the Orians and Pearson model.

Other researchers have confirmed the qualitative increase of load size with distance (Bowers and Ellis 1993) and have shown that load sizes also increase as a result of delayed access to the food source, even when distance does not change (Lair et al. 1994). The predicted increase of load size with food density was confirmed qualitatively (Kramer and Weary 1991). A decrease in load size from patches with less canopy cover and presumably more risk of aerial predation has been found in some situations (Bowers and Ellis 1993), but not in others (Otter 1994). Furthermore, three studies found a strong, but unpredicted, decrease in load size with increasing trip number (Giraldeau and Kramer 1982; Bowers and Adams-Manson 1993; Lair et al. 1994). Because these tests did not measure the loading curves, it is not clear whether the patterns arise through differences in loading and travel rates that changed the delivery rate optimum or as independent variables that produce deviations from the delivery rate optimum.

Implications   The load size decision seems relatively minor in that animals ought to be able to get by quite satisfactorily by simply filling their cheek pouches to some constant level on each trip. However, chipmunks showed rapid short-term flexibility not only to distance but also to several other environmental conditions. These patterns would not have been discovered without the a priori theory. Optimality models provide a framework for the discovery of natural patterns even when they fail to predict accurately. To illustrate, I will discuss some hypotheses to explain why chipmunks take loads smaller than the predicted optimum. These suggest future studies and ultimately stronger predictions.

First, consider possible limitations in the model resulting from potentially relevant variables that were ignored. When an animal takes smaller loads than predicted, it makes more trips per unit of foraging time than predicted and thus spends more time traveling and less time in the patch than it would if it took the predicted load sizes. Thus, we should consider alternative currencies that would recognize the benefit of more trips or more traveling relative to more time loading. Note that small loads require more of an explanation than large loads. Consider the rate of gain curves in figure 18.1D and E. An animal taking too large a load suffers less of a reduction in gain rate than an animal taking too small a load. Hence, we might expect less of a cost in responding to variables that require an increase than a decrease in load.

The most obvious simplification in this model is its currency of gross rate of food delivery. Net rate of energy gain might seem to be more appropriate. However, a 3-g load of sunflower seeds contains about 12 kcal of energy, while loading requires about 0.02 kcal/min and running requires about 0.035 kcal/min (M. Humphries, personal communication), so net rate will not differ much from gross rate in our study. Furthermore, the energy cost of running is higher than the energy cost of loading, so including energy costs will increase the discrepancy. Predation risk could be a factor if loading is more dangerous than traveling or if predation risk increases disproportionately with larger loads. Intraspecific competition might also be a factor. Small loads reduce the rate of food delivery, making an animal less effective in scramble competition for a short-lived patch. They also reduce time at the foraging site, where a dominant animal might gain an additional guarding advantage. On the other hand, if the main threat from conspecifics is pilferage from the burrow while the owner is away, taking smaller loads could improve the guarding of the hoard by increasing the frequency with which the burrow is visited. Ydenberg et al. (1986) developed several models indicating how animals could use active and passive interference as competitive tactics to increase their share of a limited ephemeral resource. This suggests that a game theory model of load size taking intraspecific competition into account might have greater predictive power.

Having considered possible limitations of the model, we should now consider possible limitations of the animal. To match the predictions of the model using behavioral flexibility, an animal has to have information about travel time, loading rate, and how they are to be integrated. Determining these from recent experience involves complex issues such as how many trips to include, how to weight more recent versus less recent trips, and how to deal with interruptions such as time vigilant, hiding from predators, grooming, or engaged in aggressive interactions with conspecifics. Furthermore, animals may measure time and mass on a different scale from scientists (Kacelnik and Bateson 1996). If a chipmunk underestimated loading rate or travel time, its loads would be too small. Most chipmunks' prior experience would involve far lower food densities than what we offered, and optimal loads, therefore, almost always would be smaller. If the animal used a simple rule of thumb, it is likely that the food type, location, or mode of presentation did not match the evolutionary or developmental environment in which the rule of thumb had been established. Models that take into account specific mechanisms by which animals determine and integrate the foraging parameters might be able to improve the precision of predictions at a considerable loss of generality.

Future Directions

Foraging studies have a long history of providing conceptual and methodological advances in behavioral ecology. This field provides an excellent example of the advantages of integrating quantitative theory with rigorous empirical testing. The number of variables influencing the "simple" load size decision suggests that there are numerous other decision rules to be discovered, possibly for load size, and certainly for the large number of other decisions involved in foraging. In fact, a few elegant models such as prey choice, patch departure (marginal value theorem), and spatial distribution (ideal free distribution) have received disproportionate research attention. For many other decisions of equal importance, clear general models remain to be discovered. Furthermore, many interesting models proposed over the last two decades have received little or no testing. This is true even for situations where simple optimality approaches apply; the investigation of social impacts on these decisions adds new challenges and should receive a strong impetus with Giraldeau and Caraco's (2000) book.

Extending the scope of foraging studies will yield fascinating insights into the links between behavior and the neuroscience of information acquisition and processing. The field of risk-sensitive foraging illustrates these connections. While there was not space in this chapter to discuss how animals deal with variance in foraging and other fitness benefits, this is currently a lively area (see Kacelnik and Bateson 1996 and other papers in the same symposium). Foraging studies can provide valuable insights for other areas of physiology. The conceptual framework of foraging is closely related to that of the acquisition of water, minerals, and oxygen, and these processes can be studied with similar approaches, even though different physiological systems are involved.

Optimal foraging arose from questions raised by community ecologists but developed with a focus on individual behavior. Our understanding of foraging is now such that the field can make a real contribution at the levels of population and community. Although ecological situations are often too complex for predictions based on simple behavioral theories, the theory and empirical base provided by foraging studies provide important insights into processes of linking behavior and populations.

Acknowledgments I thank the many undergraduate and graduate students and postdoctoral scholars who have worked with me on chipmunk behavior at McGill over the past 25 years as well as colleagues at many other institutions who have offered feedback and made constructive suggestions. D. Gidley, L.-A. Giraldeau, C. Hall, M. Humphries, P. McDougall, C. Schiffer, and H. Young provided helpful feedback on an earlier draft. M. Humphries and D. Roff provided timely help with the figures. Our research has been funded by NSERC Canada and FCAR Quebec and has been greatly aided by the staff and facilities of the Gault Nature Reserve in St. Hilaire.

19

The Evolutionary Ecology of Movement

HUGH DINGLE
MARCEL HOLYOAK

Organisms move, and their movement can take place by walking, swimming, or flying; via transport by another organism (phoresy); or by a vehicle such as wind or current (Dingle 1996). The functions of movement include finding food or mates, escaping from predators or deteriorating habitats, the avoidance of inbreeding, and the invasion and colonization of new areas. Virtually all life functions require at least some movement, so it is hardly surprising that organisms have evolved a number of structures, devices, and behaviors to facilitate it. The behavior of individuals while moving and the way this behavior is incorporated into life histories form one part of this chapter. This discussion focuses on the action of selection on the evolution of individual behavior, on how specific kinds of movement can be identified from the underlying behavior and physiology, and on the functions of the various movement behaviors.

The other major part of our discussion focuses on the consequences of movement behaviors for the ecology and dynamics of populations. The pathways of the moving individuals within it can result in quite different outcomes for a population. First, movements may *disperse* the members of the population and increase the mean distances among them. The separation may be a result of paths more-or-less randomly chosen by organisms as they seek resources, or it may be a consequence of organisms avoiding one another. In contrast to dispersing them, movement may also bring individuals to-

gether, either because they clump or *congregate* in the same habitat patch or because they actively *aggregate* through mutual attraction. Clumping can also lead to aggregation and mutually attracting social interactions. A classic example is the gregarious (aggregating) phase of the desert locust *(Schistocerca gregaria)*, in which huge swarms of many millions of individuals first congregate in suitable habitats and then develop and retain cohesion based on mutual attraction. The foraging swarms make the locust a devastating agricultural pest over much of Africa and the Middle East (Farrow 1990; Dingle 1996). It is the aggregation of locusts that makes them such destructive pests; they would be far less harmful if the populations dispersed. The desert locust is a prime example of how a specific type of individual behavior, aggregating movements, produces a distinctive population outcome different from what it would be were individuals moving apart. Thus, in studying movement it is important to recognize two distinct levels of analysis: the behavioral, which applies to individuals, and the ecological, which applies to populations (Kennedy 1985).

History

The movements of organisms take place over scales that range from a few body lengths to pathways that may span the planet. Much effort by behavioral biologists in the first half of the 20th century was expended on the short-distance movements of "lower" animals (and in some cases plants) and especially on the mechanisms of orientation. Much of this work was summarized by Gottfried Fraenkel and Donald Gunn in their book *The Orientations of Animals* (1940). After World War II, focus shifted to complex behaviors in natural environments ("ethology") rather than the simple movements observed in the laboratory that were the primary focus of Fraenkel and Gunn. The latter, however, did draw attention to two distinct kinds of orientation movements: kineses, which consist of changes in the rate of movement and turning to remain within or to locate a resource (a planarian, for example, slows down and turns less in shadow so that it tends to remain there), and taxes, which are movements directed at the source of a stimulus.

There is continuing interest in taxes by chemical ecologists studying pheromones, the air- or water-borne chemical messengers used by many organisms. Pheromones used in mating are usually called *sex attractants,* because originally it was supposed that wind-borne odor plumes were literally scent trails along which flying insects were guided chemotactically. This, however, turns out not to be how flying insects actually behave in the odor plume, because pheromones are not carried downwind in a manner that would allow "trail following." Rather, wind-borne odors occur more-or-less in puffs, so the problem is one of maintaining contact with a disappearing and reappearing source. On contacting pheromone, an insect turns upwind (amenotaxis) and performs a zigzagging behavior as long as contact with the chemical is maintained (Kennedy et al. 1981); if contact is lost, it begins wider sweeps, termed *casting,* until contact is regained. Other sensory inputs are also integrated into the amenotactic wind response. The exact nature of the responses is important because synthetic pheromone traps are used to monitor pest insects and to disrupt mating.

Large-scale movements of animals, especially migration, have been noted for centuries. Frederick II of Hohenstaufen, for example, in his treatise *The Art of Falconry,* written about 1250, made the astute observation, recently confirmed (Dingle 1996), that migratory birds use the wind to aid their journeys. Most studies of migration from Frederick to the present focus on a single taxon (usually a vertebrate), tend to be largely descriptive, and define migration in terms of the geography of pathways with return movements defining the behavior. In the 1950s, the perspective on migration began to change when some ornithologists like David Lack began studying bird migration as an evolutionary strategy that was an important component of life histories. Lack asked *why* migration occurred and first clearly outlined an approach based in evolutionary ecology. Although he was interested in birds, his ideas could be extended to other migratory organisms. Lack emphasized competition as the basis for migration, and his ideas on competition have been extended by others since.

Although birds held center stage because of their visibility, parallel changes in thinking were occurring among students of other taxa, especially insects. C. B. Williams in the 1930s had drawn attention to the fact that many insects migrate, and C. G. Johnson (1960), J. S. Kennedy (1961), and Sir Richard Southwood (1962) put studies on a solid biological footing. Johnson put migration in the context of life histories and physiology by pointing out that it usually took place prior to the onset of reproduction, a juxtaposition he called the "oogenesis-flight syndrome." In Johnson's view, migration was seen as much an episode of colonization as an escape mechanism. Kennedy demonstrated with sophisticated experiments in the black bean aphid, *Aphis fabae,* that migratory behavior involves a complex interaction between flight responses and those of settling on a new host, including larviposition (this aphid is parthenogenic and live-bearing). He provided for the first time a definition of migration independent of the pathway of the migrant (see below). Finally, Southwood provided a sound foundation in evolutionary ecology by demonstrating that temporary habitats like early plant successional stages are the ones most likely to select for migration.

Modern Concepts and Case Studies

Types of Movement

Movements can be roughly divided into two categories, those that are directed toward activities and resources on the home range, where an organism spends most of its time, and those that are not (Dingle 1996 and table 19.1). Many movements are directed at locating resources within the home

Table 19.1 Types of movement of individuals.

Movement	Characteristics	Examples
I. Movements that are home-range- or resource-directed		
A. Station keeping	Movement keeps organism in home range.	
1. Kineses	Changes in rate of movement or turning.	Planarian in shadow.
2. Taxes	Directed movement in response to a stimulus.	Upwind flight of moth in pheromone "plume."
3. Foraging	Movement in search of resources. Ceases when resource encountered.	Search for food, mate, shelter, oviposition site; parasite or parasitoid host seeking.
4. Commuting	Periodic (often daily) forages in search of resources. Ceases when resource encountered.	Albatrosses foraging; vertical "migration" of plankton.
5. Territorial behavior	Movement and agonistic behavior directed toward intruders in territory. Ceases when intruder leaves.	Many.
B. Ranging	Movement over an area so as to explore it. Ceases when new home range/territory is found.	"Dispersal" of some mammals; "natal dispersal" of birds.
II. Movements not directly responsive to resources or home range		
A. Migration	Undistracted movement with cessation primed by movement itself. Responses to resources/ home range suspended or suppressed.	Annual journeys of birds, insects, etc. Flight of aphids to new hosts. Transport of some seeds to germination sites.
B. Accidental displacement	Organism does not initiate movement. Ceases when leaves transporting vehicle.	Storm vagrancy.

range, and these we term *station-keeping movements* (Kennedy 1985). By and large, resources promote growth and reproduction and include shelter, mates, food, oviposition or nest site, suitable microclimate, and so forth. An important characteristic of these movements (and of ranging, below) is that they cease when an appropriate resource is encountered. Thus, a predator stops hunting when it kills its prey, an egg-bearing butterfly stops flying when it finds its host plant and lays eggs, and a planktonic crustacean ceases swimming downward when it reaches a suitable depth to avoid daytime predation.

We include within station-keeping movements five different patterns of behavior that are not mutually exclusive (table 19.1). The first two are kineses and taxes. Taxes are associated with some of the most complex orientation behaviors known among organisms, from bacteria to birds (Dingle 1996). Sources of information used in orientation include the sun, polarized light, star patterns, chemicals, the wind, and the earth's magnetic field, a by no means exhaustive list. Most animals use more than one system of orientation, using some to back up others. Pigeons, for example, use the sun to home when it is visible, but the magnetic field on cloudy days. The sun itself can be used in two ways. The simplest is that the organism moves toward the sun without correcting for the path across the sky. Such orientation is obviously imprecise because direction of movement will change with the sun's position. The second way to use the sun, which improves accuracy, is to compensate for its path by changing the angle of orientation. This maneuver requires a bicoordinate system with two reference points, the sun and the time of day (to know how much to compensate for the sun's position to maintain a constant direction). This bicoordinate system requires in turn an internal timing device or biological "clock." A time-compen sated sun compass has been demonstrated in a great variety of invertebrates and vertebrates.

There will always be considerable overlap between orientation and our third category, foraging, because foragers obviously orient. Foraging is covered in Kramer (this volume), but we also discuss an example here to illustrate differences between foraging and other types of movement. This is the Old World desert locust *Schistocerca gregaria* briefly mentioned in the introduction. This is one of several species of locusts and grasshoppers adapted to exploit dry environments. The desert locust displays in extreme form a phenomenon known as *phase polyphenism*, in which gregarious individuals that develop under crowded conditions are larger, darker, and of different shape and behavior than solitary forms developing at low densities (we distinguish here between "polyphenisms," where form is modified environmentally, from "polymorphisms," where forms result from gene differences). It is the behavioral differences between the two forms that concern us here (Farrow 1990; Dingle 1996).

Crowded individuals of the desert locust undergo a remarkable behavioral transformation in which they are strongly attracted to each other and form large, cohesive aggregations or swarms. Low-density individuals display no such mutual attraction. Because of attraction, locusts that leave the edge of the swarm, whether in the air or on the ground, soon turn and reenter it, maintaining swarm integrity. When feeding, most swarm members are on the ground, but many fly overhead in a more-or-less consistent direction, although the headings of flying individuals within the swarm may vary considerably. So long as wind speeds aren't excessive, the direction of movement is generally upwind (figure 19.1). This pattern of swarm behavior occurs because as individuals at the rear run out of food, they embark on flight and overfly the locusts remaining on the ground to land again at the swarm's leading edge where more food is available. To an observer viewing the swarm from outside, it appears as a rolling mass of insects moving in more-or-less the same direction. In good habitat, locusts spend most of their time feeding or resting, and the swarm moves slowly. In poor habitat, however, most locusts become airborne, and if wind velocities increase, the whole swarm may be bundled across the countryside in a downwind direction, stopping only when a suitable habitat with plenty of green plants is again encountered. Straight-line movement ensures that the swarm will not backtrack over a previously denuded area. Because appropriate resources stop the movement, most swarm behavior is best considered extended foraging (Farrow 1990) rather than migration, although in some circumstances migration by swarms does occur (Pener et al. 1997). More recent studies (Dingle 1996), especially those using insect-detecting radar, reveal that the solitary form is migratory, and that most migration takes place at night (when gregarious forms are mostly grounded).

In some organisms extended foraging may take place repetitively between a home site and a resource along a more-or-less prescribed pathway. We refer to this as commuting, by analogy with the daily movements of the human workforce. Some animal commutes are quite remarkable. Many seabirds make long-distance commutes between nesting colonies and rich foraging areas; for example, satellite tracking has revealed that wandering albatrosses make commutes of up to 33 days over distances of 3600–15,000 kilometers when their mates are incubating and for 3 days and over 300 km after the chicks have hatched (Jouventin and Weimerskirch 1990). These movements are longer than the migratory journeys of many birds (and are round-trip!) but are proximately driven by dependence on certain resources, because the outward flight ends when resources are encountered. Other examples of commuting include the daily vertical round trips made by both freshwater and marine plankters. These planktonic movements are called *vertical migration* in the literature, but we argue they are best considered commuting because, once again, they cease when a suitable resource, food or predator-free space, is encountered.

Foraging and commuting both usually take place within an area that may be extensive but is visited repeatedly and so constitutes the home range, although note that locusts are an exception because their paths almost never cover the same ground. There are also movements expressly evolved to take an organism out of its home range; these are ranging and migration (table 19.1 and Dingle 1996). Ranging involves movement over an area to explore it, and it ceases when a new suitable home range is found. Ranging behavior is often called *dispersal*, or *natal dispersal* when young prereproductive individuals are involved, but this terminology confounds the behavior of individuals with population consequences because *dispersal* also means an increase in mean distances among

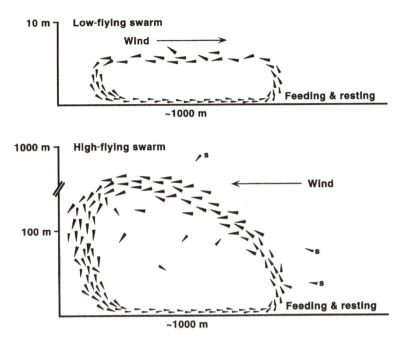

Figure 19.1 Patterns of locust swarms in low wind velocities (top) versus high wind velocities (bottom). Triangular arrows point in the general direction of movement of individuals, and s represents scattered individuals in high winds.

individuals. The point can be illustrated by the Seychelles warbler, *Acrocephalus sechellensis* (Komdeur et al. 1995). This species breeds on small islands of the Seychelles Archipelago in the Indian Ocean, islands so small that all suitable breeding territories are filled and young remain on their natal territory to assist their parents to raise more offspring, acting as "helpers." When a territory becomes available through disappearance of its previous owner, it is quickly occupied by a former helper from elsewhere. This movement to a newly available territory has been described as "dispersal," but note that the number of territories remains the same, and there is no change in mean distance among individuals. We suspect that helpers continuously range over neighboring territories or even the whole island and move into a vacant territory when they discover it, so that ranging stops when a suitable resource, a vacant territory, is encountered. Only detailed observations of individuals can determine if this scenario is correct. Such observations sometimes indicate far more ranging than previously suspected, as exemplified by supposedly sedentary brook trout, *Salvelinus*

fontinalis, that in fact move relatively long distances along Colorado mountain streams, probably in search of new feeding home ranges (Gowan and Fausch 1996).

Migration is a behavior that is distinctly different from other movements, and this difference is rooted in the underlying physiology (Kennedy 1985; Dingle 1996). First, migrants often respond to cues like photoperiod or crowding that predict habitat change, rather than simply responding to habitat change itself. Second, migratory movements tend to be persistent and straightened out, unlike the zigzagging or wandering often seen in foraging or ranging. They can be self-generated or can result from deliberate embarkation on the wind or currents, or by attaching to another organism; that is, they are not "passive" even when aided by wind or current. Third, station-keeping responses are *temporarily* suppressed or inhibited, and migrants will *not* stop when they encounter resources, such as food or potential mates, that would stop other movements. While they are migrating, wildebeest do not stop to graze, seabirds to fish, or butterflies to oviposit on suitable host plants. The extreme occurs in migrating Pacific salmon *(Oncorynchus spp.)* which lose the ability to feed because their jaws are modified for fighting for nest sites at the end of the journey. A test for whether an individual is a migrant, rather than a ranger or forager, is whether it will respond to appropriate resources. On long journeys, migrants may enter and leave the migratory phase repeatedly, foraging intermittently to replenish energy or other essential resources. Finally, the migratory behavior itself primes (disinhibits) responses to resources, so that at the end of migration, these responses are strong. A migrant aphid, for example, will ignore suitable host leaves during its flight, but especially long flights will lower the response thresholds so that the aphid will larviposit on even barely suitable leaves, and migrating male salmon will not fight other males until their journeys are complete. These characteristics define migration in terms of its behavior, not its pathway or population outcome (e.g., note that it doesn't matter whether or not migration is a round trip).

Accompanying migratory behavior is often a syndrome of other traits that increase fitness. A common characteristic of migrants from birds to insects is the use of fat as fuel because fat metabolizes to produce about twice as much energy as carbohydrate or protein and requires no water for storage; in contrast it requires 3 g of water to store 1 g of carbohydrate. Metabolic pathways shift to produce more fat storage, and the extra fat can double the body mass of migrants. Elaborate orientation and navigation systems and specific life history and morphological traits are also frequently associated with migration. Thus, migrants may be larger, have longer wings (birds, insects), reproduce earlier, or show higher fecundities. The evolution of migratory syndromes is discussed below (see "Life Histories and Movement").

Causes and Consequences of Movement

Population Dynamics So far we have discussed movement primarily from the perspective of individuals. It is important to realize that in actual fact, there is a two-way interaction between individuals and larger groups, such as populations. Individuals contribute to a wide variety of factors that influence entire populations, for example, total resource levels, biomass available to predators,

and genetic diversity. Population-level factors can, in turn, influence behaviors, traits, and fitness of individuals. In this section, we use selected examples to discuss the contribution of individual movement to population dynamics (and genetics). The links between population dynamics, population genetics, and evolutionary ecology are covered elsewhere in this book.

A common result of movement behavior is that populations of organisms change their spacing or dispersion. Untangling why individuals alter their spacing is complex because there are many potential reasons, including attraction to resources or mates, or to areas where survival is enhanced. Consider the case of mobile grazers on a plain with high- and low-resource areas. If individuals are highly mobile and frequently forage (table 19.1) over both high- and low-resource areas, they could potentially select habitat based on levels of resources. For a grazer, the number of resources available per individual is a function of both local population density and resource density within an area over which individual foraging extends. Habitat selection could result simply from individuals altering their rate of movement (kineses; table 19.1) in response to a stimulus that reflects food availability. Turning less frequently and moving more rapidly in less favorable areas will increase the net displacement per unit time, promoting movement to other areas. Such behavior often continues until a favorable stimulus is encountered, causing the individual to slow down and turn more frequently, so that it remains longer in favorable areas. Movement behavior could lead to gains in the rate of resource acquisition per individual up to the point where individuals forage optimally and resource acquisition is both equal in all individuals and maximal (see Kramer, this volume).

If individuals did not adjust their movement behavior in response to differences in available resources, overcrowding could lead to starvation of some or all individuals in areas where density exceeds available food. By contrast, with resource-dependent movement (foraging) the average rate of resource acquisition per individual may be increased, an increase promoting selection for the kineses and taxes that are involved if there are fitness benefits. In extreme cases, scramble competition for resources may lead to all individuals in overcrowded areas dying, promoting strong selection for movement away from overcrowded areas at the level of individuals and optimal foraging at the population level. The desert locust example in the previous section illustrates how feeding efficiency of individuals is increased by aggregation and travel in straight lines (see "Types of Movement"). In this case, swarms aggregate through mutual attraction, and travel in straight lines increases grazing efficiency because few individuals feed in denuded areas.

In a specialist predator and its prey, there can be strong selection for movement behavior because the predator would gain a fitness advantage by responding to prey density (just like any other consumer), and prey might increase their fitness if they can move to evade capture and consumption. Heterogeneity in the risk of prey being attacked can lead to population stability, allowing populations to persist. In many species, consumption rates are high, and consequently, strong selection for prey-avoiding predators would be expected. A wide variety of types of movement behaviors of prey in response to predators has evolved. Escape behaviors include jumping (e.g., click beetles), flight (predation is often cited as a reason for wingedness), dropping to the ground, and swimming (e.g., many frog species). Alternatively, prey may interrupt movement and rely on remaining still and crypsis to avoid predation. Prey can also alter their dispersion, thereby exploiting either predator satiation in high-density prey areas or the inability of predators to locate prey in low-density areas. Predators could also evolve counteradaptations in response to prey adaptations, including speed, prey caching (storage) to overcome satiation, and exploiting signals (e.g., chemicals, carbon dioxide) to locate sparsely distributed prey. Feedback between individual movement behavior and either population extinction or individual fitness in predators and prey could also promote selection for particular kinds of movement behavior.

Population dynamic and genetic models deal with movement in a variety of ways. A useful distinction is between models where movement occurs at a restricted rate and those where there is frequent movement between any available habitat patches. First, let us consider models with organisms that are very mobile relative to the distance between patches. High interpatch movement rates, perception of patch quality, and mechanisms leading to longer residence in higher-quality patches are the domain of foraging models, such as the ideal free distribution.[1] Individuals would then be expected to evolve to move frequently between

patches and to spend more time in more profitable patches (Parker and Stuart 1976). Foraging models also represent an extreme for population dynamics and genetics, where a single panmictic population with a single risk of extinction occupies all habitat patches over which foraging extends. These models usually assume that organisms have perfect knowledge and can select the highest-quality patch from either some or all patches of suitable habitat. While this is a useful null model, it distances researchers from how organisms select the best available habitat (termed *habitat selection*) and decide when to leave patches (see "How Good Is the Fit Between Models and Biology?").

If populations become sufficiently isolated, the amount of gene flow between them may be restricted, and population dynamics (including the risk of extinction) of different local populations may become somewhat independent (Nunney, this volume). At this point, we see a switch from the relevance of foraging and population models to so-called metapopulation models. A metapopulation consists of a suite of habitat patches that are separated by a transit habitat in which reproduction and long-term survival are impossible. The question of whether movement occurs within a population or within a metapopulation (between populations) is largely a matter of scaling. Organisms are expected to move over an area in which they can both obtain all resources necessary during their lifetime and maximize fitness through factors like taking advantage of fitness gains in vacant resource-rich habitats and by avoiding depression in fitness from inbreeding (Waser and Williams, this volume), predation (Abrams, this volume), competition (Kramer, this volume), and so on. Whether increased fitness is possible within a single patch or requires movement over an extensive network of patches is a function of the density of resources within patches and the distance between patches; it is also often the case that several distinct types of movement evolve that operate over different spatial and temporal scales (see "Types of Movement"). Models of metapopulation dynamics emphasize that the likelihood of a species surviving within a region is enhanced if extinction of local populations is followed by timely recolonization (Levins 1969). If different local populations have semi-independent probabilities of extinction, then the probability of simultaneous extinction of the entire metapopulation will be reduced. Independence in the risk of extinction across patches is a function of the distance over which factors influencing dynamics are correlated; frequent interpatch movement and regional environmental factors (e.g., drought) synchronize dynamics, whereas local factors, such as variability in population growth rates or localized physical disturbances, may promote asynchrony across patches in dynamics and within-patch extinctions. In metapopulations, interpatch movement is also the source of recolonization of local populations that have gone extinct, and it may increase the size of local populations, making extinction through chance events less likely, the so-called rescue effect (Brown and Kodric-Brown 1977). Metapopulation dynamics are also possible in a continuous habitat if rates and distances of movement are appropriate to influence dynamics without creating a single diffuse population.

In metapopulations, extinction of local populations leads to a loss of genetic variability. Following extinction of local populations, movement is important as a source of both colonization and gene flow, which helps to restore heterozygosity to habitat patches that have recently been colonized and are therefore low in heterozygosity; at the level of the metapopulation, movement may help to ameliorate loss of genetic variability through local extinctions (Hedrick and Gilpin 1996).

Natural Selection of Movement   All populations are ephemeral because either the habitat itself is ephemeral or the population has a finite risk of extinction due to factors like chance events, environmental catastrophes (floods, fires, etc.), biotic interactions (disease, predation, etc.), and loss of genetic variability. Movement is frequently hypothesized to have evolved as a way of dealing with ephemerality of populations. The general result of these studies can be illustrated with a simple probability model (Roff 1990). Imagine a species with a population in a patch that has a probability p of remaining habitable and allowing reproduction from one generation to the next. The probability that the habitat patch will persist to a second generation is p^2 and that it will persist for t generations is then p^t. Now, suppose that there are two such patches that are completely isolated, so that there is no movement between them. Extinction from both patches would then require only that each patch become unsuitable within the period t, giving a probability of persistence of $1 - (1 - p^t)^2$ without movement. Next consider what would

happen if the patches were connected by movement of individuals, so that there is instantaneous recolonization of within-patch populations following extinction of a population. Both patches would go extinct only if both of the local populations went extinct at the same time. The probability of persistence for a period t is therefore $(1 - (1 - p)^2)^t$ with movement. The significance can be seen with a numerical example. For example, with $p = 0.95$ and $t = 100$ generations, the migratory species is 66 times more likely to persist in two patches with movement than without. Adding more habitat patches that are connected by movement further reduces the chance of regionwide extinction.

A second series of models considers movement driven by resource availability within a metapopulation. Gadgil (1970) used abundance in a habitat patch (N) and the patch's carrying capacity (K) to define resource availability through *crowding* (N/K). This model gives the intuitive result that movement is selected for when there is a chance of inhabiting another habitat patch that is less crowded and therefore more favorable than the one currently occupied.

More complex models of the evolution of movement and habitat crowding consider whether a movement rate is an evolutionarily stable strategy (ESS) that maximizes fitness under the chosen conditions and therefore cannot be invaded by any other strategy. Hamilton and May (1977) developed a population model for movement of a wholly parthenogenetic species where the proportion of offspring that move, v, is determined by genotype and is therefore set by the parents. They showed that if a fraction p of offspring that move survive to reproduce, then the ESS proportion moving (v^*) was $v^* = 1/(2 - p)$. This analysis implies that an adult should always force at least half of its offspring to move. Although this conclusion seems counterintuitive, it can be explained by each patch holding a fixed number of individuals (one per patch in the case of this model), so that only one individual can survive to replace the parent, whereas moving individuals may encounter a vacant site. All else being equal, movement is selected for because it is better for offspring to compete with nonsiblings at many sites than with themselves at a single site. Hamilton and May also modified their model to consider sexual reproduction and obtained analogous results with the addition of an effect that in a species with two sexes, greater movement rates by one sex can offset low movement rates by the other sex. Other models of the evolution of movement consider other factors that influence movement, such as costs of movement or degrees of relatedness.

At low movement rates, sufficient genetic variability may be lost through inbreeding that negative effects are seen in the form of both reduced fecundity ("inbreeding depression") and a reduced ability of populations to cope with environmental change (Waser and Williams, this volume; Hedrick, this volume). Models of the evolution of movement that include the effects of relatedness of individuals, such as that by Frank (1986), show that movement is influenced by both the degree of relatedness of different (randomly chosen) individuals (ρ) and the cost of movement (c, measured in terms of fitness). Specifically, Frank showed that the ESS proportion dispersing was $v^* = (\rho - c)/(\rho - c^2)$, assuming (for mathematical reasons) that the cost of movement is less than the degree of relatedness and both are scaled between 0 and 1. Note that *disperse* is the appropriate term here because its opposite, *aggregation* or *congregation,* would lead to increased inbreeding. Figure 19.2 gives a numerical example that shows that greater movement rates will be selected for by greater degrees of relatedness and by lower fitness costs of movement.

The probability model of Roff (1990) discussed above addresses the question of why individuals should move in ephemeral habitats. Habitat ephemerality has also been considered a factor driving the evolution of the rate of movement (Southwood et al. 1974), or it has been considered a force selecting for a particular kind of movement, migration (table 19.1). First, let us consider the evolution of rates of movement with habitat ephemerality, where the driving force is local extinction. Models of the evolution of movement in metapopulations (Olivieri and Gouyon 1996) suggest that the rate of interpatch movement in metapopulations is the result of two antagonistic selective forces, which combine to give a "metapopulation effect." First, movement is selected against by mortality during movement (from factors like predation, desiccation, and starvation) and by individuals arriving in unfavorable habitat patches where survival is not possible (see also figure 19.3). In habitats consisting of mosaics of favorable and unfavorable patches, only favorable patches produce emigrants, but all patches receive immigrants. Consequently, a gene promoting movement would leave favorable patches more frequently than it is reintroduced, so it would decline in frequency. In opposition is selection for movement among local populations caused by avoidance of sib competition, deterioration of local conditions, and improved conditions elsewhere. More generally, interpatch movement is selected for by high spatiotemporal variability in population sizes (reflecting local conditions), so that a gene promoting movement increases in frequency when conditions decline in patches (and have historically declined in patches) and survival is possible only by moving to other patches where conditions are likely to be different at any point in time. Extinction of local within-patch populations is an extreme example of spatiotemporal variability that should lead to selection for movement.

The types of movement occurring in a particular population or species will depend on the ways selection acts on the individuals within it, and ultimately, that selection will depend on the distribution of resources. In the case of foraging, the sophistication of the orientation involved in the behavior is a function of the abundance of a resource and/or the difficulty in acquiring it. Thus, the skills of a predator that depends on rare prey will be greater than those of an herbivore feeding in a lush grassland (although the herbivore may well be skilled at predator avoidance). With respect to ranging, the selective factors favoring exploration for and location of new home ranges will be

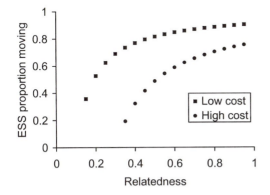

Figure 19.2 Predicted proportions of individuals moving between patches under different degrees of relatedness of individuals and different fitness costs of migration from a model by Frank (1986). Squares show the predicted ESS proportion moving, v^*, with "costs" set at $c = 0.1$ and circles for $c = 0.3$.

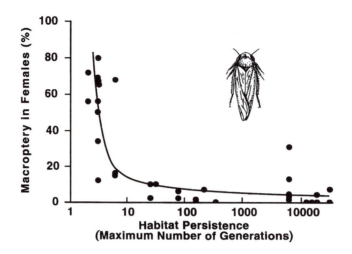

Figure 19.3 The frequency of macroptery (long wings) in female planthoppers and habitat persistence measured in generations for 35 species of planthoppers representing 41 populations. The occurrence of long-winged forms declines dramatically in persistent habitats. (Redrawn from Dingle 1996, after Denno et al. 1991.)

characteristics of the environment that favor reproduction in a new area. These factors are likely to include greater abundance of resources, greater access to resources in the absence of competition from conspecifics (parents, sibs, or others), and the ability to optimize inbreeding or outbreeding. Resources may be abundant over a fairly wide area, but access to these resources depends on local variation in competition and resource abundance.

Migration tends to be favored when habitats deteriorate and resources decline over a wide area relative to the size of the migrating organism. Habitats can be ephemeral because they are seasonal, because they are undergoing succession, because of the decline of a resource (e.g., aging of a food plant), because of unpredictable resource variation over time and space (e.g., in deserts because of low and variable rainfall), or for a combination of reasons. The overall relation between migration and habitat can be expressed as the ratio H/τ where H = duration of a favorable habitat and τ = generation time (Southwood 1977). As this ratio approaches unity, migration is increasingly favored. Seasonal migrations are perhaps the clearest example of how selection would favor a movement behavior one of whose characteristics is temporary suppression of responses to stimuli that would ordinarily stop it. If an environment were deteriorat-

ing as a consequence of seasonal change, selection would favor, first, response to a cue that would trigger departure *before* full deterioration had occurred. If the organism waited until the habitat was unsuitable, it would not be able to fuel for the journey. Indeed migration is triggered by reliable cues, such as day length, that forecast seasonal change. Second, during migration, selection would favor movement beyond favorable nearby habitats because they, too, will soon become unsuitable. Similar arguments apply when migration involves departure from a senescing host plant, as in the case of aphids, or a drying patch of desert where recent rains have stimulated plant growth, as in the case of many locusts. If one's host plant is senescing or drying out, others are probably doing likewise over a wide area. Ceasing to migrate at the first opportunity will therefore be unlikely to result in an improved environment, which is much more likely to occur farther afield.

Comparisons of the frequencies of migrant species among regional faunas suggest that different aspects of climate may be selecting for migration (Dingle et al. 2000). Most seasonal migrants in the temperate Northern Hemisphere move along a north-south axis, and birds are the most familiar example. Northern regions hold a higher proportion of migrants than southern in both eastern

North America and the western Palearctic, and statistical comparisons using regression indicate that from 85% to 98% of the variation in migration frequency is explained by latitude (table 19.2) with more migration expressed in more northerly-occurring bird species. Similarly, latitude explains 80% of the variation in the proportion of migrants among South American tyrant flycatchers. A regression analysis of Australian butterflies, however, indicated that none of the variance in migration frequency was explained by latitude. Why this failure of latitude to correlate with migration in Australian butterflies? One major reason seems to be that Australia is a very dry continent. A higher frequency of butterfly migrants is found in areas of lower soil moisture, indicating dryer conditions in a given geographic subregion. The amount of rainfall is not associated with latitude, and so latitude does not predict migration. On the other hand, if one examines butterfly faunas only along the east coast of Australia, where rainfall amounts are similar to those in the western Palearctic and eastern North America, 72% of the variation in migration is accounted for (table 19.2). It seems likely that where rainfall is relatively high, latitude is a surrogate for temperature, and indeed, temperature measures are the best predictors of migration frequency once latitude is statistically controlled. In dry climates, on the other hand, the availability of rainfall is what counts. In either case, it is migrants that can take full advantage of habitats that are only periodically or sporadically available.

How Good Is the Fit between Models and Biology? In both population and metapopulation models movement is frequently treated as "dispersal," occurring at a fixed, density-independent rate over distances that are often constant. This treatment stems from difficulty in accurately measuring the number of individuals that move and the proximate mechanisms driving movement (Ims and Yoccoz 1996). However, such treatment also creates a rift between the behavioral descriptions of different types of movement and much of the literature on population dynamics, genetics, and evolution.

While there are many population-level phenomena that can be understood without needing to understand the behavioral mechanisms, frequently this is not the case. Sometimes, it is necessary to understand the mechanisms that drive movement in order to predict how either individuals or populations will respond to change, such as habitat loss or fragmentation. Sutherland (1996) provides a graphic example of how loss of wintering sites could influence populations of shorebirds moving from site to site within their wintering grounds. Movement between wintering sites is driven by prey depletion within sites, so that removing a site or reducing its quality can lead to overwintering migrants going to other sites. This increases the densities of wintering birds in the remaining sites, accelerating the rate of prey depletion within these sites, and can produce nonadditive changes in survival. To predict how loss of particular sites will influence the number of birds surviving through each winter, it is necessary to understand how removal of particular sites will alter the patterns of movement, population density, and resource depletion in the remaining sites. Without knowledge of the movement behavior of the organisms, such attempts would be futile.

Like Sutherland's study, existing population models that include a particular mechanism controlling movement illustrate that the different types of movement behavior can have strong effects on population dynamics and evolution. We draw on selected population dynamic examples to illustrate some of these effects. Departure from patches that is dependent on the density of conspecifics can potentially stabilize populations that would otherwise show large fluctuations in size (Denno and Peterson 1995); movement can prevent overexploitation of resources and subsequent population crashes. The timing of movement relative to other events in the life cycle can also influence population stability. For example, in parasitoids that mature within host insects and have generation times similar to those of their hosts, remaining too long in host patches after emergence can lead to overex-

Table 19.2 Variance in proportion of migrants in a fauna explained by latitude for temperate zone birds and Australian butterflies (Dingle et al. 2000).

Sample	Variance explained (%)
European summer birds	97
N. American summer birds	98
S. American tyrant flycatchers	82
Australian butterflies	0
Australian butterflies (East Coast only)	72

ploitation of host populations within patches. This in turn can destabilize populations, leading to regional host and parasitoid extinction. Finally, the distance moved can have strong effects on metapopulation persistence. Fryxell and Lundberg (1993) contrasted the effects on predator and prey metapopulation persistence of having predators that can move across all patches with those that move only a short distance, and they showed that specialist predators foraging optimally for their prey could stabilize populations if movement was local, but not if movement occurred across all patches.

Life Histories and Movement

Selection acting on life histories that include major commitments to movement, and especially to migration, often results in the evolution of characteristic complexes of traits or syndromes (Dingle 2000). Migration requires two major commitments; one is the time required and the risk assumed during movement, and the second is the energy required to sustain the journey. During the time an organism is migrating, it must postpone reproduction (and increases the risk of not reproducing at all), and it must devote energy to movement that might otherwise contribute to the number and condition of offspring. At the same time, for any colonizing organism such as a migratory insect, there would be a fitness advantage to a rapid and large reproductive output on reaching a new and empty habitat (Roff, this volume; Messina and Fox, this volume). We would predict, therefore, that natural selection would produce migratory syndromes that reflect these trade-offs.

Good examples of such syndromes with obvious trade-offs are the wing polymorphisms and polyphenisms of insects in which both flying longwinged and flightless short-winged (or wingless) individuals occur in the same species. These are clear cases of plasticity maintained by selection (Pigliucci, this volume). Flight is an obvious advantage, but wings and especially wing muscles are expensive to maintain. For example, in the cricket, *Gryllus firmus,* the wing musculature of long-winged individuals comprises 9–14% of body weight compared to around 5% in short-winged individuals, and the in vitro basal respiration rate of the long-winged muscles is threefold higher (Zera et al. 1997). Small wonder, then, that the short-winged or wingless females of numerous species display

higher fecundities and more rapid development than their long-winged counterparts, because they can channel more energy into growth and reproduction (Roff and Fairbairn 1991; Dingle 1996; Zera and Denno 1997). The question of importance here is: Under what circumstances would we expect the advantages of short wings to be favored over those of retaining wings and flight?

As we saw at the beginning of this chapter, Southwood (1962) showed that migration was more frequent among insects of temporary habitats. Consistent with this finding, he also demonstrated that winglessness was more common where habitats were long-lasting, such as mature forests or large lakes where selection would favor staying put. Southwood's general result has since been confirmed with appropriate statistical tests. For example, Denno et al. (1991) examined the proportion of long-winged (macropterous) individuals in populations from 35 species of planthoppers as a function of habitat persistence (figure 19.3) and showed that there was a marked increase in macroptery in short-lived habitats.

Case Study: The Migration Syndrome of the Milkweed Bug

The notion that selection acts on migration within the context of life histories to produce adaptive syndromes suggests that the traits making up those syndromes might share pleiotropic genes, so that there are genetic correlations among the traits. If this is the case, the existence and strength of the correlations could themselves be the objects of selection. Furthermore, if the traits making up a syndrome do so because of sharing of pleiotropic genes, then selection on any one of the traits should generate a response not only in that trait, but in the other traits constituting the syndrome as well.

To test this idea, Dingle and colleagues studied the migration syndrome of the milkweed bug, *Oncopeltus fasciatus* (summarized in chapter 14 of Dingle 1996). To do this, we took advantage of the fact that this widely distributed species has both migratory and nonmigratory populations. We examined, first, a migratory population from the midwestern United States that invades northern areas in late spring and migrates south in the autumn, displaying an annual migratory cycle much like that of the monarch butterfly. Second, we stud-

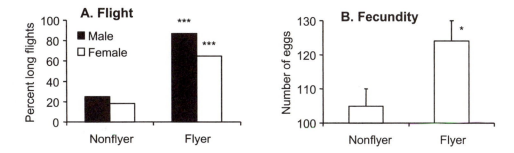

Figure 19.4 Correlated responses to selection on flight duration in migratory Iowa milkweed bugs, *Oncopeltus fasciatus*. Bugs were selected for both longer (flyer) or shorter (nonflyer) flights. (A) More males and females of the long-flying line made flights over 30 min than bugs of the short-flying line. (B) More eggs were laid per female in bugs selected for long-duration flight. Asterisks indicate levels of significance in tests conducted separately on males and females (***P < .001, *P < .01), and error bars indicate standard errors. (From Dingle 1996.)

ied a nonmigratory population that breeds year-round on the Caribbean island of Puerto Rico. We asked whether the migratory population displayed a syndrome of flight and life history traits not present in the nonmigratory population.

One of the methods chosen to address this question was artificial selection on laboratory populations derived from field-captured individuals from the two sources. Initially, we selected for wing length, because longer wings are associated with migration in many species of insect (and in birds), and in a few generations, we produced lines in each source population that had longer or shorter wings that unselected control lines. In the migratory bugs, longer wings were accompanied by higher fecundity and a greater frequency of long-duration flight, but these correlated responses were not seen in nonmigratory bugs, which did not differ among lines in either fecundity or flight. Therefore, a syndrome involving wing length, fecundity, and flight was present in the migratory population only.

To measure flight in the above selection experiment, we took advantage of the insect flight reflex: A bug could be temporarily glued to a short stick, and when lifted from the substrate, flight is initiated and its duration is recorded. To confirm a migratory syndrome, we also selected directly on flight itself in the migratory population. We produced bugs of differing flight duration by selecting individuals performing the longest and shortest flights as parents of the next generation and carrying the selection regime through two generations. The results of selection on flight are shown in figure 19.4A, where it is clear that the proportion of bugs flying for longer than 30 min is higher in the line selected for longer flight. The occurrence of a syndrome involving both flight and reproduction is indicated in figure 19.4B, which shows a statistically significant difference in fecundity (mean number of eggs laid by a female in the first 5 days of reproduction) between long and short flyers. In addition to fecundity, there was also a difference in wing length, longer flyers having longer wings as predicted from the previous experiment, where selection was on wing length.

The results from selection both on wing length and flight indicate a migratory syndrome in *O. fasciatus* that involves at least wing length, flight duration, and fecundity. This syndrome consists of traits that are influenced by the same genes because selection on one trait, flight or wing length, produces a correlated response in the other two (i.e., a genetic correlation). Selection is thus acting not only on individual traits, but also on the correlations among them. In the case of milkweed bug migration, the syndrome involves behavior (flight duration), morphology (size and wing length), and life history (fecundity), all of which would be advantageous to a colonizing migrant. This genetic relationship was not present in the nonmigratory population. In addition to avoiding the risks of migration, the nonmigrant Puerto Rico bugs devote less energy to flight. This and their smaller size give them an advantage when food is in short supply, as it often is in Puerto Rico, where milkweeds are less productive.

Genetic correlations come about through either linkage or pleiotropy, and pleiotropy seems more likely for migration syndromes because genetic linkages tend to be broken up by recombination events over time. Direct evidence for pleiotropy in an insect flight syndrome comes from a small Mediterranean bug, *Horvathiolus gibbicollis,* a species that is wing-dimorphic (Solbreck 1986). In this bug, wing length is a Mendelian trait with short wing dominant to long. The pleiotropic effect of the locus is revealed by the more rapid maturation, earlier oviposition, and larger early eggs of the short-winged morph, again revealing a syndrome of traits associated with presence or absence of flight.

Future Directions and Conclusions

In this chapter, we have considered the evolutionary ecology of movement at three levels: behavior, life histories, and population dynamics. At each level, much needs to be done both to understand the causes and consequences of movement and to integrate across this biological hierarchy. Considering behavior first, our classification of movement on the basis of proximate response, or lack of it, to resources raises important questions about how selection is acting on movement behavior. For example, what ecological circumstances might favor ranging over migration? Migration requires a major commitment of physiological resources and has attracted some attention from evolutionary ecologists. As a result, we have gained important insights, such as the role of ephemeral habitats in promoting migration, although much remains to be learned about how response to resources is both proximately and evolutionarily regulated. Ranging is quite another matter. It has gone virtually unrecognized as an important type of movement and has been confounded with population processes through confusing use of the term *dispersal*. We have pointed out that ranging, like other movements, can lead to aggregation or congregation as well as dispersal. An important question about its evolu-

tion concerns the circumstances in which it results in one or the other of these population outcomes.

There has also been little attention to the population consequences of the different kinds of movement. Most models of movement other than foraging assume a generalized sort of unidirectional movement (usually called *dispersal* because not being round-trip, it cannot be called *migration*). Once focus is shifted to the properties of movement behavior rather than the properties of pathways, questions can be raised about the consequences of different sorts of movement. Again looking at ranging versus migration, what, for example, would be the consequences for metapopulation analyses if with migration the first "hit" on a suitable patch failed to induce settling, and movement had to take place for a specified period to prime settling and lower response thresholds (a question that might well apply to, say, an aphid metapopulation)? Within subpopulations of a metapopulation, one can also model the consequences of different thresholds for departure (which will influence the proportion moving; see "Natural Selection of Movement") or investigate circumstances in which ranging might escalate to migration. These questions require an understanding of behavior to add the appropriate biological reality to models.

In reality, the evolution of movement behavior cannot be considered in isolation from life histories. An important question here concerns which traits have evolved together to produce complexes of traits resulting from shared genes, what we and others have called *syndromes*. Cost-benefit analyses have helped with understanding. Thus, the evolution of a syndrome that includes variation in wing form can be understood as a trade-off between the costs and benefits of flight and reproduction. It is also apparent, however, that some migrants have minimized those costs (Rankin and Burchsted 1992), and a fruitful direction for future inquiry should be studies of which costs are reduced by natural selection and the mechanisms by which that cost reduction is accomplished. We also know too little about how the development of movement systems is controlled, and the influence the controlling mechanisms may have on which traits will be genetically correlated in life history syndromes. Finally, syndromes, like traits, do not evolve in isolation, and the relations to life history and morphological adaptations largely remain to be explored.

There are many biological processes and their interactions which are complex enough that initial understanding will come only through the use of appropriate models. For example, what are the evolutionary consequences of having several movement behaviors (foraging, ranging, and migration) that occur in the same population? Can these movement behaviors occur independently? Do population genetics operate at different spatial and temporal scales from population dynamics, and if so, do these different scales lead to different types of movement behavior? Is short-distance movement repeated for many generations equivalent to long-distance movement for a few generations?

The answers to these questions will undoubtedly depend on the process of interest (dynamics, genetics, social structure, life history, etc.). If we are to understand the full complexity of different types of movement behavior and how these relate to the rest of evolutionary ecology, it is essential that we open our minds to the natural history around us and not be overly constrained by existing ecological and evolutionary theory!

Note

1. The distribution of individual foragers throughout a habitat so that all individuals have the same rate of consumption.

PART IV

INTERSPECIFIC INTERACTIONS

20

Ecological Character Displacement

DOLPH SCHLUTER

Ecological character displacement is phenotypic evolution wrought or maintained by resource competition between species. By resource competition, I mean the negative impact of one species (or individual) on another arising from depletion of shared resources. Character evolution driven by other mutually harmful interactions, such as intraguild predation or behavioral interference, is not included in the current definition of character displacement but perhaps should be in future. For the purposes of this chapter, however, character displacement is synonymous with the coevolution of resource competitors.

The idea that competition between species has a significant impact on character evolution has a lively history. Prior to about 20 years ago, competition was seen as one of the major factors responsible for the evolution of species differences, particularly in traits affecting resource exploitation (e.g., body size, beak shape). The idea is seen again and again in the early literature not because it was rigorously established but because it so readily accounted for observed patterns of species differences in nature. Support for the idea began to slip soon afterward, however, as alternative hypotheses were developed and as it became clear that the quality of most of the available evidence was poor. More recently, the idea has become respectable again as evidence from several systems has become more solid. My goal in this chapter is to present an overview of some of this evidence and how it has affected our understanding of the process.

I begin with a brief historical sketch of character displacement and the expectations from theory. I then present a few of the highlights emerging from observational studies of patterns suggesting character displacement, their limitations, and their implications. I follow with an overview of recent experimental work that complements studies of pattern but goes beyond them by testing novel predictions of character displacement hypotheses. I end with suggestions about where the most significant future discoveries lie.

Historical Context

The history of ideas on competition and character divergence begins with Darwin (1859), who regarded interspecific competition for resources as a fundamental and ubiquitous agent of divergent natural selection: "Natural selection . . . leads to divergence of character, for more living beings can be supported on the same area the more they diverge in structure, habitats, and constitution" (p. 105). Missing, however, were clear examples of this process among contemporary species. David Lack's (1947a) study of the Darwin's finches described several cases that finally filled the gap. Lack pointed to several pairs of closely related finch species whose beak differences were greater where they occurred together (sympatry) than where they occurred separately (allopatry). His most memorable case was of the small and medium ground finches, *Geospiza fuliginosa* and *G. fortis,* whose beak sizes are similar on islands where each occurs alone but different on islands where the two coexist. Several years later, Brown and Wilson (1956) compiled a list of similar examples from other taxa and named the phenomenon *character displacement.* These latter two studies had a profound influence on the development of ecology and evolutionary biology, and the idea that competition greatly influenced the evolution of diversity soon gained widespread approval. The role that other interactions might play in divergence was hardly mentioned.

The favorable consensus over character displacement's importance began to unravel about 20 years later, to be replaced by a more skeptical attitude in the minds of many researchers. The main factor to bring about this change of mind was a growing realization that the evidence for character displacement was not as strong as was once believed. In fact, it was downright poor. Alternative explanations were raised for patterns of greater divergence between species in sympatry than in allopatry, and it became evident that available data were not detailed enough to rule them out. For example, Grant (1975) showed that two species with partly overlapping geographic ranges might exhibit greater divergence in sympatry than allopatry even if each was responding independently to a geographical gradient in some feature of environment. Arthur (1982) pointed out that a heritable basis for character shifts, a prerequisite for evolutionary change, had been demonstrated in very few cases. Doubts were even raised over whether differences in sympatry were exaggerated at all. New statistical analyses suggested that morphological differences between sympatric species were not usually greater than those seen in randomly generated "null" communities of species (Strong et al. 1979; Simberloff and Boecklen 1981). The decline in the perceived importance of character displacement was also aided by growing doubts over the supremacy of interspecific competition in natural communities. Some researchers felt that competition was too weak and intermittent to be a significant force in divergence (Wiens 1977). Others felt that competition was frequently overruled by stronger, more direct interactions such as predation (Connell

1980). The validity of each of these criticisms was hotly debated, but regardless of the limitations of the alternative perspectives, they exposed real deficiencies in the case for character displacement.

In retrospect, one of the beneficial outcomes of these debates was the development of more rigorous standards of evidence, which researchers were then eager to apply to putative examples of character displacement both new and old. Renewed interest was also spurred by a proliferation of experimental studies showing that interspecific competition is indeed common in nature (Schoener 1983; Connell 1983; Gurevitch et al. 1992). This made the idea of character displacement even more plausible: How could a mechanism be so important in natural communities and *not* have evolutionary consequences?

The upshot was significant growth in the abundance and quality of evidence for character displacement. A recent survey of the literature found 75 hypothesized cases, many of them compelling (Schluter 2000). While much research remains to be done before we can claim to fully understand the evolutionary consequences of resource competition, at least we can say that it plays a fundamental role in divergence in many instances. Over the past 15 years, mathematical theory for character displacement has also advanced, giving us a better idea of the circumstances that should favor it. A brief outline of character displacement theory is given below. These theoretical explorations have uncovered some surprising finds, including the expectation that resource competition should often lead to convergence rather than divergence (summarized in Abrams 1996). Theory has also begun to explore the evolutionary consequences of other mutually harmful interactions such as apparent competition (competition for "enemy-free space") mediated by shared predators rather than shared resources (Brown and Vincent 1992; Abrams 2000). My summary below focuses on divergence via resource competition.

Concepts

How Character Displacement Should Work

A rich body of theory has been developed to help us understand the various evolutionary outcomes of resource competition between species. The most intuitive conceptual scenario provides a good place to begin this summary. Imagine two species that utilize resources along a continuous resource gradient. The position of an individual along this gradient and the range of resources used by that individual are determined by its measurements for a single phenotypic trait (e.g., beak size). The trait is assumed to vary continuously in the population, and some of this variation is heritable. Phenotypes compete when they share resources, that is, when their utilization curves overlap. The more similar their trait measurements, the greater the degree of resource sharing, and hence, the more intense the competition between them. For now, assume that competition is symmetric: An individual of the larger phenotype has the same competitive effect on an individual of the smaller phenotype and vice versa. Resources are assumed to be nutritionally substitutable; that is, each one of them provides all essential nutrients. For example, to a carnivore, prey of different in sizes are probably nutritionally substitutable even if they are not equally profitable. However, sunlight and soil nitrogen are not substitutable resources for a plant, and the simple scenario painted here would not apply. Finally, resource abundance is assumed to be the only factor regulating population size. A species present by itself attains a carrying capacity set by the abundance of those resources it is able to utilize. When two (or more) species are present and resources are shared, population size in each is of them is determined not just by resource availability but also by competition for these resources.

Given these assumptions, and provided that individual fitness increases with food intake, phenotypic evolution is driven by the "positive response" to resources (Abrams 1986b): The mean phenotype of a solitary species should tend to evolve toward a value enabling utilization of those resources present in greatest abundance, and away from uncommon resources. The principle is the same when two species are present (sympatry), but resource availability for one species is determined not only by environmental food supply but also by competition from the other species. In effect, the realized location of a peak in resource abundance to a given species is shifted when a competitor is present.

A convenient fitness function incorporating these elements extends the usual Lotka-Volterra equations for competition:

$$f(z) = 1 + r - \frac{r}{K(z)} \int p_i(x)\alpha(z,x)N_i dx \quad (20.1)$$

(Slatkin 1980). N_i is population density of species i, and $p_i(z)$ is its frequency distribution of phenotypes. $K(z)$ is the carrying capacity function, indirectly describing prey available to each phenotype z by the equilibrium number of identical individuals that can be supported. In the simplest renditions, $K(z)$ is gaussian; that is, it is symmetric and bell-shaped with a single peak at an intermediate phenotype. Width of the gaussian bell indirectly reflects the breadth of resources available. The constant r is the instantaneous rate of population growth in both species, assumed independent of z. $\alpha(z,x)$ is the intensity of competition between an individual of phenotype z and another individual whose value for the same trait is x. This competition function indirectly represents overlap between phenotypes in resource utilization. Typically, $\alpha(z,x)$ attains its maximum value at $x = z$, corresponding to complete resource sharing, and declines gradually for larger or smaller values of x. This decline is described by another gaussian curve, whose width symbolizes the breadth of resources utilized by each phenotype.

Resource density is not explicitly included in this type of model but is represented by total population size in relation to carrying capacity. Similarly, resource depletion is not explicate-modeled but is represented by the integral and all terms to its right, which accumulates the total competition an individual of phenotype z experiences from members of both its own and the other species. This is in contrast to another class of theory, in which the dynamics of both consumers and resources (usually just two types) are explicitly modeled (Abrams 1986b and later papers). Results from the two approaches (at least those I will concentrate on) are largely similar, and I refrain from a detailed comparison here.

Divergent character displacement occurs easily under the above conditions, the magnitude depending on details of $K(z)$, $\alpha(z,x)$, and the rate of evolution of intrapopulation genetic variance in z. In the absence of complicating factors, the mean phenotype of a solitary species exhibits the "positive response" to resources and evolves toward the peak in the resource distribution, represented by the single peak in carrying capacity. When more than one species is present, this tendency for each of them to evolve to the resource peak is counteracted by competition from the other species, which favors the use of otherwise less abundant resources located to either side of the original resource peak.

The result is a displacement in mean phenotype of both species to either side of the value expected for a solitary species. The magnitude of the shifts depends on the shape of the food abundance distribution and the breadth of resources utilized by each species. Under this simple scenario, divergence is the only possible outcome of interspecific competition.

The possible evolutionary outcomes of character displacement are more varied once we deviate from the simple scenario above. One of the least robust expectations from the simple case is that the mean phenotype of a solitary species will evolve to match the peak in carrying capacity. For example, if competition is not symmetric, the mean phenotype of a solitary species should be shifted toward whichever direction exhibits competitive dominance. If large individuals have a stronger competitive effect on small individuals than vice versa, a shift to larger size is expected. However, divergence is still expected: Sympatric species should be displaced to either side of their mean phenotype when solitary.

Less intuitively, competing species are expected to shift in parallel in sympatry or even to converge, rather than diverge, under several realistic modifications to the simple scenario. Abrams (1996) summarizes several possibilities. One situation occurs when resources are not substitutable. If addition of a second species to an environment results in depletion of a nutritionally essential resource for both of them, both are expected to increase investment in acquiring that resource. The result is convergence in the trait used to harvest the essential resource. Convergence does not mean failure to diverge. The latter may occur in the simple scenario above despite competition (e.g., if the resource distribution is narrow relative to the breadth of utilization by each species) whereas true convergence is enhanced similarity driven by competition. Another situation favoring convergence is competition for low-quality foods when high-quality food is rare and gut capacity is limited. Depletion of low-quality foods by one species will often favor greater investment in traits used to acquire those foods by a second species, resulting in convergence.

Observational Evidence

Two principal types of observation have been used to suggest that character displacement has taken place. The most common observation reported is "exaggerated divergence in sympatry," whereby differences in mean phenotype between coexisting species (sympatry) are greater than those seen between randomly chosen solitary species (allopatry). Examples include beak sizes of Galápagos ground finches (Lack 1947a), body shapes of solitary and paired threespine sticklebacks in postglacial lakes (figure 20.1), body sizes of solitary and paired *Anolis* lizards on small islands of the Lesser Antilles (Schoener 1970; Losos 1990), competitive ability of sympatric and allopatric populations of storksbill, *Erodium* (Martin and Harding 1983), and performance of fungus-eating isopods of three rivers in Italy (figure 20.2). Competition between species in sympatry is a compelling explanation for such patterns, especially since all cases involve traits used in exploiting resources.

The second type of observation is "trait overdispersion" (also known as *constant size ratios*), whereby the mean phenotypes of species making up an assemblage tend to be evenly spaced along a size or other phenotypic axis. Organisms exhibiting such a pattern include canine diameters in the *Felis* cats of Israel (figure 20.3), mandible lengths of tiger beetles (Pearson 1980), and flowering times of *Acacia* trees (Stone et al. 1996). Truly constant size-ratios are not the most likely outcome of character displacement, and perhaps for this reason, the majority of cases report a tendency toward even spacing instead.

A third and rarer type of observation is species-for-species matching, defined as an unusual degree of similarity in guild structures or phenotype distributions between sets of species that have evolved independently (Schluter 1990). Matching implies that coexisting species are assigned to niches in a consistent way. One example is the independent evolution of similar sets of ecomorphs in *Anolis* of large islands in the Greater Antilles (Williams 1972; Losos et al. 1998).

How reliable are these patterns in indicating character displacement? The answer must be that by themselves, the patterns are suggestive, but tests are needed to rule out alternative explanations. Here are some of the more realistic alternatives that have been proposed:

1. The patterns are the result of pure chance.
2. Shifts in trait means between sympatric and allopatric populations represent phenotypic

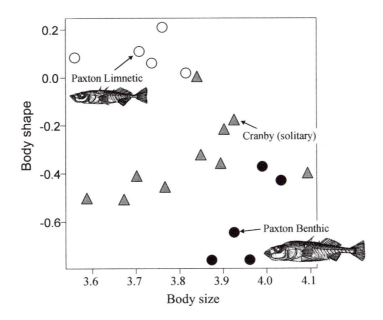

Figure 20.1 Mean phenotypes of adult male threespine sticklebacks (*Gasterosteus aculeatus* species complex) in small coastal lakes of southwestern British Columbia. Each lake is inhabited either by a single species (▲) or by a pair of species, one "limnetic" (○) and one "benthic" (●). The limnetic species forages primarily on zooplankton in the water column, whereas the benthic forages on littoral invertebrates living on or in the sediments or attached to vegetation. Solitary stickleback species occurring in nearby lakes similar in size to those containing pairs are intermediate in morphology and consume both zooplankton and benthic invertebrates. Body size represents position along a composite growth axis involving several size-related body dimensions. Body shape is a composite trait indicating variation in several size-adjusted traits. A higher value for shape indicates longer and more numerous gill rakers, slender bodies, and narrow mouths. Arrows indicate the three populations used in pond experiments. (Modified from Schluter and McPhail 1992.)

plasticity, not genetic changes between populations.

3. The patterns are the outcome of "species sorting," the biased extinction of species too similar to those already present, rather than evolutionary divergence in situ.
4. Traits measures are unconnected with resource utilization and have some other function instead.
5. Differences in phenotype between sites of sympatry and allopatry are the result of consistent differences between sites in resource availability.

6. The pattern results from another type of interspecific interaction, such as apparent competition or intraguild predation rather than resource competition.

The first hypothesis may be statistically tested in cases of exaggerated divergence in sympatry if there are replicate populations (such as in sticklebacks; figure 20.1) but not otherwise (such as in the isopods of figure 20.2). Trait overdispersion is testable statistically when there are multiple species (e.g., the *Felis* of figure 20.3). The second alterna-

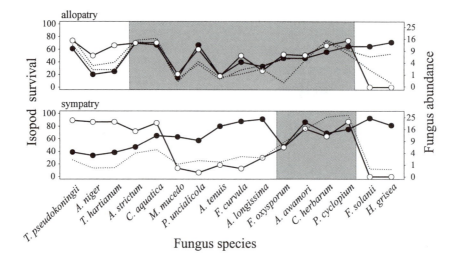

Figure 20.2 Performance (laboratory survival) of two aquatic isopods, *Asellus aquaticus* (●) and *Proasellus coxalis* (○), raised on 16 different species of fungi, their natural diet. Survival was measured on F₁ offspring of wild-caught parents; high survival reflects an evolved capacity to tolerate and metabolize fungus-specific toxins and nutrients. The upper panel measures one allopatric population of each species, whereas measurements in the lower panel are from a river where both isopods coexist. The shaded regions indicate fungi on which performance of the two species is equivalent. The dashed lines indicate abundance of the fungi in the wild, measured as number of colonies/30 petri dishes. The panels reveal that allopatric populations are adapted to exploiting all the common fungi, whereas in sympatry, only the most abundant fungi are shared and the less abundant fungi are divided between the two species. (Based on data from Rossi et al. 1983.)

tive hypothesis is ruled out if population differences persist in a common natural environment (such as the cats of figure 20.3) or in the lab (such as the isopods of figure 20.2). The third alternative hypothesis may be ruled out in cases of exaggerated divergence in sympatry if the span of trait values in sympatry exceeds that seen in solitary populations (e.g., figure 20.1). Measurements of resource utilization or performance on different resources (e.g., figure 20.2) may be used to test the fourth alternative. Measurements of food availability in sites of sympatry and allopatry (e.g., figure 20.2) may be used to test alternative 5. Finally, field experiments may be used to test whether spe

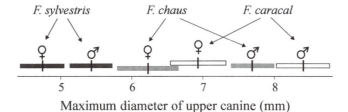

Figure 20.3 Nearly constant size-spacing of mean maximum diameter of upper canine teeth in the small wild cats of Israel *(Felis).* Vertical lines indicate means, and horizontal bars span one standard deviation to either side. (Modified from Schluter 2000 after Dayan et al. 1990.)

cies coexisting in sympatry compete for resources (alternative 6). If all six alternative hypotheses can be rejected in a specific case, then the evidence for character displacement can be regarded as solid despite being indirect.

Few of the 75 known cases of character displacement have been subjected to tests of all six alternative hypotheses, and the observational evidence is therefore incomplete. Most of the studies have rejected the first alternative hypothesis, but only about half have tested the next three. Some controls for effects of external environment (cf. hypothesis 5) have also been implemented in about half the cases, but possible differences between sites in availability of the main resources have been accounted for in only two cases exhibiting exaggerated divergence in sympatry (one of them is shown in figure 20.2). The last hypothesis has been tested least often, in only about a quarter of cases, and this means that in the majority of cases, we cannot be sure that resource competition is the mechanism underlying exaggerated divergence in sympatry, trait overdispersion, and species-for-species matching. Uncertainty over mechanism is therefore one of the biggest gaps in our understanding of these patterns in nature.

General Findings

The literature compilation of observational studies (Schluter 2000) suggests that character displacement is common in nature. Unfortunately, the existence of 75 cases, some of them very strong, cannot tell us the frequency of species assemblages in nature that show character displacement, since there is likely to be a strong publication bias against nonexamples. A small number of cases are known that looked promising at one time but later failed a test against one of the alternative hypotheses, but these are too few to be relied on. Nevertheless, an examination of the positive cases, those for which evidence to date remains consistent with the hypothesis of character displacement, might help reveal the situations in which the process occurs most readily and the kinds of patterns likely to be produced by it. Three patterns leap out.

First, every known case of character displacement represents divergence, a finding suggesting that convergent character displacement is rare. No examples are available in which two or more competing species have become more similar in their trophic characters when they occur together than when they occur separately. Examples of convergence exist for other types of interactions (e.g., Mullerian and Batesian mimicry) but not for competition. A methodological bias might contribute to this pattern, since tests of trait overdispersion are capable of detecting only divergence. Yet, comparisons of trait differences in sympatry and allopatry should be just as effective in detecting convergence as in detecting divergence in sympatry, if it exists. Possibly, the theoretical conditions known to promote convergent character displacement hardly ever occur in nature. However, an alternate possibility is that instances of convergent character displacement in nature are overlooked or, if found, are treated with skepticism and therefore not pursued because the idea that competition could lead to convergence is so counterintuitive (Abrams 1996). The theoretical possibility of convergence has been developed relatively recently, and perhaps, examples will emerge in the future now that we have a better idea of the conditions that promote it.

Another finding is that most examples of character displacement involve closely related species, implying that the process occurs most often in the relatively early stages of divergence. Of the 75 known cases, 47 (63%) involve congeneric species only, whereas most of the remaining cases (17 of 28) include two or more species from different genera within the same family. The most distantly related species showing evidence of character displacement are from different phyla: The smallest ground finch species (*Geospiza* spp.) on the Galápagos islands are even smaller on islands lacking the endemic carpenter bee, *Xylocopa darwini,* and their use of flower nectar is elevated. An alternative explanation for the shortage of cases of character displacement between distantly related species is that the process is somehow less detectable when phylogenetically unrelated species are compared, perhaps because divergence of unrelated species is likely to involve disparate traits.

The third striking pattern is the vast overrepresentation of carnivores compared with other trophic groups (figure 20.4). Herbivores (mainly granivores, but also including folivores and nectarivores) are the next most frequent trophic category. Few cases of character displacement are known among primary producers (plants), microbivores, and detritivores. Carnivores therefore appear especially prone to character displacement for reasons that are not known but are worth speculating on. One possibility is that carnivores experience stronger and more

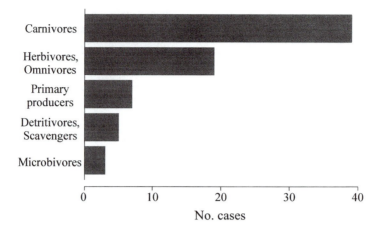

Figure 20.4 Trophic levels represented in known cases of ecological character displacement. (Modified from Schluter 2000.)

frequent interspecific competition than species at other tropic levels because they are at the top of the food web, whereas species at lower trophic levels may be limited by predation as well. However, many of the carnivores in the data set are not top predators but are intermediate-level predators that are themselves preyed on. The hypothesis also assumes that parasites of top carnivores are ineffective at reducing interspecific competition between them. Finally, this hypothesis does not explain the variability in the number of cases between other trophic categories (figure 20.4).

Alternatively, character displacement may be just as common at lower trophic levels as at high trophic levels but is less detectable there. For example, greater species richness at lower trophic levels may dilute the outcome of competition between pairs of species, reducing the magnitude of character shifts and consequently their detectability. Or foods of species at lower trophic levels may not be as nutritionally substitutable as the prey of carnivores, in which case, character displacement is more complex and may include convergence rather than divergence (Abrams 1986b, 1996). Unfortunately, there are no examples of convergent character displacement at any trophic level. Perhaps we haven't looked hard enough.

A third possible explanation of the preponderance of carnivore cases is that trait overdispersion and exaggerated divergence in sympatry are not solely the result of competition but also represent evolutionary consequences of intraguild predation,

aggressive interference, and other direct antagonisms. Such behaviors may be more prevalent among carnivores than among species in other trophic categories. If so, many of the putative cases are not of character displacement at all (at least not by the current definition, which is based on resource competition). This hypothesis may become testable once the mechanisms of divergence become identified in a larger subset of cases. Note that the hypothesis does not explain the variation in the frequency of cases among other trophic categories.

Finally, the pattern may result from detection bias. Character displacement may be most easily detected when species differ in easily measured morphological traits that map neatly onto resource utilization. Straightforward relationships between phenotype and resource use are far less frequently described for folivores, plants, and detritivores than for carnivores or even for granivores, which make up the majority of the herbivore cases. This suggests that detection bias is part of the answer even if it is not the full explanation.

Case Study

Experimental Study of Character Displacement in Sticklebacks

The above summary reveals that direct evidence linking divergence to resource competition is one of the main shortfalls of observational studies of character displacement. Some evidence of competition is available in 19 of 75 cases, of which 12 are based on field experiments rather than field observation. In at least some of these 12 cases, contest competition was detected rather than resource competition (but perhaps contest competition is itself an evolved response to resource competition). Thus, there is ample justification for further experimental studies. As well, differences between environments in resource availability are rarely accounted for, and experimental studies may eventually help overcome this deficiency, too.

However, there is another reason to conduct experiments beyond the need to measure interactions or control for food: Experiments may allow us to test stronger predictions of the hypothesis of ecological character displacement. In particular, they may allow us to determine how interaction strength has changed during the putative character displacement sequence, and to investigate how natural selection pressures on a species are altered when other species are present in the environment. Experiments on character displacement may thus take us further than the usual competition experiments.

In this section, I summarize experimental studies of character displacement with threespine sticklebacks (see figure 20.1). The experiments were designed to test three related predictions of the character displacement hypothesis: (1) Competition intensity should decline as divergence proceeds; (2) addition of a new species to an environment should alter natural selection pressures on those species already present, favoring divergence; and (3) natural selection favoring divergence should be frequency-dependent.

The experiments were carried out in a series of ponds on the campus of the University of British Columbia (Schluter 1994). Each pond is 23 m × 23 m, and its depth slopes gradually to 3 m in the center. The ponds were constructed in 1991 and seeded with plants and invertebrates from Paxton Lake on Texada Island, British Columbia (an 11-ha lake used as the source of experimental limnetic and benthic populations). Ponds are sand-lined and edged with limestone extracted from surface mines near Paxton Lake. The ponds are intended not to be identical to wild lakes but to mimic natural conditions sufficiently well that they may serve as useful tools for field studies of species interactions and their evolutionary consequences. All invertebrates found in the diets of experimental fish were characteristic of the species in the wild. Fish predators

of sticklebacks are absent in the ponds, but insect predators are common. Predation by piscivorous birds (herons, kingfishers, and mergansers) is present but sporadic. Each pond can sustain thousands of sticklebacks over multiple generations, and life cycles in the ponds are similar to those in native lakes (D. Schluter, unpublished observations). Howyever, the experiments summarized below were all short-term, lasting 7–12 weeks within single generations.

Experiments were designed around a hypothesis proposed by J. D. McPhail for the origin of the species pairs (figure 20.5). He suggested that each limnetic-benthic pair formed from two separate invasions of lakes by the marine threespine stickleback, a zooplanktivore similar in shape to the present-day limnetic. Sticklebacks in lakes invaded only once evolved an intermediate body form that made use of both littoral and open-water environments (stage 2 of figure 20.5). In those lakes colonized a second time, the intermediate form was displaced toward a benthic lifestyle, whereas the second invader remained a zooplanktivore. This scenario is consistent with the observed pattern of differences in lakes containing one and two species (figure 20.1), and makes predictions about the strength of interactions between populations differing in mean phenotype. Molecular and physiological studies offer support for the double invasion scenario.

The first experiment tested the simple prediction that intensity of competition experienced by sympatric species should have declined as divergence proceeded, yielding descendants whose present-day interactions are a "ghost" of their former strength. Pritchard and Schluter (2001) tested the prediction by re-creating historical and contemporary stages of the postulated character displacement sequence, taking advantage of the availability of the ancestral marine species in the ocean. The marine species was the target of the experiment. It was placed together with intermediate sticklebacks from Cranby Lake, Texada Island (simulating stage 3 of figure 20.5), or with benthic sticklebacks from Paxton Lake (simulating stage 5 of figure 20.5) in three divided ponds. This design held constant between treatments of the species of zooplanktivore so that any treatment effects are attributable to changes in the first invader, the form that has evolved the farthest from the ancestral state according to this scenario. The "ghost" prediction was confirmed (Pritchard and Schluter, 2001). Growth rate and degree of habitat specialization of

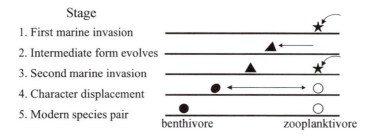

Stage

1. First marine invasion
2. Intermediate form evolves
3. Second marine invasion
4. Character displacement
5. Modern species pair

benthivore   zooplanktivore

Figure 20.5 The hypothesis of double invasion with character displacement, leading to the modern benthic (●) and limnetic (○) species pair. The marine form (★) is the ancestor of all the freshwater populations. (Modified from Pritchard and Schluter, 2001.)

the marine species were higher in the benthic treatment than in the intermediate treatment, a finding suggesting a decline in the strength of resource competition through time.

The second experiment tested whether natural selection favored divergence between sympatric stickleback species: That addition of a zooplanktivore (second invader) should alter natural selection pressures in a solitary population having an intermediate phenotype, favoring the more benthic individuals within it. The target of this second experiment was the intermediate stickleback from Cranby Lake, Texada Island. This species was manipulated by hybridization prior to introduction, to increase its levels of phenotypic variation and thereby achieve a more sensitive measurement of changing natural selection pressures. Hybridization further enabled us to mark different trophic phenotypes within the target form by using traits (armor and gill raker number) not susceptible to developmental plasticity that might otherwise have arisen from shifts in habitat and diet. The intermediate species was placed either alone or with the limnetic species from Paxton Lake (the latter treatment simulates stage 3 of figure 20.5) in two divided ponds. The density of the target was held constant between treatments to ensure that any changes in selection detected would arise solely from the addition of the zooplanktivore, but as a consequence, the total density of fish was higher in the preinvasion treatment. The purpose of the experiment was to test the prediction that phenotypes most like the added competitor should be most heavily affected by this addition. As predicted, presence of the zooplanktivore differentially depressed the growth rate

of those individuals within the target species that were closest to the added competitor in morphology and diet (figure 20.6). The effect of the addition on target phenotypes diminished gradually with increasing morphological distance away from the limnetic, with no growth depression in the most benthic phenotypes within the target species.

The third prediction tested was that natural selection stemming from interspecific competition is "frequency dependent," that is, whether the direction and strength of competition depends on the phenotype of the added competitor, not just its density (Schluter, ms.). In particular, the experiment tested whether zooplanktivory was a key component of the events that transpired after double invasion, or whether instead any competing species would have produced the same effect. The target and design of the experiment was the same as previously, except that one side of each divided pond received limnetics from Paxton Lake and the other side received benthics from the same lake. The density of the target was again held constant, but this time, total fish densities were the same between treatments since competitors were added to both sides. The results were similar to those shown in figure 20.6. Relative growth rates of different phenotypes within the target intermediate species changed between treatments. The slopes of the growth regressions differed as before, but this time mean growth was about the same in the two treatments. As a result, the regression lines from the two treatments crossed in the middle of the range of phenotypes rather than at the left end, as seen in figure 20.6. The phenotypes closest to the added competitor always suffered the greatest impact.

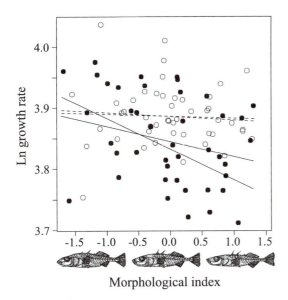

Figure 20.6 Natural selection for divergence in an experiment on character displacement. Symbols indicate growth rates of different phenotypes within an intermediate species (from Cranby Lake) in the presence (●) and absence (○) of the limnetic species from Paxton Lake, a zooplanktivore. The morphological axis distinguishes phenotypes within the target population from more benthic-like individuals on the left to more limnetic-like individuals on the right, in arbitrary units. Lines are regressions of ln(growth rate) on morphology for each pond in the presence (solid) and absence (dashed) of the limnetic species. Data combine observations from experimental and control sides of two divided ponds; each symbol is an average of 3 adjacent points. (Modified from Schluter 1994.)

This result confirms the third expectation that the impact of species addition on the target was frequency-dependent.

Future Directions

The strength of evidence for character displacement has improved a great deal over the past decade. The earlier argument that character displacement lacks empirical support is no longer tenable, and the heated debates of the past decades are unlikely to continue. Plenty of work nevertheless remains to be done, and in this final section, I highlight a few of the most critical and promising issues.

One important objective is to press on with observational studies of individual cases of character displacement. Many of the existing cases have not yet ruled out several of the most obvious alternative hypotheses for the patterns exhibited. For example, the role of differences between environments in resource availability is poorly understood in virtually all cases of exaggerated divergence in sympatry. Also badly needed are studies aimed at a better understanding of the mechanisms underlying patterns of exaggerated divergence in sympatry, trait overdispersion, and species-for-species matching. Competition experiments are now standard tools in community ecology, but few have been carried out on species exhibiting a pattern suggesting character displacement. At this point, even observational studies of species interactions (measurements of resource depletion and correlated changes in population density) would represent substantial progress.

Most needed are further experiments on character displacement, aimed at testing explicit predictions of the hypothesis beyond the requirement that species should compete for resources. In my view, this field, still in its infancy, represents the most promising future direction of research into the evolutionary consequences of competition (and species interactions generally). These remarks are not intended to underplay the role of observational studies, since experiments have their own limitations and by themselves are not enough. Indeed, it is unthinkable that an experiment should ever be carried out before a solid case has been built with observational approaches. Nevertheless, by allowing us to measure the forces of selection resulting from the interaction between coexisting species, experiments permit strong tests of the hypothesis of character displacement. Selection experiments have been carried out in only one system, the threespine sticklebacks, and studies of other systems are needed. Such experiments should also carry on beyond natural selection: The evolutionary consequences of natural selection arising from competition (i.e., heritable changes in trait means from one generation to the next) have not yet been measured in any field experiment.

For a long time, research has focused on the question of whether character displacement occurs at all, whereas focus is now shifting toward understanding the circumstances that promote or inhibit it and the form that resulting patterns of character shift should take. The database of studies to date suggests some interesting features of character displacement in nature that require further exploration: (1) Divergence is much more common than convergent character displacement; (2) character displacement occurs more often between close relatives than between more distantly related species; and (3) the likelihood of character displacement varies between trophic levels, carnivores being especially prone. Another pattern not discussed here is that the largest displacements are also the most symmetric, whereas asymmetric displacements tend to be small (Schluter 2000). A future priority is to determine whether these patterns are real or are merely outcomes of observational bias. If they are real, we need to understand their causes. These three issues hardly begin to exhaust the questions that remain concerning the circumstances that favor character displacement. Others include the role of resource dynamics, predation, environmental variability, and gene flow.

Finally, interspecific competition is not the only interaction that may generate exaggerated divergence in sympatry, trait overdispersion, and species-for-species matching. Other mechanisms that received scant attention in the past from community ecologists, but that are now being explored much more thoroughly, include intraguild predation and "apparent competition." Intraguild predation occurs when consumer species sharing resources also prey on each other, an interaction that should augment any resource competition and facilitate character displacement or should even cause character shifts by itself when resource competition is weak or absent. Apparent competition is the mutually antagonistic interaction between species that arises via shared predators rather than shared resources (Holt 1977). Theoretical studies indicate that apparent competition may drive divergence of species in sympatry as readily as competition for resources (Brown and Vincent 1992; Abrams 2000), but there are no good examples yet. Interest in these alternatives stem from more than the desire to rule out alternatives to competition in tests of character displacement. Rather, competition is plainly just one of many direct and indirect interactions between species in nature, all of which may have significant evolutionary consequences. In other words, species do not interact merely as competitors in isolation; rather, they belong to larger "webs" of interacting species. The evolutionary implications of this fact have been little explored, a state of affairs that should change dramatically in the years ahead.

21

Predator-Prey Interactions

PETER A. ABRAMS

Predation has been given many different definitions. For the purposes of this chapter, it is an interaction in which one free-living individual kills and derives resources from another organism. This definition includes finches that consume seeds but does not include fish that eat the siphons of clams that are unable to retract them quickly enough (assuming the clam usually survives the loss of tissue). Both broader and narrower definitions of predation are possible, and a variety can be found in ecology textbooks. Because broad definitions include herbivory and parasitism as forms of predation, the definition used here was chosen to minimize overlap with other chapters in this section.

Predation probably arose early in the history of life, and since then, it has been a major source of natural selection on both parties in the interaction. Given the lethal consequences of predation, it is clear that predators will usually have some effect on the rate of increase of their prey. If prey differ in their susceptibility to predators due to heritable differences in characteristics, evolutionary change in antipredator traits will ensue. Because predators must consume prey to survive and reproduce, the selective importance of predation-related traits is obvious. Predators have undergone considerable change and diversification since the first predatory protocell evolved from what was probably a scavenging ancestor. Darwin regarded some of the clearest cases of natural selection as due to the interactions between predator and prey, and that

viewpoint is also held by many current-day evolutionary biologists (e.g., Dawkins and Krebs 1979; Vermeij 1994).

Predation can be regarded as the most basic interaction between populations. Herbivory and parasitism share the basic property of predation, that one organism consumes some or all of another living organism. Many cases of competition involve predation on the same set of prey species by two or more different predator species. Even when competitors consume nonliving foods, many aspects of the consumption process are similar to consumption of prey by predators. Even mutualism frequently involves one organism eating parts or products of another.

Given this centrality of predation in the spectrum of ecological interactions and selective agents, one would think that its evolutionary consequences would now be well understood. Unfortunately, that is not the case. This chapter will try to convey some appreciation of the number and complexity of the problems faced by a biologist who would understand the evolution of this interaction.

There are many components in a predator-prey interaction. Predatory and antipredator adaptations may affect any one (or several) of the five sequential stages of detection, identification, approach, subjugation, and consumption. In most cases, a large number of traits in the predator affect one or more of these five stages in prey capture. These include traits affecting any sensory modalities that allow a predator individual to detect prey or the physical or chemical traces of prey. Traits of predators influencing their interactions with prey also include characteristics that (1) affect movement; (2) affect the predator's detectability to prey individuals; (3) allow the predator to obtain more nutritional benefits from a captured prey item; and (4) reduce the amount of time that the predator requires to subdue, ingest, and digest its prey. In addition, aspects of the life history of a predator are associated with changes in its ability to capture prey. The age and size at maturity, amount of reproductive effort, rate of senescence, and almost every other life history parameter is affected by the energy budget of the predator and is thus influenced by the level of development of any trait affecting capture. A similar variety of prey traits affect the prey's ability to either avoid encountering a predator or escape after the encounter has occurred. Escape can often be brought about if the prey can discourage the predator from attacking, by appearing dangerous, nonnutritious or unpalatable, or so proficient at escape that the attack would be futile. Life history adaptations of prey to deal with high risk of predation may include faster progress through vulnerable stages or more general adaptations to high mortality (e.g., earlier reproduction and greater reproductive investment). Understanding the evolution of a given predator-prey interaction includes understanding how all of these characteristics are likely to change over time in an environment where both species occur and interact. This is already a tall order, because a change in any one trait may affect other traits directly and generally affects selection on those other traits. For example, if a prey species is to develop and reproduce quickly, its opportunity to develop elaborate morphological defenses is restricted. Predicting the evolutionary fate of a predator-prey interaction is further complicated by the fact that the process of predation affects birth- and death rates of both species.

When predators become more adept at capturing and consuming prey, they get more to eat, and their birthrate increases, their death rate decreases, or both. When prey become more adept at avoiding or escaping predators, their own death rate should decrease; their birthrate may also increase if lower risk allows them to forage more effectively. The resulting changes in population density have further consequences. When a prey population gets better at escaping from its predator, the prey population size is likely to increase. This increased prey population density affects the cost-benefit balance of capture-related traits of the predator. For example, a costly predator trait for increasing prey capture may exhibit little or no long-term evolutionary change following the evolutionary improvement of the prey's escape ability (Abrams 1986a). The reason is that the increased number of prey offset the decreased capturability of prey individuals. Evolutionary improvements in the predator's ability may either increase or decrease the predator's population size, so the density effects may either enhance or offset the change in selection pressure due to the predator's evolutionary improvement.

Changes in population density also generally result in shifts in behavior, and the latter also feed back onto the evolutionary process. If a particular prey species gets better at escape, this may cause the predator's behavior to change so that it either increases or decreases its exploitation of another prey species. This reaction has its own effects on the population densities of both species, which may then again affect behaviors. Induced chemical and morphological defenses represent additional classes of traits that interact with genetically determined defenses in ways that are likely to be similar to behavior.

Population cycles are a potential consequence of predator-prey interactions, and both predatory and antipredator traits affect the stability of the interaction (Abrams 1992, 1997; Seger 1992; Abrams and Matsuda 1997). Moreover, almost all predator-prey interactions occur in the context of a much larger food web, and changes in population density may depend on species at higher or lower trophic levels. To understand the evolutionary time course of traits affecting predation in a given species pair, one must take into account the changes in population densities and dynamics that are an inevitable consequence of that evolution.

The difficulties mentioned above are added to the more general problems of studying evolutionary change of any type of character. There is usually at best a partial record of the change from the fossil or historical record. Laboratory systems are able to examine only a limited array of organisms in artificially simplified environments over a relatively small number of generations. The genetic architecture of most ecologically important traits is usually completely unknown.

This chapter will focus on two questions about

the evolution of the predatory interaction: (1) What is the response in the trait value in one species to a selectively favored evolutionary change in an interacting species? Here the focal species is either the predator or the prey. By determining how each species evolves in response to trait changes in the other, it may be possible to understand coevolutionary changes over longer time periods. (2) What is the response in the trait value in one species to the introduction of a species of the other type? In the case of the predator, this of necessity means an additional prey species, since the predator could not exist in the absence of prey. However, it may mean the first introduction of any predator species for a prey. This question is analogous to the question considered by Schluter in the preceding chapter: How do traits determining resource use in a focal species change following introduction of a competing species.

The presentation will ignore some of the complications alluded to above. For the most part, the theory and empirical work have focused on a single phenotypic trait in each species and have often focused on systems where it is possible to isolate a single predator and a single prey species from the rest of the biotic community. The chapter will largely ignore the role of evolutionary forces other than selection. This is more a reflection of the lack of knowledge about the roles of gene flow and drift than a reflection of any agreement that they are unimportant. The discussion begins by summarizing how the evolutionary interaction of predator and prey has been conceived of in the past.

Historical Context

Darwin was certainly not the first biologist to think about how predator-prey systems came to have their present characteristics, but he can be credited with initiating modern-day thought on the topic. He felt that predation was one of the major factors acting to prevent the indefinite increase of populations: "Each species, even where it most abounds, is constantly suffering enormous destruction at some period of its life, from enemies or from competitors for the same place and food" (Darwin 1859, p. 121). Thus, evolution of predators was the first example Darwin discussed in his chapter on natural selection: "Now, if any slight innate change of habit or of structure benefited an individual wolf, it would have the best chance of surviving and of leaving offspring. Some of its young would probably inherit the same habits or structure, and by the repetition of this process, a new variety might be formed which would either supplant or coexist with the parent-form of wolf" (Darwin 1859, p. 139).

Darwin felt that this process of natural selection on predators could lead to divergence of geographically separated predator populations encountering different prey:

> Or, again, the wolves inhabiting a mountainous district, and those frequenting the lowlands, would naturally be forced to hunt different prey; and from the continued preservation of the individuals best fitted for the two sites, two varieties might slowly be formed. . . . I may add, that, according to Mr. Pierce, there are two varieties of the wolf inhabiting the Catskill Mountains in the United States, one with a light greyhound-like form, which pursues deer, and the other more bulky, with shorter legs, which more frequently attacks the shepherd's flocks. (Darwin 1859, p. 139)

Darwin identified this discussion of evolutionary change in wolves as a "simple" example of natural selection, which he followed with a "more complex" example involving the coevolution of flowers and pollinators. It is interesting that Darwin apparently did not consider the possibility of coevolution of predators and prey. It is also somewhat ironic that we currently have far more evidence that predators have acted as selective agents that have changed the characteristics of their prey than vice versa (Endler 1986, table 5.1; Abrams 1990; Brodie and Brodie 1999b).

The realization that predators and prey might affect each other's evolution, an effect resulting in a conceptually and dynamically complicated process, was already present by mid-century, by which time Cott (1940) had formulated the concept of an evolutionary arms race. This metaphor was later revived by Dawkins and Krebs (1979). Seger (1992) credits J. B. S. Haldane with introducing modern thinking about victim-exploiter coevolution in the late 1940s, although Haldane apparently considered parasites in more detail than predators. In the following decades, most biologists continued to consider one-way evolutionary interactions between predator and prey. In the late 1960s and early 1970s, many biologists were concerned with explaining why predators did not evolve such a high efficiency that they drove their prey extinct. The possibility of predator-prey coevolution began to be discussed more seriously in the 1960s and 1970s (e.g., Schaffer and Rosenzweig 1978).

More recent theoretical work has revealed an enormous diversity of outcomes when predator and prey both undergo natural selection on traits relevant to the interaction. Most of that work has been theoretical (Saloniemi 1993; Marrow et al. 1996; Gavrilets 1997; Abrams and Matsuda 1996, 1997; Abrams 1992, 1997; Matsuda and Abrams 1994). Much of the complexity revealed by these studies results from the interaction of population dynamics and evolutionary change. Unfortunately, population densities are usually impossible to assess from the fossil record, and experimental systems where two species can be maintained for sufficiently many generations are generally limited to bacteria and phage. Such systems are generally more relevant to parasite-host coevolution than to predator-prey interactions. Few natural systems have been studied thoroughly enough and long enough for an assessment of changes in both predation-related traits and population densities over periods of many generations. This has left a rather unfortunate gap between theory and experiment. This gap is most likely to be closed by studies that are able to deduce some of the properties of the evolutionary interaction from measurements over one or a few generations.

The following section presents a largely theoretical view of natural selection on predators and prey. This is followed by a description of what is known about the outcome of this process in one system where there is good evidence of predator evolution in response to prey defenses.

Concepts

The two problem areas mentioned in the introduction were the questions of (1) evolutionary changes in one species in response to evolutionary change in the interacting species and (2) evolution in response to introduction of an interacting species. Both problems can be adequately understood only with the aid of a mathematical framework. The reason such an approach is necessary is that evolution in the context of species interactions always involves linked feedback processes of changes in traits and in population densities. The framework presented here will focus on the evolutionary tra-

jectory of a single trait in each of the two interacting species. Thus, it will ignore the potential interaction of different traits within a species that affect different aspects of the predation process. This is largely due to limitations of space and of present-day knowledge.

The basic model starts with a description of population dynamics, because population densities affect the costs and/or benefits of almost all traits involved in the interaction. I assume a differential equation model of a predator species (population size P) and a prey species (population size N). The model has the following form:

$$dN/dt = Nf(N) - CNPg(CN) \qquad (21.1a)$$

$$dP/dt = P[B(CNg(CN)) - D(P)] \qquad (21.1b)$$

where $f(N)$ gives the per capita growth rate of the prey (in the absence of predation); this rate depends on prey population density N and either may be monotonically decreasing or may increase at low N before decreasing as N becomes larger. This growth function may also depend on the prey's antipredator traits, since these will generally entail some cost. The rate at which an average nonsatiated predator individual captures prey is CN, where C is a capture rate constant that may be a function of environmental factors, prey traits influencing their vulnerability, and predator traits influencing their capture ability. The product, CN, also represents the capture rate of prey by an average predator while it is searching at its maximum potential. The realized capture rate is generally below that of a completely nonsatiated individual, and the fractional reduction is described by the function $g(CN)$. Here g is a decreasing function of the capture rate that would occur given maximal search (i.e., CN). The function B describes the rate of production of new predators, given food intake rate $CNg(CN)$. The function D represents the predator's per capita death rate. This is often influenced positively (e.g., due to group hunting) or negatively (e.g., because of aggressive interactions) by the predator's own density. The per capita deathrate is often affected by the same traits that influence the capture rate and, in those cases, will be a function of those traits. This basic population dynamics model (with C constant) is considered in textbooks and is known to be capable of having either a stable point or a limit cycle as its long-term dynamics.

The next component of a framework for studying predator-prey evolution is a model for the change in traits that accompanies the changes in population densities. Here, I adopt an approach based on quantitative genetic recursions that has been discussed in Taper and Case (1992) and Abrams et al. (1993). It assumes that traits change at a rate that is proportional to the gradient of individual fitness with respect to that trait. Here fitness, W, is measured as the expected instantaneous per capita growth rate of a population with the individual's trait value. An individual's per capita growth rate is often influenced by both its own trait and by the average trait values (or distribution of trait values) in one or both species. Because the population dynamics equations (21.1a,b) implicitly assume that mean ecological parameters can describe the system, the same will be assumed for the evolutionary dynamics. The fitness gradient is the rate of change of fitness with a unit change in the trait; dW/dz for trait z. The constant of proportionality in this relationship is the additive genetic variance of a polygenic trait. In a more general model (or if the quantitative genetic traits are not measured on the appropriate scale), the proportionality factor may change with the trait value. For example, the proportionality factor should become very small as the mean trait value approaches a minimum or maximum possible value (Abrams and Matsuda 1996). I will denote the predator's and prey's traits by x and y, with mean values x^* and y^*. The proportionality factors will be denoted $v_p(x^*)$ for the predator and $v_n(y^*)$ for the prey. The approximation used here assumes a relative small range of fitnesses, and fitness gradients exist within the population (Abrams et al. 1993; Gavrilets 1997).

Arguably, the most important parameter characterizing the interaction is C, the predator's capture rate per unit prey density when the predator is searching at its maximal rate. If both predator and prey traits affect the rate of encounter between predator and prey, an individual predator's C may often be expressed as a function of x and y^*. Increases in an individual's predatory trait x will increase the C that it experiences, while increases in the mean trait of the prey, y^*, will decrease the predator's C. The C experienced by a prey individual is a function of its own trait value y, and the mean predator trait, x^*. The mean prey trait affects the predator population's satiation function, g. Each of the two individual traits (x, y) has a

negative effect on some other component of fitness; that is, adaptations to predation have costs. In the simple context given above, it seems most likely that larger values of x would increase the predator's per capita deathrate, D, although they might also decrease the birthrate, B, or affect the shape of the satiation function, g. Larger values of y are likely to reduce the prey's per capita growth rate function, f, although it is again possible that they might affect g. The most likely alternatives lead to the following model:

$$\frac{dx^*}{dt} = v_p(x^*)\left(\frac{dW_p}{dx} \right)$$

$$= v_p(x^*)\left(N\frac{\partial C}{\partial x}(g + CNg')B' - \frac{\partial D}{\partial x} \right) \quad (21.2a)$$

$$\frac{dy^*}{dt} = v_n(y^*)\left(\frac{dW_n}{dy} \right)$$

$$= v_n(y^*)\left(\frac{\partial f}{\partial y} - Pg\frac{\partial C}{\partial y} \right) \quad (21.2b)$$

Primes denote derivatives of functions with respect to their (single) arguments.

It must be admitted that this model is a gross simplification of most genetic systems. However, it is also a reasonable approximation of adaptive evolution under many different genetic assumptions (Abrams et al. 1993). Because the population dynamics are also at best a rough approximation to any real system, the approximations involved in the evolutionary dynamics are not likely to represent the most serious departure of the model's dynamics from those of any system in nature. However, the purpose of models such as these is not to describe real systems precisely, but to help to reveal and understand the potential mechanisms behind potential evolutionary responses in natural systems. The dynamics described by equations 21.2a forms the basis for the graphical representations presented in the following section.

Responses to Evolutionary Change in the Other Species

The questions of responses to species additions or trait changes can be addressed by elaborations of equations 21.1 and 21.2. Many of the recent studies of predator-prey coevolution have been based on models that either are similar in form to these equations or have dynamics that are often approximated by a particular realization of these equations. Interested readers should consult any of the following more technical papers for details: Abrams (1986, 1990), Marrow et al. (1996), Abrams and Matsuda (1996, 1997), or Gavrilets (1997). To illustrate the use of this framework, I will consider the question: How does a predatory trait change in response to an evolutionary improvement in prey defense? The scenario is that a novel mutation improving the prey's defense has recently arisen and quickly gone to near fixation in the prey population. For purposes of this hypothetical question, we will treat the mean prey trait y^* as constant at its new (higher) level. For simplicity, the system is assumed to come to a stable equilibrium for both traits and populations.

Because evolution, as represented by equations 21.2a,b, usually maximizes the benefits minus the costs of a predatory adaptation, the outcome of the process can be represented graphically. Two different scenarios for predator evolution are represented in figures 21.1 and 21.2. In figure 21.1, as the predatory trait increases, the benefits to the predator approach a maximum set by the predator's ability to process prey. An increase in the prey's ability (y^*) to escape lowers the predator's benefit curve, favoring an increase in the predator's capture-related trait. However, this result depends on the shape of the predator's trait-related cost and benefit curves, as shown in figure 21.2. In this case, there is a linear increase in benefits with increasing prey capture, and the predator should decrease its capture-related trait (to incur lower costs) when prey become better at escape. Both of these figures are incomplete representations of the evolutionary process because they do not incorporate the change in density of the prey that follows from their increased escape ability. This compensation in prey density usually reduces the magnitude of the selectively favored response of the predator and may in some cases reverse the response.

The graphical treatment above can be extended to include density compensation and can be made more rigorous by a return to equations 21.1 and 21.2 and their use to determine how a small increase in y^* affects the equilibrium value of x^*. The mathematics approximates the nonlinear equations 21.1a,b and 21.2a with a set of linear equations and then uses linear algebra to determine how a change in a parameter (here y^*) affects the

equilibrium values of the others. See Abrams (1986a, 1990) for applications of this method to evolution of predatory traits. The question at hand is whether an increase in the prey's y^* will increase, decrease, or leave the predator's x^* unchanged. The possible outcomes can be understood in a general way by simple consideration of the linear approximation of equation 21.2a at equilibrium, and how this is changed as y^* increases. Thus, we set equation 21.2a equal to zero, differentiate with respect to y^*, and solve the resulting equation for the partial derivative $\partial x/\partial y^*$. This partial derivative gives the change in the equilibrium x (or x^*) following an increase in y^*. To simplify the analysis, predators are here assumed to have no satiation (i.e., $g = 1$). The result is the following relationship, where all derivatives are evaluated at the equilibrium point:

$$\frac{\partial x}{\partial y^*} = \frac{\dfrac{\partial N}{\partial y^*}\dfrac{\partial C}{\partial x} + N\dfrac{\partial^2 C}{\partial x \partial y^*} - \dfrac{\partial^2 D}{\partial x \partial P}\dfrac{\partial P}{\partial y^*}}{-N\dfrac{\partial^2 C}{\partial x^2} + \dfrac{\partial^2 D}{\partial x^2}} \quad (21.3)$$

Although this expression appears rather complicated, the individual terms are not difficult to understand. The denominator is the negative of the second derivative of individual predator fitness at the equilibrium value of $x = x^*$. It must be positive if the evolutionary equilibrium is one that maximizes predator fitness. Although stable evolutionary equilibria sometime minimize fitness (Abrams et al. 1993; Abrams and Matsuda 1996), we will assume that that is not the case here. Something is also known about the signs of the terms in the numerator. In the first term, $\partial N/\partial y^*$ is likely to be positive; that is, prey population size is likely to increase as its mean defensive trait y^* increases. There are some exceptions to this statement, if the defensive trait affects the prey's exploitation rate of its own biological resources. However, for the present discussion, that possibility will be ignored. The second factor in the first term, $\partial C/\partial x$, is positive by definition. Thus, the first term must be positive. The signs of the second and third terms in the numerator hinge on the three partial derivatives in those terms, and none of these derivatives has a determinate sign. The term $\partial^2 C/\partial x \partial y^*$ is the effect of a larger prey trait on the slope of the relationship between capture rate and the predator's trait. This may be positive, negative, or zero; for example, greater prey speed may mean that a given change

Cost and Benefit of Predatory Trait x

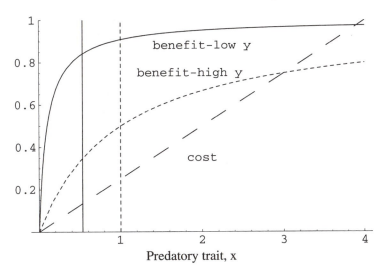

Figure 21.1 The cost and benefit curves of a costly predator trait that increases the predator's expected capture rate of prey. The value of the predator's quantitative trait is x, the cost curve reflects a higher mortality rate with greater values of the trait, and the benefit curves reflect the greater birthrate resulting from greater rates of prey capture. The predator has a maximum capture rate determined by its ability to handle prey. The short-dashed benefit line is for a situation where the prey have a greater ability to escape the predator. The optimal predator trait occurs at a trait value for which benefits minus costs are maximal; this is given by the two vertical lines for the two types of prey. As can be seen, the optimal predator trait is larger when the prey has a greater escape ability.

in predator speed has more or less of an effect on probability of capture. The term $\partial^2 D/\partial x \partial P$ may also be positive or negative; a higher predator density may make the predator's per capita deathrate more or less sensitive to its trait value. Finally, $\partial P/\partial y^*$ may also be positive (if the predator is initially overexploiting the prey) or negative (if overexploitation is not occurring at the original equilibrium). The indeterminacy of the signs of these three derivatives makes it impossible to generalize about whether natural selection will favor an increase or a decrease in the trait value of the predator. However, this analysis highlights the biological conditions that favor a decrease in the predatory trait when prey become more proficient at escape: (1) The attack rate C becomes less sensitive to the predator's trait as the result of the prey's evolutionary innovation; (2) higher predator population

makes the predator's per capita mortality rate less sensitive to its own trait value, and the prey is not overexploited; (3) higher density makes the predator's per capita mortality rate more sensitive to its trait value, and the prey is overexploited.

More specific results can be derived by assuming particular relationships between the two traits and the capture rate. For example, when the two traits affect movement rates, and capture occurs when predator and prey come within a small distance of each other, C can be expressed as a product of predator movement rate and prey movement rate, measured on an appropriate scale (Abrams 1986a). Here, prey movement is not sufficiently rapid to permit escape; higher movement rates (lower y) imply a larger encounter rate with the predator. If we further assume that the predator's population density does not affect its per capita

Cost and Benefit of Predatory Trait x

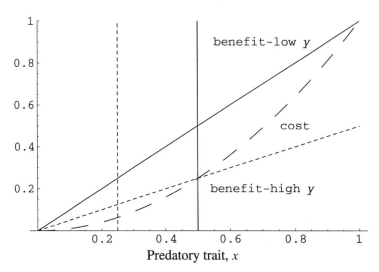

Figure 21.2 The cost and benefit curves of a costly predator trait that increases the predator's expected capture rate of prey. This figure differs from figure 21.1 in that the benefit of greater prey capture increases linearly with the capture rate, and the cost of greater values of the predator's trait increases at an accelerating rate. In this case, the optimal predator trait decreases after prey evolve a greater ability to escape.

growth rate directly, the numerator of expression (21.3) vanishes (Abrams 1986a). The increase in prey population density exactly balances the decrease in encounter with individual prey (due to their larger average y), and the net result is no change in the optimal trait value of the predator. In figures 21.1 and 21.2, this case means that the increase in prey density exactly counteracts the change in the benefit curve produced by a change in the prey's trait. Abrams (1986a, 1990) presents several results for the evolutionary response of one species following a change in its partner, derived under the assumption of a stable equilibrium point.

This example arbitrarily assumed that the prey trait underwent a rapid increase and then remained constant. To understand the full range of outcomes possible when both parties to the interaction undergo simultaneous change, the same methods may be applied, but the additional equation 21.2b makes the results more difficult to interpret. In addition, the assumption of a stable equilibrium becomes less defensible. Both coevolution and predator satiation can often lead to cycles. In such cases, there

is no way to avoid numerical methods (simulation of equations 21.1a,b and 21.2a,b or some more detailed model) to determine the outcome of the interaction. Given the range of outcomes that are possible, there are few general results about the outcomes of coevolution. Some of the results that have been derived are

1. Population dynamical cycles usually result in significantly different evolutionary outcomes when compared to stable systems. This is frequently true when individual fitness is nonlinearly related to the evolving character (Abrams 1997). For example, the scenario described above, in which the effect of prey density cancels the effect of lower encounter rate, is usually not true when there are predator-prey cycles (Abrams 1997).

2. Coevolution can drive cycles in both traits and population densities. The ability of predator-prey evolution or coevolution to drive cycles (in population density and/or trait values) has been a topic of continuing interest. Seger (1992) drew attention to the fact that parasite-host models had often predicted cycles, while most previous preda-

tor-prey models had not. However, at about that time, several studies appeared that showed that predator-prey models could also drive cycles. Abrams (1992) showed that adaptive change in predators (either behavioral or evolutionary) could drive cycles if the optimal consumption rate constant decreased as prey density increased. Another early paper that treated predator-prey cycles and coevolution was Saloniemi (1993). However, Saloniemi's model left out the demographic costs of stabilizing selection, and with these included, sustained cycles are not possible given the other assumptions of the model (Abrams and Matsuda 1996). A range of other analyses also found that sustained cycling could occur in coevolving predator-prey systems in cases where the two species would have reached a stable equilibrium in the absence of evolutionary change (Marrow and Cannings 1993; Marrow et al. 1996; Gavrilets 1997). Abrams and Matsuda (1996) showed that the most general cause of cycles was a fitness architecture that was similar to that of many parasite-host systems, where specific defenses match specific methods of attack. In the case of predator-prey systems, this architecture entails a bidirectional (or multidirectional) axis of prey escape ability. This means that the prey can reduce their probability of being eaten by either increasing or decreasing their trait value above or below a maximally vulnerable value determined by the predator's mean trait value. This type of fitness architecture can lead to a scenario where the prey's trait evolves first to larger and then to smaller values, continually "chased" by the predator's trait. Cycles are not an inevitable consequence of this type of scenario. They are most likely if the maximal rate of evolutionary change in the prey's trait is faster than that of the predator. Stable equilibria are likely if one or both species have traits under sufficiently strong stabilizing selection due to costs of trait exaggeration that increase faster than linearly with the departure of the trait from some intermediate optimal value (Abrams and Matsuda 1996). The escalation in costs as trait values depart from the equilibrium maintains the stability of the system. Cycles driven by evolution seem highly unlikely when there is a unidirectional axis of ability in the interaction, that is, when larger (smaller) values of some trait always confer greater ability to capture (in the predator) or avoid capture (in the prey).

3. The course of coevolution is greatly affected by the genetic system of trait determination when there are potential evolutionary equilibria where a trait in one or both species undergoes disruptive selection. Under some circumstances (e.g., asexual inheritance and some cases of sexual inheritance and assortative mating), disruptive selection is likely to result in splitting of the lineage undergoing disruptive selection into two separately evolving lines. However, under many types of sexual mating systems, this circumstance will result in a stable equilibrium or cycles, but no increase in the number of independently evolving units (Abrams et al. 1993).

4. Adaptive evolution can cause the extinction of the evolving species. When fitnesses are frequency-dependent, natural selection does not, in general, maximize population size. Matsuda and Abrams (1994) show that when the prey's defense entails less foraging activity, it is theoretically possible for the prey to evolve to such small population sizes that extinction is probable. Abrams and Matsuda (1997) show that (in some models) prey evolve greater vulnerability in systems with large-amplitude population cycles, because predators are either rare or satiated for much of the time. However, larger vulnerability means larger amplitude cycles, which are likely to cause extinction of predators or of both predators and prey in finite populations.

One of the topics discussed by many researchers is the question of whether there is a trend for one or the other party of the predator-prey relationship to become more successful over evolutionary time. *Success* is difficult to define, but it usually corresponds to the maximal rate at which a predator individual captures prey in some standardized circumstances. Even in the absence of information on the best measure of predatory success, there is strongly suggestive evidence in some systems that one or the other party in a predator-prey interaction has increased its success in the interaction over time. Examples include Bakker's (1983) analysis of locomotor traits in ungulates and their cursorial predators and of tooth wear in the ungulates. Both of these seem to suggest a trend toward a decrease in the risk of mortality due to predators over the past 60 million years. Similarly, the fact that the shells of some species of snails became more heavily armored while their crab predators did not alter claw morphology suggests that those snail species have improved their performance in the interaction over time (Vermeij 1994). When this is combined with the evidence from Endler's (1986, table 5.1) survey, there does appear to be an asymmetry. However, the difficulty of measuring many predatory traits and predators' longer generation times both lead to a publication bias in favor of studies on prey species. Brodie and Brodie (1999b) have recently reiterated Vermeij's (1994) suggestion that predators are more insulated from the consequences of differences in capture rate of a particular prey species, due to their ability to switch their feeding to other prey species. Unfortunately, this logic often does not work. The presence of alternative prey may simply increase predator population size (Abrams 1991). It is probably fair to say that we are still very uncertain whether predator lineages are on average less successful than prey lineages over evolutionary time. This is unlikely to be true of all lineages, given that predators are still very well represented among the earth's fauna.

Responses to the Introduction of Another Species

One area where theory and data agree, and where there are many empirical examples, is the question of how defenses of a prey change when a predator is first introduced. Unfortunately, the prediction that defenses increase when this is possible is often not of great interest. However, there are some cases where it is not immediately obvious how the trait should respond to predator introduction. That is, what constitutes a defensive trait? Such ambiguity characterizes at least some life history responses of prey to predators, where it is often not clear what change in traits represents an adaptive response to the mortality posed by predation. Reznick et al. (1990) documented a lower age and size at maturity for guppies exposed to the predatory fish *Crenicichla*. Both of these are predicted by theory under a wide range of assumptions (Abrams and Rowe 1996), although there are again exceptions. For example, predators can favor a later age of maturity if the prey reduce their growth in order to avoid predators, but they must still mature at a similar size, either because of constraints on, or the selective advantage of, larger size.

The question of how predatory traits change in response the introduction of a second (or additional) prey species has not received a great deal of attention, either theoretical or empirical. This is somewhat surprising, given the fact that evolutionary biologists concerned with competition have focused almost exclusively on the analogous question

for that interaction (see the preceding chapter). Endler's (1986) survey does not list a single study in which an introduced prey species is the agent of selection on a predator. The answer to the question clearly depends on whether the second prey can be caught by the same traits or methods as the first prey. If the prey require different capture techniques or traits, one must know whether there are trade-offs or genetic correlations between traits used to capture different species. It is often assumed that such trade-offs exist, and that introducing a second prey species will therefore reduce the traits that contribute to capturing the original prey. However, this reasoning ignores the change in population size of the original prey that occurs as the result of the introduction of the second prey. A large increase in the density of the original prey can result in increased adaptations for capturing that prey species (Abrams 1991).

General Principles

Given the current lack of knowledge about the forms of most of the functional relationships that determine the evolutionary course of predator-prey interactions, it is worth considering what principles can provide some heuristic guidance in thinking about the evolution of predatory success. One rather obvious, but nonetheless useful, idea is what Dawkins and Krebs (1979) label the "rare-enemy principle." Predators that rarely attack a given prey species, either because they themselves are rare or because they prefer other prey, are unlikely to cause a measurable response in the predation-related traits of the prey. Similarly, prey species that constitute a sufficiently small proportion of the predator's diet will not have a significant effect on the predator's capture-related traits.

A second general principle is that the evolutionary response of a species depends on the marginal benefits (or costs) of change in a given trait rather than on the absolute benefits (or costs). In other words, it is the rate of change of benefits with a change in the trait that is important in determining the evolutionary fate of the trait. If predators suddenly become more adept at capturing prey, and the prey's mortality rate increases dramatically, this does not necessarily imply that the prey's antipredator adaptations should increase (or decrease either). However, if the increase in predator ability also increases the slope of the relationship between prey survival probability and trait value, then the prey would be expected to increase their antipredator traits. It is conceivable that the predator's innovation will render all values of the prey's defensive trait ineffective, in which case, the prey's defense would decline, given that it is costly and ineffective as well. As a general rule, the evolutionary equilibrium of a continuous trait occurs where the rate of change in fitness benefits with trait value is equal to the rate of change in fitness costs.

Finally, as with any other evolving trait, the rate of change of the trait increases as the additive genetic variance in trait value increases, and it decreases as the generation time increases. Thus, when predators have a much longer generation time than prey, their main opportunity for increasing capture rate over evolutionary time is often by having a much greater genetic variance for traits that influence capture rate.

Case Studies

A major problem in empirical studies of predator-prey relationships is that it is often difficult to find species pairs in which both members have easily measurable traits influencing the interaction. Thus, there are many cases where predators have selected for more cryptic color patterns in their prey species (Endler 1986). However, the predator traits involved in detecting cryptic prey are much more difficult to measure. The fossil record provides examples of change in morphological traits that affect predation, but many of the potential adaptations of predators (increased visual acuity or larger visual or olfactory areas of the brain) would have left no trace in the record.

One of the better-studied examples where there is evidence for evolution of both predator and prey is the interaction between the garter snake, *Thamnophis sirtalis,* and the toxic newt, *Taricha granulosa.* The newts produce the highly potent neurotoxin tetrodotoxin from skin glands, and *T. sirtalis* is the only predator known to be able to survive consumption of the newt. Both species occur over a wide geographical range, and each species occurs in some areas where the other species is absent. Brodie and Brodie (1990, 1991, 1999b) have shown that there is genetically based geographic variation in the degree of resistance of the snake and geographic variation in the degree of toxicity of the newt. There is also within-population variation in the resistance of the snake. This is expected, since

tetrodotoxin resistance entails a cost in decreased locomotor performance by the snakes (Brodie and Brodie 1999a). The decreased maximal speed of resistant snakes is compensated for by their ability to partially overcome the paralytic effects of the toxin after ingesting newts. The patterns of variation for the most part exhibit positive between-species correlations (more resistant snakes occur where there are more toxic newts), but the correlation is not perfect, and good assays for newt toxicity have only recently been developed (Brodie and Brodie 1999a). Figure 21.3, based on Brodie and Brodie (1991), illustrates the difference in resistance to tetrodotoxin between two populations of *T. sirtalis,* one sympatric and one allopatric with the newt.

Toxin-antitoxin traits are very unlikely to combine multiplicatively in determining a predation rate (Abrams 1986a). An additive combination seems to be more appropriate, and models with this assumption predict that predators will increase costly detoxification traits in response to increased prey toxicity (Abrams 1986a). Unfortunately, the effects of each species on the population dynamics of the other are essentially unknown. Brodie and Brodie (1999b) speculate that the garter snake may have sufficiently limited prey-discrimination abilities that it cannot drop the newt from its diet without also reducing its consumption of many other nontoxic amphibians. This may contribute to the pattern of positive correlations between toxin and resistance.

Future Directions

Our knowledge of the evolutionary past increases gradually with the discovery of new fossils. However, almost all of the previous fossil-based scenarios for evolution of predators or prey remain controversial. A major problem is the inability to deduce demographic parameters, such as the predator-caused death rate of prey, from the morphological traits that are evident in fossilized material. Thus, macroevolutionary patterns may not shed much light on the mechanics of an interaction's evolutionary trajectory.

One of the areas where short-term research could lead to major advances is the question of the "fitness architecture" of predatory and antipredator traits. That is, how do different traits combine to determine parameters of the predator's per cap-

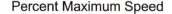

Percent Maximum Speed

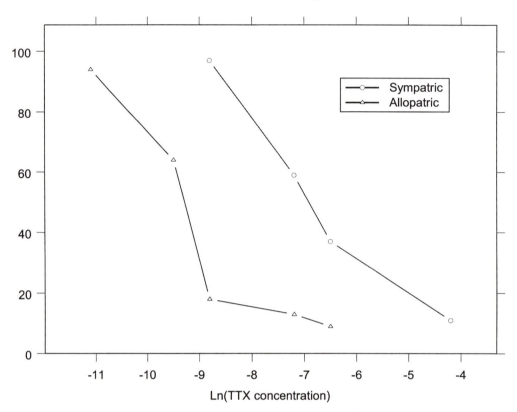

Figure 21.3 The effect of different doses of tetrodotoxin on the locomotor abilities of two populations of the garter snake, *Thamnophis sirtalis,* one sympatric and one allopatric with the normally toxic newt, *Taricha granulosa.* (After Brodie and Brodie 1991.)

ita consumption rates, as well as the birthrate and deathrates of both species? This is arguably the single most important piece of information for determining how evolutionary change in one species in a predatory interaction is likely to affect the future course of evolutionary change in the other. The previous section has mentioned some of the special properties of traits that combine multiplicatively. For a comparison of outcomes with multiplicative and additive traits, see Abrams (1990). These two possibilities are, of course, only two of the simplest end points of a wide spectrum of possible ways that traits in predator and prey could combine to determine the capture rate. The preceding sections have also mentioned the importance of knowing whether or not the relative effectiveness of a trait can be measured independently of the trait value in the other species. When prey can avoid capture by evolving larger or smaller trait values than some maximally vulnerable phenotype (a bidirectional axis of escape), then cycles driven by coevolution are particularly likely. Although bidirectional axes are very popular among theoreticians, we know almost nothing about their occurrence in nature.

Many of the other areas where research could quickly result in significant advances fall under the heading of increased complexity. These include spatially heterogeneous systems, systems in which the predator and prey are imbedded in a food web, and systems where there are several interacting traits within each species. In reality, each of these elements of complexity is likely to apply to almost every predator-prey pair in nature. What is needed is empirical and theoretical work on systems or scales that are not artificially simplified to exclude the complexity.

Evolution in metapopulations usually differs significantly from evolution in single panmictic populations. Metapopulations provide an opportunity for selection between groups, with the possible result that traits that increase the productivity of the subpopulation but decrease the relative productivity of the individual within the subpopulation may be favored. Needless to say, analyzing models of metapopulations is necessarily more complicated than analyzing models of single populations. There are correspondingly fewer examples where such models have been analyzed (but see Hochberg and van Baalen 1998). The main result found so far is that predators will have lower levels of prey exploitation in metapopulations than in a single population, because metapopulations with very efficient predators also have lower population sizes and higher probabilities of extinction (Gilpin 1975). Metapopulations are also harder to maintain in the laboratory than are single populations, although it has been done with some protists and mites. We are surprisingly ignorant about the degree to which most present-day species, let alone past species, are better approximated by metapopulations than by homogeneous populations. Spatial heterogeneity can have major affects on population dynamics, and this alone should produce some different evolutionary outcomes. Thompson (1994) has emphasized the importance of spatial heterogeneity, but most well-studied examples involve parasites and hosts rather than predators and prey.

No natural predator-prey system exists in complete isolation from other species, and the fact that a predator-prey system is embedded in a food web can have a host of evolutionary consequences that have only begun to be considered. One of the simplest examples is the potential for a predator to produce indirect evolutionary interactions between their common prey. Changes in the density or vulnerability of any one of several prey species that share common predators is likely to change the predation pressure experienced by other prey species. This will, in turn, favor shifts in the defensive traits of the remaining prey (Abrams 2000). These types of indirect predator-mediated character shifts are analogous to the shifts produced by shared resource use among competitors that was discussed by Schluter in the previous chapter. However, there are no well-studied examples, and theory is only beginning to be developed.

Similarly, we lack studies and models on interacting defensive traits of prey or interacting attack-related traits of predators. There is, for example, no good example of, or theory to describe, how life history and morphological defenses in prey should influence each other's evolutionary trajectory. Darwin's example of "simple" natural selection has proven to be far more complicated than he realized.

Acknowledgments   I thank the University of Toronto for financial support and Troy Day for discussions of coevolution.

22

Parasite-Host Interactions

CURTIS M. LIVELY

The diversity of known strategies for parasitic lifestyles is truly astonishing. Many species of parasitic worms, for example, utilize only one host species, while others cycle between two or more (as many as four) different species of hosts. Some parasites are highly virulent, seriously debilitating or even killing their hosts, while others cause only minor damage. Some parasites (such as viruses) are very small relative to their hosts and have the capacity for explosive reproduction. Others are almost as large as their hosts, and have relatively slow generation times. Therefore, parasites are difficult to categorize. Here, I use *parasite* to refer to organisms that have an obligate association with, and a negative effect on, another organism (the host).

Host strategies for dealing with parasites are equally complex. Vertebrates have highly specialized immune systems that can rapidly respond to infection and then store information that can be used to mount future responses to the same type of infection. Invertebrates lack the memory cells of true immune systems, but they do have complex self-nonself recognition systems for recognizing and killing foreign tissues. Plants also have highly specialized defenses against pathogens, and the genetic basis of these defenses is especially well known due to the work of plant pathologists on crop plants.

The myriad of details involved in the interactions between hosts and their parasites is over-whelming, but there are some shared, general aspects of these interactions that are of particular interest to evolutionary ecologists. First, parasites may attack in a frequency-dependent way. In other words, the probability of infection for a particular host genotype is expected to be, at least in part, a function of the frequency of that host genotype. This expectation has implications for sexual selection and the evolutionary maintenance of cross-fertilization (Sakai, this volume; Savalli, this volume). Second, parasites may affect the population density of their hosts, and host density may feed back to affect the numerical dynamics of the parasite. Host density may also affect natural selection on the reproductive rates of parasites, which in turn is likely to affect host fitness and host dynamics. These issues can become quite complex, but they are nonetheless important for understanding the ecology and evolution of natural populations and the emergence of new diseases. In what follows, I provide examples that illustrate some of the more important ideas that are currently emerging from studies of parasite-host interactions.

Concepts and Case Studies

Gene Frequency Dynamics: The Red Queen Hypothesis

Consider the fate of a very common host genotype in a population exposed to parasites. Assume that

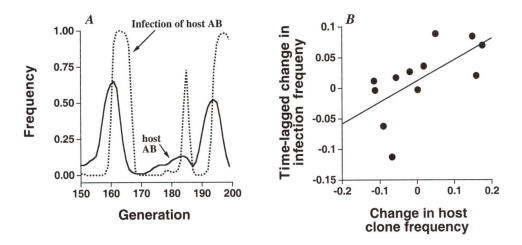

Figure 22.1 The Red Queen hypothesis. The left panel (A) shows the results from computer simulation results of gene frequency dynamics in a host-parasite interaction. The solid line gives the frequency of the *AB* genotype in a haploid host population; and the dashed line gives the frequency of infection in that same host genotype. The simulation model assumed that an exact genetic match was required for infection (e.g., parasite genotype *AB* can infect host genotype *AB* but cannot infect host genotypes *Ab*, *aB*, or *ab*), and that there were two haploid loci with two alleles each (four genotypes). The model also assumed that hosts are exposed to one parasite propagule, and that successful infections reduce the host's fitness by 60%. Note that the infection curve is shifted 90 degrees to the right of the host-frequency curve. Therefore, a change in frequency of the host genotype *AB* is correlated with a time-lagged change in frequency of infection in that genotype. This result leads to a testable prediction regarding dynamics in natural populations. Specifically, a change in frequency of a host genotype should be correlated with the time-lagged change in infection of that same genotype. The right panel (B) gives a test of this prediction. Changes in frequencies of four common snail clones in Lake Poerua (South Island, New Zealand) are plotted against the time-lagged change in infection frequency for the years 1992 through 1996. The parasite (a trematode worm) causes complete sterilization of infected snails, and it infected 3–11% of the snail population during the course of the study. The significantly positive correlation ($r = .748$; $P = .026$) indicates that an increase in a host clone between years (e.g., 1992–1993) was followed by a similar increase in infection of that clone in the following pair of years (1993–1994). Similarly, a decrease in infection of a snail clone was followed by decrease in infection in the following pair of years. This result is consistent with the Red Queen hypothesis. (Redrawn with permission from figure 4 in Dybdahl and Lively 1998.)

successful infection requires that the parasite precisely match the genotype of its host. Otherwise, the host recognizes the parasite as an invader and kills it. Parasite genotypes that can successfully infect (match) the most common host genotype will be favored by natural selection, and the associated alleles will spread in the population. If the parasite is sufficiently virulent, evolutionary change in the parasite population will result in ever-increasing selection against common host genotypes, eventually driving them down in frequency. When the frequency of the common host genotype is diminished, some new, previously rare, host genotype becomes the most common one. This then provides

a new target for the parasite population, and the newly common host genotype is expected to be attacked in the same way, but by a different parasite genotype. This kind of coevolutionary interaction could easily lead to the cycling of both host and parasite allele frequencies (figure 22.1A), and therefore it stands as a powerful mechanism for the maintenance of genetic diversity in natural populations. The technical term for this kind of interaction is *time-lagged, frequency-dependent selection*, which simply means that there is selection against common host genotypes and that this selection is lagged in time. The time lag is due to the fact that the parasites cannot instantaneously respond to changes in the host population. In fact, it may take several generations for the parasite population to "lock onto" the most common host type.

Such interaction between hosts and parasites is commonly referred to as an arms race, but this is a misleading analogy. An arms race involves an escalation in weaponry (from clubs to swords to guns to bazookas . . .), not a recycling of weapons. Yet, it is precisely a recycling of types that concerns us here, so a more useful analogy is a chase. If you are chasing me, and I dodge to the right, you are forced to also dodge right, but your adjustment to my new direction will occur with a slight delay. Once you make this adjustment, I will dodge left, forcing you to also dodge left, but with another delay. And even if you are faster, these delays mean you will stay slightly behind me. Note that neither of us changes weapons; we are just running: a leg race instead of an arms race. The important point here is that my dodge forces you, the parasite, to a change in course, which in turn causes me, the host, to change course, and so on. In an important sense, then, even though both of us are running as fast as we can, neither of us is getting anywhere, although we are covering a lot of ground. This running as fast as we can without getting anywhere is the gist of the Red Queen's famous remark to Alice in *Through the Looking Glass* (Carroll 1872): "Now here, you see, you have to run as fast as you can to stay in the same place." Hence, the idea that host-parasite coevolution can lead to oscillatory gene-frequency dynamics in both the host and the parasite has come to be known as the *Red Queen hypothesis* (following Bell 1982).

In the analogy, a change of directions is analogous to the spread of a different, but not new, genotype in a polymorphic population. Now, consider an asexual mutant in a large population of sexual hosts. Let us say that the mutant clone has an uncommon genotype, so it spreads. Because the mutant is asexual, it faithfully reproduces its genotype every generation, and it very quickly becomes common. Now, as above, there is selection on individuals in the parasite population to be able to infect this newly abundant host clone, at which point one of two things is expected: (1) The parasite drives the clone to extinction, or (2) the parasite drives the clone down to a point where it is rare, but not extinct, and then the clone begins to spread again due to its rare advantage. Either way, the parasite population has acted to keep the asexual

host clone from replacing the sexual host population. This is a special case of the Red Queen hypothesis applied to breeding-system evolution. According to this idea, parasites may impose selection on their hosts for sexual reproduction, and vice versa.

There is, however, one additional consideration. Clones, because they produce only daughters, have a much greater rate of intrinsic growth than the average of all the sexual females. In fact, the advantage is expected to be as much as twofold, meaning the clone would initially double every generation. The question now becomes: Can the parasite prevent the fixation of a clone with a twofold reproductive advantage? In theory, coevolutionary pressure from parasites can successfully prevent the clone from replacing the sexuals only if they kill or sterilize infected individuals (May and Anderson 1983). This is a lot to ask, even from a parasite. Most parasites are simply not that virulent.

Please note, however, the logical error. I implied that the theory is incorrect because most parasites don't kill or sterilize their hosts. But the theory makes no prediction about what *most* parasites do. It only requires one "sufficiently virulent" parasite or, perhaps more likely, a combination of parasites. Further, the theory does not require that the parasites themselves kill or sterilize the host, only that they lead, directly or indirectly, to a catastrophic loss of host reproductive potential. How could that happen?

William Hamilton, who originally fleshed out the Red Queen hypothesis (Hamilton 1980), has also suggested (with R. Axelrod and R. Tansey) a solution to the virulence problem (Hamilton et al. 1990). Suppose, they imagined, that parasites track common host genotypes as suggested above. Suppose further that none of the different species has much of an individual effect, but the independent effects add up, so that a host with many species of parasites is somewhat sicker than a host with few parasites. Nonetheless, all the worms, fungi, bacteria, and viruses combined do not kill the host. Finally, suppose that the host competes for resources (which should be true most of the time), and that the least infected hosts are the best competitors. If, for example, the resource will support only 80% of the individuals in the population, then it is likely that a significant fraction of the 20% that fail to reproduce are also the most infected. In this example, parasites did not directly kill their hosts; they

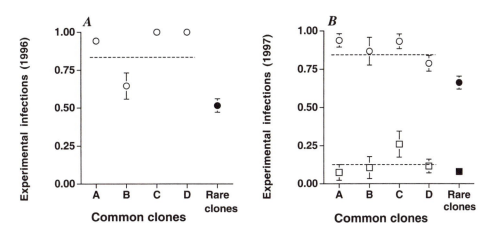

Figure 22.2 Results of an experimental infection experiment comparing common clonal genotypes with rare clonal genotypes from Lake Poerua (South Island, New Zealand). The left panel (A) gives the results from an experiment conducted in 1996. Common clones (open circles) were significantly more infected as a group than the group of 40 different rare clones (closed circles) ($X^2 = 41.39$; $df = 1$; $P < .0001$). Clones were designated as "common" if they exceeded 15% of the population during at least 1 year during the 4-year period prior to the experiment. Clones were individually designated as "rare" if they did not exceed 5% of the population during the 4-year period prior to the experiment. (Redrawn with permission from figure 5 in Dybdahl and Lively 1998.) The right panel (B) shows the results from a similar infection experiment conducted in 1997. In this repeat of the 1996 experiment, the infection rate of common clones (open symbols) was compared with the infection rate in rare clones (closed symbols) for both sympatric parasites from Lake Poerua (circles) and allopatric parasites from a distant lake (Lake Ianthe, squares). Note that the sympatric parasites infected rare and common clones more than the remote source of parasites. Note also that the local, sympatric parasites infected common clones more than rare clones (as observed in 1996) ($X^2 = 14.81$; $df = 1$; $P < .0001$), but the allopatric parasites did not infect common clones more frequently than the rare clones ($X^2 = 2.06$; $df = 1$; $P < .1517$). This result suggests that the common clones were more infected by the sympatric parasites due to coevolutionary interactions, rather than some inherent physiological characteristic of the common clones. (Redrawn with permission from figure 2 in Lively and Dybdahl 2000.) For both figures, vertical bars give one binomial standard error; and the horizontal dashed line gives the average infection rate for the four common clones.

simply rendered the hosts less able to compete in a very competitive world. Such a situation can lead to selection for sexual reproduction, even if the clones produce twice as many daughters as the sexual females. This is a very important idea, which also points to a large gap in present knowledge. We do not know the effects of infection on competitive ability for very many species.

As it stands, the Red Queen hypothesis is difficult to test, but progress has been made by examination of the assumptions. The most apparent assumption for Red Queen dynamics and the parasite theory of sex is the presence of rare advantage. Host genotypes that have been rare in the recent past should be less targeted by parasites than

host genotypes that have been common in the recent past. Happily, this is a testable idea.

Rare Advantage in Clonal Snails   A recent study of rare advantage was conducted on a clonal population of freshwater snails in New Zealand (Dybdahl and Lively 1998). We wanted to know two things: (1) Do host clones oscillate over time in a way that is consistent with the Red Queen hypothesis? And (2) is there an advantage to being rare? To address the first question, we plotted the change in frequency of clones against the time-lagged change in the prevalence of trematode infection in these same clones. So, for example, we plotted the change in host clones between years 1 and 2 against the change in infection between years 2 and 3. We plotted these changes for a 6-year study, reasoning that the Red Queen hypothesis predicted that the correlation should be positive, since parasites are expected to respond to host genotypes with a lag. We found that host clones oscillated over time, and as anticipated by the theory, host clone changes were correlated with the lagged change in infection (figure 22.1B).

To test for rare advantage, we conducted a laboratory experiment at the end of the field study. The experiment tested whether host clones that had remained rare over the course of the study were also less susceptible to infection than clones that were common over the course of the study. We found that rare clones were indeed significantly less infected following experimental exposure to the local source of parasites (figure 22.2A). We then asked whether the rare advantage was due to being locally rare per se, or whether there were correlated traits associated with rareness that also made the rare genotypes less infectable (Lively and Dybdahl 2000). We reasoned that common clones may be common as a result of a greater competitive ability, but that this ability to compete for resources might trade off with resistance to parasites. Therefore, the most common clones might be easier to infect, independent of any coevolutionary interactions. In a follow-up experiment, we used an outside source of parasites, one that was drawn from a different population of snails, which could not have been tracking the common clones in the original population we studied. We found that the outside source did not discriminate between rare and common clones (figure 22.2B). The local source of parasites, however, did again disproportionately infect the common clones. Hence, having a rare genotype, and not some correlated trait, conferred an advantage on the rare host clones, thus demonstrating frequency-dependent selection on the genes directly involved in the interaction. These results are also consistent with experiments showing local adaptation by these same trematodes (Lively 1989), and with field surveys that show a positive, significant correlation between the frequency of sexual individuals in a population and the risk of infection by trematodes (Lively 1992).

Multiple Mating and Parasite Resistance in Bee Colonies   Another recent experimental study that supports the Red Queen theory was reported by the Swiss team of Boris Baer and Paul Schmid-Hempel

(1999). In this study, they wanted to know the value to queens of mating with multiple males (polyandry). Multiple mating should produce a more genetically variable brood, which should result in fewer parasites. They worked with bumblebees, and they played the role of male bees by artificially inseminating queens with the sperm of either (1) four brothers, to give "low-diversity" broods ($N = 12$), or (2) four unrelated males, to give "high-diversity" broods ($N = 4$). They then stored the queens for one month at 6°C to simulate winter before allowing them to start their first broods in the lab. Finally, they put the developing colonies into flowering meadows in the Swiss countryside.

The results were striking (figure 22.3). The intensity of infection (number of parasites in an infected individual) and the prevalence of infection (frequency of infected individuals within a colony) were both significantly lower in the high-diversity broods than in the low-diversity broods. Apparently, the worker bees were exposed to parasites while foraging. These parasites then spread once inside the colony, and the infection spread faster in colonies where the broodmates were expected, based on treatment, to be more closely related. Importantly, this expectation for relatedness was confirmed by direct analysis of molecular markers (P. Schmid-Hempel, personal communication), and the idea that disease transmission occurs more readily among related individuals was directly demonstrated for this bee species (Bombus terrestris) in a separate experimental study (Shykoff and Schmid-Hempel 1991). Hence, taken together, the results demonstrate that polyandry (mating with multiple unrelated males) reduces the spread of disease within bumblebee colonies, and they may explain the occurrence of polyandry in many animal species.

The Baer and Schmid-Hempel (1999) study also showed that disease negatively affected the colonies' production of fertile individuals. In fact, the high-diversity colonies produced about twice as many reproductive bees as the low-diversity colonies. Thus, there is direct experimental evidence showing that genetic diversity not only reduces disease but also increases colony fitness. These results are entirely consistent with expectations under the Red Queen hypothesis, and they suggest that genetic diversification of offspring through outcrossing may be especially valuable when offspring are aggregated in colonies or litters (as suggested by Rice 1983).

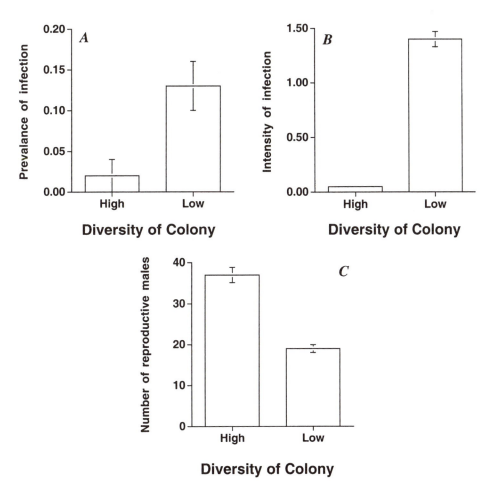

Figure 22.3 Selected results, from the bumblebee study in Switzerland, comparing infection and reproduction in colonies having high genetic diversity and low genetic diversity. Top-left panel (A) shows the
prevalence of infection by a microsporidian parasite *(Nosema)*; the difference between treatments was
highly significant ($P = .01$). A different parasite, the trypanosome *(Crithidia bombi)*, showed the same
trend, but was not significantly different between treatments ($P = .12$; data not shown). Top-right panel
(B) shows the intensity of infection for the microsporidian parasite *(Nosema)*, which was highly significantly different in the low- and high-diversity treatments ($P = .006$). Intensity of infection was significantly different and in the same direction for *Crithidia* ($P = .006$; data not shown). The bottom panel
(C) gives the mean number of males produced in the high- and low-diversity treatments, which is a
measure of the reproductive output of the colonies. The difference between these means is statistically
significant ($P = .028$). Vertical bars are one standard error about the mean. (Redrawn with permission
from figures 1 and 3 in Baer and Schmid-Hempel 1999.)

Rarity and Disease Spread   The specific effect of
host-genotype frequency on the spread of disease
was elegantly shown in a study by Shamsul
Akanda and Christopher Mundt (1996). They used
four different cultivars of wheat, each of which
was susceptible to different strains of rust (caused
by a fungus). They planted the wheat cultivars in

pairs, in five different ratios (10:90, 25:75, 50:50,
75:25, and 90:10) in a huge replicated experiment
that was repeated for 2 years. They wanted to
know how cultivar frequency would affect the
spread and severity of the rust, so they inoculated
the replicated field plots with different strains of
the disease by transplanting infected seedlings into

the plots. Here again, the results were dramatic. The study showed that the severity of the rust increased significantly with cultivar frequency (figure 22.4). There was, therefore, an advantage to having a rare genotype in the face of the spreading disease. Interestingly, the plots in which two cultivars were planted in a 50:50 ratio had the lowest average number of rust lesions per plant. This result demonstrates the value of mixing genotypes in agricultural situations, and it further suggests the value of diversifying offspring by outcrossing.

Note that I have mixed two issues above concerning rare advantage that I now want to cleanly separate. One advantage stems from having a genotype that has a recent history of rareness, as in the snail study (coevolutionary rare advantage). The other advantage stems from having a rare genotype during the spread of a disease, as in the wheat study, independent of coevolutionary effects (epidemiological rare advantage). In nature, both coevolutionary rare advantage and epidemiological rare advantage would be expected to be very important, and it is easy to imagine that epidemiological rare advantage would fuel the coevolutionary process (see also Thompson, this volume).

Parasites and Population Dynamics

In the previous section, we were interested in the effects of parasite-host interactions on gene frequencies. We turn now to conceptually similar issues, but ones that focus on population dynamics, rather than gene frequency dynamics. We are also interested in whether host density affects the transmission of parasites and the stability of parasite populations.

It seems clear that the density of susceptible hosts should affect the transfer of parasites between hosts, as in the rust study just discussed. Exactly how this should occur, however, is less intuitive. Fortunately, the process can be modeled, and the parameters of the simple model have intuitive appeal. One of these parameters is the reproductive rate of the parasite (R_0), which we can think of as the number of infections generated by each infected host. If this value is less than 1, then the parasite cannot spread in the host population; if the value is greater than 1, the parasite will spread at an exponential rate, at least initially.

Here is the basic model (taken from a review by Heesterbeek and Roberts 1995). The rate of change

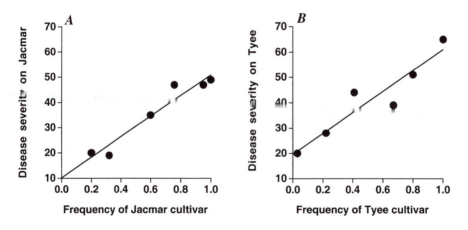

Figure 22.4 Selected results from the wheat rust study showing the relationship between the frequency of two cultivars (Jacmar in left panel, A, and Tyee in right panel, B, and the severity of rust infection in plots that contained mixtures of these same two cultivars in different frequencies. Severity was defined as the percentage of leaf area covered by rust lesions. In both cases, the relationship between cultivar frequency and disease severity was highly significant ($P < .001$). Similar results (not shown) were gained for these same two cultivars at a different site in the same year (1993) and in the following year (1994). Similar results were also gained for different pairwise comparisons of cultivars, so the result was robust to the changes in cultivar composition as well as site and year of the experiment. (Redrawn with permission from figure 1 in Akanda and Mundt 1996.)

(dI/dt) in the number of infected individuals in the population depends on the number of susceptible individuals in the population *(S)*, times the number of infected individuals in the population *(I)*, times the probability of contact between an infected and uninfected individual (β), all divided by the total host population size *(N)*. This gives

$$\frac{dI}{dt} = \frac{\beta SI}{N} \qquad (22.1)$$

If infected hosts become resistant at a certain rate due to acquired immunity, we must subtract that rate (γ) times the number of infected individuals *(I)*. This gives the rate of change in the number of infected individuals as

$$\frac{dI}{dt} = \frac{\beta SI}{N} - \gamma I \qquad (22.2)$$

If we now introduce a small number of infected individuals in a large population at time zero (I_0), then $S \approx N$, and $I = I_0$, and we get

$$\frac{dI}{dt} = \beta I_0 - \gamma I_0 = I_0(\beta - \gamma) \qquad (22.3)$$

The disease will spread if the rate of change is positive, which requires only that β > γ, or equivalently,

$$\frac{\beta}{\gamma} > 1 \qquad (22.4)$$

Hence, in this model, the basic reproductive rate, R_0, is the transmission coefficient divided by the recovery rate; the disease will spread if the former is greater than the later.

Continued spread of the disease is more restrictive because as individuals become infected, they are no longer susceptible to new infection. Thus, the current reproductive rate, R_c, given *S* susceptible hosts, $R_c(S)$, must be greater than 1 for the epidemic to continue spreading, where $R_c(S) = R_0 S / N$ (which is simply R_0 times the frequency of susceptible hosts in the population).

The recovery rate (γ) is an interesting parameter. It makes one wonder what would happen if, by vaccination, a proportion of the population were to become "recovered" even before the disease began to spread. To see this effect in the model, we simply need to weight the number of susceptible indi-

viduals *(S)* by the frequency of individuals that have not been vaccinated $(1 - v)$, where *v* is the frequency of vaccinated hosts. Thus in equation 22.2, we replace *S* with $S(1 - v)$. Now, the rate of change in the number of infected individuals becomes

$$\frac{dI}{dt} = \frac{\beta S(1 - v)I}{N} - \gamma I \qquad (22.5)$$

As above, $S \approx N$ and $I = I_0$ at the beginning. Hence

$$\frac{dI}{dt} = I_0[\beta(1 - v) - \gamma] \qquad (22.6)$$

The disease will now spread in the partially vaccinated population when

$$\frac{\beta}{\gamma}(1 - v) = R_0(1 - v) > 1 \qquad (22.7)$$

Note that the conditions for spread of the disease are now more restrictive because the parasite's reproductive rate (R_0) is multiplied by the fraction $(1 - v)$. If *v* is very high (near 1), then the disease is extremely unlikely to spread. Hence, there are two reasons to have a vaccination against the next flu: One is so that you do not get the flu, and the other is so that the pool of susceptible individuals is reduced, thereby reducing the chance of an epidemic. In the present model, we would want to vaccinate enough individuals so that

$$v > 1 - \frac{1}{R_0} \qquad (22.8)$$

in order to prevent the initial spread of infection.

Population Cycles in Red Grouse If an epidemic does occur, there are a few possible outcomes. One is that the host population is driven to extinction. Another is that the host population decreases in size but is not driven extinct. In this latter case, if the number of susceptible hosts is driven below the threshold value for parasite spread [$R_c(S) < 1$], then the parasite will begin to die out (at least locally). This die off of parasites would then allow the host population to increase again, which suggests the potential for oscillatory population dynamics, analogous to that seen in Red Queen models of gene frequency dynamics (figure 22.1A). Under this theory, such oscillations in population size are more likely if infection has a much greater effect on fecundity than on mortality (Dobson and Hudson 1992; May and Anderson 1978). Cyclical population dynamics have been observed in red grouse in Scotland since 1977, with population crashes about every 4–8 years. These crashes were associated with high levels of nematode (round worm) infections, which cause significant reductions in reproductive output (Hudson et al. 1992). The question is: Do the worms cause the cycles, or are they just along for the ride? In 1989, Peter Hudson, Andy Dobson, and Dave Newborn treated four of six populations of grouse with an anthelmintic (Hudson et al. 1998). They did not treat every individual, but they were able to treat 15–20% of the breeding populations, thereby decreasing the number of susceptible individuals. In 1993, they repeated the treatment in two of these four populations. They cleverly picked these years (1989 and 1993) because they were predicted to be crash years. If the parasites were causing the crashes of red grouse, a crash would be expected in the two control populations, but not in the four treated populations. This is exactly what they found. The two untreated control populations crashed in 1989 and 1993, exactly as expected, but the populations treated in both 1989 and 1993 did not crash (figure 22.5). One of the populations treated once (in 1989) showed a small dip in that year, but nothing of the magnitude of the two control populations. This is striking experimental evidence that parasites can regulate host population size as well as cause host populations to cycle in a predictable way.

The Evolution of Virulence

Earlier, I mentioned the possibility that parasites might decimate their host populations to the point of extinction. This assertion may seem unlikely at first, because the parasites would also become extinct in the process. Thus, it might seem that parasites should be selected to become increasingly benign, perhaps to the point of becoming commensal with their hosts, or even mutualistic. This way of reasoning formed a kind of "conventional wisdom" in the medical community until very recently (Ewald 1994).

Recent theories on the evolution of parasite virulence have shown that the conventional wisdom was incorrect. Natural selection operates not ac-

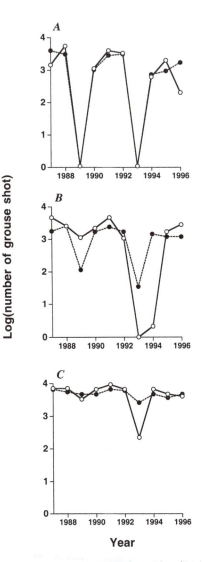

Figure 22.5 Cycling in red grouse in Scotland. The top panel (A) shows the number of red grouse in two populations that were not treated with an anthelmintic, which kills the parasitic nematodes of the grouse. Note that both populations crashed in 1989 and 1993. The middle panel (B) shows the results for two grouse populations that were treated for nematodes in 1989. Note that these two populations did not crash in 1989, as did the controls, but they did crash 4 years later in 1993. The bottom panel (C) shows the results for two populations that were treated for nematodes in both 1989 and 1993. Note that these populations did not crash at any time during the course of the study. (Redrawn with permission from figure 2 in Hudson et al. 1998.)

cording to long-range visions of what is best for the population, but by culling out genotypes that are less suited to present environmental conditions [a process that works best, in general, when the effective population size (N_e) times the selection coefficient (s) is much greater than 1: $N_e s \gg 1$]. Why should this culling be any different for parasites? Individuals that have the highest probability of successful propagation into the next generation would be favored, independent of the long-range consequences. As you might expect, selection on the rate of parasite reproduction should be affected by the density of susceptible individuals in the host population.

The first biologists to dismember the previously popular conventional wisdom were Roy Anderson, Robert May, and Paul Ewald (Anderson and May 1982; Ewald 1983). The theory that they (and their followers) have developed relies on some basic principals of life history evolution (see also Roff, this volume). The basic idea is that the intrinsic rate of growth (R_o) for a parasite depends on a trade-off between making many propagules in the host (e.g., lots of viruses) and the damage this causes to the host (figure 22.6). If, for example, the parasite propagates so fast that it kills the host, then there is selection to reduce the rate of propagule production (i.e., the parasite becomes less virulent). In this case, the conventional wisdom is right, but for the wrong reason. If, on the other hand, the pathogen produces only a few propagules, but without any effect on the host, it will surely be replaced by a strain that makes more propagules. This kind of reasoning suggests that parasite virulence should evolve to an intermediate level, but the term *intermediate* is somewhat vague. What it really means is that parasite virulence should evolve to the point where the marginal gains in fitness due to propagule production are equal to the marginal losses in fitness due to host mortality.[1] This balance should depend on the parasite's life cycle and the probability of transmission to the next host. This probability will depend on the local density of susceptible hosts.

Consider, for example, a virus of insect larvae that is transmitted only by causing the host to explode (Myers 1993). In this case, parasite transmission is coupled with host death, so the virus should pack as many virions as possible into the larva before causing the larva to splatter its contents into the environment. These contents, mostly virus, are

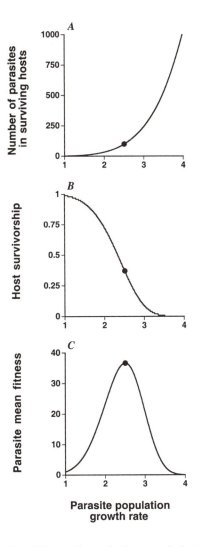

Figure 22.6 Illustration of ideas underlying the evolution of virulence. The top panel (A) shows the number of parasites in surviving hosts at time t as a function of the population growth parameter, λ. The exponential increase was calculated from the standard equation for population growth: $N_t = N_0 \lambda^t$, for $N_0 = 1$ and $t = 5$. The middle panel (B) shows host survivorship as a function N_t where the effects of parasites are independent [survivorship = $(1 - s)^{N_t}$, where s gives the effect of each individual parasite]. The bottom panel (C) shows the products of the curves in A and B, which gives parasite fitness as a function of λ. The analytical solution for λ that maximizes parasite fitness is given in each panel as a solid circle. Note that host survivorship at this maximum is quite low, a finding suggesting a highly virulent parasite at equilibrium.

picked up by other larvae, and transmission is accomplished. Viruses that did not kill the host would have a selective disadvantage. So much for the idea that parasites should evolve to be benign.

There are, of course, less extreme examples, but the basic idea holds: Parasite populations should evolve under natural selection for individuals that maximize their own reproductive success, without regard to the host's future. And in some cases, such evolutionary change could lead to the host's extinction. Along these lines, Van Valen (1973) showed that some clades have declined in species number at a constant rate over evolutionary time. One way this could happen, he suggested, is that there is a reasonably constant probability that tight coevolutionary interactions go haywire, perhaps due to the evolution of hypervirulent parasites. This reasoning lead Van Valen to the original formulation of the Red Queen hypothesis, which was concerned with the rise and fall of clades (macroevolution) rather than alleles (microevolution). It is difficult to know whether the idea is correct, but it is nonetheless very interesting. Parasite virulence would be expect to increase under selection whenever the probability of transmission increases. Does this happen at a constant rate in evolutionary time?

Virulence theory, like most evolutionary theories, is difficult to test directly, but researchers can go after the critical assumptions. The conceptual core of the current theory on the evolution of virulence rests on two critical assumptions: (1) that higher rates of replication in the parasite lead to greater reductions in host survivorship, and (2) that higher rates of replication in the parasite lead to a higher probability of transmission to the next host (figure 22.6). These assumptions were examined in an elegant study by Margaret Mackinnon and Andrew Read (1999) of Edinburgh University. They injected one of eight different isolates of rodent malaria (Plasmodium chabaudi) into different groups of inbred mice. They then measured the replication rate of the parasite, the effect of infection on each mouse, and the proportion of mosquitoes that became infected after taking a blood meal from infected mice. The results showed that replication rate was significantly different among the parasite strains, and that these differences were genetically based. In addition, replication rate was positively correlated with virulence; that is, mice infected by parasite strains having high rates of replication were also more anemic. Finally, replica-

tion rate was significantly and positively correlated with the probability that the infection was transmitted to mosquitoes. Hence, both of the critical assumptions of the model were shown to be true.

Catastrophic Disease in Natural Populations

From the above, it may seem as if only two things matter to the spread and evolution of disease: gene frequencies and population density. This, of course, is not true. Here, I illustrate, using a few examples, the effects of climate and habitat on the evolution and ecology of host-parasite interactions. This addition makes the situation more complex, but also more realistic.

One example involves a seastar (Heliaster kubiniji) in the Gulf of California. Prior to the summer of 1978, the rocky intertidal shores of this region were packed (up to $1/m^2$) with these voracious predators of barnacles, mussels, and many species of snails. This animal was already well known as a possible "keystone predator" due to its role in promoting local species diversity by eating whatever was most common. During a 2-week period in June 1978, however, Heliaster was virtually wiped out throughout the Gulf of California by a bacterial disease. What happened? How could such a common species become instantaneously so rare?

The most sensible explanation for the epidemic is that it had something to do with the exceptionally high sea-surface temperature during the summer of 1978, due to a particularly severe El Niño (Dungan et al. 1982). The very warm water may have reduced or eliminated Heliaster's ability to combat diseases on its epidermis. Alternatively, the warm water may have made the disease more virulent than it would otherwise have been. It may be dangerous to speculate about the effects of climatic change (experiments are hard here), but such widespread patterns cannot be ignored. An entire species spanning the entire 1000-km stretch of the Gulf of California disappeared virtually overnight, a circumstance suggesting at the very least that there is more to disease than genetics and host density (or the interaction between the two).

In a recent review, Jared Diamond (1998) makes a compelling historical case for the joint effects of ecology, epidemiology, and evolution in the spread of human disease to the Americas. First, he convinces the reader that diseases were as important in the conquest of Native Americans as the superior weaponry of the European invaders, or more important. The statistics he gives are staggering. In 1520, smallpox was introduced into the Aztec population of Mexico by the Spanish. At the time, there were 20 million people, none of which had been exposed to anything similar at any time in their lives, nor had their ancestors. An epidemic quickly killed about half the population, demoralizing the Aztecs. By 1618, the population was reduced to only 1.6 million, a reduction of 92%. Similarly, smallpox reduced a Mandran tribe of 2000 people in the Great Plains of North America to only 40 survivors in less than a month, and there are many other examples in Polynesia as well as the Americas.

Such devastation of human populations is heart-wrenching and difficult to fathom, but one has to wonder why it occurred. Diamond asks several interesting questions along these lines: Why did it work this way, and not the reverse? Why didn't the original inhabitants of the Americas give the Spanish invaders something equally deadly? And where did these European diseases come from? The hypothesis that Diamond gives combines ecology, epidemiology, and evolution. First, the high densities of European populations (due to the development of agriculture) led to the conditions for the spread of highly virulent diseases, which could not be maintained in small nomadic populations (but, remember, there were 20 million Aztecs, so this cannot be the whole story). Second, the emergent diseases led to strong selection on European human populations, so that genetically resistant individuals, which could then develop immunity, were more likely to survive. The Native Americans could not have been through the horrendous bouts of selection that accompanied the epidemics in Europe, and they did not have any way to develop immunity. So, when small pox was introduced to the New World, it rapidly spread in the highly social, dense populations of peoples that descended from the original colonists of the region. This new epidemic made it easier for the bearers of the disease to prevail over the peoples devastated by the disease.

But how did the Europeans first get the disease? Diamond suggests that it jumped from the domesticated animals that lived in close association with the peoples in Europe. Therefore, it would seem that the domestication of animals in Europe led indirectly to the devastation of the indigenous peoples of North and South America.

Future Directions

Parasitism is a rich and complex subject that captures many of the important ideas in evolutionary ecology (e.g., life history evolution, trade-offs, frequency- and density-dependent selection, kin selection, and breeding-system evolution). For evolutionary ecologists, there are vast opportunities (and many unanswered questions) and vast numbers of potential study systems. The most exciting part of the research to come lies in the diversity of interactions and weaving together the effects of genetics and ecology under field conditions. This weave may or may not lead to general rejection of the atomized view of infection gained by the isolated search for resistance genes, but it will nonetheless lead to important discoveries and a much better understanding of one of the important forces of evolutionary and ecological change.

Several aspects of disease would at this point seem to merit additional study. One of these aspects is the effect of infection on the competitive ability of hosts, which is important for two reasons. One is that if diseases, which might otherwise be considered avirulent, significantly reduce the ability of their hosts to compete for resources, the Hamilton et al. (1990) modification of the parasite theory of sex is greatly strengthened. In one such study of the effect of rust infection on jewelweed *(Impatiens capensis)*, we found that rust infection had no effect on the growth rate of plants in the wild when conspecific competitors were experimentally removed (Lively et al. 1995). Hence, it would seem that the disease, by itself, was not having much of an effect on plant fitness. But infected plants grew at a significantly slower rate than uninfected plants under the natural conditions of high plant density. There was thus an interaction between infection and plant density of the kind required by the Hamilton et al. (1990) model. More experiments of this kind are needed to disclose whether the result is general.

A second reason for understanding the effect of disease on the competitive ability of infected individuals relates to the structure of communities. There is a rich literature on the effects of predators on the structure of communities, especially in rocky intertidal zones, but corresponding studies of the effects of parasites are relatively rare. One example of a very nice study was recently published by Keith Clay and Jenny Holah (1999). They found in a 4-year field experiment that an association with a fungal endophyte increased the competitive dominance of a grass *(Festuca arundinacea),* which led to a decrease in local plant diversity. In other words, the presence of an endophyte in one host species altered the structure of the entire community. These kinds of effects may be very common, but at present, there are few studies of the effects of diseases on the structure of communities.

Another gap in present knowledge concerns the genetic basis for infection. The results from plant pathology suggest that gene-for-gene interactions determine whether infection will occur, but less is known about natural populations. At present, there is an interesting debate about whether the results of plant pathology can be extended to the natural world. The details of the genetic architecture turn out to be critical to models of host-parasite coevolution, but there are insufficient data to reveal the answer.

Finally, there would seem to be considerable promise to fusing the genetics of Red Queen models with the ecological perspective of the epidemiological models (following May and Anderson 1983). The mathematics of these models could get quite gnarly, but there would nonetheless seem to be considerable potential in the effort.

Acknowledgments   I thank P. Schmid-Hempel, C. Mundt, A. Dobson, L. Delph, and the members of my lab group (especially T. Frankino, M. Neiman, D. Repasky, and R. Winther) for comments on the manuscript.

Note

1. This is a simplified view because it does not take into account the variance in fitness. High variance in fitness might discount the propagule production as a way of bet hedging. Similarly, the simplified view does not take interactions with other parasites into account, which might select for higher virulence. Finally, the simplified view does not take into account any effects of parasite population structure.

23

Plant-Herbivore Interactions

MAY BERENBAUM

As is the case with most supposedly modern concepts in evolutionary biology, the idea of coevolution, or reciprocal evolutionary change between interacting species, actually goes back to Charles Darwin. In the introduction to *The Origin of Species* (1859), he wrote:

> In considering the Origin of Species, it is quite conceivable that a naturalist, reflecting on the mutual affinities of organic beings, on their embryological relations, their geographical distribution, geological succession, and other such facts, might come to the conclusion that species had not been independently created, but had descended, like varieties, from other species. Nevertheless, such a conclusion, even if well-founded, would be unsatisfactory, until it could be shown how the innumerable species inhabiting this world have been modified, so as to acquire that perfection of structure and coadaptation which justly excites our admiration. It is, therefore, of the highest importance to gain a clear insight into the means of modification and coadaptation. (p. 26)

Early on, then, Darwin pointed out the importance of interactions among organisms in determining evolutionary change, as opposed to "external conditions such as climate, food," or even "the volition" of the organism itself. Interactions among organisms, however, take many forms. Antagonistic interactions, in which one species benefits and the other is harmed, are themselves diverse. Among those interactions in which both species are animals, the gamut runs from predation, in which one species kills and consumes several individuals of the other species during its lifetime, to parasitism, in which one species merely saps the "reserves" and rarely kills its host. Intermediate and unique to the phylum Arthropoda is parasitoidism, in which one species kills its prey, as does a predator, but, like a parasite, is normally restricted to a single host individual. A comparable continuum exists for interactions between an animal and a plant species; these associations are usually referred to as forms of herbivory (with parasitoidism akin to internal seed feeders of plants).

In mutualistic interactions, both species benefit from the interaction. Mutualisms can involve interactions between animals and plants, generally in which a food reward from the plant is exchanged for mobility provided by the animal partner. Pollination and fruit dispersal are two such mutualistic processes. In that a food reward is involved in the symbiotic association, animal mutualists of plants are very often herbivorous. Thus, the degree to which an interaction is mutualistic or antagonistic depends on the fitness impact of the herbivory; the types of interactions continuously intergrade.

Taxonomically, the most prevalent form of interspecific interaction in terrestrial systems involves plants and the animals that consume them. Of these plants and animals, it has been estimated that angiosperm plants and herbivorous insects alone constitute close to half of all terrestrial species. Feeding on plants within the class Insecta occurs to a great extent only in 9 orders (out of over 20), but these orders are among the largest in the animal kingdom. Sister group comparisons of diversity between herbivorous taxa and their close, non-herbivorous relatives indicate that adoption of a plant-based diet is associated with speciation and diversification (Mitter et al. 1988). That more orders within the class Insecta are not associated with plants is remarkable considering that plant material is among the most abundant food resources and the vast bulk of it appears to go unused. In many communities, such as deciduous forests or oldfields, only 1–2% of the net primary production is consumed by herbivores (Ricklefs 1976).

Why insect herbivores are not more abundant has been a question of long-standing interest to ecologists. Hairston et al. (1960) attempted to explain the apparent paradox by asserting that, while every other life form is food-limited, herbivores are predator-limited. They reasoned that, since insects rarely demolish their food (and do so almost invariably under circumstances in which a pest is introduced into an area without its normal predators), the earth remains green because predators and parasites act to reduce insect population levels. This notion was reformulated by Bernays and Graham (1988) and has led to competing schools of thought on whether regulation of herbivore populations is top-down (predator-controlled) or bottom-up (plant-controlled).

Hairston et al. (1960) and their intellectual descendants made one major assumption in their argument, that is, that everything green is edible. This has repeatedly been shown not to be the case. Herbivores can starve in the midst of apparent plenty because there are many problems associated with eating plants. Simple morphological obstacles can hinder ingestion. Hairs, glands, spines, and wax can all act as physical barriers. Grasses are equipped with parallel rays of silica; these glasslike particles can cause abrasion even to sclerotized insect mouthparts. Specialists on grass as a consequence tend to have strongly reinforced chisel-like mandibles as well as proportionately larger heads, to accommodate the larger muscle mass needed to power the mouthparts (Bernays 1991).

Nutritional deficiencies plague herbivores as well. The bulk of plant tissue is cellulose, which,

although a polymer of glucose, is not readily utilized by herbivores: With its β 1–4 linkage, it defies metabolism by most multicellular organisms. Some insect herbivores, such as cows and other mammalian herbivores, rely on endosymbiotic microbes for cellulose digestion. Among these are some beetle grubs; rates of cellulose digestion in the scarab beetle *Pachnoda marginata* exceed 60%. Other herbivores, however, may rely on endogenously produced cellulolytic enzymes; among these are some cerambycid (longhorn) beetles and walkingsticks with extremely low numbers of microbes in their intestinal tracts (0.1–1% of numbers found in other arthropods).

Plant tissue is generally much lower in protein than animal tissue, and, because of their small size and relative immobility in immature stages, plant-feeding insects face greater challenges in acquiring adequate protein than do most other herbivores. The efficiency of conversion of ingested food, a rough index of the usability of food that measures how much ingested food is processed into insect biomass, generally increases with food nitrogen content; one consequence of this relationship is that consumption rate increases with decreasing nitrogen content. This pattern holds both within species and among species; herbivores feeding on low-nitrogen tissues generally consume more food than do herbivores feeding on high-nitrogen plant tissues and display lower efficiencies of conversion of ingested food into biomass. Thus, consumers of bark, which contains less than 1% protein, or xylem sap, which contains less than 0.04% protein, often must grow slowly or, alternatively, consume 100–1000 times their body weight per day.

Total nitrogen or protein content is a misleading concept in insect nutrition; amino acid balance is equally critical to growth and development. Insects may have a particularly high requirement of aromatic amino acids, which are used in the construction of the rigid exoskeleton. If such amino acids are in short supply, herbivorous insects may economize. Adult beetles in the family Chrysomelidae, all of which are herbivorous, contain significantly less exoskeleton relative to overall mass than beetle families containing carnivorous members. To produce this exoskeleton, a chrysomelid would require only 43% of the nitrogen required by a predatory beetle of the same size.

Water, too, is a major problem for plant feeders, which is exacerbated as water content of plant feeders decreases and the differential between food and consumer water content increases. While forb feeders such as the European cabbageworm *Pieris rapae,* with a body water content of 83–84%, consume foliage that can reach 90% in water content, tree foliage feeders such as *Telea polyphemus,* the polyphemus caterpillar, with a body water content of 90%, consumes foliage that is less than 75% water. Stored-product insects such as the rice weevil, *Sitophilus oryzae,* with a body water content of 48–50%, consume dried wheat, with a water content of 15–16%.

Micronutrients also present problems to insect herbivores. Plant tissue is extremely low in sodium, except for rare halophytic exceptions. Compounding the problem for herbivores is the fact that many plant constituents interfere with sodium availability. To some extent, plant-feeding insects compensate physiologically for low sodium intake, but no insect has completely eliminated its requirement. Butterflies, particularly males, have long been known to form aggregations or "puddle clubs" at mud puddles, bird droppings, and animal urine. Sodium ions in soil induce the formation of puddle clubs, a finding suggesting that plant feeding in the larval stage fails to provide adequate sodium for adult function. Other micronutrients, although abundant in plant tissues, may be rendered unavailable to herbivores. Iron, for example, is necessary for the function of enzymes involved in pheromone biosynthesis and toxin degradation. In seeds of legumes and other plants, iron often occurs irreversibly bound to phytic acid, a phosphorus storage compound that is a powerful mineral chelator (Berenbaum 1995). In its chelated form, iron is not readily available to herbivores.

In addition to mineral micronutrients, the vitamin content of plant tissue may influence its suitability for herbivores (Berenbaum 1995). Most herbivorous insects, for example, have an absolute requirement for vitamin C (ascorbic acid). While an absolute requirement for vitamin A has not yet been demonstrated, carotenoids (which cannot be synthesized by insects) are essential for regulating photoperiod response in some species and are important in a wide range of species for coloration and hence mate recognition and escape from predators.

While many different kinds of nutritional barriers are important to herbivores, the plant constituents that have received the most attention in the context of plant-insect interactions are allelochemicals, or plant secondary compounds. These are chemical compounds that are not known to play a role in primary metabolism (e.g., photosynthesis, respiration) but that have effects on other organisms (from the Greek *allelo,* meaning "one another"). Secondary substances can affect insect growth in several ways; they can reduce feeding rates (antifeedants), they can cause mortality outright (toxins), and they can reduce the amount of energy or protein available for growth (antinutrients). There are examples in abundance of a single compound acting in all three ways. Tannins, polyphenolic compounds present in foliage of many trees, act as feeding repellents to both insect and mammalian herbivores; they display toxicity independent of effects on consumption rate; they can create lesions in gut walls that reduce an insect's ability to process food; and they may reduce digestibility by complexing with gut proteases and with dietary proteins.

That plants are nutritionally challenging to insects is not necessarily indicative of past interactions with insects; low levels of protein, water, vitamins, minerals, and other essential nutrients may arise due to the idiosyncratic demands of plant physiology, or to ecological constraints other than herbivory. There are, however, features of plant anatomy and morphology that are difficult to account for outside the context of herbivory. One such example involves the yellow "egg-mimic" structures on the tendrils of *Passiflora* passionflower vines, which strongly resemble the eggs of *Heliconius* butterflies (Marquis 1992). These butterflies, the larvae of which are specialized on passionflower vines, will not oviposit on plants already containing eggs of conspecifics; experimental studies confirm that the plant structures resembling eggs elicit the same oviposition avoidance behavior as do actual conspecific eggs.

Historical Context

Plants present tremendous barriers to herbivores; whether these barriers are in place as a result of selection pressure exerted by herbivores over evolutionary time has been subject to speculation for over a century. One problem in recognizing coevolution has been in defining the term. To carry things to logical extremes, all organisms interact with other organisms; not all interactions, however, are coevolutionary. Reciprocity is usually regarded as the determining factor in identifying coevolution. Reciprocity or mutual evolutionary response has been most apparent in interactions that are obligatory on the part of one or both interacting species. The term *coevolution* was, in fact, coined with respect to such an interaction (Mode 1958). Pathogen-host interactions are ideal candidates for identifying reciprocal adaptive responses inasmuch as a highly host-specific parasite has a stake not only in maintaining its own life but also in maintaining the life of its host to sustain the interaction. Coexistence between pathogen and host can result from increased host resistance and decreased pathogen virulence. Reciprocity can be more difficult to document in an antagonistic interaction in which participants have no obligatory or exclusive association. This is often the case in insect-plant interactions; most plants are attacked by more than a single species of insect, and while most insect herbivores are specialized in the sense that they feed on plants in three or fewer families (Bernays and Graham 1988), their host range generally consists of multiple species.

With respect to animal-plant interactions, by the end of the 19th century, an attractive hypothesis had been advanced to account for the diversity of plant secondary substances. After studying the relations between snails and the plants they eat, Stahl (1888, cited in Fraenkel 1959) formulated a general hypothesis to account for chemical variation among plants:

> We have long been accustomed to comprehend many manifestations of the morphology [of plants] as being due to the relations between plants and animals, and nobody, in our special case here, will doubt that the external mechanical means of protection of plants were acquired in their struggle . . . with the animal world. . . . In the same sense, the great differences in the nature of chemical products [Exkrete] and consequently of metabolic processes, are brought nearer to our understanding, if we regard these compounds as means of protection, acquired in the struggle with the animal world. Thus, the animal world which surrounds the plants deeply influenced not only their morphology, but also their chemistry. (p. 1470 in Fraenkel 1959)

The importance of insects in particular as selective agents in determining plant chemistry in the "parallel evolution" (Brues 1920) of insects and plants was not emphasized until Fraenkel (1959) suggested that "reciprocal adaptive evolution" determined patterns of host plant utilization by in-

sects—that plant chemicals can act as repellents to most insects and at the same time can act as attractants for those few species of insects that can feed on each plant species. In this paper, titled "The Raison d'être of Secondary Plant Substances," Fraenkel contended that since all green plants are essentially equivalent nutritionally, differences in host plant utilization patterns must be attributable to nonnutritive factors. He then introduced the idea that the reasons for the existence of the diversity of plant chemicals not known to play a role in plant physiological processes is ecological: to repel or attract herbivores, specifically insect herbivores. In Fraenkel's words, "This reciprocal adaptive evolution . . . occurred in the feeding habits of insects and in the biochemical characteristics of plants" (p. 1466) as well as in the pigment chemistry of plants and in the behavior and morphology of pollinators. In support of this hypothesis, in addition to citing many laboratory studies documenting the various physiological effects of plant secondary substances, he cited the evolutionary conjunction of the appearance of angiosperm plants and the diversification of higher insects in the Cretaceous.

Building on Fraenkel's (1959) concept of reciprocal adaptive evolution, Ehrlich and Raven (1964) used the term *coevolution* to describe the stepwise procedure by which plants elaborate chemical defenses and insects evolve resistance or tolerance to those defenses. According to Ehrlich and Raven (1964), who based their assumptions on circumstantial evidence in the form of feeding patterns of various groups of butterflies, insect-plant coevolution is a five-step process:

1. Angiosperms produce new secondary substances as the result of mutation or recombination.
2. By chance, some of the new substances affect the suitability of the plant as food for insects.
3. Plants escape to a large extent from insect herbivory and are free to undergo evolutionary radiation.
4. By mutation or recombination, insects evolve resistance to the new compounds.
5. Insects enter a new adaptive zone and undergo evolutionary radiation, in some cases using the erstwhile toxic or repellent compounds as attractants.

Thus, this coevolutionary process, subsequently labeled the "escape-and-radiate model" (Thompson 1996), accounts not only for the biochemical diversity of angiosperm plants but also for the tremendous diversification of angiosperm plants and herbivorous insects.

Ehrlich and Raven's (1964) paper inspired decades of both theoretical and empirical work designed to document all or part of the evolutionary scenario. In one such study (Berenbaum 1983), the stepwise scenario was applied to a group of secondary chemicals called *furanocoumarins*. All known furanocoumarins derive from a single biosynthetic precursor, umbelliferone (7-hydroxycoumarin) (figure 23.1). Umbelliferone and related hydroxycoumarins are far more widespread among plant families than are the derivative furanocoumarins, occurring in approximately three dozen plant families. Among the furanocoumarins, there are two principal structural classes, derived from the activity of two different location-specific prenylating enzymes: linear (with the furan ring attached at the 6,7 positions) and angular (with the furan ring attached at the 7,8 positions). Linear furanocoumarins, known from about a dozen plant families,

are more widely distributed, in turn, than are the angular furanocoumarins, which are restricted to approximately three plant families. When tested against a generalist insect, linear furanocoumarins are appreciably more toxic than are hydroxycoumarins, and, when tested against an adapted specialist, angular furanocoumarins are more toxic than linear furanocoumarins. This progressive increase in toxicity, corresponding to biosynthetic advancement, is consistent with the Ehrlich and Raven scenario. Moreover, as Ehrlich and Raven predicted, plant taxa with more advanced coumarins are more species-rich than taxa with less advanced coumarins, and insect taxa feeding on plants with complex coumarins are more species-rich than taxa feeding on plants with simple coumarins. This increase in diversity was interpreted as evidence of the existence of Ehrlich and Raven's "adaptive zones."

If the evolution of biochemical innovations in plants and insects leads to adaptive radiation, Mitter and Brooks (1983) reasoned that coevolution between plants and animals should lead to parallel

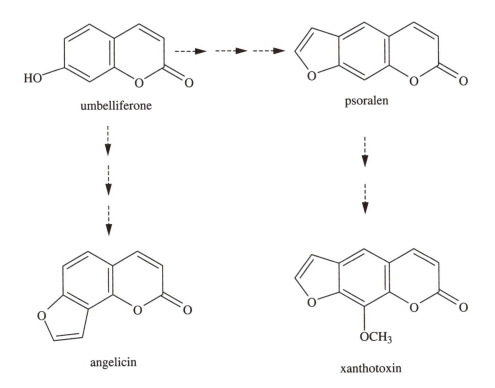

umbelliferone

psoralen

angelicin

xanthotoxin

Figure 23.1 Structure of hydroxycoumarin (umbelliferone), linear furanocoumarin (psoralen), substituted linear furanocoumarin (xanthotoxin), and angular furanocoumarin (sphondin).

cladogenesis—alternating periods of diversification in concert will be reflected in the evolutionary record as phylogenies with identical topologies. This suggestion stimulated a wave of effort among systematists to compare cladograms of herbivorous insects and their associated host plants.

The enthusiasm with which the Ehrlich and Raven (1964) scenario was greeted waned gradually in the absence of well-designed studies. Bemoaning the misuse of the word *coevolution*—specifically, its use in the absence of any evidence of reciprocity—Janzen (1980) suggested that much of what appeared to be coevolution was more properly regarded as coadaptation, the acquisition of traits in independent environments that serve as preadaptations in a new environment. Janzen then redefined coevolution in a narrow sense as genetic change in a population resulting directly from interactions with another population, which, in turn, promote genetic changes in the first population. Thus, for coevolution to occur, genetic variation must exist in the interacting populations in traits relevant to the interaction. He also proposed the term *diffuse coevolution* to emphasize interactions between plants and guilds, or suites, of herbivores that utilize their host plants in similar ways and thus exert collective selection pressure, differentiating this process from species-by-species interactions.

While it was rarely disputed that floral structures and nectar consumed by pollinators evolved in response to selection by these mutualistic partners of plants, it was difficult for many people to accept the idea that insect consumers of foliage, fruit, seeds, roots, stems, and other plant parts could indeed exert a negative selective force sufficiently strong to alter plant chemistry or morphology. Jermy (1976), an outspoken skeptic, enumerated several of the reasons behind his reluctance:

1. "Most of the phytophagous insect species are rare in natural communities. . . . Since they affect only a very small part of the host-plant's population, their rôle as selection factor must be negligible" (p. 110).
2. Plants with many insect associates are common (e.g., Gramineae), so insects cannot possibly have a "noticeable population controlling effect."
3. Plant secondary substances are involved in allelopathy and disease resistance and so cannot be involved in insect resistance.
4. Series of closely related plant species attacked

by closely related insect species should exist (parallel cladogenesis), yet they are not widely documented.
5. "Population densities of most phytophagous insect species is [*sic*] always very low compared to the copiousness of available food. . . . Consequently, competition must be negligible as a factor in the evolution of phytophagous insects" (p. 112).

Jermy proposed the term *sequential evolution* to describe the process by which plants evolve in response to a myriad of selection factors other than herbivores, prompting these herbivores to evolve in response to their changing host plants. Subsequently, Bernays and Graham (1988) resurrected the argument that herbivorous insects do not reach population densities high enough to generate selection pressure on their host plants, attributing the host specificity of herbivorous insects not to the chemical complexity of their hosts, but to selection pressure generated by insect predators. Also casting doubt on the role of stepwise adaptive radiation in generating diversity among angiosperm plants and herbivorous insects was a study of fossil insect remains that suggested that high insect diversity resulted from low extinction rates rather than high origination rates and that familial diversification began 245 million years ago, before angiosperms appeared in the fossil record (Labandeira and Sepkoski 1993).

In recent years, considerable progress has been made in evaluating competing hypotheses relating to plant-insect interactions. Specifically, studies have documented the selective impact of some herbivores on host plant fitness and on the existence of competition within certain guilds of herbivores, the selective impact of plant chemistry on insect fitness, and the presence of genetic variation in traits relevant to interactions between plants and insects. Documentation of congruent cladogenesis is still restricted to only a handful of taxa.

Concepts

Ecological Impacts of Insects on Plants

Because it is important, it is worth belaboring the fact that insects can have a demonstrable ecological impact on plant populations. In an extensive review, Marquis (1992) identified many studies demonstrating intraspecific variation in herbivore damage, intraspecific variation in plant traits associated with the damage, and association between variation in these traits and fitness; these are the conditions required for demonstrating selection by insect herbivores on plants. He highlighted several case studies in which evidence of evolutionary impacts of herbivore on plants was compelling. Among the case studies he selected was his own work with *Piper arieianum,* an evergreen shrub of the lowland wet forest of Costa Rica. Damage levels experienced by these plants vary both within and among populations; cloning experiments confirmed that this variation was at least partly genetically based. In different time periods, negative genetic correlations between growth and total leaf damage per clone indicated that selection for resistance to leaf damage had taken place over the course of the experiment.

Another case study involves the annual morning glory, *Ipomoea purpurea,* and its leaf and floral herbivores. Rausher and Simms (1989) documented directional selection for resistance to two chrysomelid beetles, *Deloyala guttata* and *Metriona bicolor.* Spraying plants with insecticide resulted in a 20% increase in seed production in comparison with plants from the same genetic families in unsprayed paired field plots, a finding indicating that levels of herbivory in the field were sufficient to exert selection.

Yet another of the case studies cited by Marquis (1992) focuses on the introduced European biennial weed *Pastinaca sativa,* wild parsnip, and its introduced oligophagous herbivore *Depressaria pastinacella,* the parsnip webworm. The plant is a monocarpic biennial (i.e., reproducing once in its life, during its second year), and the insect consumes reproductive structures: buds, flowers, and immature fruits. Thus, the potential for a selective impact is great. Parsnip plants contain at least six different furanocoumarins in their above-ground parts, which vary over the course of the season. Berenbaum et al. (1986) established that approximately 75% of the variation in resistance to herbivory in the wild parsnip is attributable to variation in two furanocoumarins in seeds and leaves. Moreover, these chemical traits are at least partly under genetic control, as determined by half-sib heritability analysis. Using quantitative genetics techniques that examine the covariance between relative fitness and a given character to determine whether or not selection has acted on that character, Berenbaum et al. (1986) demonstrated that several of the furanocoumarin traits displayed a selective response to herbivory; moreover, the chemical traits altered by herbivory responded as a result of direct selection on those traits (as opposed to indirect selection resulting from selection on a correlated trait). Indeed, populations with a long history of intense webworm herbivory had higher constitutive and inducible levels of furanocoumarins than did a population with historically lower levels of herbivory, and individual plants in a given population with the highest levels of resistance-associated furanocoumarins were less likely to be attacked by webworms (Zangerl and Berenbaum 1993). These studies, then, satisfy two of Janzen's requirements for demonstrating coevolution: that chemistry confers genetic resistance on certain plant individuals and that insect herbivory can change the genetic constitution of a plant population.

Levels of herbivory need not be consistently high to bring about differential reproduction and survival of individuals within a population. Root (1996) examined variation and cumulative effects of herbivore pressure on goldenrods *(Solidago altissima).* By varying pattern of insecticide application, Root created sets of goldenrods with varying levels of herbivore pressure. These goldenrods responded to herbivory by reducing inflorescence size and number of stems that produced flowers. Over the 6-year period of the study, even low levels of herbivory had cumulative effects, such as reduced flowering. Heavy attack occurred only about 5% of the time, but, when it did occur, it led to serious reduction in vegetative growth and complete inhibition of flowering. Such levels of herbivory are almost assuredly sufficient to exert selection on plant characters via differential reproduction of genotypes.

Whether suites of herbivores act collectively to bring about diffuse coevolution may depend on the system. In a 3-year study of the multispecies herbivore community associated with silky willow, *Salix sericia,* Roche and Fritz (1997) documented significant narrow-sense heritabilities for resistance to two species of caterpillars; *Phyllonorycter salicifoliella* and *Phyllocnistis* sp. However, the genetic and phenotypic correlation structure varied from year to year, and the absence of significant genetic correlations between species argues against any collective selective impact of the sort implicit in discussions of diffuse coevolution.

Evolutionary Responses of Plants to Insect Attack

A necessary component of the scenario by which herbivore selection pressures lead to evolutionary responses of plants is the existence of genetic variation in plant traits associated with resistance. Evidence of such variation, both direct (involving studies within species) and indirect (involving interspecies comparisons), has been found in many systems. Escape in time and space is one route by which plants have responded to insect selection pressures; that is, plants evolve mechanisms that function before feeding begins. This stands in contrast to resistance mechanisms, which function to reduce growth rate, survivorship, or fecundity of insects that do feed on the plants. Escape mechanisms result in the elimination or exclusion of certain plants from an insect's host range. One way to reduce the probability of encounter between plant and herbivore is by temporal displacement. *Quercus robur,* common oak, is the principal host of winter moth *Operophtera brumata.* Females lay their eggs on twigs in the crown; eggs hatch in spring, and newly hatched larvae can feed only on oak buds that have just opened (they cannot penetrate buds or eat leaves a few days old). Defoliation occurs only when egg hatch and bud break coincide, a rare event. Usually, larval mortality is very high due to lack of coincident timing.

Habitat location is another means by which plants escape from insect herbivores. *Phyllotreta* and *Psylliodes* flea beetles, which feed broadly on cruciferous plants, have strong habitat preferences regardless of host plant. Transplantation of *Dentaria diphylla,* a normally shady-habitat plant, into a sunny open field resulted in heavy damage by the flea beetles, all open-field species that never occur naturally on *D. diphylla.* Growing in shade, *D. diphylla* escapes from herbivory from a number of otherwise competent herbivores (Feeny 1977).

Crypsis, or camouflage, is a mechanism of morphological escape. Australian loranthaceous mistletoes, plant parasites, develop leaf shapes that resemble those of their host plants. Over 75% of Australian mistletoe species are mimics of sympatric plants. One species, *Dendrophthae shirleyi,* mimics hosts with three different leaf shapes: flattened, linear-lanceolate leaves that resemble eucalypus and acacias; thick, rounded leaves that resemble mangroves; and linear, compressed leaves that resemble *Casuarina, Greveillea,* and *Eremophila.* Ly-

caenid caterpillars, pierid caterpillars, and leaf-eating possums are thought to be selective agents in the evolution of mimicry in these plants.

Plants can escape from herbivores that orient primarily by olfaction in two ways. They can produce repellent substances; considerable evidence shows that many components of volatile oils can deter herbivores without necessitating physical contact with the plant. Some essential oil components are even toxic via fumigant action. They can also escape detection by failing to produce attractants. Plants that lack a particular chemical "token stimulus" used by the herbivore for orientation can escape detection. Feeny (1977) suggests that this phenomenon accounts for the absence of typical crucifer herbivores on field pennycress, *Thlaspi arvense.* The crucifer allelochemical sinigrin, or allyl-glucosinolate, normally hydrolyzes to allylisothiocyanate, an attractant for flea beetles. In *Thlaspi,* however, it hydrolyzes to allylthiocyanate. When vials of isothiocyanate are placed in the field adjacent to *T. arvense* plants, they are discovered and attacked by *Phyllotreta* at a much faster rate.

Resistance involves mechanisms that reduce herbivory or the effects of herbivory after contact or ingestion. Genetic variability in resistance traits of all descriptions is well documented (Berenbaum and Zangerl 1992). The modus operandi of resistance can be physical or chemical. Physical defenses include hooks, spines, hairs, or sticky glands, which act by immobilizing the insect or by puncturing the body wall so death occurs by slow starvation or loss of hemolymph. In California, populations of jimsonweed, *Datura wrightii,* are polymorphic with respect to leaf vesiture; some individuals produce a majority of glandular trichomes that produce toxins, and others produce a majority of non-glandular trichomes (Elle et al. 1999). Individuals with glandular trichomes are resistant to whiteflies, which cannot colonize the sticky surface; as well, the tobacco hornworm, *Manduca sexta,* cannot grow as rapidly on these plants. In a field test, sticky plants experienced less damage than did plants lacking glandular exudates. Plants with nonglandular trichomes may persist because the glandular trichomes are used to facilitate host plant location by a mirid bug, *Tupiocoris notatus,* which is unaffected by the toxins produced.

Chemical defenses act in either acute or chronic fashion. Acute toxins are immediately lethal and cause high mortality, whereas effects of chronic toxins are manifested over the growth period (e.g., lower fecundity, reduced digestive efficiency). Another distinction can be made in defensive strategy regarding the distribution and abundance of a chemical defense. Allelochemicals can be present at effective concentration at all times (static or constant) or synthesized de novo when the need arises (inducible). The phenomenon of induction has been long known in plant pathology: Pathogenic fungi in contact with a plant promote the production of fungitoxic substances called *phytoalexins*. Similar results have been obtained with insect damage. Gall formation, phenol and tannin production, and decreased protein and mineral nutritional suitability are characteristic responses of many tree species to insect attack. In herbaceous plants, proteinase inhibitors in tomato, cucurbitacins in squash, and terpenes in sweet potato are locally induced by insect feeding damage (Karban and Baldwin 1997). In at least one species, *Pastinaca sativa,* the degree of inducibility of defense compounds (in this case, furanocoumarins) has a genetic component and thus is available for selection.

An explanation offered both for the heterogeneous distribution of plant allelochemicals within a plant and for widespread inducibility of production is that these compounds are metabolically costly to produce; in the absence of selection pressure from herbivory, genotypes producing lower levels of defense compounds enjoy a fitness advantage. Enormous effort has been aimed at documenting such costs of defense. In general, demonstrations of cost (by documenting negative genetic correlations between the resistance trait and fitness in the absence of herbivory) have been successful in systems in which resistance mechanisms are identified and unsuccessful in systems in which the resistance mechanisms have not yet been characterized.

Another form of evolutionary adaptation to insect herbivory is to respond to insect damage simply by outgrowing the damage—compensatory growth and reproduction, so the overall effect on fitness is nil. This form of response is known as *tolerance* and is characterized by variation in the slope of full/half-sib family regressions of natural damage levels on fitness. Using rapid-cycling *Brassica rapa* plants, Stowe (1998) selected for three levels of foliar glucosinolate levels and imposed three levels of damage on these selected lines. A significant interaction between selection line and damage level was found; low-defense selection lines demonstrated less reduction in fitness in response to damage than did both control and high-

defense lines. Thus, individual plants that are less well defended chemically can sustain greater amounts of tissue loss without a reduction in fitness.

Evolutionary Responses of Insects to Plant Defenses: Response of Insects to Plant Escape

Insects have adapted to both plant escape and plant resistance. Adaptation to escape in time is illustrated in the fall cankerworm, *Alsophila pometaria*. Fall cankerworms are geometrid caterpillars that feed on a variety of deciduous trees in eastern North America (Schneider 1980). Eggs overwinter and hatch in spring; larvae feed for up to a month and drop to the ground to pupate. Adults emerge, mate, and lay eggs in November–December. In this set of characters, *A. pometaria* resembles the winter moth *Operophtera brumata*. Like the winter moth, *A. pometaria* suffers greater mortality when egg hatch and bud burst are out of synchrony. Hatching before burst results in starvation, dispersal, and mortality; hatching long after burst results in low fecundity due to decrease in the nutritional quality of leaves. There is, then, considerable selective pressure to synchronize hatch time with bud burst. The way this species has adapted to utilizing temporally variable host plants is that most populations of *A. pometaria* are actually several morphologically identical parthenogenetic clones. Differences in average date of hatch of populations of egg masses are due to differences in the distribution of clones. Oak egg masses hatch later than maple egg masses because the genetically identical oak-associated females oviposit later in the fall than the genetically identical maple-associated females. Timing differences between bud burst and egg hatch are reduced, too, by virtue of female flightlessness; females tend to oviposit on the same tree they grew up on, and trees tend to leaf out predictably at around the same time each year. The offspring, being genetically identical clones, possess the same oviposition preferences possessed by their parents and will in turn prefer the same host tree for oviposition. Thus, flightlessness and parthenogenetic reproduction allow for close, precise tracking of foliation and reduction in efficacy of plant escape in time.

That host plant phenology can act as a selective agent on herbivore populations is also suggested by Filchak et al. (1999), who examined fitness trade-offs among races of *Rhagoletis pomonella,* the apple maggot, on hosts with different phenology. A major constraint on these maggots, which complete development in fallen fruit, is the rate at which these fruits rot. On completion of development, larvae leave fruits and pupate in the soil. Because apple trees produce fruit earlier in the season, maggots associated with these fruits ("apple race") experience less drastic selection to complete development rapidly than the "hawthorn race." As well, because apple flies enter diapause earlier in the fall, they tend to develop more slowly (or experience a deeper diapause) during the winter. These two races are genetically distinct and can be characterized by differences in six allozymes. Exposing hawthorn flies to a longer prewintering period than that experienced by the apple flies leads to changes in allele frequency toward the distribution typical of apple flies, and lengthening the winter period for pupae selects as well for allozyme distributions that resemble those of apple flies.

Response of Insects to Plant Resistance

Insects have overcome plant resistance as well as plant escape. Neither physical nor chemical defenses are absolute, and demonstrations of adaptation to chemical defense abound. There are three primary means of dealing with plant allelochemicals: behavioral, biochemical, and physiological. Behavioral adaptation is perhaps least well documented. Selective feeding is the most obvious mode— simply avoiding plant parts or individuals rich in toxins. The ability to detect deterrents, or to perceive toxic substances at low concentrations and to react with an avoidance response, is highly developed in many generalized feeders. Genetic variation in responsiveness to plant allelochemicals, however, has not yet been demonstrated. For plants with allelochemicals that are activated by sunlight, leaf-rolling behavior has been proposed as a behavioral method for avoiding toxicity. This behavior is light-inducible in the caterpillar *Sparganothis reticulana* when it feeds on phototoxic plants in the Asteraceae (Aucoin et al. 1995). Another example of facultative behavior triggered by host plant defenses is trenching, a feeding behavior that involves cutting through laticifers or other plant vessels so as to eliminate influx of defensive chemicals into a portion of the plant. In the generalist feeder *Trichoplusia ni*, this facultative behavior is linked to the presence of laticifers (Doussard and Denno 1994).

Biochemical resistance is associated with many kinds of detoxification enzymes. Chief among these are cytochrome P450 monooxygenase enzymes. These enzymes convert lipophilic substrates into hydrophilic metabolites by means of oxidation; this conversion often results in reduced toxicity and increased exportability out of the body. Many allelochemicals that are substrates for P450s induce elevated enzyme activity (Brattsten 1992). But not all P450-mediated conversions are detoxifications; some allelochemicals are "bioactivated"— the hydrophilic conversion actually enhances toxicity. This is true for the naturally occurring pyrrolizidine alkaloids. Insects that feed on species containing the pyrrolizidine alkaloid monocrotaline (e.g., *Crotalaria*) do not respond biochemically by inducing P450 activity: for example, *Utetheisa ornatrix,* a specialist, has vanishingly low P450 activity levels. *Spodoptera eridania,* a broad generalist (with over 50 host plant species) has flexible midgut enzymes—induced by plant chemicals in carrot foliage, yet suppressed by chemicals in *Crotalaria* seed diet.

Interspecific comparisons suggest that host plant selection can act to alter P450 activity levels in insects. The black swallowtail *Papilio polyxenes* and fall armyworm *Spodoptera frugiperda* can both break the furan ring double bond in the furanocoumarin xanthotoxin to metabolize it, but the swallowtails, which are specialists on furanocoumarin-containing plants, can break it almost 50 times faster than the armyworms (Berenbaum 1999). In this species, comparisons among female isolines indicate that variation in P450 activity levels (up to fourfold among families) is at least partly genetically based.

Proteinase inhibitors are produced by a number of plants (e.g., Solanaceae, Leguminosae) as defenses against insect herbivores. Insects specialized on such plants can respond to the induction of these inhibitors. Mechanical damage of potatoes leads to elevation of production of cysteine and aspartate proteinase inhibitors; damage by Colorado potato beetle leads to induction of serine proteinase inhibitors. Nonetheless, Colorado potato beetles are unaffected by these plant responses. Although the activity of beetle proteinases declines in the presence of induced plant foliage, these beetles are capable of producing inhibitor-resistant proteinases in response to plant induction of proteinase inhibitors (Bolter and Jongsma 1995). Even generalists may be capable of such responses; *Spodoptera exigua* exhibits up to a threefold increase in induction of inhibitor-insensitive proteinases when reared on induced plants. Thus, induction of gut proteinases that are resistant to plant inhibitors may be a widespread adaptation among herbivorous insects.

Some highly specific detoxification processes add insult to injury and may represent "escalations" in the evolutionary arms race. Over 95% of the free amino acid nitrogen in the legume *Dioclea megacarpa* is contained in L-canavanine, a structural analogue of the amino acid arginine. Incorporated into protein by most insects, L-canavanine is highly toxic. The bruchid weevil *Caryedes brasiliensis*, however, lives entirely within the seeds of this legume. Not only does the arginyl-tRNA synthetase of *C. brasiliensis* fail to "recognize" canavanine and incorporate into protein, but the beetle also has highly active arginase and urease, which catalyze the hydrolysis of canavanine into carbon dioxide and ammonia. The ammonia can be recycled and used for protein synthesis, so the erstwhile toxin becomes a nutrient source (Rosenthal et al. 1982).

Physiological adaptation to plant chemicals has been observed in a number of instances. Tobacco hornworms *(Manduca sexta)* can tolerate the toxic alkaloid nicotine from tobacco plant because, at physiological pH in the gut of the tobacco hornworm, the compound exists primarily in its ionized form; in this polar form, it cannot cross the neural sheath (which is impermeable to ions), where it exerts its toxic effects. Gut pH can also determine to a great extent the ability of dietary tannin to complex with protein; the greater the amount of complexing, the lower the nutritional quality of the food. Insects that feed habitually on tannin-rich plants have on average much higher gut pH (range 8.5–9.0), about the pH at which tannin-protein complexes break down. In yet another example of physiological adaptation, caterpillars of the monarch butterfly have an Na/K ATPase that is insensitive to cardenolides in their milkweed host plants; cardenolides typically interfere with the function of this enzyme. A single amino acid substitution (asparagine to histidine) at a key site may make this enzyme insensitive to the erstwhile toxin.

Sequestration is perhaps the most common means of physiological adaptation to secondary chemicals. Sequestration is the deposition of secondary substances into specialized tissues or glands of an insect. The phenomenon is widespread and is documented from dozens of species from at least seven orders (Rowell-Rahier and Pasteels 1992). A wide range of allelochemicals is involved, including cardenolides, alkaloids, aristolochic acids, mustard oils, and cyanogenic glycosides. Sequestration is a function of both intrinsic physicochemical factors (e.g., size, structure, charge, and stability) and extrinsic biological factors (toxicity, solubility, and metabolizability). The material may be sequestered unaltered (and taken up passively via concentration or pH gradients), or it may be chemically altered in structure (metabolized). Although the process of sequestration can result in toxicosis rather than detoxification, as is the case of incorporation of nonprotein amino acids into proteins, generally sequestration results in detoxification and in an added benefit, insect defense, particularly if the sequestered chemicals retain their toxic properties.

Casual sequestration results primarily from solubility characteristics. Insects tend to accumulate fat-soluble materials in particular because they eat a large percentage of their body weight in food each day and because their bodies have a high fat:water ratio. Oligophagy, or specialized feeding, helps to promote sequestration, but it is by no means a prerequisite. The tiger moth caterpillar *Arctia caja*, for example, sequesters cannabinoids from *Cannabis* and pyrrolizidine alkaloids from plants in the daisy family; the grasshopper *Romalea guttata* sequesters sulfur compounds from onions and nepetalactone, a terpene, from catnip (this insect can even sequester herbicides and incorporate them into defensive secretions).

Genetic variation among individuals in the ability to sequester has not been widely demonstrated. Although variation is well documented to exist among conspecifics in amounts of plant secondary compounds sequestered, the most likely explanation is environmental; that is, consumption of food plants varying in allelochemical content leads to differential sequestration. Rowell-Rahier and Pasteels (1992) suggest that sequestration may arise in lineages, such as the chrysomelid leaf beetles, where autogenous defenses are ancestral, because the production of autogenous toxins is metabolically costly. Costs of sequestration can actually be offset by metabolic utilization of glucose moieties of sequestered glycosides.

In many cases, sequestration results in detoxification and insect defense. The large milkweed bug *Oncopeltus fasciatus* sequesters cardenolides from its milkweed hosts in dorsolateral spaces of the thorax and abdomen—very little is detectable in the hemolymph. Characteristics of the sequestration system include high sensitivity to low concentrations, stochiometric uptake, energy independence, and independent uptake of each component. The cardenolides are not sequestered in composition equal to their occurrence in the seeds—polar compounds and polar metabolites of lipophilic cardenolides are taken up preferentially. On disturbance, *O. fasciatus* releases cardenolides from its bilateral thoracic and abdominal orifices. Cardenolides have known toxic or repellent properties toward vertebrates and invertebrates. In addition to being powerful heart poisons in vertebrates, they are potent emetics when ingested and can cause sensory disturbances. Even invertebrates can be affected by cardenolide ingestion; some mantids taste and discard *O. fasciatus,* on occasion even regurgitating after tasting them.

Oncopeltus fasciatus is strikingly colored orange and black. Many of the other insects strictly associated with milkweeds in North America are conspicuously colored. Some indeed do sequester cardenolides, but many others do not. *Aposematism* is the term used to describe any sort of warning or conspicuous coloration. Not all aposematic species, however, are unpalatable; a palatable species can gain protection from predators by associating with an unpalatable species that it resembles, or mimics. Thus, the model is aposematic and unpalatable; the mimic is aposematic and palatable; and the dupe is the species (usually a predator) that does not distinguish between the two and thus forgoes a meal. This sort of system is referred to as *Batesian mimicry* and illustrates the importance of third-trophic-level selective forces in mediating plant-insect interactions.

The intellectual pitfall of establishing a dichotomous view of regulation of plant-insect interactions—top-down (predator-based) versus bottom-up (plant-based)—is well demonstrated in taxa that rely on sequestered toxins for defense against predators. In these taxa, sequestration in most cases requires adaptation to host plant toxins (i.e., evolution of resistance factors that allow toxins to be stored in the body and deployed in the presence of a predator), as well as evolution of behavioral factors that permit host plant recognition and acceptance. Once such adaptations are in place, selection pressures from predators can maintain host plant associations. Thus, regulation is not dichotomous at all; if anything, a sequential process is suggested.

Perhaps the best example of coevolution reflecting both top-down and bottom-up processes is provided by the phylogenetic study of *Tetraopes* beetles and their *Asclepias* host plants (Farrell and Mitter 1998). These orange and black aposematically colored beetles sequester cardenolides from their milkweed host plants. As larvae, they feed internally on roots; the vast majority of species are host-specific on a single *Asclepias* species. A molecular phylogenetic study of both beetles and plants reveals compelling cladogram congruence, suggesting strongly that reciprocal adaptive radiations have taken place. Moreover, basal *Asclepias* taxa produce cardenolides that are biosynthetically simpler and less toxic than derived *Asclepias* taxa, indicative of the sort of biochemical innovation described by Ehrlich and Raven (1964).

Future Directions

Whether investigators feel that Ehrlich and Raven (1964) were correct in proposing a coevolutionary mechanism to account for both the biochemical diversity of angiosperms and the overall diversity of plants and herbivorous insects is likely associated with whether they might describe a glass of water as half full or half empty. It is certainly true that many of Jermy's (1976) objections to the scenario are still, over two decades later, justified for many taxa; congruent cladograms, for example, are vanishingly rare given the spectacular diversity of these groups. Evidence of competition among herbivorous insects (Denno et al. 1995) and of selective impact of herbivores on plant fitness (Marquis 1992) is growing, but the number of studies actually documenting the genetic component of variation in resistance traits in either plant or herbivore (Janzen 1980) remains small. The apparent desire for a single all-encompassing theory to account for patterns of host plant utilization in herbivorous insects, however, may be misplaced; due to the vast diversity of the interacting taxa, it may well be that different models are applicable to different interactions. It is not at all inappropriate, then, that coevolution, in a broad sense, has contributed not only to the diversity of angiosperm plants and herbivorous insects but also to the diversity of ecological and evolutionary theories relating to interspecific interactions.

Acknowledgments I thank Arthur Zangerl for helpful discussion. Preparation of this manuscript was supported in part by NSF DEB99–03867.

24

Mutualisms

JUDITH L. BRONSTEIN

The unusual behavior of cleaner fish has at-tracted both popular and scientific curiosity since its discovery early in the 20th century. These fish apparently make their living by removing ex-ternal parasites from "host" fishes of other species (some also remove bacteria or diseased and injured tissue). When they approach cleaners, hosts assume an unusual motionless posture that allows cleaners to feed from their scales, from their gill cavities, or even inside their mouths. For their trouble, cleaner fish get a meal, and hosts get a good cleaning. The interaction between cleaner fish and their hosts is generally classified as a *mutualism,* or mutually beneficial interaction between species. Stories about this and other mutualisms have become staples of nature documentaries and the popular literature and have helped lure many students into a lifetime of studying biology.

From the perspective of evolutionary ecology, however, the cleaner-host relationship is anything but straightforward (Poulin and Grutter 1996). First, it is not at all clear that this interaction confers re-ciprocal fitness benefits. Despite several decades of effort, only one study has shown that cleaners sig-nificantly reduce hosts' parasite loads (Grutter 1999), and none has yet demonstrated that reduc-ing parasite loads increases host success. Since cleaners often gouge the host's flesh, particularly when parasites are few, the interaction is often more costly than beneficial. Second, if cleaning does *not* confer an advantage, it is not evident why

hosts should tolerate and even actively solicit cleaners' attention. In fact, sometimes hosts lure cleaners only to eat them, but the conditions under which it might be beneficial for a host to double-cross its cleaners like this remain unexplored. Third, we don't really understand how cleaning behaviors arose in the first place, considering that the first individuals that approached hosts to feed on parasites were very likely eaten. Despite this constraint, cleaning has apparently evolved multi-ple times; it is found in at least five families, in both marine and freshwater species, and in both the temperate zone and the tropics. Finally, the ecological and evolutionary parallels between cleaning in these fish and in certain shrimp, crabs, birds, and ants that also exhibit interspecific clean-ing behaviors remain obscure.

It may seem odd to begin a discussion of mutu-alism with an interaction not yet proven to be one. However, this example serves to highlight three points that I wish to emphasize in this chapter. First, a wealth of observations is available about particular interactions that appear at least superfi-cially to be mutualistic. On the other hand, consid-eration of the "big questions" about mutualisms—the range of conditions under which they confer reciprocal fitness benefits, the processes by which they originate, and the conditions under which they can persist, coevolve, and diversify—has lagged considerably behind. Finally, parallels and differ-ences among different kinds of mutualism have only recently begun to be explored. This chapter summarizes a few of the most significant generalities that evolutionary ecologists have reached so far about mutualism. I emphasize throughout, however, how many questions remain to be investigated.

Studying the evolutionary ecology of mutualism may be a very young enterprise, but it is one with a major role to play in helping to explain the diversity and diversification of life on earth. In tropical rainforests, the large majority of flowering plants depend on mutualistic pollinators and seed dispersers. Deserts are dominated by nitrogen-fixing legumes, lichens dominate tundra habitats, and most northern hardwood forest and grassland plant species depend on mycorrhizal fungi to survive and persist. In the ocean, both coral and deep-sea vent communities are rich with mutualisms; coral itself is the product of a mutualistic symbiosis. Organisms from every kingdom are involved in mutualisms, and certain kingdoms may in fact have originated as a consequence of symbiotic innovations. Other key events in the history of life on earth have been linked to mutualism as well, including the origin of the eukaryotic cell, invasion of the land by plants, and the radiation of the angiosperms. Thus, the study of mutualism cuts across habitats and kingdoms, subsumes levels of organization from the cell to the ecosystem, and involves time scales ranging from the ephemeral to deep evolutionary time.

Historical Context

Humans were familiar with mutualisms long before they had a name (Sapp 1994). For example, Herodotus discussed the cleaning mutualism in which plovers remove leeches from crocodiles' mouths ("The crocodile enjoys this, and never, in consequence, hurts the bird"). Aristotle, Cicero, Pliny, and others repeated this story and added new ones, drawing moral lessons that showed the importance of "friendships" in maintaining nature's balance. The idea that the harmony of society mirrored an underlying harmony of nature persisted until the 19th century. Just as each person was believed to have a preordained role assigned by the Creator, so was each species. Linnaeus, for example, discussed how animals were created to serve plants by both feeding on them and dispersing their seeds.

The well-ordered universe postulated by natural theology began to unravel during the intellectual upheavals of the mid-19th century. Ideas of competition and struggle as forces of progress came to pervade political, economic, and ultimately biological thought, exemplified in the life work of Charles Darwin. However, Darwin also wrote at great length about cooperation, which provided a significant challenge to his contention that organisms do not perform actions strictly to benefit others. Hence, for example, he analyzed the evolution of fruits and flowers, showing how plant traits that benefit animals function first and foremost to increase plants' own reproductive success. He also pointed to cases in which the interests of mutualists could come into conflict, leading cheating behaviors such as nectar robbing to become favored. In showing that mutualisms could emerge in nature strictly by selfish actions, limited by costs and driven by conflicts of interest between partners, Darwin laid the groundwork for our current view of these interactions.

The term *mutualism* was first used in a biological context by Pierre van Beneden, a Belgian zoologist, in 1873 ("There is mutual aid in many species, with services being repaid with good behaviour or in kind, and *mutualism* can well take its place beside *commensalism*." Albert Bernhard Frank and Anton de Bary independently coined the term *symbiosis* a few years later, in an attempt to group physiologically intimate interactions independently of their parasitic, commensal, or mutualistic nature. A tendency began quite soon thereafter to use the terms *symbiosis* and *mutualism* interchangeably. The confusion has continued to the present day. In this chapter, I retain the original usage of these terms: Some but not all mutualisms are symbiotic, and some but not all symbioses are mutualistic.

Many of the best-known mutualisms were first identified during the late 19th and early 20th centuries. By the beginning of the 20th century, hundreds of articles had been published on a wide variety of mutualisms. Despite this growing knowledge of its natural history, however, mutualism was not a prominent concept in either ecology or evolutionary biology until the last quarter of the 20th century. Evolutionists were initially concerned with the genetic basis of individual traits; they did not turn their attention to any form of interspecific interaction until mid-century. Once they did, they focused primarily on antagonistic interactions like herbivory and host-pathogen interactions, whose genetics were already under investigation by applied researchers. Early population ecologists were likewise motivated by (and were primarily funded to study) applied problems, particularly animal population fluctuations. Consequently, predator-prey dynamics and competition came under intense scrutiny while mutualisms were relatively neglected. (I should point out that many other theories have been offered for why the study of mutualism has developed so slowly. Most argue that for various reasons, all or some subset of scientists have not been attracted to studying cooperative phenomena.)

The study of mutualisms began to take off only in the late 1960s, stimulated less by conceptual issues than by a fascination with their complex and sometimes spectacular natural history. We still know far more about the details of particular mutualisms than about what different mutualisms share ecologically or evolutionarily (Bronstein 1994a). This chapter focuses on three sets of concepts that can provide an organizational framework for the study of any mutualism. Readers interested primarily in the natural history of mutualisms should see Janzen (1985).

Concepts

Benefits, Costs, and Variation

One central evolutionary question must be addressed about any form of cooperation: Why should any organism perform actions that will benefit another individual? The conditions under which cooperation may be favored *within* species are explored by Wilson in this volume. Here, I consider cooperative phenomena *between* species. To a large extent, the answer is the same: Cooperation is a selfish act that feeds back in one way or another to benefit the individual itself. Two key differences between interspecific cooperation (i.e., mutualism) and intraspecific cooperation should be kept in mind, however. First, considerations about the relatedness of cooperating partners are irrelevant in the study of mutualism, since partners belong to different species. Hence, mutualism cannot evolve via kin selection. Second, the particular advantage accrued by cooperating often differs radically between the two partner species. For example, plants and their pollinator mutualists exchange commodities as different as transportation and food.

We can start from this simple definition: *If the benefits of a given interaction outweigh its costs for both of two species, that interaction is a mutualism.* The "interaction grid" used to summarize different kinds of interspecific interactions (figure 24.1A) therefore defines mutualism as having a net outcome of plus/plus. It is also useful to think of these interactions in terms of economic exchanges. Mutualists offer their partners commodities that are cheap for them to acquire or produce. In exchange, they receive commodities that would otherwise be difficult or impossible to acquire (Noë and Hammerstein 1995).

While these definitions are a good place to start, they raise a number of difficult questions. What kinds of benefits and costs might be involved? How exactly does one measure these benefits and costs and then weigh them against one another to quantify net outcome? At what temporal and spatial scale might benefits and costs vary in magnitude, and what is the ecological and evolutionary significance of this variation? Little systematic attention has yet been paid to these basic issues. In fact, the vast majority of empirical studies examine mutualism rather qualitatively, and from the perspective of one partner only (Bronstein 1994a).

Benefits of Mutualism

The best-understood aspect of mutualism is the nature of the benefits that partner species can confer, and the usual way to identify similarities among different mutualisms is to group them according to these benefits. Three general classes of benefits appear most common. *Transportation* includes movement of partners themselves—for instance, certain anemones hitch rides to rich feeding areas on the backs of crabs—and also movement of partners' gametes, as when animals transfer pollen between flowers on different individuals. *Protection* can involve direct consumption of enemies, as in the case of cleaners that ingest their hosts' parasites, and also other forms of defense against the biotic or abiotic environment. For example, some endophytic fungi render their host plants distasteful or even fatal to herbivores. *Nutrition* involves the provision of one or more limiting nutrients to the partner. Algae provide photosynthate to their fungal associates in lichens, *Rhizobium* bacteria produce fixed nitrogen for host plants, mycorrhizal fungi provide host plants with phosphorus, and so on. Mutualisms can then be grouped further by looking at which benefits are exchanged (table 24.1; note that some mutualisms appear in two places in this table, because one partner receives two distinct kinds of benefits from its mutualist). This approach has the advantage of grouping interactions that might not otherwise be thought to have much in common. It is by no means an evolutionary-based classification, but it does point to good places to look for convergent evolution among distantly related organisms.

These classifications of benefit are purely qualitative. Until recently, in fact, biologists relied on educated deductions rather than really quantifying the existence and magnitude of benefits. Hard data can surprise us, however. For example, as I have already mentioned, it has yet to be demonstrated that host fish really accrue a fitness advantage from

associating with cleaners. Conversely, a number of interactions believed to be antagonistic were shown to be mutualisms once benefits were finally quantified. One famous example involves the interaction in which *Pseudomyrmex* ants occupy certain tropical *Acacia* trees, feeding on specialized food bodies the tree produces and attacking its herbivores. The mutualistic nature of this interaction was postulated in 1874 by Thomas Belt, who initially observed it in the field. His hypothesis was treated dismissively by ant biologists, however. William Wheeler, the leading myrmecologist of the era, sniffed that considering ants beneficial to plants was akin to considering fleas beneficial to dogs. It took a series of elegant manipulative experiments by Janzen (1966) to resolve this debate in favor of mutualism: When ants were excluded, herbivores

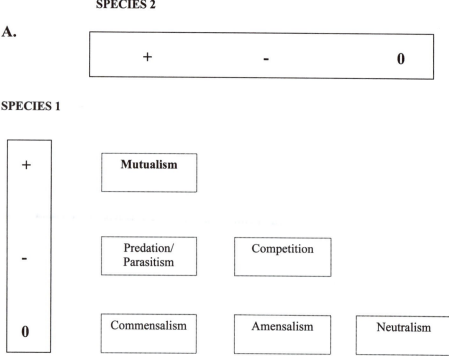

Figure 24.1 Two alternative schemes for displaying the range of interspecific interactions found in nature. (A) The "interaction grid" commonly portrayed in textbooks. Each interaction is classified according to its net effect on each of two species. (B) The "coaction compass" proposed by Haskell (1949). Different interactions grade into one another. Moving away from the center along a radius increases the magnitude but not the direction of the net effect of the interaction for one or both species; moving around the circumference shifts the direction of the net effect for one or both species.

decimated the acacias. Janzen's benchmark research provided the model for subsequent efforts to quantify the benefits of mutualism.

Before determining how to measure benefits, it is essential to decide first what we intend *benefit* to mean. Evolutionary ecologists are generally interested in whether fitness increases in the presence of putative mutualists, and if so, by how much. Hence, one usually compares success of an individual in the presence of partners with success in partners' partial or total absence, as Janzen (1966) did. Various proxies for fitness have been measured, including reproductive success, survival, growth rate, and offspring vigor. (Caveats are appropriate, of course, with regard to using any one of these measures alone to approximate fitness. For instance, an increase in growth rate may lead to a reduction in reproductive output.) By contrast, population ecologists generally choose to define the benefit of

mutualism in terms of increased population growth. For example, Breton and Addicott (1992) demonstrated that ant-tending (protection from natural enemies conferred by ants in exchange for food) increases population growth of aphids, but only at small aphid colony sizes. Unfortunately, an interaction can appear to be mutualistic in a fitness but not a population context, and vice versa. For instance, individuals with mutualists might leave relatively more offspring than those lacking mutualists, but population size may be limited by predation rather than by access to mutualists.

Costs of Mutualism

The costs of mutualism are much more poorly known than its benefits. Table 24.2 lists some currently recognized costs. Note that some of these are incurred regardless of whether an individual

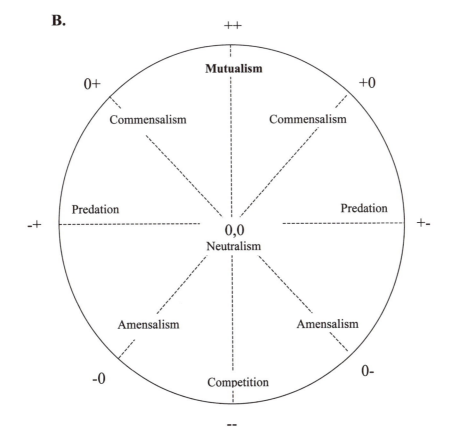

Figure 24.1 *Continued*

Table 24.1 Examples of common exchanges of mutualistic benefits.

Benefit	Transportation	Protection	Nutrition
Exchanged for Transportation	no examples?	Shell epiphytism[a]	Pollination Seed dispersal Shell epiphytism[a]
Protection		Group foraging[b] Müllerian mimicry[c]	Gut symbioses[d] Ant-tending[e]
Nutrition			Gut symbioses[d] Root symbioses[f]

[a]Transportation of certain invertebrates (e.g., anemones) to rich feeding sites on the shells of crabs; invertebrates attack predators, protecting crab.
[b]Reciprocal predator vigilance among individuals in mixed-species bird flocks.
[c]Reciprocal reinforcement of predator-avoidance behavior among effectively defended species that closely resemble one another (e.g., butterflies).
[d]Occupancy of gut of multicellular animal (e.g., mammals and termites) by unicellular organisms able to produce or acquire limited nutrients essential for host; symbiont gains both food and a highly benign environment.
[e]Protection of certain insects (e.g., aphids and lycaenid caterpillars) by ants in exchange for food secretions or excretions.
[f]Provision of limited nutrients to host plants by certain fungi (e.g., mycorrhizae) and bacteria (e.g., *Rhizobium*), which receive fixed carbon in exchange.

successfully locates mutualists, whereas others scale with the number of mutualists or antagonists attracted. For example, ant-tended lycaenid caterpillars experience fixed costs associated with the evolutionary loss of some structural defenses against natural enemies; variable costs can include slower developmental rates and smaller adult sizes when tended (Pierce 1987).

Direct measures of the costs of mutualism remain few. A quick scan of table 24.2 will probably convince readers that some costs, such as the cost of missed evolutionary opportunities, are essentially unmeasurable. The best success has been achieved in measuring the costs of producing rewards, one of the most common kinds of commodities that mutualists exchange on the "biological market" (Noë and Hammerstein 1995). The production of floral nectar, for example, has been calculated to cost 3% of a plant's total energy investment in *Pontaderia cordata* (Harder and Barrett 1992), 4–37% of photosynthate assimilated during flowering in *Asclepias syriaca* (Southwick 1984), and 50% of potential seed set in *Blandfordia nobilis* (Pyke 1991). Note the lack of consistency in the currencies used to measure costs. This inconsistency undoubtedly obstructs our ability to compare costs among different interactions and to learn why sometimes (as in the case of nectar) the same kind of cost differs so dramatically across

species. Furthermore, note that only Pyke (1991) measures costs in the same currency as benefits (number of seeds), an approach that greatly aids calculation of the net effects of mutualism.

While it is admittedly very difficult to pinpoint the costs of mutualism, there are at least four reasons why it is worth the effort. First, costs help us to understand traits that might appear puzzling if we considered mutualism as involving benefits only. For example, many species exhibit adaptations to defend themselves against antagonistic aspects of their partners' behavior, not only adaptations to attract those partners (e.g., Pierce 1987). Second, costs shed light on why mutualisms often do *not* occur at times and places and in species where one might have expected them. Certain tropical plants, for instance, rely on chemical-based defenses rather than defense by aggressive ants under nutrient conditions in which production of the food bodies on which these ants feed is excessively expensive (Davidson and McKey 1993). Third, the sensitivity of costs to ecological conditions is one major reason why the net effects of mutualisms vary so much across space and time (see below). Finally, we will see later in this chapter that the costs of mutualism help explain why "cheater" individuals and species, which gain the benefits of mutualism but invest nothing in providing commodities for exchange, arise so often.

Table 24.2 Some of the costs of mutualism.

Costs of failing to locate mutualists (or the right num-
ber of mutualists)
  Failure to survive
  Failure to grow
  Failure to reproduce
Costs of producing rewards
  Physiological costs
  Diversion of resources away from growth and sur-
  vival
Costs of visiting partner
  Physiological costs
  Missed-opportunity costs
Costs of antagonistic behaviors of mutualists
  Risk of being consumed by mutualist
  Risk of being damaged by mutualist
  Payment in offspring or gametes required to receive
  service
  Risks of having other mutualists deterred
Costs of nonmutualists
  Loss of rewards to nonmutualists
  Competitive exclusion of mutualists by nonmutualists
Evolutionary costs
  Reduced ecological breadth
  Reduced range
  Susceptibility to coextinction

Variation in Mutualism

The net benefits that any given mutualism confers
on the participants almost always varies in space
and time. For example, Cushman and Whitham
(1989) have shown that the net effect of ant-tending on one membracid (treehopper) species depends
on the local abundance of its predators. In a high-
predator year or location, membracids are decimated if not tended by ants. However, at times and
places where predators are fewer, the interaction
with ants is less mutualistic or even commensal.
That is, ants benefit just as much, but membracids
experience increasingly lower and ultimately no
benefits from being tended. In this case, variation
in the magnitude of the benefits of mutualism generates variation in its net effects. Net effects of ant-
tending can also vary with shifts in the magnitude
of costs. For example, under a fairly predictable set
of ecological conditions, ants will consume rather
than protect the aphids they tend (Sakata 1994).
This and many other interactions can therefore
range fully from mutualism through commensalism and into unilateral benefit, or antagonism.
This phenomenon has come to be called *condition-
ality* or *context dependency*.

Here is another example to point out the rich
variation that can be found within a single mutualism. Arbuscular mycorrhizal fungi increase the
availability of soil phosphorus for the host plants
with which they associate; in exchange, they receive fixed carbon from their hosts. (They also appear to protect hosts to some degree from pathogenic root fungi.) When hosts are fertilized with
phosphorus, or when they grow in phosphorus-
rich habitats, the cost of feeding the mycorrhizae
apparently comes to exceed the benefits they provide. In response, plants reduce their mycorrhizal
infections and may even exclude mycorrhizae from
their roots (Smith and Smith 1996). The net effects
of mycorrhizae on plants also may vary with ambient ecological conditions, including nutrient and
water stress, soil pH, crowding, and identity of
neighboring species, as well as with life history
stage.

Context dependency can affect both distributions and population sizes of species that rely on
mutualists (Bronstein 1994b). Its evolutionary implications are both more subtle and less investigated than its ecological consequences. However,
they are no less important: "Just as variation in
traits in populations is the raw material for the
evolution of species, variation in outcome is the
raw material for the evolution of interactions"
(Thompson 1988, p. 65). In this volume, Thompson discusses how geographic variation can influence the coevolutionary dynamics of mutualism.
Temporal variation is likewise important evolutionarily. For example, it has been argued that specialization can arise in mutualisms only when there
is long-term consistency in the magnitude of benefits that alternative partner species confer. When
the identity of the best mutualist varies over time,
however, generalization is expected (Thompson
1994).

The simple interaction grid shown in figure
24.1A reflects none of the variation in interaction
outcome that I have discussed here. In the case of
mutualism, net effects are simply categorized as
plus/plus. This is one of many weaknesses of this
highly typological view of interspecific interactions. A variety of alternative schemes have been
proposed in recent years that attempt to capture
the kinds of variability that interactions actually
exhibit. Perhaps the most successful alternative,
however, is a largely forgotten one proposed over
50 years ago by the same sociologist who conceived of the interaction grid itself, Edward Has-

kell (1949). Within his "coaction compass" (figure 24.1B), different forms of interaction grade smoothly into one another as net effects shift for one or both partners. Using this kind of scheme, one can depict fairly explicitly many different ways in which an interaction can vary in outcome over space or time. Figure 24.2 illustrates three hypothetical examples of this technique. Note that this figure maps interactions onto Cartesian coordinates rather than

onto a circle, adding a certain degree of quantitative clarity.

Cheating and the Maintenance of Mutual Benefit

Natural selection can be expected to act to increase the benefits each partner receives from mutualism, while reducing the costs it experiences. This straight-

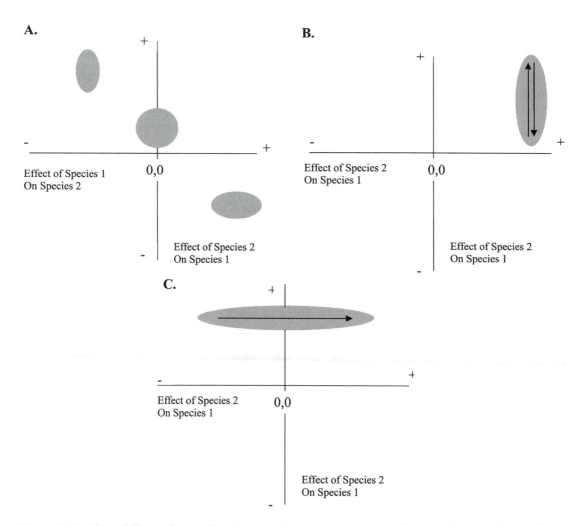

Figure 24.2 Three different forms of variation in interaction outcome. Here, variation is displayed on Cartesian coordinates rather than a compass, for quantitative clarity. In each case, the shaded area represents the region in which the net outcome of the illustrated interaction can fall; the arrows indicate temporal trajectories. (A) An interaction with distinctly different outcomes in three contrasting habitats. (B) An interaction that ranges from mutualism to commensalism; the net effect on species 2 is consistently positive, but effects on species 1 shift seasonally from positive to neutral. (C) An interaction that has shifted from strongly parasitic to strongly mutualistic over evolutionary time.

forward observation turns out to have fascinating consequences for the dynamics of mutualism, because an organism able to reduce its own costs often delivers lower benefits to its partners as a consequence. For instance, we can imagine that a pollinating butterfly that spent less time in a given flower collecting the same amount of nectar compared to a conspecific would be a more efficient forager, and hence at an evolutionary advantage. However, an incidental consequence of shorter visits may very well be the deposition and removal of less pollen, placing the plants this butterfly visits at a *dis*advantage. In the extreme case, an animal might forage on nectar most efficiently by bypassing the sexual parts of the flower entirely. When this happens, the plant can lose more than its energetic investment in nectar; it can lose its attractiveness to more beneficial visitors as well.

More generally, unless an organism has an immediate evolutionary interest in the well-being of its partner—something thought to be extremely rare in nonsymbiotic mutualisms—*selection should favor individuals that obtain what they require from other species while investing as little as possible in them.* If this process goes unchecked, mutualisms should shift toward unilateral exploitation, or antagonism. Compounding this problem, entire species with no history of providing benefits may be able to reap valuable commodities produced for mutualists. Hence, mutualisms should be easily invasible by individuals that "cheat," whether cheating arises as an alternative strategy within a mutualistic species or appears when nonmutualistic species invade the interaction.

In light of these risks, how can mutualisms possibly persist evolutionarily? This is an open question currently under very active debate, and no single answer has yet proved satisfying (Bronstein 2001). Three main hypotheses have been discussed.

First, cheating might simply not arise very often. Connor (1995), for example, argues that most mutualisms emerge from interactions that generate purely incidental ("by-product") benefits. In these cases, benefits are invested in minimally if at all, so that the costs of delivering those benefits are negligible. The advantage of cheating, and hence the risk of being cheated on, would therefore be slight. In fact, however, one of the few phenomena common to virtually all forms of mutualism, regardless of their natural history or evolutionary origins, is the presence of cheaters. For example, plant-polli-

nator mutualisms are exploited by diverse animal taxa that collect nectar but do not pollinate (Inouye 1983). Some of these are species that exclusively rob nectar, while others are individuals within generally mutualistic species that pursue either pure robbing or mixed robbing-pollinating strategies. Conversely, some plants, including many orchids, entice pollinators to visit but provide them with no reward. Ant-tending mutualisms are afflicted with ants that eat rather than tend Homoptera (Sakata 1994) and by lycaenid caterpillars that eat rather than reward ants (Pierce 1987). Host fish may eat cleaners, and cleaners may graze on their hosts; other fish species mimic the cleaners but feed exclusively on host tissue. Plants, as we have seen, can exclude mycorrhizae from their roots when the costs of maintaining them come to exceed the benefits they deliver. Phylogenetic studies are beginning to show that certain cheater species have had long evolutionary associations with mutualisms (e.g., Pellmyr and Leebens-Mack 2000). Hence, the presence of cheaters must be considered a central feature of mutualisms.

The second hypothesis acknowledges that cheaters can have a strong advantage and arise often. It suggests, however, that individual partners that cooperate can be rewarded, while individuals that cheat can be punished or have the benefits of mutualism withheld from them. Bull and Rice (1991) proposed the existence of two such recognition mechanisms, partner fidelity and partner choice. Partner fidelity is relevant when two individuals live in close association and interact repeatedly during their lifetimes. In this case, behaviors of one species toward its partner can feed back immediately to affect its own success. For instance, actions of vertically transmitted symbionts that harm hosts can rebound to reduce symbiont survival (since the host is also the symbionts' environment) as well as the likelihood of successful symbiont transmission from host parent to offspring. Cheating in this case is essentially self-regulated, because it ceases to be advantageous to the cheater itself.

Partner fidelity does not shed much light on cheater control in cases where partners are numerous and free-living and the interactions themselves relatively brief, as in most pollination and protective mutualisms. Bull and Rice (1991) argue that mutualisms of this sort can persist only if there are mechanisms for partner choice, wherein individuals can compare the quality of potential partners and then choose with whom to associate and/or how long the association will last. The best example of partner choice involves yucca plants and their obligate yucca moth mutualists. These insects both pollinate and lay eggs in yucca flowers; the caterpillars feed on developing seeds. Individuals of some yucca species are able to recognize and selectively abort developing fruits that contain particularly high numbers of the destructive caterpillars (Pellmyr and Huth 1994).

In most mutualisms, however, there is minimal evidence of traits promoting either partner fidelity or partner choice (Bronstein 2001). In many cases, evolutionary constraints clearly preclude their existence. For example, partner choice models imagine that individuals can respond quickly enough to the actions of their partners so that reward or punishment can be doled out before the interaction is severed. This assumption emerges from game theory models of within-species cooperation, in which cooperation and defection are seen as alternative behaviors available to a pair of similar, cognitively advanced organisms. It is probably much less appropriate, however, for thinking about between-species cooperation, which commonly brings together organisms from different phyla and which often involves the exchange of commodities such as nectar that must be invested in before partners even appear. The yuccas described above have a unique ability to defer their decision to reward or punish their partners until some time *after* the original encounter: They can punish their partners' offspring instead. For the vast majority of mutualisms, however, once the encounter with a partner is over, it is too late to do anything to reduce the costs of having been cheated.

A final possibility for why cheating does not extinguish mutualism is that adaptations are present that effectively protect mutualisms from the impact of cheaters. For example, many floral traits, such as long and/or thickened corollas and nectar toxins, are thought to reduce attack by nectar robbers (Inouye 1983). While many of these traits may in fact function in this context, we can often identify common cheaters able to cope quite well with these apparent adaptations. In fact, certain cheater species show elaborate adaptations for circumventing apparent defenses against them. Letourneau (1990) has described one such cheater on a tropical ant-plant mutualism. The plant, a species of *Piper*, produces costly food rewards when it detects the presence of its specific ant-defender species. However, a clerid beetle sometimes invades the plant and destroys the ant colony. The beetle then somehow fools the plant into continuing to produce food bodies for it.

Defenses against cheaters are likely to favor the evolution of adaptations by cheaters to overcome them, much as plant defenses against herbivory have selected for diverse herbivore counterdefenses. Thus, coevolutionary races between mutualists and cheaters are possible, similar to those in better-studied antagonistic interactions (see chapters 19–22). Interestingly, under certain conditions this kind of escalation should feed back to alter the coevolutionary dynamics of mutualism itself. These processes remain virtually unstudied.

Coevolution, Specificity, and Intimacy of Mutualism

Calling an interaction mutualistic does not imply that it has undergone or is currently undergoing coevolution. In fact, to limit the concept of mutualism this way would eliminate many interactions that exhibit all of the characteristics that I have discussed so far (benefits that generally exceed costs, benefits and costs that vary in space and time, and the problem posed by cheating). Consider the range of evolutionary histories exhibited by plant-pollinator interactions. The obligate, species-specific relationship between figs and fig wasps clearly has undergone extensive coevolution (see below). In contrast, most plant-pollinator interactions are the product of selection on the plant only. In essence, plants have evolved traits, including rewards like nectar and attractants like colorful petals, that allow them to co-opt existing animal foraging behaviors for their own benefit. Other plant-pollinator interactions have involved no detectable evolutionary change on the part of either partner. Some of the most interesting are cases in which one partner is an invasive or introduced species. If the alien is similar enough ecologically to a resident species' mutualists, it may be able to join or replace them. Honeybees, introduced to the Western Hemisphere from Africa around 400 years ago, now visit the flowers of diverse native plants; they generally do confer some pollination service, although they tend to be relatively poorer mutualists than native visitors. Mutually beneficial interactions can also arise between two invaders from different regions of the world, as occurs in North America when honeybees pollinate weeds of European origin. Episodes of evolution and coevolution may follow the formation of these novel mutualisms, however, at least in theory. These evolutionary phenomena would have potentially serious implications for efforts to conserve native species and native interactions (see also Hedrick, this volume).

We can find a similarly broad array of evolutionary histories represented in many forms of mutualism. This observation highlights the fact that subsets of mutualisms can easily be identified that are characterized by phenomena that cut across particular benefits that partners exchange (the method by which we grouped examples earlier in this chapter). For example, regardless of the nature of the benefits, the large majority of mutualisms are rather generalized. That is, each species can obtain the commodities it requires from a wide range of partner species (Thompson 1982, 1994). Furthermore, most mutualisms are facultative: At least some of the commodity acquired from mutualists can be obtained from abiotic sources, a possibility allowing a degree of success even in the total absence of mutualists. However, many extremely specialized mutualisms do exist. They are species-specific (i.e., there is a single mutualist species that can provide the necessary commodity) and may be obligate as well (i.e., individuals have zero fitness in the absence of mutualists). The degree of specificity is often asymmetrical within a mutualism (Thompson 1994). For instance, many orchid species can be pollinated by a single species of orchid bee, whereas these bee species visit many different orchids as well as other plants.

Thompson (1982) argued that the evolutionary origins of highly specialized mutualisms and more generalized mutualisms are somewhat divergent. Species-specific obligate mutualisms, he contended, tend to originate as species-specific obligate parasitic interactions. In contrast, generalized mutualisms tend to evolve from grazing interactions, in which one species inflicts a low level of harm on a variety of other species by feeding on them. This has selected for prey traits to reduce the costs of, or ultimately even to benefit from, the grazers' actions. For example, lycaenid caterpillars show diverse chemical adaptations to appease ants, allowing lycaenids to direct ant aggression toward their enemies rather than toward themselves (Pierce 1987). These theories argue that the degree of specificity and obligacy of an interaction may often persist for long periods of evolutionary time, even as the net outcome (from antagonism to mutualism) shifts dramatically. One important implication, to which

I will return at the end of this chapter, is that it is probably artificial to study mutualism as a phenomenon fundamentally different from the antagonistic interactions described in earlier chapters of this book.

I will mention one more way that mutualisms can fruitfully be divided up. In symbiotic mutualism, there is intimate physical contact and often physiological integration between the partners for most or all of their lives; all other mutualisms are nonsymbiotic. Most mutualistic symbioses involve either one-way or reciprocal transfer of nutrients between partners. Well-known examples include mutualisms between plants and *Rhizobium* bacteria, plants and mycorrhizal fungi, animals and their gut flora, corals and algae, and the fungi and algae that form lichens. The unique physiology of these interactions has been under scrutiny for well over a century; their evolution and ecology, on the other hand, have only recently begun to attract attention. Many mutualistic symbioses show clear signals of having originated as parasitic symbioses, providing evidence for Thompson's (1982) argument presented in the previous paragraph, as well as for Bull and Rice's (1991) partner fidelity model for the maintenance of cooperation discussed earlier. A number of evolutionary problems are unique to this group of mutualisms. For example, there has been intense speculation about the conditions under which hosts and symbionts might lose their autonomy to become a single integrated organism. It is now well accepted that both mitochondria and chloroplasts originated by this process. Current evidence suggests that each of these events occurred only once, the implication being that the conditions favoring such events are exceedingly restrictive. However, these evolutionary innovations have clearly changed the face of life on earth.

Case Study

The Fig Pollination Mutualism

The mutualism between monoecious fig trees (about 300 of the 750 *Ficus* species) and their pollinator wasps (family Agaonidae) provides a good illustration of the main points I have covered in this chapter. It is not at all a typical mutualism, since it is obligate, largely species-specific, and tightly coevolved. However, this extreme mutual dependence generates a number of advantages for investigating how mutualisms function, evolve, and persist (Bronstein 1992; Anstett et al. 1997; Herre 1999).

Most fig species are pollinated exclusively by a single species of fig wasp, which in turn is associated with a single fig species. The mutualism is summarized schematically in figure 24.3. Female fig wasps are attracted to the trees by species-specific volatile attractants released by receptive inflorescences. The wasps enter the enclosed fig inflorescences or syconia (illustrated in figure 24.4), spread pollen within them, and then deposit their eggs in a subset of the hundred or more flowers. Females generally die at this point, trapped within the fig. Each offspring feeds for the next several weeks within a single flower, at the expense of one developing seed. When the wasps and seeds are mature, the wasps mate, still within their natal fig. Females then collect pollen and depart through an exit hole chewed by the males, in search of an oviposition site. Trees within a fig population generally flower in tight within-tree synchrony, but out of synchrony with each other, forcing the wasps to depart their natal tree in order to find receptive inflorescences. Adult lifespan is exceedingly brief and survival during the adult dispersal phase extremely low, but it is evidently sufficient to have made figs ecological dominants in tropical and subtropical habitats worldwide.

This interaction provides an unusually good model for studying the net effects of mutualism, since both costs and benefits to the fig can be measured in the same currency. Specifically, net seed production is about equal to the number of seeds a fig initiates as a consequence of pollen transport by the wasps (the benefit), minus the number of seeds consumed by the wasps' offspring (the cost).

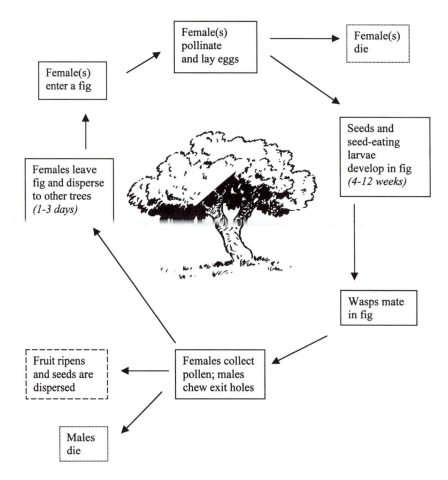

Figure 24.3 The pollination mutualism between fig wasps and monoecious fig trees.

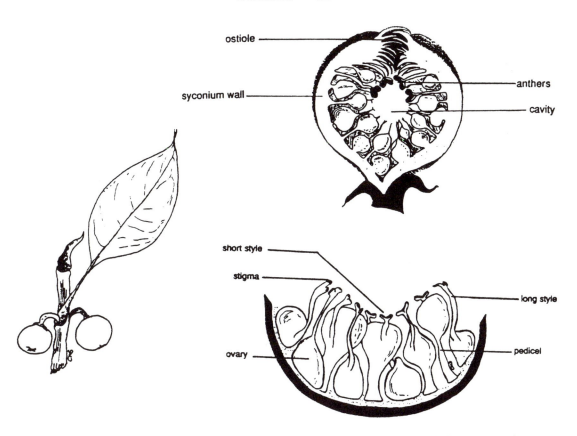

Figure 24.4 Structure of the inflorescence (syconium) of a monoecious fig tree. Pollinator females enter the syconium through the tight, bract-covered pore (the ostiole); they generally become trapped inside after pollinating and laying eggs. In contrast, most nonpollinator wasps lay eggs through the syconium wall, directly into the ovaries. Note that both female and male flowers are present in the syconium; female flowers are receptive several weeks before the anthers have matured. Female flowers vary greatly in style length, a trait often believed to regulate conflicts between figs and wasps. (From J. L. Bronstein, *Insect/Plant Interactions Volume IV,* 1992, E. A. Bernays, ed. Reproduced by permission, CRC Press, LLC.)

This cost is extremely high: Half or more of the seeds are commonly consumed. Both these costs and the associated benefits do vary substantially, however. For instance, entry by relatively few pollinators into each fig generally confers a large net positive effect on the plant: Each wasp carries a large amount of pollen, so seed initiation is high, but seed consumption is fairly low. As more and more pollinators enter a given inflorescence, however, the cost of seed consumption continues to rise, while the benefits begin to saturate because all ovaries have received pollen. Hence, late-entering wasps function more as fig antagonists than as mutualists.

The production of intact seeds by figs does not appear to be in the short-term evolutionary interests of the wasps. (Put in terms of figure 24.3, the fig wasp life cycle is not broken if there are no intact seeds left in the fruits from which pollen-loaded females depart.) Rather, seed production is an incidental consequence of the fact that wasps don't mature from many of the fig flowers that have received pollen. If a mutant wasp arose that were able to parasitize every flower successfully, it would be a pure seed predator rather than a mutualist (despite transferring a great deal of pollen between plants). In the short term, mutants should enjoy a significant advantage, since they would leave far more offspring than less exploitive wasp lineages. This is a particularly intriguing evolutionary problem because fig wasps undergo perhaps 300 generations to every generation of the fig. Hence, if a trait permitting increased exploitation arose, it could sweep through the wasp population before a single episode of selection had taken place in the fig. In the long run, of course, this behavior would be as disastrous for the wasp as for the figs. If fig seed production ground to a halt, fig populations would dwindle to the point where the wasps themselves could not persist.

It is possible that lineage selection has taken place; that is, the only fig–fig-wasp mutualisms present today might be those in which such evolution toward extinction has not yet occurred or is not yet complete. Alternatively, figs may have evolved adaptations to control this kind of overexploitation. A variety of candidate adaptations have been suggested in recent years. The most widely known of these is the style-length polymorphism that can be easily observed within fig inflorescences (figure 24.4). A fig wasp lays an egg only when her ovipositor can extend down the full length of a style and into the ovary below. Some flowers have style lengths clearly exceeding the length of the wasp ovipositor. It was long believed that these "long-styled flowers" are protected from oviposition, a protection guaranteeing that at least some uneaten seeds would always mature. However, this hypothesis has fared badly in recent years and has largely been set aside. Essentially, data from a number of species show that wasps have access to the large majority of fig flowers, but that seeds nevertheless develop in many of them. Much remains to be discovered about how this and a variety of other evolutionary conflicts between partners in this mutualism are mediated (Anstett et al. 1997; Herre 1999). The challenge, of course, is to explain how, despite the easily quantifiable conflicts of interest between figs and fig wasps, their mutualism has managed not only to persist since the mid-Cretaceous, but also to explosively diversify and spread worldwide.

Figs are exploited not only by their mutualists; a variety of other wasps oviposit within inflorescences, mostly through the outer fig wall, but they never transfer pollen (Herre 1999). These wasps, many of which are as species-specific as the pollinators themselves, are ancient associates of the mutualism, having radiated in parallel with it. Over 30 species of nonpollinators have been found associated with a single fig species; they can make up the large majority of wasp individuals developing within an inflorescence. At least some of them can inflict major costs on the mutualism by interfering with maturation of seeds and/or pollinator offspring. It is not obvious how figs limit the impact of nonpollinators. Inflorescences of some fig species have extremely thickened walls, apparently protecting the ovaries from externally ovipositing wasps. However, the nonpollinator species that specialize on these figs have exceptionally elongated ovipositors, allowing them to gain access to the ovaries quite successfully. Here, we may be seeing a coevolutionary arms race in progress.

The fig pollination mutualism is highly unusual, but it does share features with a wide array of other interactions. In particular, striking similarities have long been noted between the independently derived fig and yucca pollination mutualisms. The yucca–yucca-moth interaction is summarized schematically in figure 24.5. A comparison with figure 24.3 will make clear the many points of ecological similarity between the two mutualisms. Both involve a highly specialized insect that exhibits elaborate pollination behaviors in association with a single host plant species. Both insects then lay enough eggs to destroy a significant proportion of the seeds that their actions have initiated. Both mutualisms appear to have originated when highly specific seed predators acquired pollination behaviors, guaranteeing that their offspring would not starve during development. The ancestral fig and yucca flowers were presumably open and accessible to diverse visitors, but both now exhibit floral modifications that exclude alternative pollinators. The evolutionary mystery is why and how each of them eventually closed to exclude all but a single species, and an extremely costly one at that. These questions are currently much better resolved for yuccas than for figs (Pellmyr and Leebens-Mack 2000), in part because the phylogeny is better understood. (There is an order of magnitude fewer yucca than fig species, distributed on a single continent rather than worldwide.) For example, studies of *Lithophragma*, a saxifrage whose open inflorescences currently attract a pollinating seed predator closely related to yucca moths as well as more typical pollinating insects, has been shedding light on why yuccas might have evolved to severely restrict pollinator access (Pellmyr et al. 1996; see also Thompson, this volume). This work is likely to generate

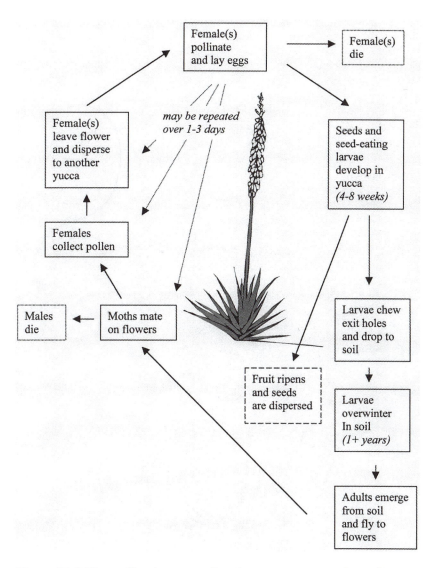

Figure 24.5 The pollination mutualism between yucca moths and yucca plants.

hypotheses for how this process might have occurred in the ancestral fig as well.

The differences between the yucca and fig pollination mutualisms are equally intriguing. In particular, yuccas exhibit an array of traits that allow them a fair degree of control over the degree to which their mutualists exploit them. Parallel traits in figs have proven elusive. For instance, unlike yuccas, figs abort almost no developing fruits at all, let alone ones that have been overexploited by their mutualists (Bronstein 1992). One simple constraint may help us understand this difference. After fig wasps have matured and mated, adult females immediately depart with pollen in search of another inflorescence (figure 24.3). Therefore, a fig that reduced wasp survival in any way would immediately reduce its own success as a pollen donor. In contrast, once yucca moth larvae have finished consuming yucca seeds, they drop to the soil and remain there for a year or more (figure 24.5). They appear no more likely to transfer pollen of their natal yucca than of its neighbors when they emerge as adults. Hence, punishment of exploiters does not carry the same kind of reproductive cost for yuccas.

Future Directions

A perennial risk of discussing mutualism is how easy it is to emphasize what we still don't know about it, at the expense of clarifying what has already been learned. Hopefully, in this chapter, I have been able to make both points.

Decades of natural history studies have generated a rich source of information for investigating conceptual issues about mutualism (Bronstein 1994a). It is particularly useful for detecting broad ecological patterns: the nature of benefits and costs and how they vary in space and time, the levels of specificity, obligacy, the intimacy exhibited by different kinds of mutualism, and the kinds of cheating present within them. To address evolutionary questions, however, more carefully directed and designed studies are necessary. Great thought still needs to be given to how benefits and costs should be defined and measured in a fitness context. Fur-

ther phylogenetic studies are needed to illuminate evolutionary transitions between mutualism and antagonism and between specificity and generalization, and to reveal which mutualists share a coevolutionary past. The genetics of traits that regulate mutualism remain largely unexplored.

It is important to realize, however, that none of these gaps in our knowledge is really unique to *mutualistic* interactions. In fact, if there is a takeaway lesson from this chapter, it is that in many ways, mutualisms are not a class of interactions fundamentally distinct from antagonisms. Many, if not most, had their evolutionary origins in antagonisms and have the apparent potential to revert to antagonisms. Many, if not most, involve significant costs, often as a consequence of antagonistic aspects of their mutualists' actions. Geographic variation seems as ubiquitous, and as evolutionarily significant, as in antagonisms (see also Thompson, this volume). These ideas are perhaps best exemplified in the coaction compass (figure 24.1B), in which different forms of interaction smoothly grade into one another. The barriers that have separated the study of different forms of interspecific interactions may ultimately dissolve as well.

25

The Geographic Dynamics of Coevolution

JOHN N. THOMPSON

Coevolution is reciprocal evolutionary change in interacting species driven by natural selection. It is a pervasive evolutionary process that has shaped many of the major events in the history of life, including the origin of the eukaryotic cell, the origin of plants, the evolution of coral reefs, and the formation of lichens, mycorrhizae, and rhizobia, all of which are crucial in the development of terrestrial communities. Just as important, evidence is increasing that coevolution is an important ongoing ecological process, continually shaping and reshaping interactions among species, sometimes over time spans of only a few decades.

This chapter is an evaluation of coevolution as an ongoing process shaped by the geographic structure of interactions among species. It is an analysis of what we have learned recently as we have taken a broader geographic view of how coevolution continually remolds the relationships among taxa. The first mathematical models of geographically structured coevolution were developed only in the past few years, and there are still fewer than a dozen empirical studies that have analyzed any aspects of coevolutionary structure and dynamics across geographic landscapes. Nevertheless, these theoretical and empirical studies have together suggested that coevolution is very likely a much more dynamic process than suggested by the previous several decades of study in evolutionary ecology.

Concepts

The Geographic Mosaic Theory of Coevolution

Coevolution is a hierarchical process. Local populations of species interact with one another and sometimes coevolve. These local populations are in turn connected through gene flow to populations in other communities, and this geographic structuring adds another level to the coevolutionary process. Local geographic clusters of populations may show metapopulation dynamics, and yet broader geographic groupings of populations may show considerable genetic differentiation in the traits of interacting species. Only a subset of locally or regionally coevolving traits will eventually sweep through all populations. Hence, coevolution as seen in comparisons of interacting phylogenetic lineages will show only a small fraction of the coevolutionary dynamics found at the population, metapopulation, and broader geographic scales.

Within this hierarchical structure of coevolution, many of the dynamics may occur above the level of local populations and below the level of the fixed traits of species for three reasons: Many species are collections of genetically differentiated populations, the outcomes of species interactions commonly differ among communities, and interacting species often do not have identical geographic ranges. The single clearest result from the

331

past 30 years of research in population biology and, more recently, molecular ecology is that most species are geographically structured. These kinds of phylogeographic results suggest that coevolving species may be constantly tugged and pulled in different directions across geographic landscapes. Any theory of coevolutionary dynamics must therefore take into account this geographic structuring of most taxa and interactions.

For example, in the northern Rockies the prodoxid moth *Greya enchrysa* is attacked by a specialist braconid wasp *Agathis* n. sp. Analyses of geographic differentiation in these two species using mitochondrial DNA sequence data indicate that each of them is a cluster of geographically differentiated populations (Althoff and Thompson 1999). Moreover, the two species differ in how their populations are geographically clustered (figure 25.1). Consequently, these two species may differ in where and how they show local adaptation to each other across the geographic range of the interaction. Another interacting pair of species in the same genera—*Greya subalba* and *Agathis thompsoni*—also show differing structures across the same geographic landscapes, and the composite structures differ somewhat from those seen in *G. enchrysa* and *Agathis* n. sp. (Althoff and Thompson 1999).

If interspecific interactions commonly show this kind of geographic structuring, then we should expect that they will exhibit three biological properties that together comprise an evolutionary hypothesis of the general structure of the coevolutionary process: (1) There is a selection mosaic among populations, favoring different evolutionary trajectories in different populations; (2) there are coevolutionary hotspots, which are the subset of communities in which reciprocal selection is actually occurring; and (3) there is a continual geographic remixing of the range of coevolving traits, resulting from the selection mosaic, coevolutionary hotspots, gene flow, random genetic drift, and the extinction and recolonization of populations. Across landscapes, selection may differ slightly and clinally among communities, or it may differ greatly (e.g., antagonism *versus* mutualism) in a patchwork pattern (figure 25.2).

This tripartite evolutionary hypothesis in turn suggests three ecological predictions: (1) Populations will differ in the traits shaped by an interaction; (2) traits of interacting species will be well-matched in some communities and mismatched in others, sometimes resulting in local maladaptation; and (3) there will be few coevolved traits that spread across all populations, because few traits will be favored across all communities.

This tripartite evolutionary hypothesis and its predictions, called the *geographic mosaic theory of coevolution*, are amenable to modeling and experimental studies within natural populations (Thomp-

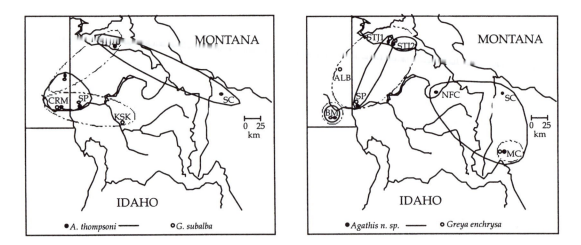

Figure 25.1 Population groupings for the prodoxid moth *Greya enchrysa* and a specialist braconid parasitoid *Agathis* n. sp. based on analysis of molecular variance (AMOVA) of differences among populations in mitochrondrial DNA sequence of cytochrome I and II subunits. (Redrawn from Althoff and Thompson 1999.)

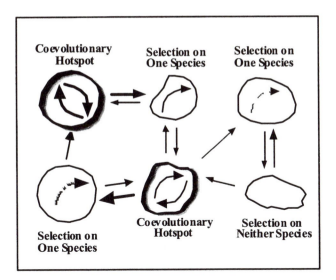

Figure 25.2 Hypothetical example of a complex geographic mosaic in a coevolving interaction. Arrows within circles (biological communities) indicate natural selection on one or both interacting species. Different arrow directions in different circles represent different (co)evolutionary trajectories. Arrows between communities indicate gene flow. (Redrawn from Thompson 1999.)

son 1994, 1999). So far, the formal models and field studies have developed mostly in parallel, exploring how selection mosaics, coevolutionary hotspots, and continual trait remixing across landscapes may shape coevolutionary trajectories. The models, however, are now sufficiently developed that they can provide predictions for tests on the spatial patterns of coevolutionary outcomes in real interactions. Concomitantly, recent empirical studies have provided the kinds of data needed for developing the next generation of models. The following sections summarize where things stand currently.

Formalizing the Geographic Mosaic Theory of Coevolution

We already know from related ecological and evolutionary models that spatial structure can stabilize some interactions or at least increase their local persistence over longer periods of time (e.g., Leonard 1998). Gandon et al. (1996), for example, explored a metapopulation model that lacks selection mosaics and coevolutionary hotspots but includes extinction and colonization. They found that metapopulation structure tends to stabilize coevolving parasite-host interactions across landscapes.

Recent explicit models of the geographic mosaic theory of coevolution have begun to evaluate formally how the components of the geographic mosaic interact, how coevolutionary hotspots develop, and how local mismatches of traits can result from the coevolutionary process. These models have incorporated, to different degrees, metapopulation structure, broader geographic structure, and selection mosaics.

The simplest genetic model of the geographic mosaic of coevolution is one in which each of two interacting species has two alleles. Allele A in species 1 matches allele B in species 2, and allele a in species 1 matches allele b in species 2. If the interaction is antagonistic (as in predator-prey or parasite-host interactions), natural selection favors parasites that matches the complementary allele in their hosts (e.g., A-B), but it favors hosts that are mismatched to the parasites (A-b). That is, natural selection favors parasites that are adapted to their host, and it favors hosts that harbor alleles to

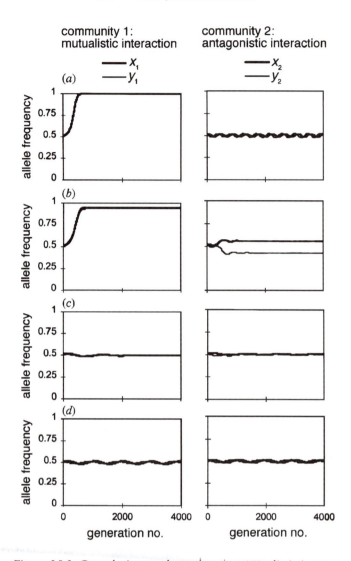

Figure 25.3 Coevolutionary dynamics of a mutualistic interaction linked to a stronger antagonistic interaction. In this genetic model, mutualism favors selection toward fixation of matched alleles in the two interacting species, whereas antagonism favors fluctuating polymorphisms as the exploiter tries to "match" alleles with the victim and the victim evolves to escape matching of alleles. Gene flow increases from (a = no gene flow) to (d = 0.05). (From Nuismer et al. 1999.)

which the parasite is not adapted. The result is fluctuating allele frequencies, in which negative frequency-dependent selection favors hosts that are the rarer genotype. In contrast, if the interaction is mutualistic, natural selection favors matches of alleles in both species through positive frequency-dependent selection. Such models seem biologically oversimplified, but they can provide important insights into the range of coevolutionary dynamics that may occur between interacting species.

Using this genetic system, Nuismer et al. (1999) explored a geographically structured model to evaluate the interplay between selection mosaics, coevolutionary hotspots, and gene flow. As in most matching-allele models of parasite-host interactions that rely on frequency-dependent selection, the allele frequencies tend to oscillate with greater amplitudes over time and never become stabilized (figure 25.3A). By comparison, mutualistic interactions rapidly become fixed for one set of alleles in both species through selection to match alleles (figure 25.3A). When an antagonistic interaction is linked by gene flow to a mutualistic interaction, however, very different results occur. If the interaction coevolves in both communities (i.e., both are hotspots), but selection on the antagonistic interaction is stronger ("hotter") than selection on the mutualistic interactions, gene flow can result in an oscillating polymorphism in both communities (figure 25.3B,C,D).

What this means is that an evolutionary ecologist studying coevolution in a local mutualistic community could document local mutualistic selection, predict selection toward fixation of matching alleles, and be dead wrong, because the overall evolutionary trajectory is actually toward fluctuating polymorphisms driven by connections to neighboring communities where selection is antagonistic. Alternatively, if selection is stronger in the mutualistic community than in the antagonistic community, allele frequencies would evolve toward fixation along a trajectory that looks as if it is driven by positive frequency-dependent selection and mutualism across both communities.

These results caution strongly against any broad interpretation of coevolutionary dynamics and the patterns they produce from a study of selection on a particular interaction within a single local community. Yet, such broad interpretations have been common in studies within evolutionary ecology. This simple model suggests that the structure of selection mosaics among communities, the degree of asymmetry in coevolutionary hotspots, and the level of gene flow across geographic landscapes will all contribute importantly to the ongoing dynamics of coevolution between species.

Moreover, if the interaction is distributed across many populations, rather than just between two as in this original model, the geographic structure and dynamics of coadaptation become even more complex. The coevolving interaction can produce many different degrees of local maladaptation, distributed in a clinal pattern across the landscape (Nuismer et al. 2000). Because many species occur as collections of genetically connected populations, clinal patterns of maladaptation may be a common outcome of geographically structured coevolution.

Using the same overall model structure, Gomulkiewicz et al. (2000) showed that coevolutionary selection need not be ubiquitous for coevolution to have important effects on the overall evolutionary trajectory of an interaction. Local but strong coevolutionary hotspots can have large influences on the evolutionary dynamics of interactions, if populations are connected by gene flow. Hotspots can maintain regional polymorphisms in coevolving alleles, and they can create novel dynamics in coldspots. Asymmetric gene flow can create situations in which the interacting species experience maximal fitness outside the hotspot or inside the hotspot, depending on the structure of selection and the pattern of gene flow.

These genetic models do not include the effects of demography on coevolutionary dynamics. They track only allele frequencies. Hochberg and van Baalen (1998), however, explored how coevolutionary hotspots may develop and become distributed across landscapes when predators and prey differ geographically in productivity (i.e., birthrates). They began with the observation that some environments support higher reproductive rates than others, and they envisioned a gradient in predator and prey productivity from the center to the edge of the geographic ranges of both species. In the model, productivity is highest at the center of species ranges, decreasing gradually outward to the edge, where the populations formed demographic sinks. These authors' analyses suggested that coevolutionary selection for investment in the interaction (i.e., level of defense in the prey and counterdefense in the predator) will be highest at the center of the range, where prey productivity is highest. Because central populations produce more offspring than peripheral populations, gene flow will be mostly from the center to the edge. Consequently, the addition of gene flow results in edge populations showing more coevolutionary investment than would be predicted from local coevolutionary selection alone.

As other models of the geographic mosaic of coevolution continue to develop, we will be able to refine our understanding of how geographic structure shapes coevolutionary dynamics across landscapes. Some recent models have indicated that time lags in local coevolutionary responses (Lively 1999) and historical events such as which genotypes colonize an area first (Parker 1999) are additional important components of coevolutionary dynamics.

Overall, the recent models of the geographic mosaic of coevolution suggest that selection mosaics, coevolutionary hotspots, and trait remixing may all have important effects on coevolutionary dynamics. Moreover, the coevolutionary process itself may generate selection mosaics across landscapes. Six points stand out in the models. First, metapopulation structure may increase the long-term persistence of coevolving interactions, at least for interactions governed by major genes. This conclusion reinforces the results of ecological models, showing increased persistence of some interactions with the addition of metapopulation dynamics (Hanski 1998). What is still unknown is how various forms of broader geographic structure in coevolving interactions shape long-term persistence. Metapopulation models are built on continual extinction and recolonization among habitats in which all genotypes may get established. Over broader geographic areas, selection mosaics create different opportunities for establishment of genotypes and different mixes of evolutionary hotspots and coldspots.

Second, population dynamics may contribute importantly to the geographic structure and dynamics of coevolution. Population density may influence whether a particular habitat becomes a coevolutionary hotspot. Consequently, studies of selection on life history evolution (e.g., timing of reproduction and other factors affecting the intrinsic rate of population increase) will be important in linking demographic and genetic approaches to the study of coevolutionary hotspots. In addition, the results of these models suggest the need for more dialogue between coevolutionary biologists and ecosystem biologists who study geographic patterns of productivity gradients.

Third, selection mosaics coupled with gene flow may commonly result in some degree of local maladaptation in coevolving interactions. Over 20 years ago, Gould and Lewontin (1979) accused many biologists of viewing organisms as perfectly adapted to their environments in a Panglossian world. The adaptations of organisms are indeed often remarkably good, but the past few decades of research have done much to highlight just how jury-rigged is the process of adaptation at the organismal level. Studies of population structure and the recent models of coevolutionary dynamics have indicated that populations are often at least somewhat lo-

cally maladapted in their interactions with one another. Continual local selection may be thwarted to varying degrees by the distribution of coevolutionary hotspots and by patterns of gene flow and random genetic drift during periods of low population numbers. The stronger the selection mosaic across geographic landscapes and the higher the gene flow, the greater the likelihood that connected populations will show local maladaptation. Such local maladaptation may be an important driving force in ongoing coevolution dynamics.

Fourth, the models suggest a world view of coevolutionary dynamics closer to Wright's shifting-balance theory of the structure of evolutionary change than to Fisher's large-population-size theory (Thompson 1994). As Wade and Goodnight (1998) argued, Wright viewed the essential problem of evolution to be the origin of adaptive novelty in a constantly changing world. In his view, the major evolutionary processes driving change are a combination of local natural selection (involving epistatis and pleiotropy), random genetic drift, gene flow, and interdemic selection. These processes occur in an ecological context that often includes a small, subdivided population and alleles whose effects vary among environments. In contrast, Fisher emphasized the continuing refinement of adaptation in large panmictic populations within stable or slowly changing environments. Mutation and selection are the dominant evolutionary processes, acting on the additive effects of alleles. Like Wright's models, the current geographic models of coevolution emphasize the dependence of the selective value of particular alleles on both the environment and the genotypes in which they occur. They indicate that deviations from panmixis can have important effects on the rates and trajectories of evolutionary change. Moreover, they suggest that geographically structured coevolution may partially generate the continual change in environment that was a cornerstone of Wright's view.

Fifth, the models indicate that ongoing, often fluctuating, coevolution across landscapes may be a common consequence of the geographic structuring of species. In fact, fluctuating coevolutionary selection may be one of the most important ways in which coevolution affects the ecological dynamics of communities. Through constantly reshaping interactions, by favoring first one set of traits and then another set, it may play an important role in molding demographic patterns across landscapes. The process may sometimes prevent regional and global extinction of some taxa. It will, however, require detailed studies of some model interactions to gain a better idea of the importance of geographically structured coevolution in the local and biogeographic dynamics of species.

Finally, the geographic models are suggesting that too many variables interact to allow a prediction of the long-term trajectories of coevolution between a particular pair or group of species. That complexity does not mean that the analysis of coevolutionary dynamics is impossible, and it does not mean that we should abandon searching for patterns. Instead, it suggests that the patterns we need to understand are not whether particular combinations of traits in two or more species will coevolve over the long term in particular ways. Rather, we need to understand the large-scale, ongoing process that results from the geographic mosaic of coevolution. These are some of the major questions:

1. How do different forms of the geographic mosaic of coevolution shape geographic patterns of genetic diversity in species?
2. How are coevolved polymorphisms maintained and at what geographic scales?
3. How commonly is local maladaptation a result of geographically structured coevolution, and how important is it as a driving force in ongoing coevolution?
4. How does the geographic mosaic of coevolution determine the forms of coevolution that are likely to dominate an interaction (e.g., an escalating arms races versus fluctuating polymorphisms)?
5. How important is coevolution in organizing the overall geographic structure of species and, more generally, biodiversity?

What we do not know from the models is as important as what we do know. None of the current models of coevolution has come to grips with complex geographic mosaics. Imagine landscapes with strong selection mosaics among communities, coevolutionary hotspots embedded within a matrix of coevolutionary coldspots, differing degrees of gene flow among populations, differing geographic structures in the interacting species, local population fluctuations that change the intensity of selection and the rate of genetic drift, and occasional local extinction of interactions. That is probably the common condition of many coevolving interactions. Although not as sophisticated as real interac-

tions with all this complexity, the models are helping us understand the process by evaluating some of these components and the relationships among them. Those efforts are gaining increased importance as the results of coevolutionary studies are applied to conservation issues, because the most important part of conservation biology is preservation of the full complexity of ecological and evolutionary processes.

Case Studies

Evidence for Geographic Mosaics in Evolving Interactions

The models indicate that selection mosaics, coevolutionary hotspots, and trait remixing may all be important whenever they occur within natural communities. Only a few interactions, however, have been studied from a coevolutionary perspective across broad geographic landscapes. Consequently, we are only now beginning to understand the actual geographic structure of the coevolutionary process. Moreover, none of the studies has directly tested any specific predictions that come from the models. Instead, the case studies so far have been devoted to defining how selection mosaics and coevolutionary hotspots are distributed across landscapes and how they interact with patterns of gene flow and fragmentation to shape coevolutionary dynamics.

I will highlight here three examples from western North America to illustrate the kinds of geographically structured data needed to evaluate the ongoing geographic mosaic of coevolution and eventually link the models with actual interactions. Some additional interactions that are developing as model systems include the ongoing detailed studies of Australian wild flax and flax rust (Burdon et al. 1999), wild parsnip and the parsnip webworm (Berenbaum and Zangerl 1998), wild *Drosophila melanogaster* and its parasitoids across Europe (Kraaijeveld and Godfray 1999), *Brassica* plants and *Phyllotreta* beetles in Denmark (Jong and Nielsen 1999), *Amphicarpaea bracteata* legumes and *Bradyrhizobium* in eastern North America (Parker 1999), *Silene* plants and *Ustilago* rust (Antonovics et al. 1998), *Potamopyrgus* snails and *Microphallus* trematodes in New Zealand (Lively 1999), and competing *Cnemidophorus* lizards (Radtkey et al. 1997).

Geographically Structured Interactions

The prodoxid moth *Greya politella* feeds on the flowers of a group of closely related plant species in the genera *Lithophragma* and *Heuchera* (Saxifragaceae). It is one of the more widespread interactions in western North America, with *Greya* populations distributed from British Columbia to Southern California and east to the Rockies. Long-term studies of this interaction have been used to evaluate whether the major components of the geographic mosaic theory of coevolution occur within natural populations. These studies have indicated that the interaction has a complex geographic structure in patterns of genetic connectedness among populations, in the distribution of phenotypic traits in the species, and in the local outcome of the interaction.

Molecular analyses of *Greya politella* populations have shown that these moths are grouped geographically into northern and southern clades and are further differentiated within those clades (figure 25.4). The northern clade, in Oregon and Washington, has less genetic variation than the southern clade, in California, and may have originated through a post-Pleistocene expansion of populations northward (Brown et al. 1997). Populations recently found farther east in the Rockies suggest additional geographic differentiation among the populations.

The plants show a similar, complex geographic structure in genetic connections among populations. *Lithophragma* is a small genus with about 10 described species, but recent work has suggested much more complex relationships among populations. Hybridization, polyploidy, and clinal variation have all interacted to create a complex mosaic of genetically differentiated populations with many uncertainties about species limits (Soltis et al. 1992; Kuzoff et al. 2000). The genus is best viewed currently as a complex set of populations/species harboring several clades. The *L. parviflorum* clade and clade 2 differ significantly in floral traits. The moths have colonized both of these plant clades (figure 25.4). The northern clade of moths (the W moth clade) oviposits into the *L. parviflorum* clade. The southern clade of moths (the C moth clade) oviposits into plant clade 2 and also, in some areas, into the *L. parviflorum* clade. Some of the remaining *Lithophragma* populations have a mix of traits from the *L. parviflorum* clade and clade 2.

Recent analyses using nuclear DNA markers have indicated that some of these populations have arisen through hybridization between other populations/species within *Lithophragma* (Kuzoff et al. 2000).

The moths, which are close relatives of yucca moths (Pellmyr et al. 1996), oviposit into the flowers of *Lithophragma* by inserting the abdomen into the corolla. Female moths have an abnormally long seventh abdominal segment, which allows them to reach the plant ovary when ovipositing through the corolla. While ovipositing into a flower, a female passively pollinates it with pollen that has adhered to her abdomen from previous flowers she has visited. Female and male moths also nectar on the flowers, but nectaring results in little or no pollination. Hence, *G. politella* moths have the potential to be mutualists (pollinators) with their hosts only when they are antagonists (floral parasites). Whether the interaction is locally mutualistic, commensalistic, or antagonistic depends in part on the abundance of local copollinators, especially bombyliid flies and some bee species (Thompson and Pellmyr 1992; Thompson, unpublished data).

In addition, moths in some areas of the northern Rockies have colonized *Heuchera*, a saxifrage genus close to *Lithophragma*. In northern Idaho, local autopolyploid populations of *H. grossulariifolia* sustain higher population levels of the moths than sympatric or parapatric diploid populations (Thompson et al. 1997). Whether the moths on *Heuchera* are major pollinators is not known, but analyses of visitation rates suggest that bumblebees and some solitary bees are potentially the major pollinators of at least one set of diploid and polyploid *H. grossulariifolia* populations (Segraves and Thompson 1999).

Overall, the interaction between *Greya politella* and its host plants is a complex, geographically structured relationship that is diversifying in different directions in different regions. Over the millennia, the geographic ranges of these insect and plant lineages have undoubtedly expanded and contracted repeatedly, creating the complex pattern evident today in the mitochondrial data, the chloroplast data, and the uncertain species limits. Local mutualistic coevolutionary hotspots are possible where copollinators are rare, and local antagonistic hotspots may occur where moths (and their seed-feeding larvae) are abundant and copollinators are also abundant. Such hotspots may blink in and out over time, and they may rarely lead to long-term direc-

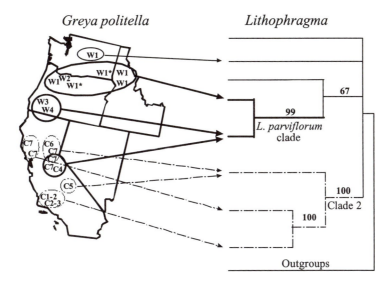

Greya politella   *Lithophragma*

——— Hosts have franciscan stigmas, mostly inferior ovaries, and smooth seed coats. Clinals occur among some populations.

—·— Hosts have nonfranciscan stigmas, ovaries less than half inferior, and rough seed coats. Some populations may be hybrids.

——— Hosts are possible hybrid between *L. parviflorum* and another species outside the *L. parviflorum* clade of populations.

Figure 25.4 Geographic distributions of RFLP haplotypes in *Greya politella* and their match to a cpDNA phylogeny of *Lithophragma*. The moths are divided into a northern group of populations (W haplotypes; asterisks indicate multiple populations) and a southern group (C haplotypes). The moths use two well-resolved clades within *Lithophragma* that differ in floral traits. Relationships to other *Lithophragma* populations and species limits are unresolved by this analysis, possibly as a result of cases of hybridization and polyploidization. Most W-clade moths use plants within the *L. parviflorum* plant clade (solid ovals, arrows, and branches on cladogram); most C-clade moths use plants within plant clade 2 (dashed ovals, arrows, and branches). (Data from Soltis et al. 1992; Brown et al. 1997; and Thompson, unpublished data. Redrawn from Thompson 1999.)

tional selection on the traits of the interacting populations. But those hotspots may be of major ecological importance, because they may buffer the populations from local extinction. The interaction is locally obligate for the moths, but not always so for the plants. When copollinators are rare, natural selection is likely to favor plants that are better at attracting moths and maximizing seed set during oviposition. When copollinators are abundant, selection may act to restrict access by the moths. Local fluctuating selection may therefore be common in these interactions, with the balance between mutualistic and antagonistic selection varying geographically.

Selection Mosaics and Coevolutionary Hotspots

Across the same western landscapes of North America, red crossbills feed on the seeds of conifers.

These crossbills, once thought to be one species, are now known to be a collection of genetically differentiated populations or species specialized to different conifers. Benkman's (1999) long-term studies have suggested that one of these specialist taxa, which feeds on the cones of lodgepole pine *(Pinus contorta)* may exhibit coevolutionary hotspots and coldspots in its interactions with these pines. Both the birds and the pines vary geographically in traits important to the interaction. In the northern Rocky Mountains, the cones of lodgepole pines have traits adapted to defense against red squirrels *(Tamiasciurus),* and coevolution between these two taxa dominates in this central region of the geographic range of the pines (Benkman 1999). The squirrels preferentially harvest cones that are narrow at the base, have more seeds than average, and have a higher ratio than average of kernel mass to cone mass. To the east and south of the Rockies, however, are some regions that lack squirrels. Here, the pines have larger, more cylindrical cones with thicker scales, which make seed extraction more difficult for crossbills with small bills. Resident birds in these outlying regions in Idaho and Alberta have converged on larger, stouter bills, which increase the efficiency with which these cones can be harvested (figure 25.5).

The overall geographic pattern of traits in crossbills and lodgepole pines suggests that there is a selection mosaic in these interactions, favoring different traits in different geographic regions. Moreover, the interaction between the birds and the pines shows regions of coevolutionary hotspots and coldspots, governed by the presence or absence of red squirrels. These results confirm the need to incorporate selection mosaics and coevolutionary hotspots into models and analyses of the coevolutionary process. Moreover, as in the models, the results show that many locally coevolved traits may undergo only limited spread throughout the geographic range of an interaction. Benkman (1999) has argued that the hotspots may have favored such divergent traits from other populations that some of these crossbill populations may be either separate species or on their way to becoming new species. Hence, through diversifying coevolution, where local specialization and trait matching driven by coevolution create barriers to hybridization with other populations, some of these hotspots may therefore be spinning off new taxa of lodgepole pine specialists.

Local Trait Matching and Mismatching

Yet a third set of interactions in western North America shows strong evidence of a coevolutionary geographic mosaic. Moreover, this interaction shows some local mismatching of traits, which is predicted by the theory and models. *Thamnophis sirtalis* garter snakes are distributed broadly over the western United States and southern Canada. Some far western populations are sympatric with toxic newts in the genus *Taricha.* These newts produce tetrodotoxin (TTX), which is a powerful neurotoxin. The toxin is particularly potent in *Taricha granulosa,* which occurs from southern British Columbia to northern California. Garter snake populations differ geographically in their resistance to TTX (figure 25.6). Those outside the range of *Taricha* are nonresistant, whereas those within the range of *Taricha* differ geographically in their degree of resistance. Other snakes sympatric with *Taricha* lack high levels of resistance to TTX. Only *Thamnophis sirtalis* has effectively countered the powerful defenses of this salamander.

Taricha also vary geographically, showing higher levels of toxicity in some populations than in others. The exact cause of differences in toxicity levels is not yet known, but toxicity levels in the newts and resistance levels in the snakes broadly match. For example, some island populations of newts in British Columbia lack TTX, and garter snakes on these islands are not resistant to TTX (Brodie and Brodie 1999a).

Unlike the populations shown in figure 25.6, not all populations of the salamanders and snakes are closely matched in defense and counterdefense. Much remains to be understood about how the overall geographic mosaic in traits in these predators and prey is maintained. Recent work has indicated that resistance to TTX comes at a cost to the snakes (Brodie and Brodie 1999a). Those snakes with high levels of resistance are slower than nonresistant snakes, so they are more vulnerable to their own predators. Hence, the degree of local matching in the traits of *Taricha* and *Thamnophis* may be shaped partly by other interactions between the snakes and their own predators.

Future Directions

The continuing challenge in the study of species interactions is to understand the scale at which many

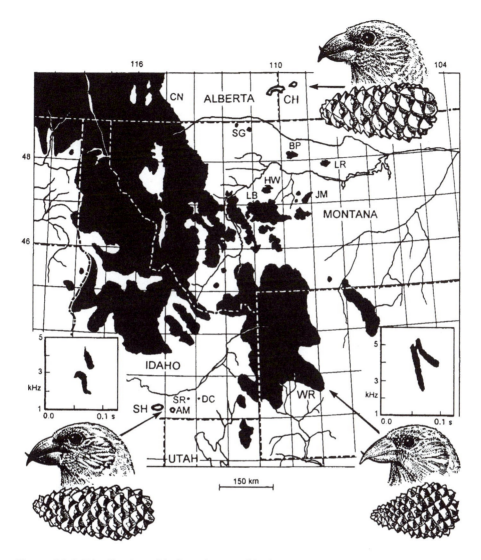

Figure 25.5 Distribution of lodgepole pine (black regions) in the Rocky Mountains of the northern United States and southern Canada and in outlying areas. Within the black areas, lodgepole pines coevolve with squirrels, and red crossbills have bill morphologies adapted to opening cones that have been shaped by pine defenses against squirrels. In the outlying areas of the Cypress Hills (CH) and South Hills (SH), squirrels are absent. In these areas red crossbills and lodgepole pines show evidence of local coevolution. The crossbills in the South Hills show other evidence of significant divergence from those in the Rockies, including flight calls (representative sonograms shown). (Reprinted from Benkman 1999.)

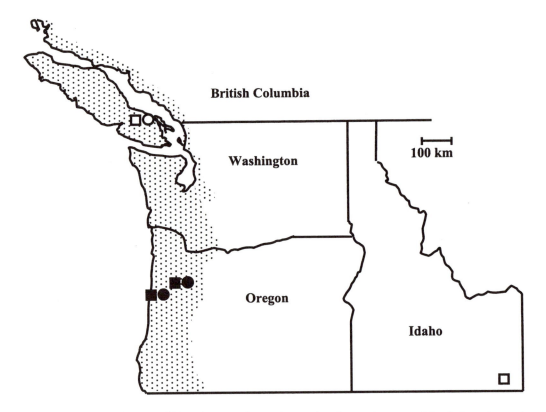

Figure 25.6 Distribution of *Taricha granulosa* salamanders in northwestern North America and level of resistance to tetrodotoxin (TTX) in sympatric and allopatric *Thamnophis sirtalis* garter snake populations. Stippled area is northwestern range of *Taricha granulosa*. Open circle on Vancouver Island represents *T. granulosa* populations lacking TTX; black circles are populations with TTX. Open squares are *T. sirtalis* populations not resistant to TTX; black squares are resistant populations. The Idaho *T. sirtalis* occurs outside the geographic range of *T. granulosa*. (Redrawn from Brodie 1999b.)

of the ecological and evolutionary dynamics occur. Without that understanding, we are left with little biological basis for interpreting the selection pressures, short-term evolutionary fluctuations, and patterns of trait matching that we so carefully document in studies of local dynamics. The kinds of studies we need can no longer be done by individual investigators working alone. Evaluating the geographic mosaic of coevolution requires evolutionary ecologists, population geneticists, molecular ecologists, and systematists working together on a group of model interactions. No single model interaction will be sufficient. But there is more to be gained at the moment from working on the half dozen or so interactions whose geographic structure has already been studied in some detail than in starting from scratch on yet another interaction.

It will be quite some time before any of these model interactions has reached the point where any new study is likely to provide only trivial refinement to our understanding. Instead, we are at the opposite point: We know just enough now about the geographic and phylogenetic structure of some of the model interactions that any study that takes up one of them and uses it for context is likely to generate results that add real depth to our conceptual understanding of the geographic structure of evolving interactions.

We also need increased refinement in our theoretical analyses of the geographic mosaic of coevolution. We now have a few models that have begun to formalize the theory and refine the predictions, but those studies have only laid the groundwork. It is the interplay between models and field studies that will advance our understanding of coevolution by pointing to the evolutionary and ecological forces most important in driving the dynamics.

The practical need for these studies is increasing. Most of the earth's biodiversity is contained not in its species but in the network of continually changing interactions that connect genetically differentiated populations across complex landscapes. The fragmentation of landscapes by human activities is rapidly reshaping those landscapes and the geographic structure of species and interactions they harbor. Our guesses about how these activities affect biodiversity have mostly been about the ways in which fragmentation affects extinction rates, loss of genetic diversity, and ecosystem services. But the real glue to biodiversity is the network of interactions among species, and that network is geographically structured. Hence, we need to understand how increased fragmentation in the geographic mosaic of interactions and the coevolutionary process directly contributes to patterns of extinction, the dynamics of genetic diversity, and the overall dynamics of regional and global population diversity.

Acknowledgments   I thank Bradley Cunningham, Derek Roff, Kurt Merg, and Scott Nuismer for helpful discussions or comments on the manuscript. This work was supported by NSF grants DEB 9707781, DEB 0073911, DEB 0083548, and by the National Center for Ecological Analysis and Synthesis (Coevolution Working Group) NSF grant DEB 9421535.

PART V

ADAPTATION TO ANTHROPOGENIC CHANGE

26

Pesticide Resistance

JOHN A. McKENZIE

Biological control, sterile insect release, autocidal control and genetically modified crops have made, and will continue to make, important contributions to specific programs of integrated pest management. However, at least into the immediate future, the effective management of agricultural ecosystems will depend on the judicious use of chemical pesticides to control fungal pathogens, weeds, nematodes, or arthropods that damage crops or livestock and lead to lower productivity. Similar conclusions can be drawn with respect to the control of insect pests that play key roles as vectors in the transmission of diseases that have devastating impact on the health of humans and animals, particularly in the developing countries of Africa and Asia.

If pesticides are used inappropriately, their effectiveness can be short-lived, and the residues of the chemicals can be harmful to the environment. Typically, resistance to the pesticide develops, often resulting in increased chemical usage at higher concentrations. This, in turn, produces higher levels of pesticide residues in the environment, with greater deleterious effect on nontargeted species through direct, unintentional exposure or through the incorporation of chemical residues into food chains. Unfortunately, this outcome has not been uncommon. The list of pests and the chemicals to which they have developed resistance is depressingly impressive (Georghiou 1986; Bergelson and Purrington 1996; Denholm et al. 1999).

The development of resistance causes significant problems. The phenomenon does, however, provide a rare opportunity: the chance to study natural selection where fundamental research on ecology, genetics, molecular, and developmental biology and physiology can be integrated. An understanding of the microevolutionary processes that lead to the development of resistance enables the derivation of better strategies of pesticide usage that minimize the evolution of resistance to future pesticides. The task of measuring selection in natural populations is not, however, trivial (Fairbairn and Reeve, this volume).

In essence, to demonstrate unambiguously that selection is occurring we must:

1. Identify the selective agent(s).
2. Mechanistically associate the action of the selective agent on the phenotype(s) with the product(s) of the genotype(s).
3. Gain predictable results after using our knowledge of the mechanism to manipulate experimental populations.

Often, the identification of the selective agent has proved the stumbling block (Clarke 1975). This is not the case in the evolution of pesticide resistance. The selective agent, the pesticide, has a clear association with phenotypic change, the development of resistance (McKenzie 1996). When combined with other advantages (table 26.1), the evolution of pesticide resistance offers an excep-

Table 26.1 The advantages of pesticide resistance systems in studying selection in natural populations.

The selective agent, the pesticide, is known.
The physiological/biochemical/molecular basis of selection is defined.
Relative fitness differences between phenotypes are frequently sufficiently large to be measured.
Change of gene frequencies is rapid.

Source: Modified from McKenzie (1996).

tional opportunity to study microevolution experimentally on a manageable time scale.

Historical Context

Home remedies and various toxic substances have been used throughout history in attempts to control species that humans perceived as pests. Melander (1914) is credited as the first to pose the question: "Can insects become resistant to sprays?" His question related to the response of San José scale, *Quadraspidiotus perniciosus,* to sulfur-lime, but in subsequent years, the question would be asked of other organisms and pesticides as the use of chemicals to control pests became widespread. From the 1940s, the common occurrence of resistance closely followed. Several hundred species have become resistant to at least one pesticide. Some have become resistant to several different groups. A notable recidivist is the Colorado potato beetle, *Leptinotarsa decemlineata,* where some populations in the eastern United States of America have developed resistance to almost every chemical available for control (Georghiou 1986).

The development of resistance may be viewed from a number of perspectives. A farmer perceives resistance as a loss of pest control, a toxicologist as a shift in a concentration mortality line (figure 26.1), a geneticist as a change in allele frequency at one or a number of loci, a biochemist as an associated change in the gene product, and a molecular biologist as a change in DNA sequence. Over many of the last 50 years, the focus of the study of resistance has tended to be discipline-specific, with a distinction between fundamental and applied approaches. As in many instances of studies of ecol-

ogy and evolution, the better studies have been multidisciplinary and have adapted advances in technology to the methodology and philosophy of the analysis of resistance (ffrench-Constant and Roush 1990; McKenzie 1996). A significant challenge is to explain the relative rates at which resistance to particular pesticides evolves.

The rate of the evolution of resistance is influenced by a complex interaction of biological, genetic, and operational factors. The rate is ultimately limited by the flexibility of the genetic system, population dynamics, the phenotypic variation in the population, and the way in which that variation is selectively screened (Roush and McKenzie 1987; McKenzie 1996; 2000; Groeters and Tabashnik 2000). A first step in unraveling the complex of interactions is to understand the genetics of resistance. We can then hope to estimate the relative fitness of resistance phenotypes associated with known genotypes.

Concepts

The Genetics of Resistance

The relative importance of monogenic or polygenic responses during adaptation has been a matter of debate in evolutionary theory (Orr and Coyne 1992). It may be argued that during adaptation genetic outcomes depend on whether the variation selected lies within an already-present continuous distribution or whether the variation, on which selection acts, results from a new gene mutation that generates phenotypes outside the initial distribution.

Let us apply these principles to the evolution of resistance. Assume that at the time a pesticide is applied, the population consists of susceptible individuals. These susceptible individuals will not, however, have an identical resistance phenotype because of differences in their size, age, or physiological state. Thus, there will be a viability distribution within the susceptible population (figure 26.1). A normal distribution of viability indicates that environmental, genetic, and interactive components contribute to the resistance phenotype. Genetic differences between individuals of a continuous normal distribution are typically polygenic (Orr and Coyne 1992). Therefore, responses to selection at concentrations of pesticide that lie within the viability distribution

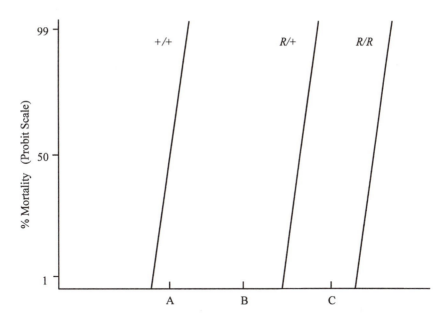

Figure 26.1 Idealized concentration mortality relationships demonstrating partial dominance for resistance phenotypes determined by a single gene [susceptible (+/+), intermediate resistant (R/+), resistant (R/R)]. Uniform application of insecticide at concentrations B or C makes resistance dominant or recessive with respect to survivorship, respectively. Similar relationships apply for polygenically determined resistance. The straight lines indicate underlying normal distributions of viability within each resistance class. Uniform application of concentration A, within the viability distribution of the susceptible population, selects for the more resistant phenotypes within the susceptible distribution, as approximately 50% of susceptibles would die at this concentration. (Adapted from McKenzie 1996.)

of susceptibles (figure 26.1, concentration A) are expected to be polygenically based.

Conversely, selection at concentrations above those that kill 100% of the susceptible population (figure 26.1, concentration B) will screen for rare mutations that produce a nonoverlapping shift between the susceptible and resistant phenotypic distributions. A monogenic response is expected in the case of such adaptations (McKenzie and Batterham 1994, 1998; McKenzie 1996). There are three general ways in which specific mutations might cause resistance: gene amplification (or duplication), mutations that alter protein structure or the amount of enzyme produced, or mutations to regulatory genes that also achieve the previous re-

sult. Examples of each occur in the literature (McKenzie 1996; Taylor and Feyereisen 1996).

Both polygenic and monogenic responses are observed in natural populations of different pest species. Although exceptions occur, the genetic mechanisms observed are in general accord with the theory discussed above (Roush and McKenzie 1987; McKenzie 1996). The systems that have proved more tractable to ecological and population genetics analysis are those in which resistance is under monogenic control. For that reason, our discussion will focus on these. It should be noted, however, that the general principles considered also apply to resistance systems that are under polygenic control.

Measuring Relative Fitness: The Theory

The relative fitness of resistance phenotypes and genotypes is a function of the pesticide concentration to which a population is exposed. With respect to inheritance, the resistant phenotype is frequently partially dominant. That is, the phenotype of the heterozygote is intermediate to those of resistant and susceptible homozygotes (figure 26.1). With respect to survivorship, and therefore potentially relative fitness, resistance may be either recessive, partially dominant, or completely dominant, depending on the concentration of pesticide to which the population is exposed.

If a population of resistance phenotypes is exposed to concentration B of figure 26.1, resistance is dominant with respect to relative fitness as heterozygotes and resistant homozygotes survive equally well while all susceptibles die. If, however, the population is exposed to concentration C (figure 26.1), resistance is recessive, as only resistant homozygotes survive. At concentrations between B and C, resistance is partially dominant with respect to survivorship. It is worthwhile restating that while the inheritance of resistance is usually partially dominant, relative fitness is concentration-dependent, and therefore, selection for resistance may be recessive, partially dominant, or dominant. This conclusion will be important to our understanding of the rates at which resistance evolves.

When monogenic resistance first occurs, it will typically be because rare heterozygotes are selected in a population largely consisting of susceptibles. In these circumstances, we expect resistance to evolve more rapidly if selection for resistance is dominant. As we will discuss in the case study, this prediction of population genetics theory is fulfilled (McKenzie 1996).

In the absence of the pesticide, resistant phenotypes are usually at a selective disadvantage relative to susceptibles because of the pleiotropic effects of the resistant allele. These pleiotropic effects occur as a consequence of the biochemical, molecular, and physiological changes associated with resistance. The capacity to withstand the toxic influences of the pesticide are thus seen as a trade-off in components of fitness such as longevity, viability, and fertility, the "cost" being observed in the absence of the pesticide (Roush and McKenzie 1987; Bergelson and Purrington 1996). It should be noted that the cost may vary from very large to

extremely small, sometimes essentially unmeasurable, depending on the pesticide and the pest system in which it is used (Bergelson and Purrington 1996; McKenzie 1996).

Evolutionary theory tells us that if there is an initial fitness cost, subsequent selection may select modifiers that ameliorate that cost (Maynard Smith et al. 1985; Orr and Coyne 1992). Thus, coadaptation may occur. The application of the principle to resistance systems is of considerable importance as, if coadaptation is common, there are significant consequences for the management of resistance and pesticide usage.

The pesticide literature frequently does assume that after resistance has evolved, the ongoing use of the pesticide results in the enhancement of relative fitness of resistant phenotypes (Taylor 1986). However, in seven appropriately controlled experiments, while genetic background is found to be important to relative fitness in four cases, only two, resistance to warfarin in rats (Smith et al. 1993) and resistance to diazinon in the Australian sheep blowfly, *Lucilia cuprina* (McKenzie 1993), provide clear evidence of coadaptation (McKenzie 1996). The *L. cuprina* example will be discussed in the case study.

The basic relative fitness relationships we have discussed are summarized in table 26.2. Our task is to now consider the ways in which we attempt to measure selection coefficients and use these estimates to explain the relative rates of evolution of resistance to different pesticides.

Table 26.2 General relative fitness relationships between genotypes (++, susceptible; $R+$, intermediate resistance; RR, resistant), where resistance is inherited as a partially dominant character controlled by a single genetic locus.

Pesticide Absent	Pesticide Present
No coadaptation $++ > R+ > RR$	Resistance dominant $++ < R+ = RR$
Coadaptation $++ = R+ = RR$	Resistance partially dominant $++ < R+ < RR$
	Resistant recessive $++ = R+ < RR$

Note: Relative fitness relationships are for the absence or presence of pesticide usage and for the influence of coadaptation in the former case.

Experimental Approaches: Measuring Selection for Resistance

Endler (1986) and Fairbairn and Reeve (this volume) have provided an excellent overview of the experimental methods that can be used to measure natural selection. Several of these approaches—correlation of phenotypes/genotypes with environmental factors, longitudinal studies over several generations, comparison among age classes or life history stages, comparisons between closely related species, agreement of specific fitness models with independently collected data, and departures from general models such as the Hardy-Weinberg equilibrium and perturbation of natural populations—have been used to study the evolution of resistance (McKenzie 1996). We will briefly consider examples of the first three.

Correlation of Phenotypes/Genotypes with Pesticide Concentration

The general association between patterns of insecticide usage and the evolution of resistance in the literature is quite well documented. The number of studies in which it is possible to relate pesticide concentration and selection intensity to specific genotypes is much more limited (McKenzie 1996). A field study in the mosquito *Anopheles culicifacies* (Rawlings et al. 1981) provides an example of the power of the approach. The survivorship of *A. culicifacies* adults was tested in huts that had been treated with one of four concentrations of the insecticide dieldrin. On several occasions, posttreatment, homozygous resistant and susceptible mosquitoes, and heterozygotes of a cross between them, were released into the huts and scored for survivorship. Mosquitoes of each genotype were marked with a different dust so that genotypes could be unambiguously identified.

At the three highest concentrations (530, 700, and 870 mg m^{-2}), mosquitoes of all genotypes were killed for up to 4 weeks after treatment. The survivorship of resistant homozygotes increased after this time, but heterozygotes and susceptibles did not survive at all until releases 12 weeks after treatment. Pesticide degradation after treatment indicates dieldrin concentrations would be much lower at this time. This is supported by the observations at the lowest concentration (270 mg m^{-2}). At this concentration, resistant homozygotes began to survive after 1–2 weeks; however, susceptibles were still killed for 12 weeks, and heterozygotes had only limited survival up to this time.

These data allow us to conclude that selection for resistance occurs over a range of concentrations and is essentially recessive, with respect to survivorship, as resistant homozygotes survive better than heterozygotes and susceptibles, which show similar survivorship over much of that range. The data also suggest that the original selection for resistance, a function of the relative fitness of heterozygotes and susceptibles, occurred at low concentrations of dieldrin, as differential survival of these genotypes is observed only many weeks after treatment at the lowest experimental concentration. The results of the experiment therefore provide information to assist our understanding of the evolution of resistance to dieldrin by *A. culicifacies*.

Longitudinal Studies

Longitudinal studies have been an extremely helpful tool in studying the evolution of pesticide resistance. The approach allows a baseline to be established, trends to be compared against competing resistance management models, and changes in the relative fitness of genotypes to be considered as rates of change of resistance are monitored within pest populations (McKenzie 1996). A classic example is the evolution of pyrethroid resistance in the pest of cotton *Helicoverpa armigera* in Australia, where resistance was first observed in the Emerald Irrigation District of Queensland in the early 1980s and in other populations soon after. Regular sampling and testing of the populations has tracked the evolution of resistance. It was found that resistance increased relatively regularly over time (figure 26.2) (Forrester et al. 1993; McKenzie 1996).

The data of the longitudinal study have enabled predictions about the selective process to be made and then tested experimentally. For example, some of the trends of resistance frequencies observed within and between years indicated that selection for pyrethroid resistance must occur at both larval and adult stages of the life cycle (Forrester et al. 1993). Subsequent experimentation confirmed this prediction.

The availability and intensive analysis of longitudinal data and the results of the experiments they suggest have allowed the fine tuning and modification of an insecticide usage strategy that has enabled *H. armigera* to be continued to be controlled in the face of pyrethroid resistance. Without longi-

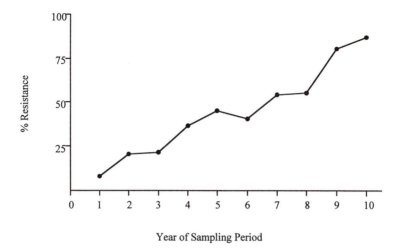

Figure 26.2 The percentage of *Helicoverpa armigera* of the Emerald district of Queensland resistant to pyrethroid insecticide over yearly samples for a 10-year period. Sampling began in 1985–1986. (Derived from Forrester et al. 1993; McKenzie 1996.)

tudinal data, resistance management would not have been as effective.

Comparison among Age-Classes

A further experiment suggested by the longitudinal study in *H. armigera* was to test if selection for resistance was dominant as the rate of change of resistance frequency was consistent with this hypothesis. The hypothesis was confirmed in a series of experiments that showed that selection was not only dominant but also age-, class-, and concentration-dependent (Daly et al. 1988). Daly et al. (1988) found that on freshly sprayed cotton leaves, survivorship of resistant and susceptible eggs and larvae, under 4 days of age, is similar. From 4 days onward, resistant larvae survive better than susceptibles on freshly treated leaves. Daly and her colleagues have subsequently been able to demonstrate that hybrids between resistant and susceptibles show survivorship similar to that of pure-breeding resistants. Thus, resistance is dominant (McKenzie 1996).

At the time the chemical is applied, selection for resistance is maximized if older and larger larvae are exposed. As the chemical decays, selection for resistance occurs at all larval stages but not at the egg stage of the life cycle (Daly et al. 1988). The effect is again dominant.

The results in *H. armigera* demonstrate that the intensity of selection for pyrethroid resistance is dependent on both the concentration of insecticide to which the pest is exposed and the life cycle stage at the time of exposure. A similar conclusion occurs for studies of permethrin resistance in the Colorado potato beetle, *L. decemlineata*, a pest that has shown an extraordinary capacity to develop resistance to a broad spectrum of different pesticides (Georghiou 1980).

Irrespective of resistance genotype, young larvae of *L. decemlineata* generally survive less well than older larvae of the same genotype after exposure to permethrin. Selection for resistance occurs across all larval age classes, but relative survivorship varies across classes. Resistance is dominant in larvae of 1–2 days of age, while partial dominance is observed for larvae aged 4–6 days (Follett et al. 1993).

Controlled experiments such as those described for the resistance systems of *A. culicifacies, H. armigera,* and *L. decemlineata* expose the complex interactions that can occur in resistance systems when resistance is selected for. The situation may be no less complex when one is attempting to estimate the "cost" of being resistant in the absence of the pesticide.

Experimental Approaches: Measuring Selection against Resistant Phenotypes

As with the methodology for measuring selection for resistance, several approaches have been used to test whether resistant phenotypes are at a selective disadvantage relative to susceptibles. These approaches include field observations where populations of a pest are monitored, after pesticide use has ceased, to observe if the frequency of resistance declines. A decline is indicative of the relative selective advantage of susceptibles in the absence of the pesticide. An example is provided by dieldrin resistance in *L. cuprina* (McKenzie 1996).

Dieldrin resistance developed rapidly in this pest after the introduction of the insecticide for blowfly control. Within 2 years treatment, the chemical failed to provide adequate protection to sheep from the blowfly. Dieldrin was therefore withdrawn as a control agent in 1958. At the time

the chemical was withdrawn, approximately 60% of blowflies were resistant. Since then, there has been a consistent decline in the frequency of the resistant phenotype, so that current frequencies are less than 1% in populations of *L. cuprina* in sheep areas of Australia (figure 26.3).

A second method, laboratory population cage experimentation, employs a philosophy similar to that of field observations: to monitor changes in resistance frequencies across generations and relate these changes to the relative fitness of resistant and susceptible phenotypes. An example comes from the study of the German cockroach *Blattella germanica* using 18 strains collected from areas of the United States (Cochran 1994).

The initial frequency of the allele that conferred resistance on the pesticide bendiocarb ranged from 0.15 to 0.99 in these strains. During the experiment, the strains were maintained under standard laboratory conditions, without pesticide, for between 3 and 25 generations. The results were heterogeneous. In four strains, the frequency of the resistant allele decreased significantly; in two, it increased significantly, and in the remaining strains,

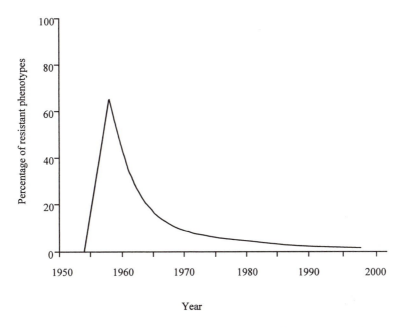

Figure 26.3 The change in the percentage of dieldrin-resistant phenotypes in natural populations of *Lucilia cuprina* from sheep areas of Australia prior to the introduction of the chemical (1955) and after its removal (1958) for blowfly control. (After McKenzie 1996.)

the initial and final frequencies did not vary significantly. The reasons for the strain differences were not investigated, but clearly, estimates of relative fitness are dependent on genetic background. Any generalizations must therefore be made with extreme caution. There may be fitness costs to resistant individuals, but the relationship may not be universal.

This point is emphasized by an example of the third methodology used to measure fitness differences of resistance phenotypes in the absence of the pesticide, estimation of components of fitness in single-generation studies. Using this approach, Minkoff and Wilson (1992) observed relatively subtle differences between the relative fitness estimates of genotypes of *Drosophila melanogaster* resistant or susceptible to methoprene. However, in population cage analysis of the same resistance system, a rapid decrease in the frequency of the resistant allele occurred. This result suggested significant selection against resistant phenotypes in the absence of the pesticide. Thus, different predictions of changes of resistance frequency in natural populations would be made depending on which laboratory estimate was used.

The studies discussed indicate that evidence for selection against resistant phenotypes in the absence of the pesticide depends on the particular resistance system, the methodology used to measure fitness, and the genetic background of resistant and susceptible strains. The latter may confound the influence of pleiotropy and linkage on fitness (McKenzie 1996). One elegant attempt to overcome these difficulties is a study of the herbicide chlorsulfuron using transgenic *Arabidopsis thaliana* as a model system (Purrington and Bergelson 1997).

Purrington and Bergelson (1997) estimated the effect of the resistant gene on lifetime seed production in control and fertilizer-enriched field environments. Their experiments controlled for genetic background, the influence of the vector plasmid used in transformations, insertion position, and the nature of the resistance mutation. The results showed a significant reduction of seed production in transgenic resistant plants, in both the control and fertilizer-enriched environments, when lifetime seed production was compared with that of susceptibles. Susceptible plants had higher seed production than nontransgenic resistant plants in the control environment, but there was no significant difference between these phenotypes in the enriched en-

vironment. Seed production was enhanced for all plants in the fertilizer-enriched environment. The results are important in the context of fitness comparisons, and also in the context of the performance of genetically engineered resistance plants. They again emphasize the difficulty of making general statements of resistant costs without having data about specific pesticides, specific environments, and specific systems (McKenzie 1996).

Summary of the Concepts

Our overview of the concepts that underpin the evolution of pesticide resistance allows us to make the following conclusions (McKenzie 1996):

1. Resistance may be under polygenic or monogenic control. The mode of inheritance is a function of how the variation is screened.
2. Resistance is usually inherited as a partially dominant character. The relative fitness of resistant phenotypes may be dominant, partially dominant, or recessive depending on the concentration to which the population is exposed.
3. The "cost" of being resistant in the absence of the pesticide varies from extremely large to essentially neutral. Outcomes depend on the pesticide, the system, the genetic background, and the environment in which estimates are made.
4. Coadaptation is possible but uncommon in resistance systems.

With this background, we are now well placed to consider our case study, the evolution of resistance to the insecticides dieldrin (a cyclodiene) and diazinon (an organophosphorus insecticide) by the Australian sheep blowfly, *Lucilia cuprina*.

Case Study: The Evolution of Insecticide Resistance in the Australian Sheep Blowfly

Selection for Resistant Phenotypes

Resistance to dieldrin occurred in *L. cuprina* 2 years after the insecticide was introduced for blowfly control in 1955. The genetic control of resistance is by allelic substitution at a single locus *(Rdl)* on chromosome V and is inherited as a partial dominant. Resistance to dieldrin occurs as the result of modifications in a GABA gated chloride channel.

Diazinon resistance was first observed in 1965, 8–9 years after first being used to control sheep blowfly. Resistance is partially dominant and essentially controlled by a gene *(Rop-1)* on chromosome IV. Allelic substitution at the *Rop-1* locus results in an altered carboxylesterase. There is no cross-resistance between dieldrin- and diazinon-resistant blowflies (McKenzie 1996; McKenzie and Batterham 1998).

The different rates of the evolution of resistance to the two insecticides is explained by comparing the relative viabilities of the different genotypes at each resistance locus. It is known that selection for resistance occurs mainly at the larval stage of the life cycle, when larvae are exposed to the insecticide in larval-initiated wounds (myiases) on the sheep (McKenzie 1993). Sheep were therefore treated with one or other insecticide or left as untreated controls. At regular times after treatment, artificial myiases were established on the sheep, and the relative viability of homozygotes and heterozygotes of each resistance system was estimated by exposure to appropriately treated sheep or untreated controls.

For dieldrin resistance, it was observed that relative viability was partially dominant for 30 weeks after treatment as the heterozygote *(Rdl/+)* had intermediate viability to the selectively favored resistant homozygote *(Rdl/Rdl)* and the disadvantaged susceptibles (+/+) (McKenzie 1996).

The relative viability of diazinon-resistant phenotypes was found to be recessive for some 20 weeks after treatment of sheep with diazinon. Heterozygotes *(Rop-1/+)* only have a significant advantage over susceptibles after this time. At the low concentrations of diazinon, between 20 and 30 weeks after treatment, when all genotypes have quite high survival levels, there is some indication of heterozygote advantage (McKenzie 1993, 1996).

The initial selection for resistance depends on the relative advantage of heterozygotes over susceptible homozygotes, as resistant homozygotes will be exceedingly rare at that time. *Rdl/+* flies have a selective advantage over susceptibles (+/+) soon after treatment (when dieldrin concentrations are high) until dieldrin concentrations have significantly decreased some 30 weeks later. The data therefore indicate that selection for dieldrin resistance is intense over a large range of concentra-

tions. Flies resistant to dieldrin, raised as larvae in the presence of dieldrin, excrete dieldrin, accumulated during development, as adults. The excreted dieldrin is lethal to +/+ flies again selecting against susceptibles (Davies et al. 1992). Given this intense selection in both larvae and adults, it is hardly surprising that resistance to dieldrin evolved so rapidly.

It is equally unsurprising that resistance to diazinon developed relatively more slowly than resistance to dieldrin in *L. cuprina*. Heterozygotes *(Rop-1/+)* are at a selective advantage over susceptibles (+/+) for a limited concentration range. The selective window between relative advantage and disadvantage, which initially occurred in the absence of the insecticide, is only just open (McKenzie 1993). It is of interest to note that initial selection for resistance to diazinon occurred at low concentrations. Such concentrations provide little protection to sheep from even susceptible blowflies. This may seem counterintuitive, but similar conclusions can be drawn from our previously discussed examples in *A. culicifacies* and *H. armigera*. Selection for resistance may therefore be maximized at high or low concentrations, depending on the specific system. In *L. cuprina*, selection for resistance to dieldrin is partially dominant over a range of concentrations. Selection for resistance to diazinon is largely recessive. Therefore, the relative rates of evolution of resistance to these two insecticides are in accord with population genetics theory, where recessive selection is expected to produce less rapid change in allele frequency (Roush and McKenzie 1987).

The observation of selection at low concentrations affords the chance of subtle frequency- and density-dependent interactions. This is realized in the resistance systems of *L. cuprina* (McKenzie 1993, 1996).

The subtlety of the interaction is exemplified at low concentrations of diazinon in a series of laboratory trials comparing *Rop-1/+* and +/+ genotypes on media containing 0%, $1.3 \times 10^{-4}\%$, and $2 \times 10^{-4}\%$ (W/V) diazinon. Susceptibles do not survive in pure culture at the latter concentration but do when in mixed culture with heterozygotes (McKenzie 1993).

The viability of both genotypes decreases with increasing concentration. Heterozygote viability is independent of frequency in mixed culture. However, at the highest concentration, the viability of susceptibles is enhanced at frequencies of 20% or less, relative to other frequencies at this concentration, in mixed culture. Egg to adult viability of susceptibles is independent of frequency in mixed culture on 0% medium but positively correlated with frequency on the $1.3 \times 10^{-4}\%$ medium. Therefore, while frequency-dependent selection occurs at both 1.3 and $2 \times 10^{-4}\%$ (W/V) concentrations, the interaction between Rop-1/+ and +/+ is competitive at the former concentration and facilitative at the latter. Media-conditioning experiments involving Rop-1/Rop-1, Rop-1/+, and +/+ genotypes show that the different results are due to concentration-dependent toxicological associations (McKenzie 1993).

I predict that if other resistance systems are investigated, similar subtle intergenic interactions will be observed. Resistance systems may therefore provide the opportunity to investigate frequency- and density-dependent selection in mechanistic terms. This opportunity has been more limited in traditional studies of these processes.

Selection against Resistant Phenotypes

The overwintering stage of the life cycle is frequently ignored in resistance studies. In both dieldrin and diazinon resistance systems of L. cuprina, it is of particular importance. We have already observed the selective disadvantage of resistant phenotypes in the absence of dieldrin as the frequency of resistance decreased in blowfly populations once the chemical was withdrawn for blowfly control (figure 26.3). Much of this selection occurs during arrested development of prepupae during the winter months (McKenzie 1990).

The population biology of L. cuprina involves wandering larvae dropping from sheep, after larval development is complete, to pupate in the soil. Under permissive conditions, development is continuous, but during winter, larvae arrest development at the prepupal stage or die. Mortality during the winter period may exceed 90%, so the potential for selection is considerable.

Wandering F_2 larvae of an initial cross between Rdl/Rdl and +/+ strains were held in the laboratory as controls or placed in containers in the ground. Experiments were conducted throughout the year to measure selective effects during continuous and arrested development. During continuous development in the field, the proportion of larvae that eventually emerge as adults is approximately 75% of those that emerge in the laboratory controls.

The frequency of the Rdl allele under continuous development conditions is close to 0.5 in both field and laboratory control populations. This is the expected allele frequency in the F_2 generation if Rdl/Rdl, Rdl/+, and +/+ genotypes have similar relative viabilities. The differences in mortality between the field laboratory control experiments is therefore not selective with respect to resistance status (McKenzie 1990).

The frequency of Rdl remains at close to 0.5 for the laboratory control populations when the field populations are exposed to winter ground temperatures. However, the frequency of Rdl in the latter populations averages less than 0.10. Therefore, severe selection against resistants occurs during winter developmental arrest. Temperature is shown to be a critical selective agent. Results similar to those observed in the field are observed when prepupae are held in the laboratory at 8°C, the ground temperature during winter, before being returned to permissive developmental temperatures (McKenzie 1990).

Similar conclusions are drawn for diazinon-resistant genotypes if a fitness modifier (see the next section) is not present. In the absence of the modifier, phenotypes resistant to diazinon overwinter less successfully than susceptibles. In the presence of the modifier, the overwintering success of all genotypes is similar (McKenzie 1996). It should therefore be noted that differential physiological cost to resistant individuals during overwintering may be influenced by genetic background. It should also be noted that while the effect of arrested development is clear in L. cuprina resistance systems, evidence for similar selection in other species is equivocal in the handful of other studies that have been conducted (McKenzie 1996). As diapause, quiescence, and dormancy are vital components of the population biology of many organisms, it is desirable to resolve if the results for L. cuprina are atypical. The effects of selection on resistance phenotypes during arrested development is an area of research worthy of more intensive investigation.

Diazinon Resistance: Evidence for Coadaptation

For coadaptation to be demonstrated, the relative fitness of the resistant phenotypes must improve over time, with any initial physiological cost ameliorated (table 26.2). Such changes result from modification of the genetic background, although there is rigorous debate about how important coadaptation is in the general evolutionary process (Orr and Coyne 1992) or in resistance systems (Taylor 1986; McKenzie 1996). The most convincing experiments involve comparisons in which relative fitness estimates are made at regular intervals, while the supposedly coadapted genome is disrupted by repeated backcrossing to an unrelated strain. In the context of resistance studies, a resistant strain from the field is backcrossed to a susceptible laboratory strain, and either single generation and/or population cage experiments are conducted to assess the relative fitness of the resistance phenotypes (Roush and McKenzie 1987; McKenzie 1996). Coadaptation in the diazinon-resistance system was investigated in this manner, and fitness changes of resistant phenotypes were shown to be due to a single modifier locus on chromosome III (McKenzie 1993, 1996).

It was initially observed in the *L. cuprina* system that if the relative fitness of diazinon-resistant flies collected in the late 1970s was compared to that of susceptibles in population cages, the values were similar (figure 26.4). This conclusion was drawn as the proportion of susceptibles in the cages remained at close to 25% over generations, the frequency expected when cages are initiated from *Rop-1/+* heterozygotes.

After nine generations of backcrossing of the resistant strain to a susceptible laboratory strain, the relative fitness comparisons were repeated. In this case, there was a consistent increase in the proportion of susceptibles in the population over generations (figure 26.4) indicating that resistant phenotypes are at a disadvantage in this genetic background.

The genetic basis of this fitness modification was determined in a series of experiments summarized by McKenzie (1993, 1996). The modifier was eventually mapped to a region of chromosome III very tightly linked to the *scalloped-wings (Scl)* locus by chromosome substitution line analysis, intrachromosomal mapping, and the use of lines of defined modifier genotype (figure 26.4). The results were confirmed by single-generation fitness estimates, which also demonstrated that the resistant genotype was relatively more fit with the modifier than without it in the absence or presence of diazinon (McKenzie 1993). This last observation is important as the modifier was selected under these conditions as diazinon has been used to help to

control blowfly since resistance developed in the mid-1960s.

This continued use of diazinon has undoubtedly been important to the selection of the modifier that is expected to occur once the resistance allele is common in the population. Thus, it is not surprising that coadaptation is not observed for dieldrin-resistance in *L. cuprina* (McKenzie 1996), as the insecticide was withdrawn for blowfly control soon after resistance developed.

Coadaptation is typically expected to be under polygenic control (Maynard Smith et al. 1985; Orr and Coyne 1992). Clearly, this expectation has not occurred in this case. The effect of the modifier is dominant. Its influence is observed at different stages of the life cycle, including overwintering (McKenzie 1993, 1996). The modifier also ameliorates a phenotypic manifestation of the developmental perturbation associated with the initial evolution of diazinon resistance, increased phenotypic asymmetry of resistant flies.

Relative Fitness, Relative Asymmetry

Evolutionary developmental biology is an emerging field that integrates developmental processes, relative fitness, and adaptation within a framework of developmental constraint (Maynard Smith et al. 1985). Changes in symmetry of organisms that are usually bilaterally symmetric is a method used to estimate developmental perturbation (Palmer 1996). The pleiotropic effects related to the evolution of resistance suggest developmental perturbation (McKenzie 1997).

The substitution of the *Rdl* or *Rop-1* allele for a susceptible allele results in a significant increase in asymmetry. The effect is dominant. As already noted, the modifier of the diazinon-resistance system returns the asymmetry score of resistant phenotypes to that of susceptibles. The effect of the modifier is also dominant (McKenzie 1993, 1997). For both dieldrin- and diazinon-resistance systems, there is a negative correlation between asymmetry score and relative fitness (McKenzie 1993, 1997).

The resistance systems of *L. cuprina* represent some of the few in which the association between asymmetry and relative fitness is unambiguously defined (Palmer 1996). The nature of the association suggests it may be possible to partition the genotypic and environmental components of the asymmetry phenotype and relate these to the developmental profile of the organism (McKenzie 1997).

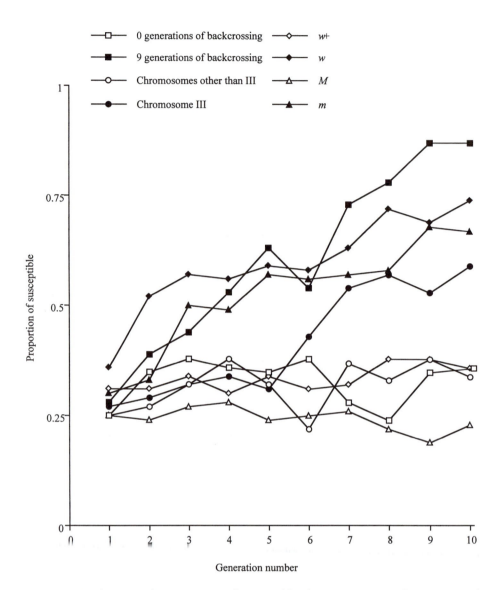

Figure 26.4 Change in the proportion of susceptible phenotypes in population cages of *Lucilia cuprina* segregating for *Rop-1* and + alleles. An increase in the proportion of susceptibles over generations indicates that resistant phenotypes are relatively less fit than susceptibles. Relative constancy of proportions around 0.25 indicates similar relative fitness of all phenotypes.

The comparisons show a fitness modifier was in the population in the late 1970s when the resistant strain was collected from the field (0 generations of backcrossing) but may not have been present when resistance first arose (9 generations of backcrossing to a laboratory susceptible strain). The modifier is localized to chromosome III (chromosomes other than III derive from the laboratory strain versus chromosome III alone from the laboratory strain) and is linked to the white locus (w^+ versus w). The data are consistent, the modifier being a single gene (M versus m). (The figure is from McKenzie 1996, reprinted with the permission of the publisher.)

In this context, these, and other, resistance systems may have an important role to play in the field of evolutionary developmental biology. This thought leads us to consider other possible future areas of evolutionary research to which resistance studies can contribute, as well as areas where current gaps in resistance studies occur.

Future Directions

Studies of pesticide resistance are frequently in specialized discipline areas. Understandably the emphasis is often on applied research. The most effective applied approaches are underpinned by fundamental research involving biochemical, molecular, genetic, population, biological, and ecological analyses. Resistance, or preferably susceptibility, management strategies benefit from interdisciplinary inputs, unified by a philosophy of the basic principles of ecological and evolutionary genetics (McKenzie 1996). In turn, resistance systems provide rare study opportunities to investigate natural selection in action. This interface offers great potential over the next decade.

Reflection on the material in this chapter allows us to identify a number of issues that will be important areas for future fundamental and applied research. Advances in molecular biology make it possible to define the basis of a resistance mutation (Taylor and Feyreisen 1996). Transformation and transgenic systems allow the mechanistic basis of resistance to be better understood and exploited (McKenzie 1996). However, to result in better pesticide usage and resistance management practice, the information must be integrated with data on the population biology of the pest. Similar arguments are equally valid for the use of biopesticides such as *Bacillus thuringiensis* and the baculoviruses (Uri 1998), no matter what the mode of delivery, if integrated pest management is to be enhanced by pesticide usage. Let us consider two examples of how approaches based on evolutionary principles may benefit future programs.

Predicting Resistance

Imagine the advantage of being able to predict the type of resistance mechanism that is likely to evolve before a pesticide is released for pest control. If predictive resistant phenotypes were generated in the laboratory, we could investigate if there

was a physiological "cost" to being resistant, the range of concentrations over which resistant genotypes were at a selective advantage, whether resistance was dominant with respect to fitness, whether one pesticide was more likely than another to select for resistance, and whether different delivery systems would influence the rate at which resistance developed. McKenzie and Batterham (1998) reported on a series of studies on the evolution of resistance to dieldrin and diazinon by *L. cuprina* to demonstrate that it is possible to do this.

A susceptible strain of the sheep blowfly was exposed to the mutagen ethyl methanesulphonate (EMS), and progeny were selected for resistance to either dieldrin or diazinon by screening above the lethal concentration of susceptibles. The genetic and molecular bases of the resultant resistant strains proved to be identical to the resistance that evolved in natural populations. Thus, in both the laboratory and the field, dieldrin resistance was controlled by the *Rdl* locus and was due to an amino acid substitution of serine[320] for alanine in the gene product. Resistance to diazinon resulted from a mutation at the *Rop-1* locus, causing an amino acid substitution of glycine[137] for aspartic acid (McKenzie and Batterham 1998). If this work had been done prior to the introduction of the insecticides, and if the relative fitness estimates of the different genotypes had been made, the relatively more rapid rate of the evolution of resistance to dieldrin would have been predicted and the mechanism of resistance identified. Similar experiments with chemicals to which resistance has yet to develop, as well as studies using *Drosophila melanogaster* as a model system, have produced informative results (McKenzie and Batterham 1998).

The strategy of anticipating resistance therefore seems promising. It is necessary, however, to extend the methodology to other organisms and resistance systems to see how generally that promise can be realized in resistance management programs.

Selection of Resistant Natural Enemies

A second example of how an understanding of evolutionary principles may assist in integrated control programs is through the selection of pesticide-resistant predators. The evolution of resistance is uncommon in natural populations of predators of pests. Laboratory selection of resistant variants that would survive pesticides, used against target species in an integrated pest management program, is therefore necessary. Thus far, with the exception of responses in some phytosecid mites, the results of laboratory selection on resistance to strains to be released in the field have been disappointing (Hoy 1990).

Two general hypotheses are advanced to explain these results. First, it is suggested that the evolution of natural predators has been such that they lack the capacity to detoxify pesticides. Second, the lack of variation in laboratory populations is argued to limit the capacity of these populations to respond to artificial selection (Hoy 1990).

It is more likely that the outcomes are better explained by exposure history in natural populations and by programs of laboratory selection by pesticides within the distribution of susceptibles. In the later instance, where there has been a limited response to selection, the results are consistent with the expected polygenic control. To generate resistant predators suitable for use in natural populations may require allelic substitution at a single genetic locus (figure 26.1). Initial selection of heterozygotes at pesticide concentrations above those lethal to susceptibles is a more appropriate method. The probability of generating a resistant allele that is screened in the selection program would be enhanced by prior mutagenesis of the susceptible populations. Selection of resistant predators by such means is potentially a fruitful area for future research, research that would be contemplated only if resistance is considered within an evolutionary genetics framework (McKenzie 1996; 2000).

Acknowledgments   The research discussed in the case study was supported by the Australian Research Council.

27

Predicting the Outcome of Biological Control

JUDITH H. MYERS

The movement of humans around the earth has been associated with an amazing redistribution of a variety of organisms to new continents and exotic islands. The natural biodiversity of native communities is threatened by new invasive species, and many of the most serious insect and weed pests are exotics. Classical biological control is one approach to dealing with nonindigenous species. If introduced species that lack natural enemies are competitively superior in exotic habitats, introducing some of their predators (herbivores), diseases, or parasitoids may reduce their population densities. Thus, the introduction of more exotic species may be necessary to reduce the competitive superiority of nonindigenous pests.

The intentional introduction of insects as biological control agents provides an experimental arena in which adaptations and interactions among species may be tested. We can use biological control programs to explore such evolutionary questions as: What characteristics make a natural enemy a successful biological control agent? Does coevolution of herbivores and hosts or predators (parasitoids) and prey result in few species of natural enemies having the potential to be successful biological control agents? Do introduced natural enemies make unexpected host range shifts in new environments? Do exotic species lose their defense against specialized natural enemies after living for many generations without them?

If coevolution is a common force in nature, we

expect biological control interactions to demonstrate a dynamic interplay between hosts and their natural enemies. In this chapter, I consider biological control introductions to be experiments that might yield evidence on how adaptation molds the interactions between species and their natural enemies. I argue that the best biological control agents will be those to which the target hosts have not evolved resistance.

Historical Context

Classical biological control is the movement of natural enemies from a native habitat to an exotic habitat where their host has become a pest. This approach to exotic pests has been practiced since the late 1800s, when Albert Koebele explored the native habitat of the cottony cushion scale, *Icrya purchasi,* in Australia and introduced Vadalia cardinalis beetles (see below) to control the cottony cushion scale on citrus in California. This control has continued to be a success. Control of prickly pear cactuses, *Opuntia inermis* and *O. stricta,* in Australia by the South American moth, *Cactoblastis cactorum,* is the first great success story for biological control of weeds and has been the basis for continued optimism about this technique.

Although biological control seems to make ecological sense, the success rate of control programs is not very high. For biological control of insects, success is approximately 10–15% (Hawkins 1993), and for weeds, it is similar (Myers 1992), although some estimates reviewed by McFadyen (1998) are higher: 33% overall, 26% for South Africa, and 33% for Hawaii. For these latter estimates success is defined as situations in which no other type of control is necessary. Unfortunately, very few quantitative studies have been done on the impact of control agents on the densities of pests. Therefore, success is quite arbitrary and might simply imply that no more effort is made to control the pest because there is nothing else to do.

The relatively low success rate of biological control suggests two interpretations: (1) Most natural enemies do not greatly reduce the densities of their hosts, or (2) a complex of natural enemies is required to reduce host densities. One way of looking at these alternative hypotheses is to review successful biological control programs and ask if success was due to a complex of introduced agents or to just one. Although there are a number of potential biases associated with evaluating hypotheses from the biological control literature, for biological control of both weeds and insects a majority of successes are attributed to a single species of agent even when several agents have been introduced (Myers 1984; Myers et al. 1989). One interpretation of this result is that most plant or insect hosts are well adapted to their natural enemies and that only certain species are capable of reducing their host's densities. If we knew why some natural enemies have impacts on the density of their host while others do not, we would perhaps be able to predict the characteristics of effective biological control agents. We would also understand much more about the evolution of hosts and natural enemies.

Concepts and Case Studies

How Can We Select Effective Biological Control Agents?

While some consider that in the last decade theoretical ecology has increased our understanding of the dynamics of single populations (chaos or stability, or stage-structured dynamics), theory has done little to help identify if particular natural enemies regulate populations or what characteristics to look for in potentially effective control agents.

Considerable effort in population ecology has been directed at determining if natural enemies are density-dependent. However, for successful biological control, it may not be necessary that the attack by an agent is density-dependent. It may simply be more important that the agent increase the total level of mortality. Because the introduction of each new, exotic control agent has an associated possibility of indirect and nontarget impacts (Cory and Myers 2000), minimizing the number of species introduced will reduce the risks. Therefore, there is great incentive to be able to select agents with good potential to control their hosts.

Will Common or Rare Parasitoids or Herbivores Have More Impact?

Careful study of an organism in its native habitat could help to identify a good control agent for introduction to the exotic habitat. However, it is necessary to know what characteristics to look for. It is generally held that good biological control agents will be common and widespread in their native habitats. But it is possible that the opposite is true. I propose that natural enemies that are relatively rare in their native habitats will have the best potential as biological control agents because their hosts will lack resistance mechanisms—the biological tolerance hypothesis (figure 27.1). Several studies allow evaluation of this hypothesis.

Biological Control of Winter Moth in Canada   Winter moth, *Operophtera brumata,* was the focus of a classical insect population study (Varley et al. 1973). In Britain, where winter moth is native, the tachnid fly parasitoid, *Cyzenis albicans,* caused low and density-independent mortality to winter moth. Predation of moth pupae in the ground was related to moth density, and most of the variation in moth. Population densities was associated with "winter disappearance," including poor synchrony between leaf development and egg hatching. Based on these observations, Varley and Gradwell predicted that *C. albicans* would be a poor control agent, unable to respond to the high density of its host in the exotic habitat.

This prediction proved to be wrong, however. Following the introduction of *C. albicans* to both eastern and western Canada, population densities of the exotic winter moth declined. Released from its own natural enemies or competitors, *Cyzenis*

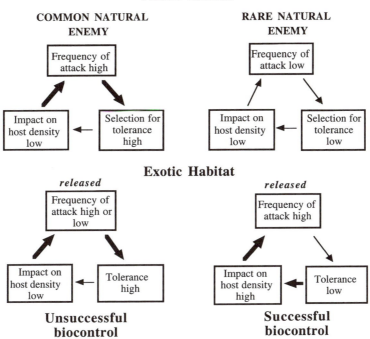

Figure 27.1 Diagram of the "biological tolerance hypothesis," which predicts that hosts will not have evolved tolerance or resistance to natural enemies that are rare in the native environment and therefore will be more susceptible to them if they reach high densities when released in the exotic habitat. Width of arrows indicates the strength of the interactions.

attained higher levels of parasitization in Canada (usually around 4% parasitism in Britain and 30–60% parasitism in Canada; Horgan, et al. 1999). In association with density-dependent mortality to winter moth pupae by ground predators, beetles, and ants, *Cyzenis* has been a successful biological control agent.

This example indicates that a natural enemy that parasitizes only a low proportion of its host in the native habitat can have a larger impact in an exotic habitat. There might have been a clue that *C. albicans* would be successful since the highest densities of this parasitoid occurred in Britain as host populations were declining (figure 27.2). However, the typically low parasitization level will not have selected for resistance in the winter moth, and therefore, high levels of parasitization occurred in Canadian populations.

Biological Control of Cottony Cushion Scale and Larch Casebearer   Although based on less work in the native habitat, supporting evidence for this hypothesis can be found with the biological control of cottony cushion scale. In 1888–1889, Albert Koebel explored cottony cushion scale populations in Australia for potential biological control agents for introduced populations of the scales in California. He found that the scales were attacked by parasitic flies, green lacewings, and the lady-bird beetle, *Vadalia cardinalis.* Koebel was able to collect and ship to the United States over 10,000 individuals of the parasitic fly *Cryptochaetun iceryae,* but only 129 individuals of the Vadalia cardinalis beetle (Doutt 1964). It was the Vadalia cardinalis beetle that was the successful control agent, rare at home but dynamite abroad. More recently, larch casebearer, *Coleophora laricella,* was successfully

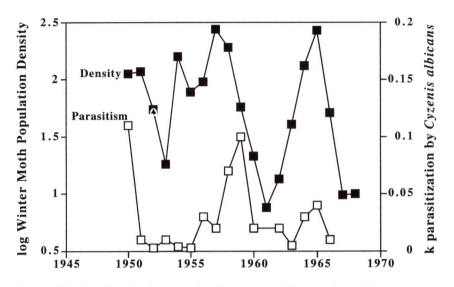

Figure 27.2 Relationship between density and parasitization (k = difference in log density before and after parasitization) of winter moth by the Tachinid fly *Cyzenis albicans*. Although levels of parasitization are not high, they are highest during periods of population decline of winter moth. (Data from Varley et al. 1973.)

controlled in the western United States by a braconid wasp, *Agathis pumila*. This parasitoid is weakly, inversely density-dependent in Europe, where it is native (Ryan 1997). Six other species of parasitoids were also introduced but remained at low density and had no impact on the host.

Biological Control of Weeds   Several examples also occur in weed biological control that indicate that herbivores that are relatively rare in the native habitat can reach high densities in the exotic habitat. Zwölfer (1973), in a study of potential biological control agents for nodding thistle, *Carduus nutans*, concluded that an insect species maintained at low density by competitors and/or predators in the native habitat would be adapted for rapid population increase when released into an exotic habitat. The seed-feeding weevil, *Rhinocyllus conicus*, is one of several species of insects that feed on seeds in thistle flower heads, but it was a poor competitor with the other species. Following introduction to North America, *R. conicus* became a successful control agent (McFayden 1998).

Another successful biological control agent, the mealy bug, *Hypogeococcus festerianus*, occurs in scattered patches in Argentina but rapidly expanded and achieved good control of Harrisia cactus, *Eriocereus martinii*, in Australia (McFadyen and Tomley 1981). Populations of the cinnabar moth, *Tyria jacobaea*, were the target of several studies in Europe, where it is native. Here, populations periodically cause severe defoliation of their food plants, tansy ragwort, *Senecio jacobaeae*. Introduced populations of cinnabar moths showed similar population outbreaks in North America, but they were not successful control agents. The plants were able to compensate for their attack. A small flea beetle, *Longitarsus jacobaeae*, on which little information is available from European populations, has successfully controlled tansy ragwort in several areas in North America (McEvoy et al. 1993).

Unfortunately, it is often difficult to find information on the status of biological control agents in the native habitat. The introduction of herbivorous insects into new habitats has demonstrated that species suppressed in their native habitats by predators, parasitoids, diseases and competitors have the potential to erupt in an exotic habitat, while species that are common in their native habitats may sometimes be difficult to introduce, do not persist, or have little impact in exotic habitats.

Force (1972) suggested that the best biological control agents would be those that are not dominant in undisturbed habitats, are widely spread geographically, and are capable of exploiting high host density associated with disturbance or newly available habitats. These suggestions have received little attention by biological control practitioners.

Parasitoid Diversity, Host Feeding Niche, and Biological Control Success

The diversity of parasitoids varies with the feeding niche of the insect hosts (Hawkins 1993). Leaf miners have the highest average number of parasitoid species, and insects that feed in protected sites, such as in roots and flower heads, have the lowest diversity of parasitoids. Also, both the establishment of parasitoids and the proportion of introductions of parasitoids in which some biological control success was achieved were associated with host feeding niche and therefore, by implication, with parasitoid diversity. Success was highest (approximately 18%) for programs against leaf-mining insects and lowest (approximately 7–8%) for those against root feeders and concealed insects (Hawkins 1993). Patterns such as these are difficult to interpret, since surveys do not necessarily distinguish between the number of parasitoids attacking single populations of hosts or the number recorded over the host's range. However, it is assumed that additional adaptations are required to parasitize concealed hosts and that this need for special adaptations reduces the number of parasitoid species attacking concealed hosts compared to those parasitizing exposed hosts. If concealed hosts have fewer species of parasitoids, they may be better able to adapt to those they do have. Therefore, they may be more resistant to introduced parasitoids in a biological control attempt. On the other hand, species of exposed hosts with more numerous natural enemies may be less able to adapt to any particular species of parasitoid. They may therefore be more vulnerable to biological control agents.

Host Resistance to Parasitoids   A good biological control agent is able to reduce the density of its host. By implication, the host will not have evolved resistance mechanisms against the attack of that agent. A biological control strategy using agents to which pests were not resistant was proposed by Hokkanen and Pimentel (1989). They suggested that new associations of parasitoids or herbivores and hosts based either on geographic isolation or on the "exploiter" being maintained at low density by high mortality or competition would permit greater impact on the "victim" when the species

were brought together. While the initial analysis supported this hypothesis, this idea has been controversial. Another analysis of available case histories by Waage (1990) found just the opposite, that success with old associations was better than with new associations. It is likely that there is a continuum from associations that are "too new" and result in little more than a minor attack level because the parasitoid is not adapted to the host, to those that are old and involve highly evolved adaptations, including resistance by the host to the parasitoid. Testing this hypothesis is difficult without careful consideration of the details of each case.

Resistance of insect hosts to parasitoids could be associated with physical refuges, or it could be related to physiological mechanisms such as the ability of the host to encapsulate parasitoid eggs. The coevolution of host resistance and parasitoid virulence has been recently reviewed (Kraaijeveld et al. 1998). Much of the work on variation in host resistance and parasitoid virulence has been done on *Drosophila*. Populations of *Drosophila* differ geographically in their resistance to parasitoids, and parasitoids vary in their virulence. But temporal studies have not been done to determine if these factors change with host density. Kraaijeveld et al. (1998) conclude that Red Queen coevolution of virulence and resistance is not occurring with parasitoids. Therefore, it is possible that hosts that are the target of biological control introductions will differ in their susceptibility to parasitoids.

One way of looking at the susceptibility of a biological control target species to parasitism by introduced parasitoids is to evaluate the levels of parasitism reached in the exotic habitat. A review of 73 parasitoid species that were introduced and became established in biological control programs showed that 59 (81%) reached a maximum level of parasitization above 40%. Some species were recorded to have parasitized 100% of the sample of hosts (Hawkins et al. 1994). What actually determines the levels of parasitization in these cases is not known. Since maximum levels were recorded, it is not known if the high levels were typical or exceptional. These data are biased because information for parasitoids that never establish or have only low rates of attack are less likely to be published.

It would be useful to evaluate interactions between hosts and parasitoids by comparing levels of attack by parasitoid species in native and exotic habitats. This would need to be done over a range of host densities. A comparison of this sort would help one to evaluate if resistance to parasitoids is a characteristic of the host species or the ability of the parasitoids to reach high densities. These types of comparisons are perhaps more readily done with plants and herbivores (see below and table 27.1).

Resistance of Plants to Herbivores   Plants frequently have large numbers of herbivore species that attack them at some place and time. A well-studied example is golden rod, *Solidago altissima*, in New York, on which 138 species of insect have been recorded. However, only 7 of these insect species reached densities at which their biomass exceeded 0.1% of the leaf biomass. Yellow star thistle, *Centaurea solstitialis*, has 6 seed-feeding insects alone that have been introduced and become widely established in western North America. At least 36 species of oligophagous insects have been found in Europe to attack either spotted, *Centaurea maculosa*, or diffuse, *C. diffusa*, knapweed, and they include at least 6 species feeding on the flower heads and another 5 species feeding in the roots. Fireweed, *Epilobium angustifolium*, has at least 10 species of insects feeding on it at one location in British Columbia (Hicks, personal observation). Plants have many species of insects that attack them.

A big difference between parasitoids and plant herbivores is that the former kill their hosts and the latter usually only damage their hosts. In fact, plants are able to compensate for considerable lev-els of insect damage. For example, cinnabar moth, *Tyria jacobaea*, can heavily defoliate stands of tansy ragwort, *Senecio jacobaeae*, without causing a long-term reduction in plant density (Myers 1980). In British Columbia, a 95% reduction of seed production was not sufficient to reduce the density of diffuse knapweed because compensatory survival of seedlings buffered plant densities from this level of insect attack (Myers 2000). A simple simulation model using relationships measured in the field for density-related seedling survival and seed production showed that the type of agent with the best potential to reduce plant density is one that attacks and kills the rosette stage of the plant (figure 27.3). Although in field observations the survival of the rosette stage of knapweed was not related to plant density, populations at two of three sites were able to compensate over one summer for removal of 80% of the rosette plants (Myers et al. 1988). Removal of up to 100% of the leaves from lantana plants *(Lantana camara)* three times in 1 year did not significantly reduce plant growth or survival, although defoliated plants allotted more biomass to reproduction than did undefoliated plants (Broughton 2000).

This ability of plants to compensate for insect attack is undoubtedly related to the relatively low rate of success of biological control. In a review of biological control of weeds in Canada, Harris showed that only about 10% of introduced insect species had any impact on host plant density. It seems that successful biological control agents are those able to kill their host plants (Myers 2000)

Table 27.1 Tests of the evolution of increased-competitive-ability hypothesis based on results and studies reviewed by (Willis et al. 1999).

Plant species	Characteristics	Indigenous	Introduced
Many	Vigor, reproduction	Less	Greater
Lythrum salicaria	Leaf phenols	Greater	Less
	Growth of specialist herbivore	Same	Same
	Survival of specialist herbivore	Same	Same
	Growth of generalist herbivore	Same	Same
	Survival and growth of root weevil	Less	Greater
	Beetle host preference	Less	Greater
Spartina alterniflora	Leaf hopper host choice	Less	Greater
	Herbivore impact	Less	Greater
	Intrinsic growth rate	Greater[a]	Less

[a]Contradicts evolution of increased competitive ability hypothesis

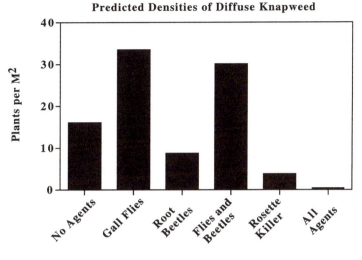

Figure 27.3 Densities of diffuse knapweed predicted by a model that incorporates density-related plant survival, reproduction, and different types of damage from biological control agents. Biological control agents include gall flies, which reduce seed production; root-feeding beetles, which reduce seed production and development rate of plants; and a hypothetical rosette-killing agent. (Based on Myers 2000.)

and in that way act in a manner similar to parasitoids attacking their insect hosts. If plants such as the thistles and the knapweeds have lots of seeds, they also seem to have lots of seed predators. This may indicate that they are resilient to the loss of seed production. One interpretation is that plants are able to compensate for the attack of their common herbivores.

Host Specificity

Host Selection: Plant and Herbivore Phylogenies The specificity and stability of host selection are important characteristics for potential biological control agents and are important attributes for reducing the possibility of introduced species feeding on nontarget host plants. Host selection by an insect can be constrained by phylogeny. If variation in host choices occurs among biotypes or species of insects, comparisons of the phylogenies of the herbivore and the hosts may help to elucidate host choices and can contribute to the study their evolu-

tion. Farrell and Mitter (1990) found strong evidence for parallel diversification between beetles in the genus *Phyllobrotica* and their host plants that indicated that the host choice of the insects had been constrained by plant speciation events. Futuyma and McCafferty (1990) also found phylogenetic constraints in host shifts of oligophagous beetles in the genus *Ophraella,* but on host plant use these constraints did not parallel the evolution of the plants. For host shifts to occur, genetic variability must exist for oviposition, feeding preferences, and larval survival on different plant species. Polyphagous insects have greater genetic variability for characteristics determining host selection than do monophagous species (Futuyma, this volume). Therefore, the patterns of relationships observed between the phylogenies of plants and herbivores may vary with the degree of speciality of the feeding choices of the insects.

Briese (1996) studied the phylogenies of two groups of insects that are biological control agents on thistles. Weevils in the genus *Larinus* attack thistles in the genera *Onopordum* and *Cynara*. In the eastern Mediterranean, *L. latus* occurs only on *Onopordum*, while in the western Mediterranean, *L. cynarae* has biotypes on *Onopordum* species in France and northern Spain, on *Cynara humulis* in southern Spain, and on *C. cardunculus* in Italy and Greece. Comparing the distributions and host choices of the insects to the distributions of host plants led Briese to conclude that host shifts occurred after host plant speciation rather than in parallel with it.

Another weevil, *Rhinocyllus conicus,* is considered to be a single species, but biotypes have been identified based on host plant use and climate. A Mediterranean climate type uses thistles in the genus *Silybum*, as well as in the genera *Carduus* and *Notobasis*. A continental climate type is restricted to the plant genera *Carduus* and *Cirsium*, and an oceanic climate group in western Europe specializes on *Cirsium*. Briese concluded that for both *Larinus* and *Rhinocyllus*, allopatry was necessary for shifts in host affiliation. He interpreted phylogenetic conservatism to indicate that sudden shifts in host plant use will not occur after agents are introduced to new areas.

Introducing Biotypes   Three biotypes of *R. conicus* were introduced to North America as biological control agents for three thistles: nodding or musk thistle, *Carduus nutans;* Italian thistle, *C. pycncephalus;* and milk thistle, *Silybum marianum*. These biotypes differed genetically and also in their propensity to attack different thistle host species. The current attack of native thistles by *R. conicus* calls into question the introduction of biotypes of insects that vary in their host plant choice.

Another way in which the introduction of different strains or biotypes has been considered is as a means of increasing genetic variability of the biological control agents. By expanding the genetic diversity of the initial population, the probability of adaptation to the climate or host type may be increased. To determine if the introduction of multiple provenances is a useful strategy for biological control releases, Clarke and Walter (1995) reviewed cases of multiple introductions of conspecific populations of biological control agents. Of 178 projects involving 417 importations of insects, 29 projects (16%) involved introductions of two strains at the same time. Whether success was due to one or the other of the strains could not be determined because they lacked distinguishing markers. The success rate of these projects was 11.8%,

which is comparable to the overall success rate for "normal" classical biological control introductions of 12–16%. In only 11 cases (6.2%) was success achieved after a second importation of the same morphologically defined species of agent. These successful cases included 5 host-related strains that are likely to have been cryptic species.

Clarke and Walter also reviewed cases that have been recorded to demonstrate success of multiple provenance introductions. These included introductions of the egg parasitoid *Trissolucus basalis* for control of the bug *Nezara viridula* in Australia, of the wasp *Trioxys pallidus* to control walnut aphid in California, of the wasp *Aphytis maculicornis* to control olive scale in California, and of the wasp *Comperiella bifasciata* to control diaspine scales. In all cases, the geographic provenances of the parasitoid were discovered retrospectively to be two species. Clark and Walter conclude that introducing two or more populations of a biological control agent can result in the introduction of cryptic species, and that control is not apparently increased by introducing several populations of the same species.

Introducing biological control agents from different source populations to increase genetic variability and the potential for adaptation has not been shown to be a useful strategy for biological control. Because biotypes can vary in their host plant acceptance and introducing different biotypes can result in the introduction of cryptic species not tested for host specificity, introducing multiple biotypes should be avoided in responsible biological control programs.

Host Specificity and Shifts to Nontarget Hosts   The greatest fear in introducing biological control agents is that they will shift onto other species of hosts and reduce population densities of native species (Cory and Myers 2000). Two recent analyses have looked at the use of nontarget hosts by biological control agents, one for biological control of weeds and the other for biological control of insects.

To determine the threat of introduced biological control agents to native plants, Pemberton (2000) analyzed data on the use of host plants by 112 insects, 3 fungi, a mite, and a nematode. He analyzed data from three areas: Hawaii, the continental United States, and the Caribbean. Of the agents introduced, approximately 13% (15 insect species) used 41 native plant species in addition to their targeted host plant. However, these cases did not indicate major host plant shifts, since 36 of the native plants attacked were congeneric with the target weeds, and 4 others were in closely allied genera. In only one case was a plant unrelated to the target weed used by the biological control agents. This analysis indicates considerable stability of food plant use by insects. The best way to avoid the use of nontarget host plants is to introduce biological control agents only to areas in which the target weed lacks close relatives.

Host specificity has been less of an issue for biological control of insects. In fact, generalist predators and parasitoids with obligatory second or third generations dependent on nontarget hosts have often been introduced in past biological control programs. Shifts in host range or in levels of attack on a species could have a genetic basis or could be environmentally controlled. Although there are few studies, a genetically based change in host specificity has not been shown for any biological control agent (Van Klinken 1999). On the other hand, native parasitoids frequently attack introduced biological control agents and reduce their potential effectiveness. An analysis of these new associations between parasitoids and hosts could be used to investigate the factors that influence host specificity of parasitoids. Are there phylogenetic linkages, or are ecological conditions most important?

Loss of Resistance or Competitive Ability in Introduced Pests Plants growing in exotic areas frequently are larger, produce more seeds, and are more competitively fit than those from native habitats. This increased vigor could be the direct result of the plants having few natural enemies. An alternative or additional factor proposed by Blossey and Nötzold (1995) is that invasive plants that lack natural enemies shift the allocation of resources from herbivore defense to competitive ability. This has been called the *evolution of increased competitive ability hypothesis*. This hypothesis predicts that plants from populations that have existed for some time in exotic areas without natural enemies will have increased vegetative growth compared to plants from populations still in the native range. In addition, it predicts that herbivores will colonize the plants from exotic areas in preference to those from the native area. The assumption behind this hypothesis is that plant defenses are costly and that therefore, when plants

grow in the absence of particular herbivores, they will lose their defensive mechanisms. Showing the cost of plant defenses has been difficult, and so this comparison between indigenous and introduced populations of plants provides a good opportunity to evaluate whether costly herbivore defenses are lost.

The results of several tests of this hypothesis that have been done are summarized in table 27.1. These studies show some supporting evidence, but not all results are as predicted. A recent comparison of the size of European plants growing in Europe compared to the same species growing in North America failed to find the predicted larger size of exotic plants (Thébaud and Simberloff 2001). The rate of change in the characteristics of introduced plants will depend on the cost of maintaining defenses in the absence of herbivores and the strength of selection for characteristics that benefit competitive ability in the new environment. More tests of this hypothesis will be required before the generality of this mechanism is known. However, the great number of plants that have been spread from native to exotic habitats provides ample opportunities for further experimentation. Introduced insect pests might also change in their resistance to parasitoids, and these provide another arena for experimentation.

Future Directions

Host resistance, plant defenses, plant tolerance, competition, coevolution, costs of resistance, and host finding and specificity are all characteristics of biological organisms on which evolution can act. These are also characteristics of potential biological control agents and their target hosts that may change with the environment and with varying selection pressure. Whether or how rapidly these characteristics change may also influence the success and environmental safety of biological pest reduction. Biological control is more than exploration, host testing, and release of agents. Consideration of the evolutionary history of potential agents with the target species may facilitate the prediction of the impacts and safety of biological control agents. Although there is some evidence that target species can change after many generations without the selection pressure of most of their natural enemy complex, rapid changes in host specificity have not been observed in biological control programs. Much further work is needed on evaluating the resistance and susceptibility of target species to particular species of natural enemies to make the success rate of biological control introductions more predictable.

Dedication

This chapter is dedicated to the memory of Samantha Hicks, who was to be a coauthor before her sudden death in September 1999. Sam was attempting to test some of the ideas presented here using the insect complex on fireweed, *Epilobium angustifolium,* as a model system. It is my hope that this chapter will stimulate other students to use biological control systems to investigate the evolutionary changes between natural enemies and their hosts. The research opportunities created by the redistribution of species around the globe are enormous.

28

Evolutionary Conservation Biology

PHILIP W. HEDRICK

Conservation biology as a discipline focused on endangered species is young and dates only from the late 1970s. Although conservation of endangered species encompasses many different biological disciplines, including behavior, ecology, and genetics, evolutionary considerations always have been emphasized (e.g., Frankel and Soule 1981). Many of the applications of evolutionary concepts to conservation are ones related to genetic variation in small or subdivided populations. However, the critical status of many endangered species makes both more precision and more caution necessary than the general findings for evolutionary considerations. On the other hand, the dire situations of many endangered species often require recommendations to be made on less than adequate data. Overall, one can think of the evolutionary aspects of conservation biology as an applied aspect of the evolution of small populations with the important constraint that any conclusions or recommendations may influence the actual extinction of the populations or species under consideration.

From this perspective, all of the factors that influence continuing evolution (i.e., selection, inbreeding, genetic drift, gene flow, and mutation; e.g., Hedrick 2000) are potentially important in conservation. The evolutionary issues of widest concern in conservation biology—inbreeding depression and maintenance of genetic variation—can be seen in their simplest form as the joint effects of inbreeding and selection, and of genetic

drift and mutation, respectively. However, even in model organisms such as *Drosophila*, the basis of inbreeding depression and the maintenance of genetic variation are not clearly understood. In addition, findings from model laboratory organisms may not provide good insight into problems in many endangered species, the most visible of which are generally slowly reproducing, large vertebrates with small populations.

Here we will first focus on introductions to two important evolutionary aspects of conservation biology: the units of conservation and inbreeding depression. Then, we will discuss studies in two organisms as illustrations of these and related principles—an endangered fish species, the Gila topminnow, and desert bighorn sheep—to illustrate some evolutionary aspects of conservation. In the discussion, we will mention some of the other evolutionary topics that are relevant to conservation biology.

Concepts

Units of Conservation

The first step in conservation is the identification of the taxon or unit that should be the object of conservation. Although in some instances this may be straightforward, determining the unit of conservation for an endangered species may often be both difficult and controversial. Some conservation biologists suggest the species as the appropriate unit; however, most others consider the unit of conservation an evolutionarily significant unit (ESU), that is, a group that is on independent evolutionary trajectories, whether it is a population, a stock, a subspecies, or a species (e.g., Moritz 1994; Waples 1995). Moritz (1994) suggested that ESUs could be identified as groups whose mtDNA were reciprocally monophyletic; in other words, lineages within each group share more recent common ancestors than with any lineages from other groups. However, because there is no recombination in mtDNA, information from mtDNA is equivalent to that for only one gene and makes general recommendations somewhat risky because different genes may have different histories over groups. Waples (1995) suggested a more comprehensive approach to identifying ESUs, an approach followed by most researchers, and he recommended that other important historical, ecological, distributional, and genetic data be included to provide an overall perspective.

Moritz (1994) suggested that some groups that do not qualify as ESUs may be significant for conservation as management units (MUs). MUs "represent populations connected by such low levels of gene flow that they are functionally independent" and can be identified as groups "that have diverged in allelic frequencies" (Moritz 1994, p. 374). Notice that both the ESU and the MU categories are developed with management considerations in mind, but that they are based on evolutionary perspectives. We will discuss application of these concepts to both Gila topminnows and desert bighorn sheep below. However, let us slightly digress now to give a perspective on the use of modern genetic data because this application is fundamental in understanding these evolutionary-based conservation categories.

Endangered species often have low genetic variation when assayed by allozymes or restriction fragment length polymorphisms (RFLPs) of mtDNA, and as a result, there has been the assumption that statistically significant differences between groups imply biological significant differences. Recently, the wide application of highly variable genetic markers and DNA sequencing has allowed documentation of genetic variation within and between many groups found by older techniques to be invariant or undifferentiated. However, the statistical power using these new molecular genetic markers is often high, and for example, all popula-

tions except very similar ones may be statistically significantly different. As a perspective on evaluating such molecular data, let us examine the connection of statistical significance and biological meaningfulness of comparisons of groups (Hedrick 1999).

First, there may be both no significant statistical and no meaningful biological difference between groups. Second, there may be a significant statistical difference between groups, and this difference may reflect a meaningful biological difference. In both these cases, statistics based on genetic markers result in an appropriate evaluation of the real biological situation.

However, problems result when statistical significance does not reflect biological meaningfulness, a conflict that can occur in two basic forms. First, there may be no statistical significance when there is an actual biologically meaningful difference between groups. For allozymes, it was often the case that there was very low statistical power to detect real differences between groups because of the low number of polymorphic loci or the low sample size. However, when highly variable loci are used, this has become much less of a problem because most organisms are variable for a number of loci and sample sizes can often be larger because of less intrusive sampling of individuals.

In some cases, there may be no significant difference based on genetic markers, but other, adaptively important loci may be highly differentiated between populations. For example, in Scots pine (*Pinus sylvestris*) from Finland, molecular markers such as allozymes, random amplified polymorphic DNA (RAPD), and microsatellites all show very little differentiation between northern and southern populations (Karhu et al. 1996). Howyever, a number of important adaptive quantitative traits, such as date of budset, show high levels of genetic differentiation between these populations in common experimental environments. In this case, the molecular data appear to be adequately reflecting the high level of gene flow in Scots pine. However, the selective forces between populations are so strong (leading to such low fitness in transplants that it is generally recommended that transplants not be over 100 km in latitude) that they overcome the effects of gene flow and result in large adaptive genetic differences between populations. Both selection and gene flow in other instances may not be as strong but could result in similar adaptive differences between populations. In this case, the error is not a typical "false negative" because the result is correct for the neutral nuclear markers. The error results from not knowing (assaying) the genes involved in adaptation, not an easy problem to overcome in most instances.

Second, there may be statistical significance between groups when there is no meaningful biological difference. This conflict portends to become a major concern in both evolutionary and conservation biology as large numbers of highly variable markers become available in many species. For example, thousands of highly polymorphic microsatellite loci are available in humans, and with the statistical power provided by so many highly variable loci, very small differences between groups would be statistically significant. This is not a typical "false positive" because the differences detected are real but they may not reflect a biologically meaningful difference.

For endangered species, these considerations may, on the one hand, lead to the conclusion that many different populations are significantly different and should be maintained as ESUs or MUs. On the other hand, they may lead to the conclusion that there is no biological difference between groups (no different ESUs or MUs) because molecular markers show homogeneity, but these neutral markers reflect only gene flow and do not detect important adaptive differences between groups. Along with other considerations, such perspectives need to be a part of ESU and MU discussions.

Inbreeding Depression

Although the detrimental effects of close inbreeding on traits related to fitness have long been known (Waser and Williams, this volume), the negative effects of inbreeding in endangered species were first documented by Ralls et al. (1979; see review by Hedrick and Kalinowski 2000). They compared juvenile survival in inbred and outbred captive vertebrate populations and found that the average survival of inbred animals was lower than that of noninbred animals in 41 of 44 species examined. In a follow-up survey, Ralls et al. (1988) showed that 36 out of 40 captive populations exhibited decreased juvenile viability in inbred animals, although there was extensive variation among estimates including populations with nonsignificant inbreeding depression. Recognition of the potential negative effect of inbreeding on fitness made inbreeding depression a concern in small-popula-

tion conservation and inbreeding avoidance a priority in captive breeding programs.

Let us introduce the approach generally used to determine the extent of inbreeding depression influencing survival in a pedigreed population, that is, a population in which the inbreeding coefficients for the different individuals can be calculated from their known pedigree. If loci determining fitness have independent, multiplicative effects, then fitness is expected to decline exponentially with the inbreeding coefficient f as

$$\overline{w}_I = \overline{w}_R e^{-Bf} \qquad (28.1)$$

where $\overline{w}_I$ and $\overline{w}_R$ are the mean fitness for an inbred population and a random mating population, respectively, and B is a constant, characteristic of a population. Notice that if there is inbreeding depression, $B > 0$, then the expected inbred fitness declines as inbreeding increases. This model is most useful for examining the relationship between inbreeding and viability, and in this case, $2B$ is approximately equal to the number of lethal equivalents in a diploid genome. Lethal equivalents are a unit that can be used to quantify the effects of genes on survival. One lethal equivalent is defined as a set of alleles that, if dispersed in different individuals, would, on average, be lethal in one individual of the group. In the survey of Ralls et al. (1988) of 40 captive populations, the median number of lethal equivalents was 3.14.

Generally only survival has been examined in endangered species because of the greater ease in collecting reliable data. However, other components of fitness, such as reproduction and mating, may be detrimentally affected by inbreeding as much as or more than survival. In addition, inbreeding depression for a given trait may be either population-specific or environmentally specific. For example, under stress conditions, inbreeding depression may be larger than under more benign conditions. A good example is from the white-footed mouse study by Jimenez et al. (1994), which we discuss below, where the number of lethal equivalents in juvenile mice in the laboratory was 0.45 based on 0.879 survival of noninbred mice and 0.822 survival of inbred mice ($f = 0.25$) from birth to weaning at day 20. On the other hand, for the initial 3-week release period in the wild, adult survival of inbreds was very low at 0.040, and for noninbreds, it was 0.194. In terms of the survival data in nature, the estimated number of lethal equivalents is 12.64, illustrating the extreme environmental dependence of inbreeding depression in this case. Finally, some attributes of the data set may make it difficult to detect inbreeding depression even though it may be present. For example, if the number of animals is small, then there may be low statistical power to detect inbreeding depression, or if the distribution over inbreeding classes is in a narrow range of inbreeding categories, it may be difficult to detect inbreeding depression.

In the past few years, there have been a series of studies consistent with negative effect of inbreeding in situations related to conservation. Three different experiments have shown that gene flow from outside populations can seemingly restore the fitness in a population with low fitness that was apparently reduced by past genetic drift, resulting in fixation of detrimental alleles. In these instances, the mean fitness of the F_1 individuals appears to be higher than that in the ancestral population. Examination of inbreeding depression in these populations before infusion of outside individuals using the traditional approach given above may not reveal inbreeding depression because all the individuals are similar genetically for detrimental alleles. Only by crossing to individuals from outside the population is the fitness restored to the same level as before the effect of genetic drift.

For example, the Florida panther (Felis concolor coryi) has a suite of traits—high frequency of cryptochordism (monochordism or unilaterally undescended testicles), deformed sperm, kinked tail, and cowlick—that are found only in high frequency in the Florida panther and are unusual in other puma subspecies (Roelke et al. 1993). The Florida panther has been isolated in southern Florida since the early 1900s, and the effective population size in recent decades appears to be 25 or less. As a result, a program to release females from the closest natural population from Texas was initiated in 1995 to genetically restore fitness in this population (Hedrick 1995). The introduced Texas females have bred with resident Florida panther males and 14 F_1 offspring have been reproduced. Of these, none have a kinked tail and only one had a cowlick (Land et al. 1999). Although the sample size is small, the frequency of these detrimental traits has been greatly reduced.

Also, an isolated population of an adder (Vipera berus) in southern Sweden declined in total estimated population size to about 10 males in 1992 (Madsen et al. 1999). During this decline, a high proportion of deformed or stillborn offspring were observed, and very low genetic variability was documented. At that time, 20 males from another population were captured and released for 3 years into the site. In 1996, the first year that new adult male adders were expected to be observed, an increased recruitment of F_1 individuals was observed. The number continued to increase, and in 1999, 32 male adders were observed, the maximum number found from 1981 to the present.

Finally, a remnant population of the greater prairie chickens (Tympanuchus cupido pinnatus) in Illinois decreased from around 2000 individuals in 1962 to fewer than 50 in 1994 (Westemeier et al. 1998). Fitness, as measured by fertility and hatching rates, declined over this period, and an estimate of genetic variation in the population was low compared both to populations from other states and historical specimens from Illinois. In 1992, translocation of 271 birds from large populations in other states was carried out, and nests that were monitored after the translocation appeared to have restored fertility and hatching rates.

Three other studies also appear to document the detrimental effects of inbreeding in nature. The population of song sparrows (Melospiza melodia) on Mandarte island, British Columbia, undergoes periodic crashes, probably due to severe winter weather (Keller et al. 1994). One such crash occurred in 1989 in which only 11% of the adult animals survived. Because there was an extensive program to mark individuals in this population over several generations, the inbreeding coefficient for most of the individuals was known before the crash. After the crash, the inbreeding coefficient for the 10 survivors was 0.0065 (only 3 had known inbreeding), while the inbreeding coefficient for 206 birds that died was significantly higher at 0.0312. In addition, all the birds with inbreeding coefficients of 0.0625 or higher, approximately 14% of the population, died during the crash. Later, high inbreeding depression was shown in the population, consistent with the differential survival observed between inbred and noninbred individuals.

Saccheri et al. (1998) reported that the rate of extinction was negatively correlated with heterozygosity in 42 small populations of fritillary butterflies (Melitaea cinxia). Over the summer of 1996, 7 of these populations went extinct. The heterozygosity, as determined by seven allozyme loci and one microsatellite locus, was significantly lower in the populations that went extinct than in the surviving populations. In addition, from laboratory studies, there appears to be substantial inbreeding depression in this species from one generation of full-sib mating. This amount of inbreeding is similar to that probably found in some of the isolated populations studied that may have been founded by a single female.

Jimenez et al. (1994) examined the survival of adult noninbred and inbred white-footed mice *(Peromyscus leucopus noveboracensis)*. The stock for the experiment was captured from the natural study site, brought into the laboratory, and bred to produce individuals with inbreeding coefficients of 0.00 or 0.25 (from full-sib matings). Almost 800 mice, nearly equally split between noninbred and inbred, were released during three different periods. The area used had a low number of mice during the release, suggesting that the environment was harsh because of some unknown cause. By an approach that utilizes capture-recapture data, the weekly survival of the inbred mice was 56% and that of the noninbred mice. In addition, inbred male mice lost body mass throughout the experiment, while noninbred male mice and non-inbred and inbred female mice did not differ.

Recognition that inbreeding depression was detrimental to captive populations was soon followed by an apparently successful attempt to purge inbreeding depression from the captive population of the endangered Speke's gazelle (Templeton and Read 1983). Although purging has not become an accepted strategy for managing small populations, the Speke's gazelle captive breeding program of Templeton and Read has remained a prominent case study in the inbreeding-depression and conservation-biology literature. The original captive Speke's gazelle population descended from three females and one male and therefore was soon faced with unavoidable half-sib or parent-offspring matings. Templeton and Read selected mating pairs for the population for 3 years and documented that the second and third generations of inbred births had significantly higher viability than the first generation of inbred gazelles. Specifically, they suggested that inbred gazelles with inbred parents had higher survival than inbred gazelles with noninbred parents.

Recently, Kalinowski et al. (2000) argued that the prominence of the Speke's gazelle breeding program has been based on mischaracterization of when viability of inbred births increased. They show that the viability of inbred births improved halfway through the first generation of inbreeding, around 1976, and then remained approximately constant through 1982 (figure 28.1). Before this time, the only inbred births were to noninbred parents, and they had low survival. The increase in viability occurred similarly for inbred births from both inbred and noninbred parents and cannot be explained by a theoretical examination of a purging model. If this interpretation is correct, the Speke's gazelle captive breeding program is a better example of the complexity of inbreeding depression, apparently related to unknown changes in the environment over time, than of purging. In either case, the Speke's gazelle captive breeding illustrates that populations with significant inbreeding depression can be successfully founded from a small number of individuals if their reproductive potential is large enough.

The Speke's gazelle captive breeding program fueled interest in purging as a conservation strategy, and some experimental work has examined the purging process in detail. A 10-generation study of inbreeding depression by Lacy and Ballou (1998) found that each of three subspecies of deer mice responded to continued inbreeding differently. In general, fitness measures declined in each of the three subspecies during the first generation of inbreeding. Subsequent generations of inbreeding were accompanied by improvements for one of the subspecies, another subspecies showed consistent decreases in fitness with subsequent inbreeding, and inbreeding depression in the third subspecies was exacerbated by further inbreeding.

Overall, deliberate inbreeding to purge inbreeding depression is a risky strategy in endangered species that have low reproductive potential. The added reduction in fitness may result in extinction, and during a purging process, detrimental alleles may become fixed permanently, lowering the population fitness (Hedrick 1994). Overall, the generally accepted approach in captive populations is to manage inbreeding depression by avoiding close inbreeding and not to attempt purging.

Case Studies

Gila Topminnows

The Gila topminnow *(Poeciliopsis occidentalis occidentalis)* is a small live-bearing fish of the family

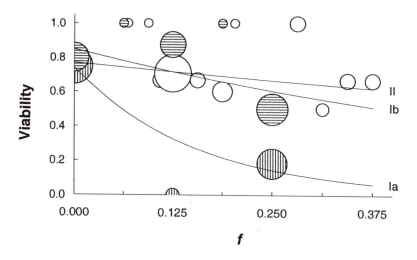

Figure 28.1 Observed (circles) and fitted (lines) viability as a function of the inbreeding coefficient *(f)* for Speke's gazelles for offspring of noninbred gazelles born prior to 1976 (vertically crosshatched circles and line Ia), offspring of noninbred gazelles born in 1976 or later (horizontally crosshatched circles and line Ib), and offspring of inbred gazelles (open circles and line II).

Poeciliidae that was once considered among the most abundant fishes in the lower Gila River basin in Arizona. They are now federally endangered and persist at only a few watersheds in southeastern Arizona, primarily because of loss and fragmentation of adequate shallow-water habitat and the widespread introduction of the nonnative western mosquitofish *(Gambusia affinis)*. Gila topminnows have been the subject of extensive genetically related research, first by Vrijenhoek and his colleagues (Vrijenhoek et al. 1985; Quattro and Vrijenhoek 1989) and more recently by me and my colleagues (Sheffer et al. 1997, 1999; Parker et al. 1999). One of the striking comparisons between these studies is that many of the conclusions from the research carried by Vrijenhoek and from our work are not consistent. Such differences suggest that conservation recommendations should be made cautiously, and when appropriate and if all possible, more than one research group should investigate a particular conservation situation.

For example, Vrijenhoek et al. (1985) found more allozyme variation in a sample from Sharp Spring than in a sample from Monkey Spring and higher values of traits potentially related to fitness (table 28.1) in laboratory experiments for Sharp Spring (Quattro and Vrijenhoek 1989) and recom-

mended the "the Sharp Spring stock currently offers the best choice from stocking the Gila River system" (p. 492). As a result, the U.S. Fish and Wildlife Service stopped using Monkey Spring fish for reintroductions and has since used only Sharp Spring stock. These studies have become one of the most cited, positive associations between heterozygosity and fitness in an endangered species.

We examined the amount of genetic variation using highly variable genetic markers and were unable to confirm that Sharp Spring fish had higher genetic variation for either microsatellite loci (Parker et al. 1999) or a major histocompatibility complex (MHC) locus than other samples. For example, the sample from Cienega Creek had approximately the same level of genetic variation for microsatellite loci (table 28.1), and a sample analyzed recently from the lower Santa Cruz River had the highest genetic variation of all samples examined so far.

Further, Sheffer et al. (1997) found that measures of traits related to fitness in fish from all four watersheds were generally similar. Differences between our data and those of Vrijenhoek and colleagues are not trivial. Under our laboratory conditions, Sharp Spring had neither higher survival, less bilateral asymmetry, nor larger size than samples

Table 28.1 Measures of expected heterozygosity (H) and fitness correlates by (a) Vrijenhoek and others and (b) Hedrick and others for four population of the endangered Gila topminnow where * indicates significantly statistical heterogeneity at $P < .05$ level for the fitness correlates.

	Monkey Spring	Sharp Spring	Bylas Spring	Cienega Creek
(a) Vrijenhoek and others				
H (allozymes)	0.000	0.037	0.000	0.000
Egg number*	5.8	13.3	—	—
Bilateral asymmetry*	0.51	0.15	—	—
Viability*	0.45	0.56	—	—
Body length			—	—
Female*	19.5	24.2		
Male*	19.3	23.0		
(b) Hedrick and others				
H (microsatellites)	0.101	0.140	0.038	0.132
Brood size				
Wild-caught*	12.1	15.8	12.8	17.8
Captive-reared	6.0	5.5	6.1	6.6
Bilateral asymmetry	0.05	0.06	0.09	0.10
Viability	0.93	0.96	0.90	0.96
Body length				
Female*	27.5	26.5	26.8	26.5
Male*	24.6	22.7	22.3	21.6

from the three other populations (table 28.1). Even for fecundity, in which the Sharp Spring sample of wild-caught fish had a higher value than that from the allozymically monomorphic Monkey Spring, the sample from Cienega Creek had slightly higher fecundity levels. These fecundity differences disappeared, however, in the next laboratory-raised generation.

Sheffer et al. (1997) concluded that the differences in the results were most likely attributable to the much more stressful laboratory environment used by Vrijenhoek in New Jersey than ours in Arizona. The difference is exemplified by an eightfold higher mortality over the first 12 weeks in their fish than in ours and bilateral asymmetry that is an order of magnitude higher in their Monkey Spring sample than in ours. Obviously, ability to cope with stressors is important for endangered species, but in this case, the stressors in New Jersey are unknown and are probably unrelated to any stressors that would be encountered in natural populations in Arizona.

Parker et al. (1999) surveyed samples of Gila topminnows from the four major watersheds, but recently, we surveyed the six other known natural populations (Hedrick et al. 2001). In addition, we sampled a population that is a natural recolonization (Santa Cruz River), a population that is the source of many of the early transplants (Boyce-Thompson Arboretum), and one transplant population of unknown origin (Watson Wash). Figure 28.2 gives a neighbor-joining tree for these data from the microsatellite loci we have examined.

The natural populations appear to fall into five groups based on these microsatellite data and the MHC data that are not included in this tree. In addition, other distributional and ecological attributes of these populations support these groupings (see Parker et al. 1999 for discussions of the ecological, life history, and other attributes of the sites). First, the two populations Bylas Spring 1 and Bylas Spring 2, the only samples from the mainstem Gila River, are very similar to each other and different from others. They also have low genetic variation as measured by both average expected heterozygosity ($H = 0.059$) and average observed number of alleles ($n = 1.2$). The Bylas populations are isolated by at least 580 km of stream channel from the closest other populations with very extensive dry reaches.

Second, the Monkey and Cottonwood Springs samples nearby in the upper Sonoita Creek watershed are similar to each other and different from all other natural populations. They also both have intermediate levels of genetic variation (average of $H = 0.207$ and $n = 2.3$). Monkey Spring appeared

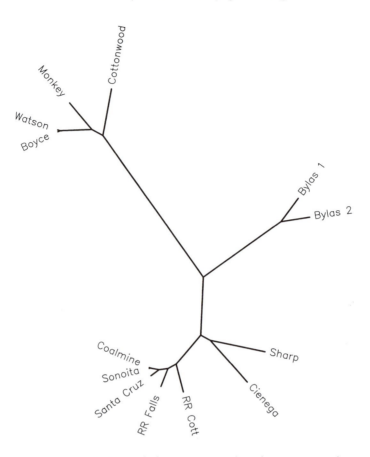

Figure 28.2 UPMGA phylogenetic tree based on genetic distances for microsatellite loci in different Gila topminnow samples.

to be the most distinctive sample in the Parker et al. survey, and this site appears to have been isolated from Sonoita Creek, into which it flows, through formation of a travertine dam, perhaps 10,000 years ago. As indicators of the long-term isolation of this site from immigration, it was occupied by a now-extinct species of pupfish and a now-extinct, morphologically distinct, form of the Gila chub. In addition, we found a substantial life history difference in that male development time was 50% longer in Monkey Spring males than in males from other sites. As a result, of the totality of these findings, Parker et al. concluded that Monkey Spring should be considered a separate ESU, and with the genetic similarity of the Cottonwood sample, it could be included in this category.

Third, the other samples from the Sonoita Creek drainage, Sonoita Creek, Coalmine Canyon, Red Rock at Cott tank, and Red Rock Falls cluster together, Coalmine Canyon and Sonoita Creek being very similar. Except for Red Rock at Cott tank, which is variable for only one locus, these populations are intermediate in genetic variation. The sample from the recolonized Santa Cruz River population clusters with these nearby Sonoita Creek samples. The Santa Cruz sample is the most variable sample analyzed with the highest heterozygosity and number of alleles ($H = 0.473$ and $n = 4.4$).

The Cienega Creek and Sharp Spring samples form the fourth and fifth groups. Although they do not appear to be as different on the microsatellite phylogenetic tree as the other groups, they have nearly nonoverlapping sets of alleles for two microsatellite loci and do not share any alleles at the MHC locus (not included in the tree). Sharp Spring has higher than average genetic variation ($H =$

0.283 and $n = 3.8$), and Cienega Creek has intermediate genetic variation. Based on these data and the discussions of other factors in Parker et al. (1999), we recommend that there be two ESUs, one composed of Monkey and Cottonwood Springs, and one composed of the remaining eight natural populations, which itself should be subdivided into four MUs. Management in this manner should allow conservation of genetic variation and evolutionary differences and also provide an approach to the long-term persistence of the Gila topminnow.

Quattro and Vrijenhoek (1989) compared several traits potentially related to fitness in the laboratory for a Tule Spring sample of Yaqui topminnows, *P. o. sonoriensis,* a related subspecies from far southeastern Arizona (and Sonora, Mexico), as an example of intermediate heterozygosity, in addition to those from Monkey and Sharp Springs. In table 28.1, we show the lack of concordance with the fitness results of Quattro and Vrijenhoek and those of Sheffer et al. (1997) within Gila topminnows, and we have now also looked at molecular variation from a sample of Yaqui topminnow from Tule Spring and at seven microsatellite loci and a MHC locus (Hedrick et al. 2001). This sample shares only 2 out of 25 microsatellite alleles (and no MHC alleles) with any of our Gila topminnow samples, and for three of the microsatellite loci, the number of repeats is greatly different for these two taxa. For example, at one microsatellite locus, the allele size in the Gila topminnow is between 143 and 153, while Yaqui topminnow is fixed for an allele that is over 50 dinucleotide repeats greater in length. As a result of this great divergence in different genetic markers, it appears that these two subspecies should be considered for ranking as separate species. Further, using microsatellite loci in common to the two taxa, Tule Spring sample has $H = 0.443$, and Sharp Spring has $H = 0.202$. Overall, then, the comparison by Quattro and Vrijenhoek (1989) used a sample from what is most certainly a different species, and the ranking of genetic variation based on allozyme loci is not consistent with that found for more variable markers.

The Gila topminnow provides an unusual example of an endangered species, one that can be bred and evaluated in captive situations with replicates, simultaneous controls, and the other attributes that make model organisms useful. For example, Sheffer et al. (1999) examined populations from the four major watersheds for the presence of inbreeding or outbreeding depression for several traits potentially related to fitness. These results are summarized in table 28.2 and show that there was generally high survival, similar body size, and little bilateral asymmetry, for all the inbred and outbred matings; that is, there was no evidence of either inbreeding or outbreeding depression for these traits. Similarly, there was no evidence of inbreeding or outbreeding effects for fecundity or sex ratio except for the sample from Monkey Spring, which had highly skewed (female-biased) sex ratios and low fecundity after one generation of inbreeding. In addition, in inbred lines maintained by full-sib mating, the fecundity for Sharp Spring fish declined over time. Even between these groups that we have suggested are different ESUs or MUs, there does not appear to be a large detrimental effect of outbreeding.

Desert Bighorn Sheep

With the western spread of Europeans in North America, the four putative subspecies of desert bighorn sheep *(Ovis canadensis),* which occupied the desert mountain ranges in California, Arizona, New Mexico, and Nevada in the United States and Sonora and Baja California in Mexico, were greatly reduced in both distribution and abundance. Although the numbers of bighorn sheep before Europeans is not known exactly and may have numbered over 1 million, only approximately 25,000 desert bighorn now remain (Lee 1993). A number of factors, such as overhunting, habitat loss and modification, competition with livestock for food and water, and predation, appear to have been important, but the single most important factor reducing bighorn sheep numbers appears to have been disease transmission from domestic livestock (Lee 1993). In recent decades, domestic livestock have been removed from much of the remaining bighorn sheep range, many waterholes have been developed in desert mountains, and transplantation has resulted in reestablishment of populations that had gone extinct. However, numbers of desert bighorn continue to decline in some areas, and for example, the desert bighorn in the Peninsular Ranges of Southern California were listed as endangered in 1998.

Traditionally, the four subspecies of the desert bighorn have been identified by male skull and horn measurements. Genetic studies using allozymes and mtDNA have proved inconclusive in re-

Table 28.2 The average measure of five traits potentially related to fitness for inbred, outbred, and simultaneous controls over four populations of Gila topminnows for the F_1 and F_2 generations (Sheffer et al. 1999) where * indicates statistical significance at $P < .05$ level compared to the simultaneous control.

Trait	F₁			F₂		
	Inbred	Outbred	Control	Inbred	Outbred	Control
Brood size	5.2*	8.8	9.9	—	—	—
Bilateral symmetry	0.05	0.03	0.04	0.07	0.07	0.05
Viability	0.95	0.91	0.91	0.95	0.91	0.93
Body length						
Females	30.7	30.4	29.8	30.3*	26.3*	25.0
Males	23.4	24.9	24.1	23.3	23.2	23.3
Sex ratio	0.35*	0.38*	0.52	0.40	0.51	0.42

solving the relationships between these subspecies, making highly variable loci, such as microsatellite loci, important (Gutiérrez-Espeleta et al. 2000). Although samples were not obtained from all the representative populations, we examined genetic variation within and between nine populations of desert bighorn sheep and two populations of Rocky Mountain sheep.

The data for the microsatellite loci can be summarized in two different ways. One would expect that different populations of the same subspecies would cluster together in a phylogenetic tree of the different populations. Although this is true of neighboring populations of the same subspecies, not all populations of the same putative subspecies cluster together. For example, the three *O. c. nelsoni* samples from northwest Arizona cluster together but away from the three Southern California *O. c. nelsoni* populations (Gutiérrez-Espeleta et al. 2000).

Another approach to understanding these data is to plot the relationship of genetic distance and geographic distance (top of figure 28.3). There is a statistically significant relationship between genetic and geographic distance, but there appears to be an asymptote of genetic distance around 0.6 at approximately 250 km. Such saturation of genetic distance over space for microsatellite loci, which often have substantial "back" mutation, has also been found in other species. In addition, if the genetic distance is plotted among the populations that are less than 300 km apart (bottom of figure 28.3), and if the within-species and between-species distances are compared, there is no significant difference. This is contrary to the expectation that

given a certain geographic distance apart, a comparison between subspecies would have a higher genetic distance than a comparison within subspecies. As a result, it appears that genetic distance between populations is primarily the result of their geographic separation and not a function of their subspecies designation.

Determining ESUs and MUs for desert bighorn sheep has been a controversial issue, including suggestions that all desert bighorn sheep are one polytypic subspecies and other suggestions that nearly all ewe bands on parts of a mountain range should be maintained. From our genetic data, geographically nearby populations, whether they are called the same subspecies or different subspecies, appear to be closely related genetically and could be considered a single MU. On the other hand, it seems appropriate to classify northern Arizona populations (all considered *O. c. nelsoni*) and southern Arizona populations (all considered *O. c. mexicana*), which are about 250 km apart, as different MUs. Because we have not analyzed all the extant populations in detail, it is difficult to consider other classifications, but presumably, the same principles would prove useful.

There has been the suggestion that desert bighorn sheep are highly inbred and that one of the causes of their decline is inbreeding depression. From our survey, the level of microsatellite heterozygosity is nearly as high as that observed in Rocky Mountain bighorn sheep or domestic sheep, a finding suggesting that there has not been a substantial decline in genetic variation. However, Sausman (1984) compared 6-month survival of noninbred and inbred captive-born bighorn sheep for 172 an-

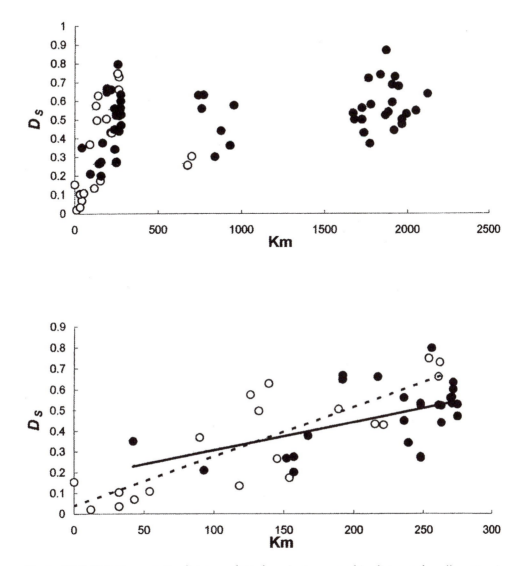

Figure 28.3 Pairwise genetic distance plotted against geographic distance for all comparisons between bighorn samples (above). Comparisons within and between populations of putative subspecies are indicated by open and closed symbols, respectively. The populations < 300 km apart (below) and the linear regression slopes for within-putative-subspecies comparisons (broken line) and between-subspecies comparisons (solid line).

imals born in the 1970s and early 1980s. She found that survival of the noninbred animals was 78% while that of inbred was only 46%, a finding suggesting that inbreeding depression was quite important in this species.

Since Sausman's study, a large number of additional bighorn sheep have been captive-born, and recently, Kalinowski and Hedrick (2001) examined the survival data for all of the 589 captive-born bighorn sheep from 1976 through 1995. The maximum likelihood estimate of the number of lethal equivalents B is only 0.23 and is not significantly different from zero, with a 95% confidence interval of (0, 0.79). In other words, there is no evidence of a decline in survival with increasing inbreeding coefficient. Kalinowski and Hedrick (2001) then categorized animals as noninbred ($f = 0$) or inbred ($f > 0$) to determine whether other factors may make obvious contributions to survival. First, figure 28.4 gives the survival as a function of date of birth, and for the first 4 years, 1976 to 1979, the survival of inbred sheep was much reduced, and significantly different, from that for noninbred sheep. In this period, the percentage survival for noninbred sheep was 67% and for inbred sheep was only 38%. On the other hand, there is no significant difference in survival for noninbred and inbred sheep born from 1980 to 1995 (77% survival for noninbred sheep and 73% for inbred sheep).

One possible explanation is that in these early years, the environment was different in some way that influenced inbred but not noninbred animals. However, such an effect that is specific to environment cannot be easily verified when the potentially important environmental differences are not known or obvious. Remember, from the discussion earlier, that lack of detection of inbreeding depression does not necessarily mean that inbreeding depression does not exist in the population or species under investigation and that in other conditions or for other traits, there may be significant inbreeding depression.

Future Directions

We have focused on two of the major evolutionary topics in conservation biology: determining the unit of conservation and the amount of inbreeding depression. In both of these areas, there is still a great deal to learn, and investigating new species often introduces new problems that have not been considered. General evolutionary guidelines are helpful in understanding particular situations and provide a flexible context for incorporating new findings in a coherent manner. However, we do not appear close to universal recommendations for either the unit of conservation or an understanding of the impact of inbreeding depression in endangered species. This may disappoint many but, in fact, reflects biological reality in that nearly every taxon appears to have its own peculiar aspects that need to be considered. As a result, it seems prudent to be cautious in recommendations, so that if they prove incorrect in some manner, they can be revised appropriately.

A number of other evolutionary topics have had considerable impact in conservation biology. For

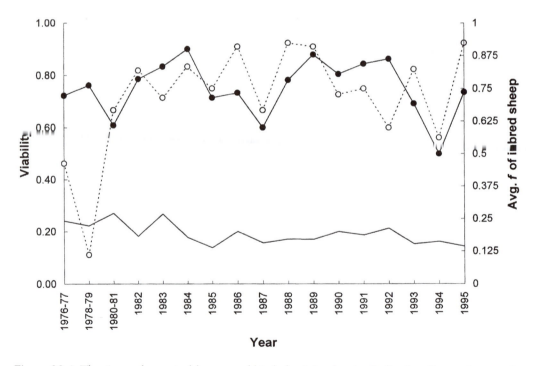

Figure 28.4 The 6-month survival by year of birth for inbred animals (broken line and open circles) and noninbred animals (solid line and closed circles). Also given is the average inbreeding of the inbred sheep for different years.

example, a major topic in both evolution and conservation in recent years has been the structure and size of the population. Nunney (this volume) has given an introduction to population structure, including metapopulations, which are often the result of fragmentation and isolation in endangered species. Generally, it is difficult to determine the present-day population structure, and it is even more difficult to determine the past structure, but new molecular techniques using DNA sequence data appear promising.

In addition, the effective size of a population (Nunney, this volume) is particularly significant because it influences both the detrimental effect of inbreeding and the general maintenance of genetic variation. The ratio of the effective population size to the adult number is also important because this ratio may be less than unity, the result being a relative loss of genetic variation and a reduction in the potential for adaptive evolution. Endangered species either may be currently going through bottlenecks or founding events or may have recently gone through them. Identifying bottlenecks has been the focus of a number of recent studies, most of which have used molecular data to document the signature of the past population size reduction.

It has been suggested that an effective population size of 50 is necessary to prevent the detrimental effects of inbreeding, and that an effective population size of 500 is necessary to prevent loss of the genetic variation necessary for continued evolution. Although this 50/500 rule has been of some general value, it has been qualified in a number of ways (Hedrick and Miller 1992). For example, there are many cases of endangered species (e.g.,

California condor, black-footed ferret, and Przewalski's horse) that have had founder numbers much less than 50 and now have rebounded in numbers without major apparent effects of inbreeding. In addition, the 500 rule was based on the balance of neutral mutational input increasing variation and genetic drift reducing variation. What the extent of adaptive input is and what other factors influence the standing genetic variation are a current topic of intense research.

Finally, determining the fitness of a population and its potential for adaptation in a changing environment is a critical unanswered question in most cases. In some cases, there appears to be a correlation with the amount of molecular genetic variation and the fitness of the population (however, see Hedrick and Miller 1992). This may provide a suggestion of the adaptive potential but may not signify specific adaptive variants, such as those involved in pathogen resistance. Obviously, it would be most appropriate to quantify the extent of different types of genetic variation, such as neutral molecular variation that can give insight into the history of the population, quantitative traits that can give insight into past selective effects and the general potential for future change, and specific adaptive loci, such as in the MHC, which can give some idea of potential adaptation to specific threats.

Acknowledgments I appreciate the contributions of G. Gutiérrez-Espeleta, S. Kalinowski, W. Minckley, R. Sheffer, and K. Parker to various aspects of this research and to NSF, Arizona Game and Fish, U.S. Fish and Wildlife, and the Ullman Professorship for financial support.

References

Abrams, P. A. 1986a. Adaptive responses of predators to prey and prey to predators: The failure of the arms race analogy. Evolution 40: 1229–1247.

Abrams, P. A. 1986b. Character displacement and niche shift analyzed using consumer-resource models of competition. Theor. Pop. Biol. 29: 107–160.

Abrams, P. A. 1990. The evolution of antipredator traits in prey in response to evolutionary change in predators. Oikos 59: 147–156.

Abrams, P. A. 1991. The effects of interacting species on predator-prey coevolution. Theor. Pop. Biol. 39: 241–262.

Abrams, P. A. 1992. Adaptive foraging by predators as a cause of predator-prey cycles. Evol. Ecol. 6: 56–72.

Abrams, P. A. 1993. Does increased mortality favor the evolution of more rapid senescence? Evolution 47: 877–887.

Abrams, P. A. 1996. Evolution and the consequences of species introductions and deletions. Ecology 77: 1321–1328.

Abrams, P. A. 1997. Evolutionary responses of foraging-related traits in unstable predator-prey systems. Evol. Ecol. 11: 673–686.

Abrams, P. A. 2000. Character shifts of species that share predators. Am. Nat. 156: S45–S61.

Abrams, P. A., and H. Matsuda. 1996. Fitness minimization and dynamic instability as a consequence of predator-prey coevolution. Evol. Ecol. 10: 167–186. (Reprinted with corrections, 1997, Evol. Ecol. 11: 1–20.)

Abrams, P. A., and H. Matsuda. 1997. Prey evolution as a cause of predator-prey cycles. Evolution 51: 1740–1748.

Abrams, P. A., Matsuda, H., and Harada, Y. 1993. Evolutionarily unstable fitness maxima and stable fitness minima in the evolution of continuous traits. Evol. Ecol. 7: 465–487.

Abrams, P. A., and Rowe, L. 1996. The effects of predation on the age and size of maturity of prey. Evolution 50: 1052–1061.

Ågren, J., and L. Ericson. 1996. Population structure and morph-specific fitness differences in tristylous *Lythrum salicaria*. Evolution 50: 126–139.

Akanda, S. I., and C. C. Mundt. 1996. Effects of two-component wheat cultivar mixtures on stripe rust severity. Phytopathology 86: 347–353.

Althoff, D. M., and J. N. Thompson. 1999. Comparative geographic structures of two parasitoid-host interactions. Evolution 53: 818–825.

Altmann, S. A. 1998. Foraging for survival: Yearling baboons in Africa. University of Chicago Press, Chicago.

Anderson, R. M., and R. M. May. 1982. Coevolution of host and parasites. Parasitology 85: 411–426.

Andersson, M. 1982. Female choice selects for extreme tail length in a widowbird. Nature 299: 818–820.

Andersson, M. 1994. Sexual selection. Princeton University Press, Princeton, N.J.

Andersson, M., and Y. Iwasa 1996. Sexual selection. Trends Ecol. Evol. 11: 53–58.

Andersson, S. 1991. Bowers on the savanna: display courts and mate choice in a lekking widowbird. Behav. Ecol. 2: 210–218.

Andersson, S. 1992. Female preference for long tails in lekking Jackson's widowbirds: Experimental evidence. Anim. Behav. 43: 379–388.

Anholt, B. R. 1991. Measuring selection on a population of damselflies with a manipulated phenotype. Evolution 45: 1091–1106.

Anstett, M. C., M. Hossaert-Mckey, and F. Kjellberg. 1997. Figs and fig pollinators: Evolutionary conflicts in a coevolved mutualism. Trends Ecol. Evol. 12: 94–99.

Antonovics, J., P. H. Thrall and A. M. Jarosz. 1998. Genetics and the spatial ecology of species interactions: The *Silene-Ustilago* system. Pp. 158–180 *in* D. Tilman and P. Kareiva (eds.), Spatial ecology: The role of space in population dynamics and interspecific interactions. Princeton University Press, Princeton, N.J.

Appel, H. M., and M. M. Martin. 1992. Significance of metabolic load in the evolution of diet specificity of *Manduca sexta*. Ecology 73: 216–228.

Arendt, J. D. 1997. Adaptive intrinsic growth rates: an integration across taxa. Quart. Rev. Biol. 72: 149–177.

Arnold, S. J., and M. J. Wade. 1984a. On the measurement of natural and sexual selection: Applications. Evolution 38: 720–734.

Arnold, S. J., and M. J. Wade. 1984b. On the measurement of natural and sexual selection: Theory. Evolution 38: 709–719.

Arnqvist, G., and F. Johansson. 1998. Ontogenetic reaction norms of predator-induced defensive morphology in dragonfly larvae. Ecology 79: 1847–1858.

Arthur, W. 1982. The evolutionary consequences of interspecific competition. Adv. Ecol. Res. 12: 127–87.

Aucoin, R., G. Guillet, C. Murray, B. J. R. Philogene, and J. T. Arnason. 1995. How do insect herbivores cope with the extreme oxidative stress of phototoxic host plants? Arch. Insect Biochem. Physiol. 29: 211–226.

Avise, J. C. 1994. Molecular markers, natural history and evolution. Chapman and Hall, New York.

Axelrod, R., and Hamilton, W. D. 1981. The evolution of cooperation. Science 211: 1390–1396.

Azevedo, R. B. R, V. French, and L. Partridge. 1996. Thermal evolution of egg size in *Drosophila melanogaster*. Evolution 50: 2338–2345.

Baer, B., and Schmid-Hempel, P. 1999. Experimental variation in polyandry affects parasite loads and fitness in a bumble-bee. Nature 397: 161–164.

Baer, C. F. 1998. Species-wide population structure in a Southeastern US freshwater fish, *Heterandria formosa*: Gene flow and biogeography. Evolution 52: 183–193.

Baer, C. F., and J. Travis. 2000. Direct and correlated responses to artificial selection on acute thermal stress tolerance in a livebearing fish. Evolution 54: 238–244.

Bakker, R. T. 1983. The deer flees, the wolf pursues: Incongruencies in predator-prey coevolution. Pp. 350–382 *in* D. J. Futuyma and M. Slatkin (eds.), Coevolution. Sinauer, Sunderland, MA.

Balmford, A., I. L. Jones, and A. L. R. Thomas. 1994. How to compensate for costly sexually selected tails: The origin of sexually dimorphic wings in long-tailed birds. Evolution 48: 1062–1070.

Barnard, C. J. and R. M. Sibly. 1981. Producers and scroungers: A general model and its application to captive flocks of house sparrows. Anim. Behav. 29: 543–555.

Barrett, S. C. H. 1992. (ed.) Evolution and function of heterostyly. Monographs on Theoretical and Applied Genetics 15. Springer-Verlag, Berlin.

Barrett, S. C. H., and L. D. Harder. 1996. Ecology and evolution of plant mating. Tr. Ecol. Evol. 11: 73–79.

Basolo, A. L. 1994. The dynamics of Fisherian sex-ratio evolution: Theoretical and experimental investigations. Am. Nat. 144: 473–490.

Bell, G. 1982. The masterpiece of nature: The evolution and genetics of sexuality. University of California Press, Berkeley, Calif.

Bell, G., and V. Koufopanou. 1986. The cost of reproduction. *In* R. Dawkins and M. Ridley (eds.), Oxford surveys in evolutionary biology. Oxford University Press, Oxford.

Benkman, C. W. 1999. The selection mosaic and diversifying coevolution between crossbills and lodgepole pine. Am. Nat. 153: S75–S91.

Berenbaum, M. R., 1983. Coumarins and caterpillars: a case for coevolution. Evolution 37: 163–179.

Berenbaum, M. R. 1991. Comparative processing of allelochemicals in the Papilionidae (Lepidoptera). Arch. Insect Biochem. Physiol. 17: 213–222.

Berenbaum, M. R. 1995. Turnabout is fair play; secondary roles for primary metabolites. J. Chem. Ecol. 21: 125–139.

Berenbaum, M. R. 1999. Animal-plant warfare: Molecular basis for cytochrome P450-mediated natural adaptation. Pp. 553–571 *in* A. Pugu and K. B. Wallace (eds.), Molecular biology of the toxic response. Philadelphia: Taylor and Francis.

Berenbaum, M. R., and A. R. Zangerl. 1992. Genetics of secondary metabolism and resistance in plants. Pp. 415–438 *in* G. A. Rosenthal and M. R. Berenbaum (eds.), Herbivores: Their interactions with secondary plant substances. Academic Press, New York.

Berenbaum, M. R., and A. R. Zangerl. 1998. Chemical phenotype matching between a plant and its insect herbivore. Proc. Natl. Acad. Sci. USA 95: 13743–13748.

Berenbaum, M. R., A. R. Zangerl, and J. K. Nitao. 1986. Constraints on chemical coevolution: Wild parsnip and the parsnip webworm. Evolution 40: 1215–1228.

Bergelson, J., and C. B. Purrington. 1996. Surveying patterns in the cost of resistance in plants. Am. Nat. 148: 536–558.

Bernays, E. A., 1991. Evolution of insect morphology in relation to plants. Philos. Trans. R. Soc. Lond. B. Biol. Sci. 333: 257–264.

Bernays, E. A., and R. F. Chapman. 1977. Deterrent chemicals as a basis of oligophagy in *Locusta migratoria* (L.). Ecol. Ent. 2: 1–18.

Bernays, E. A., and D. J. Funk. 1999. Specialists make faster decisions than generalists: Experiments with aphids. Proc. Roy. Soc. Lond. B. 266: 151–156.

Bernays, E. A., and M. Graham. 1988. On the evolution of host specificity in phytophagous arthropods. Ecology 69: 886–892.

Bernays, E. A., and W. T. Wcislo. 1994. Sensory capabilities, information processing, and resource specialization. Quart. Rev. Biol. 69: 187–204.

Berrigan, D., and J. C. Koella. 1994. The evolution of reaction norms: simple models for age and size at maturity. J. Evol. Biol. 7: 549–566.

Bertin, R. I. 1989. Pollination biology. Pp. 23–86 *in* W. G. Abrahamson (ed.), Plant-animal interactions. McGraw-Hill, New York.

Bierzychudek, P. 1987. Patterns in plant parthenogenesis. Pp. 197–217 *in* S. C. Sterns (ed.), The evolution of sex and its consequences. Birkhauser Verlag, Basel.

Birdsell, J., and C. Wills. 1996. Significant competitive advantage conferred by meiosis and syngamy in the yeast *Saccharomyces cerevisiae*. Proc. Natl. Acad. Sci. U.S.A. 93: 908–912.

Birkhead, T. R., and A. P. Møller (eds.) 1998. Sperm Competition and Sexual Selection. Academic Press, San Diego.

Bisazza A., and G. Marin. 1995. Sexual selection and sexual size dimorphism in the eastern mosquitofish *Gambusia holbrooki* (Pisces: Poeciliidae). Ethol. Ecol. and Evol. 7: 169–183.

Blossey, B., and R. Nötzold. 1995. Evolution of increased competitive ability in invasive non-indigenous plants: A hypothesis. J. Ecol. 83: 887–889.

Böhning-Gaese, K., and R. Oberrath. 1999. Phylogenetic effects on morphological, life-history, behavioural and ecological traits of birds. Evol. Ecol. Res. 1: 347–364.

Bolter, C. J., and M. A. Jongsma. 1995. Colorado potato beetle *(Leptinotarsa decemlineata)* adapt to proteinase inhibitors induced in potato leaves by methyl jasmonate. J. Insect Physiol. 41: 1071–1078.

Bowers, M. A., and R. H. Adams-Manson. 1993. Information and patch exploitation strategies of the eastern chipmunk, *Tamias striatus* (Rodentia: Sciuridae). Ethology 95: 299–308.

Bowers, M. A., and A. Ellis. 1993. Load size variation in the eastern chipmunk, *Tamias striatus:* The importance of distance to burrow and canopy cover. Ethology 94: 72–82.

Boyce, M. S., and C. M. Perrins. 1987. Optimizing great tit clutch size in a fluctuating environment. Ecology 68: 142–153.

Boyd, R., and Richerson, P. J. 1985. Culture and the evolutionary process. University of Chicago Press, Chicago.

Bradshaw, A. D. 1965. Evolutionary significance of phenotypic plasticity in plants. Adv. Gen. 13: 115–155.

Brakefield, P. M., F. Kesiseke, and P. B. Koch. 1998. The regulation of phenotypic plasticity of eyespots in the butterfly *Byciclus Anynana*. American Naturalist 152: 853–860.

Brattsten, L. B. 1992. Metabolic defenses against plant allelochemicals. Pp. 176–242 *in* G. A. Rosenthal and M. R. Berenbaum (eds.), Herbivores: Their interactions with secondary plant substances. Academic Press, New York.

Breton, L. M., and J. F. Addicott. 1992. Density-dependent mutualism in an aphid-ant interaction. Ecology 73: 2175–2180.

Briese, D. 1996. Phylogeny: Can it help us to understand host choice by biological weed control agents? Pp. 63–70 *in* V. Moran and J. Hoffman (eds.), Proc. IX. Intern. Symp. Biological Control of Weeds. University of Cape Town, Stellenbosch.

Briggs, D., and S. M. Walters. 1997. Plant variation and evolution, 3rd ed. Cambridge University Press, Cambridge.

Brodie, E. D., III, and E. D. Brodie, Jr. 1990. Tetrodotoxin resistance in garter snakes: An evolutionary response of predators to dangerous prey. Evolution 44: 651–659.

Brodie, E. D., III, and E. D. Brodie, Jr. 1991. Evolutionary response of predators to dangerous prey: Reduction of toxicity of newts and resistance of garter snakes in island populations. Evolution 45: 221–224.

Brodie, E. D. I., and E. D. J. Brodie. 1999a. Costs of exploiting poisonous prey: Evolutionary trade-offs in a predator-prey arms race. Evolution 53: 626–631.

Brodie, E. D. I., and E. D. J. Brodie. 1999b. Predator-prey arms races. Bioscience 49: 557–568.

Bronstein, J. L. 1992. Seed predators as mutualists: Ecology and evolution of the fig/pollinator interaction. Pp. 1–44 in E. Bernays (ed.), Insect-plant Interactions, Vol. 4. CRC Press, Boca Raton.

Bronstein, J. L. 1994a. Our current understanding of mutualism. Q. Rev. Biol. 69: 31–51.

Bronstein, J. L. 1994b. Conditional outcomes in mutualistic interactions. Trends Ecol. Evol. 9: 214–217.

Bronstein, J. L. in press. The exploitation of mutualisms. Ecology Letters 4.

Brooks, D. R., and D. A. McLennan. 1991. Phylogeny, ecology, and behavior: A research program in comparative biology. University of Chicago Press, Chicago.

Broughton, S. 2000. Review and evaluation of lantana biocontrol programs. Bio. Cont. 17: 272–286.

Brown, A. H. D. 1990. Genetic characterization of plant mating systems. Pp. 145–162 in A. H. D. Brown, M. T. Clegg, A. L. Kahler, and B. S. Weir (eds.), Plant population genetics, breeding, and genetic resources. Sinauer, Sunderland, Mass.

Brown, J. H., and A. Kodric-Brown. 1977. Turnover rates in insular biogeography: Effect of immigration on extinction. Ecology 58: 445–449.

Brown, J. M., J. H. Leebens-Mack, J. N. Thompson, O. Pellmyr, and R. G. Harrison. 1997. Phylogeography and host association in a pollinating seed parasite Greya politella (Lepidoptera: Prodoxidae). Mol. Ecol. 6: 215–224.

Brown, J. S., and T. L. Vincent. 1992. Organization of predator-prey communities as an evolutionary game. Evolution 46: 1269–1283.

Brown, W. L., Jr., and E. O. Wilson. 1956. Character displacement. Syst. Zool. 5: 49–64.

Brues, C. T. 1920. The selection of food-plants by insects, with special references to lepidopterous larvae. Am. Nat. 54: 313–332.

Bryant, E., and L. Meffert. 1993. The effect of serial founder-flush cycles on quantitative genetic variation in the housefly. Heredity 70: 122–129.

Bull, J. J., and E. L. Charnov. 1988. How fundamental are Fisherian sex ratios? Oxford Surv. Evol. Biol. 5: 96–135.

Bull, J. J., and W. R. Rice. 1991. Distinguishing mechanisms for the evolution of cooperation. J. Theor. Biol. 149: 63–74.

Burdon, J. J., P. H. Thrall, and A. H. D. Brown. 1999. Resistance and virulence structure in two Linum marginale–Melampsora lini host-pathogens metapopulations with different mating systems. Evolution 53: 704–716.

Burton, R. S. 1987. Differentiation and integration of the genome in populations of the marine copepod Tigriopus californicus. Evolution 41: 504–513.

Byers, D. L., and D. M. Waller. 2000. Do plant populations purge their genetic load? Effects of population size and mating history on inbreeding depression. Annu. Rev. Ecol. Syst. 30: 479–513.

Campbell, D. R. 1998. Variation in lifetime male fitness in Ipomopsis aggregata: Tests of sex allocation theory. Am. Nat. 152: 338–353.

Campbell, D. R. 2000. Experimental tests of sex allocation theory in plants. Tr. Ecol. Evol. 15: 227–232.

Campbell, D. R., and N. M. Waser. 1987. The evolution of plant mating systems: Multilocus simulations of pollen dispersal. Am. Nat. 129: 593–609.

Capy, P., J. R. David, R. Allemand, P. Hyytia, and J. Rouault. 1983. Genetic properties of North African Drosophila melanogaster and comparison with European and Afrotropical populations. Genet. Sel. Evol. 15: 185–200.

Carlson, A. 1983. Maximizing energy delivery to dependent young: A field experiment with red-backed shrikes (Lanius collurio). J. Anim. Ecol. 52: 697–704.

Carroll, L. 1872. Through the looking glass and what Alice found there. Macmillan, London.

Carroll, S. B., S. D. Weatherbee, and J. A. Langeland. 1995. Homeotic genes and the regulation and evolution of insect wing number. Nature 375: 58–61.

Carter, P. A. 1997. Maintenance of the Adh polymorphism in Ambystoma tigrinum nebulosum (tiger salamanders). 1. Genotypic differences in time to metamorphosis in extreme oxygen environments. Heredity 78: 101–109.

Case, A. L., E. P. Lacey, and R. G. Hopkins. 1996. Parental effects in Plantago lanceolata L.: 2. Manipulation of grandparental temperature and parental flowering time. Heredity 76: 287–295.

Charlesworth, B. 1994. Evolution in age-structured populations, 2nd ed. Cambridge University Press, Cambridge.

Charlesworth, B., and D. Charlesworth. 1978. A model for the evolution of dioecy and gynodioecy. Am. Nat. 112: 975–997.

Charlesworth, D. 1993. Why are unisexual flowers associated with wind pollination and unspecialized pollinators? Am. Nat. 141: 481–490.

Charlesworth, D., and B. Charlesworth. 1987. Inbreeding depression and its evolutionary consequences. Annu. Rev. Ecol. Syst. 18: 237–268.

Charnov, E. L. 1976. Optimal foraging, the marginal value theorem. Theor. Pop. Biol. 9: 129–136.

Charnov, E. L. 1982. The Theory of Sex Allocation. Princeton University Press, Princeton, N.J.

Charnov, E. L., and S. W. Skinner. 1984. Evolution of host selection and clutch size in parasitoid wasps. Florida Entomologist 67: 5–21.

Charnov, E. L., J. Maynard Smith, and J. J. Bull. 1976. Why be a hermaphrodite? Nature 263: 125–126.

Cheverud, J. M., T. T. Vaughn, L. S. Pletscher, K. King-Ellison, J. Bailiff, E. Adams, C. Erickson, and A. Bonislawski. 1999. Epistasis and the evolution of additive genetic variance in populations that pass through a bottleneck. Evolution 53: 1009–1018.

Clarke, A., and G. Walter. 1995. "Strains" and the classical biological control of insect pests. Can. J. Zool. 73: 1777–1790.

Clarke, B. 1975. The contribution of ecological genetics to evolutionary theory: Detecting the direct effects of natural selection on particular polymorphic loci. Genetics 79: 101–113.

Clausen, C. P. 1939. The effect of host size upon the sex ratio of hymenopterous parasites and its relation to methods of rearing and colonization. J. New York Ent. Soc. 47: 1–9.

Clay, K., and J. Holah. 1999. Fungal endophyte symbiosis and plant diversity in successional fields. Science 285: 1742–1744.

Clutton-Brock, T. H. 1991. The evolution of parental care. Princeton University Press, Princeton, N.J.

Cochran, D. G. 1994. Changes in insecticide resistance gene frequencies in field-collected populations of the German cockroach during extended periods of laboratory culture (Dictyoptera : Blattellidae). J. Econ. Entomol. 87: 1–6.

Cockerham, C. C. 1973. Analysis of gene frequencies. Genetics 74: 679–700.

Cohen, D. 1966. Optimizing reproduction in a randomly varying environment. J. Theor. Biol. 12: 119–129.

Cohen, D. 1971. Maximising final yield when growth is limited by time or by limiting resources. J. Theor. Biol. 33, 299–307.

Connell, J. H. 1961. The influence of interspecific competition and other factors on the distribution of the barnacle *Chthamalus stellatus*. Ecology 42: 710–723.

Connell, J. H. 1980. Diversity and the coevolution of competitors, or the ghost of competition past. Oikos 35: 131–138.

Connell, J. H. 1983. On the prevalence and relative importance of interspecific competition: Evidence from field experiments. Am. Nat. 122: 661–696.

Connor, R. C. 1995. The benefits of mutualism: A conceptual framework. Biol. Rev. 70: 427–457.

Conover, D. O., and S. W. Heins. 1987. Adaptive variation in environmental and genetic sex determination in a fish. Nature 326: 496–498.

Conover, D. O., and D. A. Van Voorhees. 1990. Evolution of a balanced sex ratio by frequency-dependent selection in fish. Science 250: 1556–1558.

Cory, J. S., and J. H. Myers. 2000. Direct and indirect ecological effects of biological control. Tr. Ecol. Evol. 15: 137–139.

Cott, H. B. 1940. Adaptive coloration in animals. Methuen, London.

Coyne, J. A., N. H. Barton, and M. Turelli. 1997. Perspective: A critique of Sewall Wright's shifting balance theory of evolution. Evolution 51: 643–671.

Crews, D., J. Sakata, and T. Rhen. 1998. Developmental effects on intersexual and intrasexual variation in growth and reproduction in a lizard with temperature-dependent sex determination. J. Comp. Bioch. Phys. C 119: 229–241.

Cronin, H. 1991. The ant and the peacock. Cambridge University Press, Cambridge.

Crozier, R. H., and P. Pamilo. 1996. Evolution of social insect colonies: Sex allocation and kin selection. Oxford University Press, New York.

Cubas, P., C. Vincent, and E. Coen. 1999. An epigenetic mutation responsible for natural variation in floral symmetry. Nature 401: 157–161.

Culley, T. M., S. G. Weller, A. K. Sakai, and A. E. Rankin. 1999. Inbreeding depression and selfing rates in a self-compatible hermaphroditic species, *Schiedea membranacea* (Caryophyllaceae). Am. J. Bot. 86: 980–987.

Cunningham, E. J. A., and A. F. Russell. 2000. Egg investment is influenced by male attractiveness in the mallard. Nature 404: 74–77.

Curio, E. 1976. The ethology of predation. Springer-Verlag, Berlin.

Daly, J. C., J. H. Fisk, and N. W. Forrester. 1988. Selective mortality in field trials between strains of *Heliothis armigera* (Lepidoptera: Noctuidae) resistant and susceptible to pyrethroid: functional dominance of resistance and age class. J. Econ. Entomol. 81: 1000–1007.

Darwin, C. 1859. On the origin of species by means of natural selection or the preservation of favored races in the struggle for life. Murray, London.

Darwin, C. 1868. The variation of animals and plants under domestication. Murray, London.

Darwin, C. 1871. The descent of man and selection in relation to sex. John Murray, London.

Darwin, C. 1876. The effects of cross and self fertilisation in the vegetable kingdom. Murray, London.

Davidson, D. W., and D. McKey. 1993. The evolutionary ecology of symbiotic ant-plant relationships. J. Hymenoptera Res. 2: 13–83.

Davies, N. B. 1991. Mating systems. Pp. 263–294 in J. R. Krebs and N. B. Davies (eds.), Behavioural ecology: An evolutionary approach, 3rd ed. Blackwell, Oxford.

Davies, N. B. 1992. Dunnock behavior and social evolution. Oxford University Press, Oxford.

Davies, A. G., P. Batterham, and J. A. McKenzie. 1992. Fatal association between dieldrin-resistant and susceptible Australian sheep blowfly, *Lucilia cuprina*. Proc. R. Soc. Lond. B. 247: 125–129.

Dawkins, R. 1976. The selfish gene, 1st ed. Oxford University Press, Oxford.

Dawkins, R., and J. R. Krebs. 1979. Arms races between and within species. Proc. Roy. Soc. Lond. B 202: 489–511.

Dayan, T., D. Simberloff, E. Tchhernov, and Y. Yom-Tov. 1990. Feline canines: Community wide character displacement among the small cats of Israel. Am. Nat. 136: 39–60.

de Jong, G., and A. J. van Noordwijk. 1992. Acquisition and allocation of resources: Genetic (co)variances, selection, and life histories. Am. Nat. 139: 749–770.

Delph, L. F., L. F. Galloway, and M. L. Stanton. 1996. Sexual dimorphism in flower size. Am. Nat. 148: 299–320.

Denholm, I., J. A. Pickett, and A. L. Devonshire (eds.) 1999. Insecticide resistance: From mechanisms to management. CABI, Oxford University Press, Wallingford, UK.

Denno, R. F., and M. A. Peterson. 1995. Density-dependent dispersal and its consequences for population dynamics. Pp. 113–130 in N. Cappuccino, and P. W. Price (eds.), Population dynamics: New approaches and synthesis. Academic Press, San Diego.

Denno, R. F., G. K. Roderick, K. L. Olmstead, and H. G. Dobel. 1991. Density-related migration in planthoppers (Homoptera: Delphacidae): The role of habitat persistence. Am. Nat. 138: 1513–1541.

Denno, R. R., M. McClure, and J. R. Ott. 1995. Interspecific interactions in phytophagous insects: Competition reexamined and resurrected. Annu. Rev. Entomol. 40: 297–231.

DeWitt, T. J., A. Sih, and D. S. Wilson. 1998. Costs and limits of phenotypic plasticity. Trends Ecol. Evol. 13: 77–81.

Diamond, J. 1998. Guns, germs, and steel. W. W. Norton, New York.

Diehl, S. R., and G. L. Bush. 1989. The role of habitat preference in adaptation and speciation. Pp. 345–365 in D. Otte and J. A. Endler (eds.), Speciation and its consequences. Sinauer, Sunderland, Mass.

Dingle, H. 1996. Migration: The biology of life on the move. Oxford University Press, New York.

Dingle, H. 2000. The evolution of migratory syndromes in insects. In I. P. Woiwod, C. D. Thomas, and D. R. Reynolds (eds.), Insect movement: Mechanisms and consequences. Oxford University Press, Wallingford, UK.

Dingle, H., W. A. Rochester, and M. P. Zalucki. 2000. Relationships among climate, latitude, and migration: Australian butterflies are not temperate zone birds. Oecologia 124: 196–207.

Dobson, A. P., and P. J. Hudson. 1992. Regulation and stability of a free-living host-parasite system, *Trichostrongylus tenuis* in red grouse. 2. Population models. J. Anim. Ecol. 61: 487–500.

Doussourd, D. E., and R. F. Denno. 1994. Host range of generalist caterpillars: Trenching permits feeding on plants with secretory canals. Ecology 75: 69–78.

Doutt, R. 1964. The historical development of biological control. In D. DeBach (ed.), Biological control of insect pests and weeds. Van Nostrand-Reinhold, Princeton, N.J.

Dudley, S. A. 1996. The response to differing selection on plant physiological traits: (evolution for local adaptation). Evolution 50: 103–110.

Dudley, S. A., and J. Schmitt. 1996. Testing the adaptive plasticity hypothesis: Density-dependent selection on manipulated stem length in *Impatiens capensis*. Am. Nat. 147: 445–465.

Dugatkin, L. A. and H. K. Reeve (eds.) 1998. Game theory and animal behavior. Oxford University Press, New York.

Dungan, M. L., T. E. Miller, and D. A. Thomson. 1982. Catastrophic decline of a top carnivore in the Gulf of California rocky intertidal zone. Science 216: 989–997.

Dunson, W. A., C. J. Paradise, and R. L. Van Fleet. 1997. Patterns of water chemistry and fish occurrence in wetlands of hydric pine flatwoods. J. Freshwater Ecol. 12: 553–565.

Dunson, W. A., and J. Travis. 1991. The role of abiotic factors in community organization. Am. Nat. 138: 1067–1091.

Dybdahl, M. F., and C. M. Lively. 1998. Host-parasite coevolution: evidence for rare advantage and time-lagged selection in a natural population. Evolution 52: 1057–1066.

Dykhuizen, D., and M. Davies. 1980. An experimental model: Bacterial specialists and generalists competing in chemostats. Ecology 61: 1213–1227.

Eberhard, W. G. 1996. Female control: Sexual selection by cryptic female choice. Princeton University Press, Princeton, N.J.

Eckert, C. G., and S. C. H. Barrett. 1993. Clonal reproduction and patterns of genotypic diversity in *Decodon verticillatus* (Lythraceae). Am. J. Bot. 80: 1175–1182.

Eckert, C. G., and S. C. H. Barrett. 1995. Style morph ratios in tristylous *Decodon verticillatus* (Lythraceae): Selection vs. historical contingency. Ecology 76: 1051–1066.

Edwards, A. W. F. 1998. Natural selection and the sex ratio: Fisher's sources. Am. Nat. 151: 564–569.

Ehrlich, P. R., and P. H. Raven. 1964. Butterflies and plants: A study in coevolution. Evolution 18: 586–608.

Einum, S., and I. A. Fleming. 2000. Highly fecund mothers sacrifice offspring survival to maximize fitness. Nature 405: 565–567.

Ekbom, B. 1998. Clutch size and larval performance of pollen beetles on different host plants. Oikos 83: 56–64.

Elle, E., N. M. van Dam, and J. D. Hare. 1999. Cost of glandular trichomes, a "resistance" character in *Datura wrightii* Regel (Solanaceae). Evolution 3: 22–35.

Emlen, J. M. 1991. Heterosis and outbreeding depression: a multilocus model and an application to salmon production. Fisheries Res. 12: 187–212.

Emlen, S. T., and L. W. Oring. 1977. Ecology, sexual selection, and the evolution of mating systems. Science 197: 215–223.

Endler, J. A. 1980. Natural selection on color patterns in *Poecilia reticulata*. Evolution 34: 76–91.

Endler, J. A. 1986. Natural selection in the wild. Princeton University Press, Princeton, N.J.

Epperson, B. K. 1999. Gene genealogies in geographically structured populations. Genetics 152: 797–806.

Eshel, I., and M. W. Feldman. 2001. Optimality and evolutionary stability under short-term and long-term selection. Pp. 161–190 *in* S. H. Orzuck and E. Sober (eds.), Adaptationism and optimality. Cambridge University Press, New York.

Ewald, P. W. 1983. Host-parasite relations, vectors, and the evolution of disease severity. Annu. Rev. Ecol. Syst. 14: 465–485.

Ewald, P. W. 1994. Evolution of infectious disease. Oxford University Press. Oxford, UK.

Ewald, P. W. 2000. Plague time. Free Press, New York.

Ewert, M. A., D. Jackson, and C. Nelson. 1994. Patterns of temperature-dependent sex determination in turtles. J. Exp. Zool. 270: 3–15.

Fairbairn, D. J., and R. F. Preziosi. 1996. Sexual selection and the evolution of sexual size dimorphism in the water strider, *Aquarius remigis*. Evolution 50: 1549–1559.

Fairbairn, D. J., and D. E. Yadlowski. 1997. Coevolution of traits determining migratory tendency: Correlated response of a critical enzyme, juvenile hormone esterase, to selection on wing morphology. J. Evol. Biol. 10: 495–513.

Falconer, D. S., and T. F. C. MacKay. 1996. Introduction to quantitative genetics, 4th ed., Longman Group, Essex, UK.

Farrell, B. D. 1998. "Inordinate fondness" explained: Why are there so many beetles? Science 281: 555–559.

Farrell, B. D., and C. Mitter. 1990. Phylogenesis of insect/plant interactions: Have *Phyllobrotica* leaf beetles (Chrysomelidae) and the Lamiales diversified in parallel? Evolution 44: 1389–1403.

Farrell B. D., and C. Mitter. 1998. The timing of insect/plant diversification—might *Tetraopes* (Coleoptera, Cerambycidae) and *Asclepias* (Asclepiadaceae) have co-evolved? Biol. J. Linn. Soc. 63: 553–577.

Farrow, R. A. 1990. Flight and migration in acridoids. Pp. 227–314 *in* R. F. Chapman and A. Joern (eds.), Biology of grasshoppers. Wiley, Somerset, N.J.

Feder, M. E., A. F. Bennett, and R. B. Huey. 2000. Evolutionary physiology. Annu. Rev. Ecol. Syst. 31: 315–341.

Feeny, P. 1977. Defensive ecology of the Cruciferae. Ann. Mo. Bot. Gard. 64: 221–234.

Fenster, C. B., and L. F. Galloway. 2000. Inbreeding and outbreeding depression in natural populations of *Chamaecrista fasciculata* (Fabaceae): Consequences for conservation biology. Conservation Biology 14: 1406–1412.

Ferguson I. M., and D. J. Fairbairn. 2000. Sex-specific selection and sexual size dimorphism in the waterstrider, *Aquarius remigis*. J. Evol. Biol. 13: 160–170.

ffrench-Constant, R. H., and R. T. Roush. 1990. Resistance detection and documentation: The relative roles of pesticidal and biochemical assays. Pp. 4–38 *in* R. T. Roush and B. E. Tabashnik (eds.), Pesticide resistance in arthropods. Chapman and Hall, New York.

Filchak, K. E., J. L. Feder, J. B. Roethele, and U. Stolz. 1999. A field test for host-plant dependent selection on larvae of the apple maggot fly, *Rhagoletis pomonella*. Evolution 53: 187–200.

Fisher, R. A. 1915. The evolution of sexual preference. Eugenics Rev. 7: 184–192.

Fisher, R. A. 1925. Statistical methods for research workers. Hafner, New York.

Fisher, R. A. 1930. The genetical theory of natural selection. Clarendon Press, Oxford.

Flores, D., A. Tousignant, and D. Crews. 1994. Incubation temperature affects the behavior of adult leopard geckos *(Eublepharis macularius)*. Phys. Behav. 55: 1067–1072.

Follett, P. A., F. Gould, and G. G. Kennedy. 1993. Comparative fitness of three strains of Colorado potato beetle (Coleoptera : Chrysomelidae) in the field: Spatial and temporal variation in insecticide selection. J. Econ. Entomol. 86: 1324–1333.

Forbes, L. S. 1991. Optimal size and number of offspring in a variable environment. J. Theor. Biol. 150: 299–304.

Force, D. C. 1972. r- and k-strategists in endemic host-parasitoid communities. Bull. Entomol. Soc. Am. 18: 135–137.

Ford, E. B. 1971. Ecological genetics. Chapman and Hall, London.

Forrester, N. W., M. Cahill, L. J. Bird, and J. Layland. 1993. Management of pyrethroid and endosulphan resistance in *Helicoverpa armigera* (Lepidoptera : Noctuidae) in Australia. Bull. Entomol. Res. Suppl. No. 1: 1–132.

Forsgren, E., C. Kvarnemo, and K. Lindström. 1996. Mode of sexual selection determined by resource abundance in two sand goby populations. Evolution 50: 646–654.

Forsman, A. 1995. Opposing fitness consequences of colour pattern in male and female snakes. J. Evol. Biol. 8: 53–70.

Forster-Blouin, S. 1989. Genetic and environmental components of thermal tolerance in the Least Killifish, *Heterandria formosa*. Ph.D. dissertation, Florida State University, Tallahassee.

Fox, C. W. 1993. A quantitative genetic analysis of oviposition preference and larval performance on two hosts in the bruchid beetle, *Callosobruchus maculatus*. Evolution 47: 166–175.

Fox, C. W. 2000. Natural selection on seed beetle egg size in the field and the lab: Variation among environments. Ecology 81: 3029–3035.

Fox, C. W., and M. E. Czesak. 2000. Evolutionary ecology of progeny size in arthropods. Annual Review of Entomology 45: 341–369.

Fox, C. W., M. E. Czesak, T. A. Mousseau, and D. A. Roff. 1999. The evolutionary genetics of an adaptive maternal effect: Egg size plasticity in a seed beetle. Evolution 53: 552–560.

Fox, C. W., and T. A. Mousseau. 1996. Larval host plant affects the fitness consequences of egg size in the seed beetle *Stator limbatus*. Oecologia 107: 541–548.

Fox, C. W., M. S. Thakar, and T. A. Mousseau. 1997. Egg size plasticity in a seed beetle: An adaptive maternal effect. Am. Nat. 149: 149–163.

Fraenkel, G. S. 1959. The raison d'être of secondary plant substances. Science 129: 1466–1470.

Fraenkel, G. S., and D. L. Gunn. 1940. The orientation of animals. Oxford University Press, Oxford.

Frank, S. A. 1985. Hierarchical selection theory and sex ratios. 2. On applying the theory, and a test with fig wasps. Evolution 39: 949–964.

Frank, S. A. 1986. Dispersal polymorphisms in subdivided populations. J. Theor. Biol. 122: 303–310.

Frank, S. A. 1994. Genetics of mutualism: The evolution of altruism between species. J. Theor. Biol. 170: 393–400.

Frank, S. A. 1996. Models of parasite virulence. Quart. Rev. Biol. 71: 37–78.

Frank, S. A. 1998. Foundations of social evolution. Princeton University Press, Princeton, N.J.

Frankel, O. H., and M. E. Soule. 1981. Conservation and evolution. Cambridge University Press, Cambridge.

Franklin, F. C. H., M. J. Lawrence, and V. E. Franklin-Tong. 1995. Cell and molecular biology of self-incompatibility in flowering plants. International Review of Cytology 158: 1–64.

Fretwell, S. D. 1972. Populations in a seasonal environment. Princeton University Press, Princeton, N.J.

Fry, J. D. 1990. Tradeoffs in fitness on different hosts: Evidence from a selection experiment with a phytophagous mite. Am. Nat. 136: 569–580.

Fry, J. D. 1993. The "general vigor" problem: can antagonistic pleiotropy be detected when genetic covariances are positive? Evolution 47: 329–333.

Fryxell, J., and P. Lundberg. 1993. Optimal patch use and metapopulation dynamics. Evol. Ecol. 7: 379–393.

Fryxell, J. M., and P. Lundberg. 1997. Individual behavior and community dynamics. Chapman and Hall, London.

Futuyma, D. J. 1998. Evolutionary biology. 3rd ed. Sinauer, Sunderland, Mass.

Futuyma, D. J. 1999. Evolution, science and society: Evolutionary biology and the national research agenda. Office of University Publications, Rutgers University, New Brunswick, N.J.

Futuyma, D. J., M. C. Keese, and D. J. Funk. 1995. Genetic constraints on macroevolution: The evolution of host affiliation in the leaf beetle genus *Ophraella*. Evolution 49: 797–809.

Futuyma, D. J., and S. McCafferty. 1990. Phylogeny and the evolution of host-plant affiliations in the leaf beetle genus *Ophraella* (Chrysomelidae: Galerucinae). Evolution 44: 1885–1913.

Futuyma, D. J., and C. Mitter. 1996. Insect-plant interactions: The evolution of component communities. Proc. Royal Soc. Lond. B 351: 1361–1366.

Futuyma, D. J., and G. Moreno. 1988. The evolution of ecological specialization. Annu. Rev. Ecol. Syst. 19: 207–223.

Futuyma, D. J., and T. E. Philippi. 1987. Genetic variation and covariation in responses to host plants by *Alsophila pometaria* (Lepidoptera: Geometridae). Evolution 41: 269–279.

Gadgil, M. 1970. Dispersal: Population consequences and evolution. Ecology 52: 253–261.

Gandon, S., Y. Capowiez Y. Dubios Y. Michalakis, and I. Olivieri. 1996. Local adaptation and gene-for-gene coevolution in a metapopulation model. Proc. R. Soc. Lond. B 263: 1003–1009.

Garland, T., Jr., and P. A. Carter. 1994. Evolutionary physiology. Annu. Rev. Physiol. 56: 579–621.

Gavrilets, S. 1997. Coevolutionary chase in exploiter-victim systems with polygenic characters. J. Theor. Biol. 186: 527–534.

Geber, M. A., T. E. Dawson, and L. F. Delph. 1999. Gender and sexual dimorphism in flowering plants. Springer-Verlag, Berlin.

Georghiou, G. P. 1986. The magnitude of the resistance problem. Pp. 14–43 *in* National Academy of Sciences (ed.), Pesticide resistance: Strategies and tactics for management. National Academy Press, Washington.

Ghiselin, M. T. 1974. The economy of nature and the evolution of sex. University of California Press, Berkeley.

Ghiselin, M. T. 1988. The evolution of sex: A history of competing points of view. Pp. 7–23 *in* R. E. Michod and B. R. Levin (eds.), The evolution of sex. Sinauer, Sunderland, Mass.

Gilbert, L. I., and E. Frieden 1981. Metamorphosis: A problem in developmental biology, 2nd ed. Plenum Press, New York.

Gilchrist, G. W., R. B. Huey, and L. Partridge. 1997. Thermal sensitivity of *Drosophila melanogaster*: Evolutionary response of adults and eggs to laboratory natural selection at different temperatures. Physiol. Zool. 70: 403–414.

Gilpin, M. L. 1975. Group selection in predator-prey communities. Princeton University Press, Princeton, N.J.

Giraldeau, L.-A., and T. Caraco. 2000. Social foraging theory. Princeton University Press, Princeton, N.J.

Giraldeau, L.-A., and D. L. Kramer. 1982. The marginal value theorem: A quantitative test using load size variation in a central place forager, the eastern chipmunk, *Tamias striatus*. Anim. Behav. 30: 1036–1042.

Giraldeau, L.-A., D. L. Kramer, I. Deplanes, and H. Lair. 1994. The effect of competitors and distance on central place foraging eastern chipmunks, *Tamias striatus*. Anim. Behav. 47: 621–632.

Godfray, H. C. J. 1987. The evolution of clutch size in parasitic wasps. Am. Nat. 129: 221–233.

Gomulkiewicz, R., J. N. Thompson, R. D. Holt, S. L. Nuismer, and M. E. Hochberg. 2000. Hot spots, cold spots, and the geographic mosaic theory of coevolution. Am. Nat. 156: 156–174.

Goodnight, C. J. 1990. Experimental studies of community evolution. 1. The response to selection at the community level. Evolution 44: 1614–1624.

Goodnight, C. J. 1995. Epistasis and the increase in additive genetic variance: Implications for phase I of Wright's shifting balance process. Evolution 49: 501–511.

Goodnight, C. J., J. M. Schwartz, and L. Stevens. 1992. Contextual analysis of models of group selection, soft selection, hard selection, and the evolution of altruism. Am. Nat. 140: 743–761.

Gosselin, L. A., and P.-Y. Qian. 1997. Juvenile mortality in benthic marine invertebrates. Marine Ecology Progress Series 146: 265–282.

Gould, F. 1979. Rapid host range evolution in a population of the phytophagous mite *Tetranychus urticae* Koch. Evolution 33: 791–802.

Gould, S. J., and R. C. Lewontin. 1979. The spandrals of San Marco and the Panglossian paradigm: a critique of the adaptationist programme. Proc. R. Soc. Lond. B 205: 581–598.

Gould, S. J., and E. S. Vrba. 1982. Exaptation—A missing term in the science of form. Paleobiology 8: 4–15.

Goulson, D., and J. S. Cory. 1995. Responses of *Mamestra brassicae* (Lepidoptera: Noctuidae) to crowding: Interactions with disease resistance, colour phase and growth. Oecologia 104: 416–423.

Gowan, C., and K. D. Fausch. 1996. Mobile brook trout in two high-elevation Colorado streams: Re-evaluating the concept of restricted movement. Can. J. Fish. Aq. Sci. 53: 1370–1381.

Grant, B. S. 1999. Fine tuning the peppered moth paradigm. Evolution 53: 980–984.

Grant, P. R. 1975. The classic case of character displacement. Evol. Biol. 8: 237–237.

Grant, P. R., and B. R. Grant. 1995. Predicting microevolutionary responses to directional selection on heritable variation. Evolution 49: 241–251.

Gray, R. D. 1987. Faith and foraging: A critique of the "paradigm argument from design." Pp. 69–140 *in* A. C. Kamil, J. R. Krebs, and H. R. Pulliam (eds.), Foraging behavior. Plenum Press, New York.

Grether, G. F. 1996. Sexual selection and survival selection on wing coloration and body size in the rubyspot damselfly *Hataerina americana*. Evolution 50: 1939–1948.

Groeters, F. R., and B. E. Tabashnik. 2000. Roles of selection intensity, major genes, and minor genes in evolution of insecticide resistance. J. Econ. Entomol. 93: 1580–1587.

Gross, M. R. 1996. Alternative reproductive strategies and tactics: Diversity within sexes. Trends Ecol. Evol. 11: 92–98.

Grutter, A. S. 1999. Cleaner fish really do clean. Nature 398: 672–673.

Gurevitch, J., L. L. Morrow, A. Wallace, and J. S. Walsh. 1992. A meta-analysis of competition in field experiments. Am. Nat. 140: 539–572.

Gutiérrez-Espeleta, G. A., S. T. Kalinowski, W. M. Boyce, and P. W. Hedrick. in press. Genetic variation and population structure in desert bighorn sheep: Implications for conservation. Cons. Genet.

Gutzke, W. H. N., and D. Crews. 1988. Embryonic temperature determines adult sexuality in a reptile. Nature 332: 832–834.

Gwynne, D. T., and I. Jamieson. 1998. Sexual selection and sexual dimorphism in a harem-polygynous insect, the alpine weta (*Hemideina maori,* Orthoptera: Stenopelmatidae). Ethol. Ecol. Evol. 10: 393–402.

Hairston, N. G., F. E. Smith, and L. B. Slobodkin. 1960. Community structure, population control, and competition. Am. Nat. 94: 421–425.

Haldane, J. B. S. 1949. Suggestions as to quantitative measurement of rates of evolution. Evolution 3: 51–56.

Hall, B. K., and M. H. Wake. 1999. The origin and evolution of larval forms. Academic Press, San Diego.

Hamilton, W. D. 1964. The genetical evolution of social behavior. 1 and 2. J. Theor. Biol. 7: 1–52.

Hamilton, W. D. 1966. The moulding of senescence by natural selection. J. Theor. Biol. 12: 12–45.

Hamilton, W. D. 1967. Extraordinary sex ratios. Science 156: 477–488.

Hamilton, W. D. 1975. Innate social aptitudes in man, an approach from evolutionary genetics. *In* R. Fox (eds.), Biosocial anthropology. Malaby Press, London.

Hamilton, W. D. 1980. Sex versus non-sex versus parasite. Oikos 35: 282–290.

Hamilton, W. D., R. Axelrod, and R. Tanese. 1990. Sexual reproduction as an adaptation to resist parasites (A review). Proc. Natl. Acad. Sci. U.S.A. 87: 3566–3573.

Hamilton, W. D., and R. M. May. 1977. Dispersal in stable habitats. Nature 269: 578–581.

Hamrick, J. L., and M. J. W. Godt. 1996. Effects of life history traits on genetic diversity in plant species. Phil. Trans. Roy. Soc. Lond. B 351: 1291–1298.

Hanski, I. 1998. Metapopulation dynamics. Nature 396: 41–49.

Harder, L. D., and S. C. H. Barrett. 1992. The energy cost of bee pollination for Pontederia cordata (Pontederiaceae). Funct. Ecol. 6: 226–233.

Hardin, G. 1968. The tragedy of the commons. Science 162: 1243–1248.

Hardy, I. C. W. 1997. Possible factors influencing vertebrate sex ratios: An introductory overview. Appl. Anim. Behav. Sci. 51: 217–241.

Harper, D. G. C. 1982. Competitive foraging in mallards: "Ideal free" ducks. Anim. Behav. 30: 575–584.

Harshman, L. G., and A. A. Hoffmann. 2000. Laboratory selection experiments using *Drosophila:* What do they really tell us? Tr. Ecol. Evol. 15: 32–36.

Hart, M. W., M. Byrne, and M. J. Smith. 1997. Molecular phylogenetic analysis of life-history evolution in asterinid starfish. Evolution 51: 1848–1861.

Harvey, P. H., and S. Nee. 1997. The phylogenetic foundations of behavioural ecology. Pp. 334–349 *in* J. R. Krebs and N. B. Davies (eds.), Behavioural ecology: An evolutionary approach, 4th ed. Blackwell, Oxford.

Harvey, P. H., and M. D. Pagel. 1991. The comparative method in evolutionary biology. Oxford University Press, Oxford.

Haskell, E. F. 1949. A clarification of social science. Main Currents in Modern Thought 7: 45–51.

Haskins, C. P., E. G. Haskins, J. J. A. McLaughlin, and R. E. Hewitt. 1961. Polymorphism and population structure in *Lebistes reticulata*, a population study. *In* W. F. Blair (ed.), Vertebrate speciation. University of Texas Press, Austin.

Hawkins, B. A. 1993. Parasitoid species richness, host mortality, and biological control. Am. Nat. 141: 634–641.

Hawkins, B. A., M. B. Thomas, and M. E. Hochberg. 1994. Refuge theory and biological control. Science 262: 1429–1432.

Hayes, T. B. 1997. Amphibian metamorphosis: An integrative approach. Am. Zool. 37: 121–207.

Hedgecock, D. 1986. Is gene flow from pelagic larval dispersal important in the adaptation and evolution of marine invertebrates? Bull. Mar. Sci. 39: 550–564.

Hedrick, P. W. 1994. Purging inbreeding depression and the probability of extinction: Full-sib mating. Heredity 73: 363–372.

Hedrick, P. W. 1995. Gene flow and genetic restoration: The Florida panther as a case study. Cons. Biol. 9: 996–1007.

Hedrick, P. W. 1999. Perspective: Highly variable loci and their interpretation in evolution and conservation. Evolution 53: 313–318.

Hedrick, P. W. 2000. Genetics of populations, 2nd ed. Jones and Bartlett, Boston.

Hedrick, P. W., and M. E. Gilpin. 1996. Genetic effective size of a metapopulation. Pp. 165–181 *in* I. Hanski and M. E. Gilpin (eds.), Metapopulation dynamics: Ecology, genetics and evolution. Academic Press, San Diego.

Hedrick, P. W., and S. T. Kalinowski. 2000. Inbreeding depression. Annu. Rev. Ecol. Syst. 31: 139–162.

Hedrick, P. W., and P. S. Miller. 1992. Conservation genetics: Techniques and fundamentals Ecol. Applic. 2: 30–46.

Hedrick, P. W., K. M. Parker, and R. Lee. 2001. Genetic variation in the endangered Gila and Yaqui topminnows: Microsatellite and MHC variation. Molec. Ecol. (in press).

Heesterbeek, J. A. P., and M. G. Roberts. 1995. Mathematical models for microparasites of wildlife. Pp. 90–122 *in* B.T. Grenfell and A.P. Dobson (eds.), Ecology of infectious diseases in natural populations. Cambridge University Press, Cambridge.

Heisler, I. L., and J. Damuth. 1987. A method for analyzing selection in hierarchically structured populations. Am. Nat. 130: 582–602.

Herre, E. A. 1985. Sex ratio adjustment in fig wasps. Science 288: 896–898.

Herre, E. A. 1999. Laws governing species interactions? Encouragement and caution from figs and their associates. Pp. 209–237 *in* L. Keller (ed.), Levels of selection in evolution. Princeton University Press, Princeton, N.J.

Herre, E. A., E. G. Leigh, Jr., and E. A. Fischer. 1987. Sex allocation in animals. Pp. 219–244 *in* S. C. Stearns (ed.), The evolution of sex and its consequences. Birkhäuser, Basel.

Herre, E. A., S. A. West, J. M. Cook, S. G. Compton, and F. Kjellberg. 1997. Fig wasps: Pollinators and parasites, sex ratio adjustment and male polymorphism, population structure and its consequences. Pp. 226–239 *in* J. Choe and B. Crespi (eds.), Social competition and cooperation in insects and arachnids: 1. Evolution of mating systems. Cambridge University Press, Cambridge.

Hines, W. G. S. 1987. Evolutionarily stable strategies: a review of basic theory. Theor. Pop. Biol. 31: 195–272.

Hochberg, M. E., and M. van Baalen. 1998. Antagonistic coevolution over productivity gradients. Am. Nat. 152: 620–634.

Hoffmann, A. A., and P. A. Parsons. 1991. Evolutionary genetics and environmental stress. Oxford University Press, Oxford.

Hokkanen, H. M. T., and D. Pimentel. 1989. New associations in biological control: Theory and practice. Canad. Entomol. 121: 829–840.

Holliday, R. 1989. Food, reproduction and longevity: Is the extended lifespan of calorie-restricted animals an evolutionary adaptation? BioEssays 10: 125–127.

Holt, R. D. 1977. Predation, apparent competition, and the structure of prey communities. Theor. Pop. Biol. 12: 197–229.

Holt, R. D., and M. S. Gaines. 1992. Analysis of adaptation in heterogeneous landscapes: Implications for the evolution of fundamental niches. Evol. Ecol. 6: 433–447.

Holt, R. D., and M. A. McPeek. 1996. Chaotic population dynamics favors the evolution of dispersal. Am. Nat. 148: 709–718.

Horgan, F. G., J. H. Myers, and R. van Meel. 1999. *Cyzenis albicans* (Diptera: Tachinidae) does not prevent the outbreak of winter moth (Lepidoptera: Geometridae) in birch stands and blueberry plots on the Lower Mainland of British Columbia. Environ. Entomol. 28: 96–107.

Houston, A., and J. McNamara. 1999. Models of adaptive behaviour. Cambridge University Press, Cambridge.

Hoy, M. A. 1990. Pesticide resistance in arthropod natural enemies: Variability and selection responses. Pp. 203–236 in R. T. Roush and B. E. Tabashnik (eds.), Pesticide resistance in arthropods. Chapman and Hall, New York.

Hudson, P. J., A. P. Dobson, and D. Newborn. 1998. Prevention of population cycles by parasite removal. Science 282: 2256–2258.

Hudson, P. J., D. Newborn, and A. P. Dobson. 1992. Regulation and stability of a free-living host-parasite system, Trichostrongylus tenuis in red grouse. 1. Monitoring and parasite reduction experiments. J. Anim. Ecol. 61: 477–486.

Hudson, R. R., D. D. Boos, and N. L. Kaplan. 1992. A statistical test for detecting geographic subdivision. Mol. Biol. Evol. 9: 138–151.

Humphries, M. M., and S. Boutin. 2000. The determinants of optimal litter size in free-ranging red squirrels. Ecology 81: 2867–2877.

Hunt, H. L., and R. E. Scheibling. 1997. Role of early post-settlement mortality in recruitment of benthic marine invertebrates. Mar. Ecol. Prog. Ser. 155: 269–301.

Hurst, L. D., A. Atlan, and B. O. Bengtsson. 1996. Genetic conflicts. Quart. Rev. Biol. 71: 317–364.

Hurst, L. D., and W. D. Hamilton. 1992. Cytoplasmic fusion and the nature of sexes. Proc. Roy. Soc. Lond. Ser. B 247: 189–194.

Hurst, L. D., and J. R. Peck. 1996. Recent advances in understanding the evolution and maintenance of sex. Tr. Ecol. Evol. 11: 46–52.

Hutchings, J. A. 1993. Adaptive life histories effected by age-specific survival and growth rate. Ecology 74: 673–684.

Hutchinson, G. E. 1957. Concluding remarks. Cold Spring Harbor Symposia on Quantitative Biology 22: 415–427.

Ims, R. A., and N. G. Yoccoz. 1996. Studying transfer processes in metapopulations: Emigration, migration and colonization. Pp. 247–265 in I. Hanski and M. E. Gilpin (eds.), Metapopulation dynamics: Ecology, genetics and evolution. Academic Press, San Diego.

Inouye, D. W. 1983. The ecology of nectar robbing. Pp. 153–173 in B. Bentley and T. Elias (eds.), The biology of nectaries. Columbia University Press, New York.

Irwin, R. E. 1994. The evolution of plumage dichromatism in the New World blackbirds: Social selection on female brightness? Am. Nat. 144: 890–907.

Istock, C. A. 1967. The evolution of complex life cycle phenomena: An ecological perspective. Evolution 21: 592–605.

Ives, A. R. 1989. The optimal clutch size of insects when many females oviposit per patch. Am. Nat. 133: 671–687.

Jablonski, D., and R. A. Lutz. 1983. Larval ecology of marine benthic invertebrates: Paleobiological implications. Biol. Rev. 58: 21–89.

Jaenike, J. 1990. Host specialization in phytophagous insects. Annu. Rev. Ecol. Syst. 21: 243–273.

Jägersten, G. 1972. Evolution of the metazoan life cycle: A comprehensive theory. Academic Press, New York.

Janz, N., and S. Nylin. 1997. The role of female search behaviour in determining host plant range in plant feeding insects: A test of the information processing hypothesis. Proc. Roy. Soc. Lond. B 264: 701–707.

Janzen, D. H. 1966. Co-evolution of mutualism between ants and acacias in Central America. Evolution 20: 249–275.

Janzen, D. H. 1980. When is it coevolution? Evolution 34: 611–612.

Janzen, D. H. 1985. The natural history of mutualisms. Pp. 40–99 in D. H. Boucher (ed.), The biology of mutualism. Oxford University Press, New York.

Janzen, F. J., and G. L. Paukstis. 1991. Environmental sex determination in reptiles: Ecology, evolution, and experimental design. Quart. Rev. Biol. 66: 149–179.

Janzen, F. J., and H. S. Stern. 1998. Logistic regression for empirical studies of multivariate selection. Evolution 52: 1564–1571.

Jarrett, J. N., and J. A. Pechenik. 1997. Temporal variation in cyprid quality and juvenile growth capacity for an intertidal barnacle. Ecology 70. 1262–1265

Jermy, T. 1976. Insect-host-plant relationship—Coevolution or sequential evolution? Symp. Biol. Hung. 16: 109–113.

Jimenez, J. A., K. A. Hughes, G. Alaks, L. Graham, and R. C. Lacy. 1994. An experimental study of inbreeding depression in a natural habitat. Science 266: 271–273.

Johannsen, W. 1911. The genotype conception of heredity. Am. Nat. 45: 129–159.

Johnson, C. G. 1960. The basis of a general system of insect migration and dispersal by flight. Nature 186: 348–350.

Johnstone, R. A., J. D. Reynolds, and J. C. Deutsch 1996. Mutual mate choice and sex differences in choosiness. Evolution 50: 1382–1391.

Jones, J. S., B. H. Leith, and P. Rawlings. 1977. Polymorphism in Cepaea: A problem with too many solutions? Annu. Rev. Ecol. and Syst. 8: 109–143.

Jong, P. W. D., and J. K. Nielsen. 1999. Polymorphism in a flea beetle for the ability to use an atypical host plant. Proc. R. Soc. Lond. B 266: 103–111.

Joshi, A., and L. D. Mueller. 1993. Directional and stabilizing density-dependent natural selection for pupation height in *Drosophila melanogaster*. Evolution 47: 176–184.

Joshi, A., and J. N. Thompson. 1995. Trade-offs and the evolution of specialization. Evol. Ecol. 9: 82–92.

Jouventin, P., and H. Weimerskirch. 1990. Satellite tracking of Wandering Albatrosses. Nature 343: 746–748.

Juenger, T., and J. Bergelson. 1998. Pairwise vs. diffuse natural selection and the multiple herbivores of Scarlet Gilia, *Ipomopsis aggregata*. Evolution 52: 1583–1592.

Kacelnik, A., and M. Bateson. 1996. Risky theories—The effects of variance in foraging decisions. Am. Zool. 36: 402–434.

Kalinowski, S., and P. W. Hedrick. 2001. Inbreeding depression in captive bighorn sheep. Anim. Cons. (in press).

Kalinowski, S. T., P. W. Hedrick, and P. S. Miller. 2000. A close look at inbreeding depression in the Speke's gazelle captive breeding program. Cons. Biol. 14: 1375–1384.

Kambysellis, M. P., and W. B. Heed. 1971. Studies of oogenesis in natural populations of Drosophilidae. 1. Relation of ovarian development and ecological habitats of the Hawaiian species. Am. Nat. 105: 31–49.

Karban, R., and I. T. Baldwin. 1997. Induced responses to herbivory. University of Chicago Press, Chicago.

Karhu, A., P. Hurme, M. Karjalainen, P. Karvonen, K. Karkkainen, D. Neale, and O. Savolainen. 1996. Do molecular markers reflect patterns of differentiation in adaptive traits of conifers? Theoret. Appl. Genet. 93: 215–221.

Kawecki, T. J., N. H. Barton, and J. D. Fry. 1997. Mutational collapse of fitness in marginal habitats and the evolution of ecological specialisation. J. Evol. Biol. 10: 407–429.

Kawecki, T. J., and S. C. Stearns. 1993. The evolution of life histories in spatially heterogeneous environments: Optimal reaction norms revisited. Evol. Ecol. 7: 155–174.

Keese, M. C. 1998. Performance of two monophagous leaf feeding beetles (Coleoptera: Chrysomelidae) on each other's host plant: do intrinsic factors determine host specialization? J. Evol. Biol. 11: 403–419.

Keller, L. F., P. Arecese, J. M. N. Smith, W. M. Hochachka, and S. C. Stearns. 1994. Selection against inbred song sparrows during a natural population bottleneck. Nature 372: 356–357.

Kelley, S. T., and B. D. Farrell. 1998. Is specialization a dead end? The phylogeny of host use in *Dendroctonus* bark beetles (Scolytidae). Evolution 52: 1731–1743.

Kennedy, J. S. 1961. A turning point in the study of insect migration. Nature 189: 785–791.

Kennedy, J. S. 1985. Migration, behavioral and ecological. Pp. 5–26 *in* M. A. Rankin (ed.), Migration: Mechanisms and adaptive significance. Institute of Marine Science, University of Texas at Austin, Port Aransas.

Kennedy, J. S., A. R. Ludlow, and C. J. Sanders. 1981. Guidance of flying male moths by wind-borne sex pheromone. Physiol. Entomol. 6: 395–412.

Kimura, M., and G. H. Weiss. 1964. The stepping stone model of population structure and decrease in genetic correlation with distance. Genetics 49: 561–576.

King, R. B. 1987. Color pattern polymorphism in the Lake Erie water snake, *Nerodia sipedon insularum*. Evolution 41: 241–255.

Kingsolver, J. G., and D. W. Schemske. 1991. Path analyses of selection. Trends Ecol. Evol. 6: 76–280.

Kirkpatrick, M., and N. H. Barton. 1997. Evolution of a species' range. Am. Nat. 150: 1–23.

Kirkpatrick, M., and L. A. Dugatkin. 1994. Sexual selection and the evolutionary effects of copying mate choice. Behav. Ecol. Sociobiol. 34: 443–449.

Kisdi, E., and S. A. H. Geritz. 1999. Adaptive dynamics in the alleles space: Evolution of genetic polymorphism by small mutations in a heterogeneous environment. Evolution 53: 993–1008.

Koenig, W. D., and S. S. Albano. 1987. Lifetime reproductive success, selection, and the opportunity for selection in the white-tailed skimmer *Plathemis lydia* (Odonata: Libellulidae). Evolution 41: 22–36.

Komdeur, J., A. Huffstadt, W. Prast, G. Castle, R. Mileto, and J. Wattel. 1995. Transfer experiments of Seychelles Warblers to new islands: Changes in dispersal and helping behaviour. Anim. Behav. 49: 695–708.

Kozlowski, J., and R. G. Wiegert. 1987. Optimal age and size at maturity in annuals and perennials with determinate growth. Evol. Ecol. 1: 231–244.

Kraaijeveld, A. R., and H. C. J. Godfray. 1999. Geographic patterns in the evolution of resistance and virulence in *Drosophila* and its parasitoids. Am. Nat. 153: S61–S74.

Kraaijeveld, A. R., J. J. M. van Alphen, and H. C. Godfray. 1998. The coevolution of host resistance and parasitoid virulence. Parasitology 116: S29–S45.

Krackow, S. 1995. Potential mechanisms for sex ratio adjustment in mammals and birds. Biol. Rev. 70: 225–241.

Kramer, D. L., and W. Nowell. 1980. Central place foraging in the eastern chipmunk, *Tamias striatus*. Anim. Behav. 28: 772–778.

Kramer, D. L., and D. M. Weary. 1991. Exploration versus exploitation: A field study of time allocation to environmental tracking by foraging chipmunks. Anim. Behav. 41: 443–449.

Krebs, J. R., and N. B. Davies. 1987. An introduction to behavioural ecology. Second edition. Blackwell Scientific, Oxford.

Kuzoff, R. K., D. E. Soltis, L. Hufford, and P. S. Soltis. 1999. Phylogenetic relationships with *Lithophragma* (Saxifragaceae): Hybridization, allopolyploidy and ovary diversification. Systematic Botany 24: 598–615.

Labandeira, C. C., and J. J. Sepkoski. 1993. Insect diversity in the fossil record. Science 261: 310–315.

Lack, D. 1947a. Darwin's finches. Cambridge University Press, Cambridge.

Lack, D. 1947b. The significance of clutch size. Ibis 89: 302–352.

Lacy, R. C., and J. D. Ballou. 1998. Effectiveness of selection in reducing the genetic load in populations of *Peromyscus polionotus*. Evolution 50: 2187–2200.

Lahn, B. T., and D. C. Page. 1999. Four evolutionary strata on the human X chromosome. Science 286: 964–967.

Lair, H., D. L. Kramer, and L.-A. Giraldeau. 1994. Interference competition in central place foragers: The effect of imposed waiting on patch-use decisions of eastern chipmunks, *Tamias striatus*. Behav. Ecol. 5: 237–244.

Land, D., M. Lotz, D. Shindle, and S. K. Taylor. 1999. Florida panther genetic restoration and management: Annual performance report 1998–1999. Florida Fish and Wildlife Conservation Commission, Naples, FL.

Lande, R. 1979. Quantitative genetics of multivariate evolution applied to brain-body size allometry. Evolution 33: 402–416.

Lande, R., and S. J. Arnold. 1983. The measurement of selection on correlated characters. Evolution 37: 1210–1226.

Lande, R., and D. W. Schemske. 1985. The evolution of self-fertilization and inbreeding depression in plants. 1. Genetic model. Evolution 39: 24–40.

Law, R. 1979. Optimal life histories under age-specific predation. Am. Nat. 114: 399–417.

Lee, E. T. 1992. Statistical methods for survival data analysis, 2nd ed. Wiley, New York.

Lee, R. M. (ed.) 1993. Desert bighorn sheep. Arizona Game and Fish Department, Phoenix.

Lefebvre, L. 2000. Feeding innovations and their cultural transmission in bird populations. Pp. 311–328 in C. M. Heyes, L. Huber, and A. Heschel (eds.), The evolution of cognition. MIT Press, Cambridge, Mass.

Leigh, E. G., E. L. Charnov, and R. R. Warner. 1976. Sex ratio, sex change and natural selection. Proc. Nat. Acad. Sci. 73: 3656–3660.

Leips, J., and J. Travis. 1999. The comparative expression of life-history traits and its relationship to the numerical dynamics of four populations of the Least Killifish, *Heterandria formosa*. J. Anim. Ecol. 68: 595–616.

Leips, J., J. Travis, and F. H. Rodd. 2000. Genetic differentiation in life-histories: Effects on population dynamics in the Least Killifish. Ecol. Mon. 70: 289–309.

Leonard, J. L. 1993. Sexual conflict in simultaneous hermaphrodites—Evidence from serranid fishes. Env. Biol. Fishes 36: 135–148.

Leonard, K. J. 1998. Modelling gene frequency dynamics. Pp. 211–230 in I. R. Crute, E. B. Holub, and J. J. Burdon. (eds.), The gene-for-gene relationship in plant-parasite interactions. CAB International, Oxford University Press, Wallingford, UK.

Leroi, A., A. K. Chippindale, and M. R. Rose. 1994. Long-term laboratory evolution of a genetic life-history trade-off in *Drosophila melanogaster*. 1. The role of genotype-by-environment interaction. Evolution 48: 1244–1257.

Letourneau, D. K. 1990. Code of ant-plant mutualism broken by parasite. Science 248: 215–217.

Levene, H. 1953. Genetic equilibrium when more than one ecological niche is available. Am. Nat. 87: 331–333.

Levins, R. 1968. Evolution in Changing Environments. Princeton University Press, Princeton, N.J.

Levins, R. 1969. Some demographic and genetic consequences of environmental heterogeneity for biological control. Bull. Entomol. Soc. Am. 15: 237–240.

Lewis, W. M., Jr. 1983. Interruption of synthesis as a cost of sex in small organisms. Am. Nat. 121: 825–834.

Lewontin, R. C. 1974. The analysis of variance and the analysis of causes. Am. J. Hum. Gen. 26: 400–411.

Lithgow, G. L., and T. B. L. Kirkwood. 1996. Mechanisms and evolution of aging. Science 273: 80.

Lively, C. M. 1989. Adaptation by a parasitic trematode to local populations of its snail host. Evolution 43: 1663–1671.

Lively, C. M. 1992. Parthenogenesis in a freshwater snail: Reproductive assurance versus parasitic release. Evolution 46: 907–913.

Lively, C. M. 1999. Migration, virulence, and the geographic mosaic of adaptation by parasites. Am. Nat. 153: S34–S47.

Lively, C. M., and M. F. Dybdahl. 2000. Parasite adaptation to locally common host genotypes. Nature 405: 679–681.

Lively, C. M., S. G. Johnson, L. F. Delph, and K. Clay. 1995. Thinning reduces the effect of rust infection on jewelweed *(Impatiens capensis)*. Ecology 76: 1851–1854.

Lloyd, D. G. 1980. Benefits and handicaps of sexual reproduction. Evol. Biol. 13: 69–111.

Lloyd, D. 1984. Gender allocations in outcrossing cosexual plants. Pp. 277–300 *in* R. Dirzo and J. Sarukhán, (eds.), Perspectives on plant population ecology. Sinauer, Sunderland, Mass.

Losos, J. B. 1990. A phylogenetic analysis of character displacement in Caribbean *Anolis* lizards. Evolution 44: 1189–1203.

Losos, J. B., T. R. Jackman, A. Larson, K. de Queiroz, and L. Rodríguez-Schettino. 1998. Contingency and determinism in replicated adaptive radiations of island lizards. Science 279: 2115–2118.

Lovelock, J. E. 1979. Gaia: A new look at life on earth. Oxford University Press, Oxford.

Lynch, M., and T. J. Crease. 1990. The analysis of populations survey data on DNA sequence variation. Mol. Biol. Evol. 7: 377–394.

Lynch, M., and B. Walsh. 1998. Genetics and analysis of quantitative traits. Sinauer, Sunderland, Mass.

Lyon, B. E. 1998. Optimal clutch size and conspecific brood parasitism. Nature 392: 380–383.

Mackenzie, A. 1996. A trade-off for host plant utilization in the black bean aphid, *Aphis fabae*. Evolution 50: 155–162.

Mackinnon, M. J., and A. F. Read. 1999. Genetic relationships between parasite virulence and transmission in the rodent malaria *Plasmodium chabaudi*. Evolution 53: 689–703.

Madsen, T., R. Shine, M. Olsson, and H. Wittsell. 1999. Restoration of an inbred adder population. Nature 402: 34–35.

Mangel, M. 1987. Oviposition site selection and clutch size in insects. J. Math. Biol. 25: 1–22.

Mangel, M., and C. W. Clark. 1988. Dynamic modeling in behavioral ecology. Princeton University Press, Princeton, N.J.

Margulis, L. 1970. Origin of eukaryotic cells. Yale University Press, New Haven.

Margulis, L. 1996. Archaeal-eubacterial mergers in the origin of Eukarya: Phylogenetic classification of life. Proc. Natl. Acad. Sci. U.S.A. 93: 1071–1076.

Marquis, R. J. 1992. Selective impact of herbivores. Pp. 392–425 *in* R. Fritz and E. Simms (eds.), Plant resistance to herbivores and pathogens: Ecology, evolution and genetics. University of Chicago Press, Chicago.

Marrow, P., U. Dieckmann, and R. Law. 1996. Evolutionary dynamics of predator-prey systems: An ecological perspective. J. Math. Biol. 34: 556–578.

Martin, G. M., S. N. Austad, and T. E. Johnson. 1996. Genetic analysis of ageing: Role of oxidative damage and environmental stresses. Nat. Genet. 13: 25–34.

Martin, M. M., and J. Harding. 1981. Evidence for the evolution of competition between two species of annual plants. Evolution 35: 975–987.

Martin, T. E., P. R. Martin, C. R. Olson, B. J. Heidinger, and J. J. Fontaine. 2000. Parental care and clutch sizes in North and South American birds. Science 287: 1482–1485.

Maruyama, T., and M. Kimura. 1980. Genetic variability and effective population size when local extinction and recolonization of some populations of frequent. Proc. Natl. Acad. Sci. 77: 6710–6714.

Matapurkar, A. K., and M. G. Watve. 1997. Altruist cheater dynamics in Dictyostelium: Aggregated distribution gives stable oscillations. Am. Nat. 150: 790–797.

Mathews, S., and R. A. Sharrock. 1997. Phytochrome gene diversity. Plant, Cell and Envir. 20: 666–671.

Matsuda, H., and P. A. Abrams. 1994. Timid consumers: Self-extinction due to adaptive change in foraging and anti-predator effort. Theor. Pop. Biol. 45: 76–91.

May, R. M., and R. M. Anderson. 1978. Regulation of stability of host-parasite population interactions. 2. Destabilizing processes. J. Anim. Ecol. 47: 249–267.

May, R. M., and R. M. Anderson. 1983. Epidemiology and genetics in the coevolution of parasites and hosts. Proc. Roy. Soc., Lond. B 219: 281–313.

Mayhew, P. J., and J. J. van Alphen. 1999. Gregarious development in alysiine parasitoids evolved through a reduction in larval aggression. Anim. Behav. 58: 131–141.

Maynard Smith, J. 1964. Group selection and kin selection. Nature 201: 1145–1147.

Maynard Smith, J. 1982. Evolution and the theory of games. Cambridge University Press, Cambridge.

Maynard Smith, J., R. Burian, S. Kauffman, P. Alberch, J. Campbell, B. Goodwin, R. Lande, D. Raup, and L. Wolpert. 1985. Developmental constraints and evolution. Quart. Rev. Biol. 60: 266–287.

Maynard Smith, J., and E. Szathmary. 1995. The major transitions of life. W. H. Freeman, New York.

Mayr, E. 1963. Animal Species and Evolution. Harvard University Press, Cambridge.

Mazer, S. J. 1987. The quantitative genetics of life history and fitness components in *Raphanus raphanistrum* L. (Brassicaceae): Ecological and evolutionary consequences of seed weight variation. Am. Nat. 130: 891–914.

Mazer, S. J., and V. A. Delesalle. 1996. Floral trait variation in *Spergularia marina* (Caryophyllaceae): Ontogenetic, maternal family, and population effects. Heredity 77: 269–281.

Mazer, S. J., V. A. Delesalle, and P. R. Neal. 1999. Responses of floral traits to selection on primary sexual investment in *Spergularia marina*: The battle between the sexes. Evolution 53: 717–731.

Mazer, S. J., and G. LeBuhn. 1999. Genetic variation in life history traits: Heritability within and geographic differentiation among populations. *In* Timo Vuarisalo and Pia Mutakainen (eds.), Life history evolution in plants. Kluwer, The Netherlands.

Mazer, S. J., and C. T. Schick. 1991a. Constancy of population parameters for life history and floral traits in *Raphanus sativus* L.: 1. Norms of reaction and the nature of genotype by environment interactions. Heredity 67: 143–156.

Mazer, S. J., and C. T. Schick. 1991b. Constancy of population parameters for life-history and floral traits in *Raphanus sativus* L. 2. Effects of planting density on phenotype and heritability estimates. Evolution 45: 1888–1907.

Mazer, S. J., and L. M. Wolfe. 1998. Density-mediated maternal effects on seed size in wild radish: Genetic variation and its evolutionary implications. Pp. 323–343 *in* T. A. Mousseau and C. W. Fox (eds.), Maternal effects as adaptations. Oxford University Press, New York.

McCormick, M. I. 1999. Delayed metamorphosis of a tropical reef fish *(Acanthurus triostegus)*: A field experiment. Mar. Ecol. Prog. Ser. 176: 25–38.

McEdward, L. 1995. Ecology of marine invertebrate larvae. CRC Press, Boca Raton, Fla.

McEvoy, P., N. Rudd, C. Cox, and M. Huso. 1993. Disturbance, competition and herbivory effects on ragwort, *Senecio jacobaea* populations. Ecological Monographs 63: 55–75.

McFadyen, R., and A. J. Tomley. 1981. Biological control of Harrisia cactus, *Eriocereus martinii*, in Queensland by the mealybug, *Hypogeoccoccus festerianus*. Pp. 589–594 *in* Proc. V. Intern. Symp. Biological control of weeds,

1980, E. Delfosse (ed.), CSIRO, Canberra, Australia.

McFadyen, R. E. 1998. Biological control of weeds. Annu. Rev. Entomol. 43: 369–393.

McKenzie, J. A. 1990. Selection at the dieldrin resistance locus in overwintering populations of *Lucilia cuprina* (Wiedemann). Aust. J. Zool. 38: 493–501.

McKenzie, J. A. 1993. Measuring fitness and intergenic interactions: The evolution of resistance to diazinon in *Lucilia cuprina*. Genetica 90: 227–237.

McKenzie, J. A. 1996. Ecological and evolutionary aspects of insecticide resistance. R. G. Landes/Academic Press, Austin.

McKenzie, J. A. 1997. Stress and asymmetry during arrested development of the Australian sheep blowfly. Proc. R. Soc. Lond. B 264: 1749–1756.

McKenzie, J. A. 2000. The character or the variation: The genetic analysis of the insecticide-resistance phenotype. Bulletin of Entomological Research 90: 3–7.

McKenzie, J. A., and P. Batterham. 1994. The genetic, molecular and phenotypic consequences of selection for insecticide resistance. Trends Ecol. Evol. 9: 166–169.

McKenzie, J. A., and P. Batterham. 1998. Predicting insecticide resistance: Mutagenesis, selection and response. Phil. Trans. R. Soc. Lond. B. 353: 1729–1734.

McLaren, I. A. 1966. Adaptive significance of large size and long life of the chaetognath *Saggitta elegans* in the Arctic. Ecology 47: 852–856.

Meagher, T. R. 1992. The quantitative genetics of sexual dimorphism in *Silene latifolia* (Caryophyllaceae): 1. Genetic variation. Evolution 46: 445–457.

Medawar, P. B. 1952. The uniqueness of the individual. Dover, New York.

Medawar, P. B. 1955. The definition and measurement of senescence. CIBA Foundation Colloquia on Ageing 1: 4–15.

Melander, A. L. 1914. Can insects become resistant to sprays? J. Econ. Entomol. 7: 167–173.

Messina, F. J. 1991. Life-history variation in a seed beetle: Adult egg-laying vs. larval competitive ability. Oecologia 85: 447–455.

Messina, F. J. 1998. Maternal influences on larval competition in insects. Pp. 227–243 *in* T. A. Mousseau and C. W. Fox (eds.), Maternal effects as adaptations. Oxford University Press, New York.

Michod, R. E. 1999. Darwinian dynamics. Princeton University Press, Princeton, N.J.

Minkoff, C., III, and T. G. Wilson. 1992. The competitive ability and fitness components of the *methoprene-tolerant* (Met) *Drosophila* mutant resistant to juvenile hormone analog insecticides. Genetics 131: 91–97.

Mitchell-Olds, T., and R. G. Shaw. 1987. Regression analysis of natural selection: Statistical inference and biological interpretation. Evolution 41: 1149–1161.

Mitteldorf, J., and D. S. Wilson. 2000. Population viscosity and the evolution of altruism. J. Theor. Biol. 204: 481–496.

Mitter, C., and D. R. Brooks. 1983. Phylogenetic aspects of coevolution. Pp. 65–98 *in* D. Futuyma and M. Slatkin (eds.), Coevolution. Sinauer, Sunderland, Mass.

Mitter, C., B. D. Farrell, and B. Wiegmann. 1988. The phylogenetic study of adaptive zones: Has phytophagy promoted insect diversifiction? Am. Nat. 132: 107–128.

Mitton, J. B. 1997. Selection in natural populations. Oxford University Press, Oxford.

Mock, D. W., and G. A. Parker. 1998. Siblicide, family conflict and the evolutionary limits of selfishness. Anim. Behav. 56: 1–10.

Moczek, A. P. 1998. Horn polyphenism in the beetle *Onthophagus taurus:* Larval diet quality and plasticity in parental investment determine adult body size and male horn morphology. Behav. Ecol. 9: 636–641.

Mode, C. J. 1958. A mathematical model for the coevolution of obligate parasites and their hosts. Evol. 12: 158–165.

Mogensen, H. L. 1996. The hows and whys of cytoplasmic inheritance in seed plants. Am. J. Bot. 83: 383–404.

Møller, A. P. 1994. Sexual selection and the barn swallow. Oxford University Press, Oxford.

Møller, A. P., and R. V. Alatalo. 1999. Good-genes effects in sexual selection. Proc. R. Soc. Lond. B 266: 85–91.

Møller, A. P., R. Dufva, and J. Erritzøe. 1998. Host immune function and sexual selection in birds. J. Evol. Biol. 11: 703–719.

Møller, A. P., and J. P. Swaddle. 1997. Asymmetry, developmental stability, and evolution. Oxford University Press, Oxford.

Monaghan, P., and R. G. Nager. 1997. Why don't birds lay more eggs? Trends Ecol. Evol. 12: 270–274.

Mongold, J. A., A. F. Bennett, and R. E. Lenski. 1999. Evolutionary adaptation to temperature. VII. Extension of the upper thermal limit of *Escherichia coli*. Evolution 53: 386–394.

Moore, A. J., E. D. Brodie III, and J. B. Wolf. 1997. Interacting phenotypes and the evolutionary process. 1. Direct and indirect genetic effects of social interactions. Evolution 51: 1352–1362.

Moritz, C. 1994. Defining "evolutionarily significant units" for conservation. Trends Ecol. Evol. 9: 373–375.

Morris, D. W. 1998. State-dependent optimization of litter size. Oikos 83: 518–528.

Mousseau, T. A., and C. W. Fox (eds.) 1998. Maternal effects as adaptations. Oxford University Press, New York.

Mueller, L. D. 1988. Density-dependent population growth and natural selection in food-limited environments: The *Drosophila* model. Am. Nat. 132: 786–809.

Murphy, M. T. 2000. Evolution of clutch size in the Eastern kingbird: Tests of alternative hypotheses. Ecol. Monogr. 70: 1–20.

Myers, J. 1980. Is the insect or the plant the driving force in the cinnabar moth-tansy ragwort system? Oecologia 47: 16–21.

Myers, J. 1984. How many insect species are necessary for successful biocontrol of weeds? Pp. 77–82 *in* Proc. VI International Symposium on Biological Control of Weeds, E. Delfosse (ed.). Agriculture Canada, Ottawa.

Myers, J. 1992. Plant-insect interactions and the biological control of weeds. Pp. 31–36 *in* E. Fontes (ed.), Proc. II Symposium on Biological Control. Pesq. Agropec. Bras, Brasilia.

Myers, J. H. 1993. Population outbreaks in forest Lepidoptera. Am. Scient. 81: 240–257.

Myers, J. 1998. Struggling with knapweed, a persistent, exotic invader. *In* P. Kranitz (ed.), Antelope-brush ecosystem symposium. Canadian Wildlife Service, Ottawa, Canada.

Myers, J. H. 2000. Why reduced seed production is not necessarily translated into successful biological weed control. *In* N. R. Spencer (ed.), Proc. X Intern. Symp. Biological Control of Weeds. Montana State University, Bozeman, Mont.

Myers, J. H., C. Higgins, and E. Kovacs. 1989. How many insects are necessary for the biological control of insects? Environ. Entomol. 18: 541–547.

Myers, J., C. Risley, and R. Eng. 1988. The ability of plants to compensate for insect attack: Why biological control of weeds with insects is so difficult. Pp. 67–73 *in* E. Delfosse (ed.), Proc. VII Intern. Symp. Biological Control of Weeds. Inst. Sper. Patol. Veg., Rome, Italy.

Nager, R. G., P. Monaghan, R. Griffiths, D. C. Houston, and R. Dawson. 1999. Experimental demonstration that offspring sex ratio varies with maternal condition. Proc. Nat. Acad. Sci. 96: 570–573.

Nei, M. 1977. F-statistics and analysis of gene diversity in subdivided populations. Ann. Hum. Genet. 41: 225–233.

Newman, R. A. 1992. Adaptive plasticity in amphibian metamorphosis. BioScience 42: 671–678.

Noë, R., and P. Hammerstein. 1995. Biological markets. Trends Ecol. Evol. 10: 336–339.

Nuismer, S. L., J. N. Thompson, and R. Gomulkiewicz. 1999. Gene flow and geographically structured coevolution. Proc. R. Soc. Lond. B 266: 605–609.

Nuismer, S. L., J. N. Thompson, and R. Gomulkiewicz. 2000. Coevolutionary clines across selection mosaics. Evolution 54: 1102–1115.

Nunney, L. 1985. Group selection, altruism and structured-deme models. Am. Nat. 126: 212–230.

Nunney, L. 1999. The effective size of a hierarchically-structured population. Evolution 53: 1–10.

Nunney, L. 2000. The limits to knowledge in conservation genetics: The predictive value of effective population size. *In* M. T. Clegg (ed.), The limits to knowledge in evolutionary genetics. Plenum Press, New York.

Nylin, S., K. Gotthard, and C. Wiklund. 1996. Reaction norms for age and size at maturity in *Lasiommata* butterflies: Predictions and tests. Evolution 50: 1351–1358.

Nylin, S., and L. Svard. 1991. Latitudinal patterns in the size of European butterflies. Hol. Ecol. 14: 192–202.

Ohlsson, R., K. Hall, and M. Ritzen. 1995. Genomic imprinting: Causes and consequences. Cambridge University Press, Cambridge.

Olivieri, I., and P.-H. Gouyon. 1996. Evolution of migration and other traits: The metapopulation effect. Pp. 293–323 *in* I. P. Hanski and M. E. Gilpin (eds.), Metapopulation dynamics: Ecology, genetics, and evolution. Academic Press, San Diego.

Olivieri, I., and P. H. Gouyon. 1997. Evolution of migration rate and other traits: The metapopulation effect. Pp. 293–323 *in* I. P. Hanski and M. E. Gilpin (eds.), Metapopulation biology: Ecology, genetics, and evolution. Academic Press, San Diego.

Olson, R. 1983. Ascidian-prochloron symbiosis: The role of larval photoadaptations in midday larval release and settlement. Biol. Bull. 165: 221–240.

Orians, G. H., and N. E. Pearson. 1979. On the theory of central place foraging. Pp. 154–177 *in* D. J. Horn, R. D. Mitchell, and G. R. Stairs (eds.), Analysis of ecological systems. Ohio State University Press, Columbus.

Orr, H. A., and J. A. Coyne. 1992. The genetics of adaptation: A reassessment. Am. Nat. 140: 725–742.

Orzack, S. H. 2002. Using sex ratios: The past and the future. *In* I. Hardy (ed.), Sex ratio

handbook. Cambridge University Press, New York.

Orzack, S. H., E. D. Parker, Jr., and J. Gladstone. 1991. The comparative biology of genetic variation for conditional sex ratio adjustment in a parasitic wasp, *Nasonia vitripennis*. Genetics 127: 583–599.

Orzack, S. H., and E. Sober. 1994. Optimality models and the test of adaptationism. Am. Nat. 143: 361–380.

Otter, K. 1994. The impact of potential predation upon the foraging behaviour of eastern chipmunks. Can. J. Zool. 72: 1858–1861.

Packer, C., M. Tatar, and A. Collins. 1998. Reproductive cessation in female mammals. Nature 392: 807–811.

Palmer, A. R. 1996. Waltzing with asymmetry. BioScience 46: 518–532.

Parker, G. A., and M. Begon. 1986. Optimal egg size and clutch size: Effects of environment and maternal phenotype. Am. Nat. 128: 573–592.

Parker, G. R., and R. A. Stuart. 1976. Animal behaviour as a strategy optimizer: Evolution of resource assessment strategies and optimal emigration thresholds. Am. Nat. 110: 1055–1076.

Parker, K. M., R. Sheffer, and P. W. Hedrick. 1999. Molecular variation and evolutionarily significant units in the endangered Gila topminnow. Cons. Biol. 13: 108–116.

Parker, M. A. 1999. Mutualism in metapopulations of legumes and rhizobia. Am. Nat. 153: S48–S60.

Partridge, L., and N. H. Barton. 1993. Optimality, mutation and the evolution of aging. Nature 362: 305–311.

Partridge, L., and N. H. Barton. 1996. On measuring the rate of aging. Proc. Roy. Soc. Lond. B 263: 1365–1371.

Partridge, L., and M. Mangel. 1999. Messages from mortality: The evolution of death rates in the old. Tr. Ecol. Evol. 14: 438–442.

Partridge, L., N. Prowse, and P. Pignatelli. 1999. Another set of responses and correlated responses to selection on age at reproduction in *Drosophila melanogaster*. Proc. Roy. Soc. Lond. B 266: 255–261.

Pawlik, J. R. 1990. Natural and artificial induction of metamorphosis of *Phragmatopoma lapidosa californica* (Polychaeta: Sabellariidae), with a critical look at the effects of bioactive compounds on marine invertebrate larvae. Bull. Mar. Sci. 46: 512–536.

Pearson, D. L. 1980. Patterns of limiting similarity in tropical forest tiger beetles (Coleoptera: Cicindelidae). Biotropica 12: 195–204.

Pechenik, J. A. 1979. Role of encapsulation in invertebrate life histories. Am. Nat. 114: 859–870.

Pechenik, J. A. 1980. Growth and energy balance during the larval lives of three prosobranch gastropods. J. Exp. Mar. Biol. Ecol. 44: 1–28.

Pechenik, J. A. 1990. Delayed metamorphosis by larvae of benthic marine invertebrates: Does it occur? Is there a price to pay? Ophelia 32: 63–94.

Pechenik, J. A. 1999. On the advantages and disadvantages of larval stages in benthic marine invertebrate life cycles. Mar. Ecol. Prog. Ser. 177: 269–297.

Pechenik, J. A. 2000. Biology of the invertebrates, 4th ed. McGraw-Hill, New York.

Pechenik, J. A., and L. S. Eyster. 1989. Influence of delayed metamorphosis on the growth and metabolism of young *Crepidula fornicata* (Gastropoda) juveniles. Biol. Bull. 176: 14–24.

Pechenik, J. A., T. Gleason, D. Daniels, and D. Champlin. In press. Influence of larval exposure to salinity and cadmium stress on juvenile performance of two marine invertebrates (*Capitella* sp. I and *Crepidula fornicata*). J. Exp. Mar. Biol. Ecol.

Pechenik, J. A., D. E. Wendt, and J. N. Jarrett. 1998. Metamorphosis is not a new beginning. BioScience 48: 901–910.

Pellmyr, O., and C. J. Huth. 1994. Evolutionary stability of mutualism between yuccas and yucca moths. Nature 372: 257–260.

Pellmyr, O., and J. Leebens-Mack. 2000. Adaptive radiation in yucca moths and the reversal of mutualism. Am. Nat. 156: S62–S76.

Pellmyr, O., J. N. Thompson, J. M. Brown, and R. G. Harrison. 1996. Evolution of pollination and mutualism in the yucca moth lineage. Am. Nat. 148: 827–847.

Pemberton, R. W. 2000. Predictable risk to native plants in weed biological control. Oecologia 125: 489–494.

Pener, M. P., A. Ayali, and E. Golenser. 1997. Adipokinetic hormone and flight fuel related characteristics of density-dependent locust phase polymorphism: A review. Comp. Biochem. Physiol. 117B: 513–524.

Perrin, N., and J. F. Rubin. 1990. On dome-shaped norms of reaction for size-at-age at maturity in fishes. Func. Ecol. 4: 53–57.

Peters, R. H. 1976. Tautology in evolution and ecology. Am. Nat. 110: 1–12.

Pettifor, R. A., C. M. Perrins, and R. H. McCleery. 1988. Individual optimization of clutch size in great tits. Nature 336: 160–162.

Pettifor, R. A., C. M. Perrins, and R. H. McCleery. 2001. The individual optimization of fitness: Variation in reproductive output, including clutch size, mean nestling mass and offspring recruitment, in manipulated broods of great tits *Parus major*. J. Anim. Ecol. 70: 62–79.

Phillips, P. C., and S. J. Arnold. 1989. Visualizing multivariate selection. Evolution 43: 1209–1222.

Pianka, E. R. 1994. Evolutionary ecology. 5th ed. HarperCollins, New York.

Pierce, N. E. 1987. The evolution and biogeography of associations between lycaenid butterflies and ants. Oxford Surv. Evol. Biol. 4: 89–116.

Pigliucci, M. 1996. How organisms respond to environmental changes: From phenotypes to molecules (and vice versa). Tr. Ecol. Evol. 11: 168–173.

Pigliucci, M. 1998. Ecological and evolutionary genetics of *Arabidopsis*. Tr. Plant Sci. 3: 485–489.

Pigliucci, M. 2000. Beyond nature vs. nurture: The genetics, ecology and evolution of genotype-environment interactions. Johns Hopkins University Press, Baltimore, Md.

Pigliucci, M. 2001. Phenotypic plasticity: Beyond nature and nurture, Johns Hopkins University Press, Baltimore, Md.

Pigliucci, M., and J. Schmitt. 1999. Genes affecting phenotypic plasticity in *Arabidopsis*: Pleiotropic effects and reproductive fitness of photomorphogenic mutants. J. Evol. Biol. 12: 551–562.

Pletcher, S. 1999. Model fitting and hypothesis testing for age-specific mortality data. J. Evol. Biol. 12: 430–440.

Policansky, D. 1981. Sex choice and the size advantage model in jack-in-the-pulpit (*Arisaema triphyllum*). Proc. Nat. Acad. Sci. 78: 1306–1308.

Policansky, D. 1982. Sex change in plants and animals. Annu. Rev. Ecol. Syst. 13: 471–495.

Pollard, H., M. Cruzan, and M. Pigliucci. in press. Comparative studies of reaction norms in *Arabidopsis*. 1. Evolution of response to daylength. Evol. Ecol. Res.

Poulin, R., and A. S. Grutter. 1996. Cleaning symbioses: Proximate and adaptive explanations. Bioscience 46: 512–517.

Preziosi, R. F., and D. J. Fairbairn. 1996. Sexual size dimorphism and selection in the wild in the waterstrider *Aquarius remigis*: Body size, components of body size and male mating success. J. Evol. Biol. 9: 317–336.

Preziosi, R. F., and D. J. Fairbairn. 1997. Sexual size dimorphism and selection in the wild in the waterstrider *Aquarius remigis*: Lifetime fecundity selection on female total length and its components. Evolution 51: 467–474.

Preziosi, R. F., and D. J. Fairbairn. 2000. Lifetime selection on adult body size and components of body size in a waterstrider: Opposing selection and maintenance of sexual size dimorphism. Evolution 54: 558–566.

Price, G. R. 1970. Selection and covariance. Nature 227: 520–521.

Price, M. V., and N. M. Waser. 1979. Pollen dispersal and optimal outcrossing in Delphinium nelsoni. Nature 277: 294–297.

Prins, H. H. T. 1996. Ecology and behaviour of the African buffalo. Chapman and Hall, London.

Pritchard, J. R., and D. Schluter. 2001. Declining interspecific competition during character displacement: Summoning the ghost of competition past. Evol. Ecol. Res. 3: 209–220.

Promislow, D. E. L. 1991. Senescence in natural populations of mammals: A comparative study. Evolution 45: 1869–1887.

Promislow, D. E. L., and M. Tatar. 1998. Mutation and senescence: Where genetics and demography meet. Genetica 102/103: 299–314.

Promislow, D. E .L., M. Tatar, S. Pletcher, and J. R. Carey. 1999. Below threshold mortality: Implications for studies in evolution, ecology and demography. J. Evol. Biol. 12: 314–328.

Provine, W. B. 1971. The origins of theoretical population genetics. University of Chicago Press, Chicago.

Purrington, C. B., and J. Bergelson. 1997. Fitness consequences of genetically engineered herbicide and antibiotic resistance in Arabidopsis thaliana. Genetics 145: 807–814.

Purugganan, M. D., and J. I. Suddith. 1999. Molecular population genetics of floral homeotic loci: Departures from the equilibrium neutral model at the APETALA3 and PISTILLATA genes of Arabidopsis thaliana. Genetics 151: 839–848.

Pyke, G. H. 1984. Optimal foraging theory: A critical review. Ann. Rev. Ecol. Syst. 15: 523–575.

Pyke, G. H. 1991. What does it cost a plant to produce floral nectar? Nature 350: 58–59.

Quattro, J. M., and R. C. Vrijenhoek. 1989. Fitness differences among remnant populations of the endangered Sonoran topminnow. Science 245: 976–978.

Queller, D. C. 1992a. Does population viscosity promote kin selection? Tr. Ecol. Evol. 7: 322–324.

Queller, D. C. 1992b. Quantitative genetics, inclusive fitness, and group selection. Am. Nat. 139: 40–558.

Queller, D. C., and K. F. Goodnight. 1989. Estimating relatedness using genetic markers. Evolution 43: 258–275.

Radtkey, R. R., S. M. Fallon, and T. J. Case. 1997. Character displacement in some Cnemidophorus lizards revisited: A phylogenetic analysis. Proc. Natl. Acad. Sci. U.S.A. 94: 9740–9745.

Raff, R. A. 1996. The shape of life: Genes, development, and the evolution of animal form. University of Chicago Press, Chicago.

Ralls, K., J. D. Ballou, and A. R. Templeton. 1988. Estimates of lethal equivalents and the cost of inbreeding in mammals. Cons. Biol. 2: 185–193.

Ralls, K., K. Brugger, and J. Ballou. 1979. Inbreeding and juvenile mortality in small populations of ungulates. Science 206: 1101–1103.

Rankin, M. A., and J. C. A. Burchsted. 1992. The cost of migration in insects. Annu. Rev. Entomol. 37: 533–560.

Rausher, M. D. 1988. Is coevolution dead? Ecology 69: 898–901.

Rausher, M. D. 1992. The measurement of selection on quantitative traits: Biases due to environmental covariances between traits and fitness. Evolution 46: 616–626.

Rausher, M. D., and E. L. Simms. 1989. The evolution of resistance to herbivory in Ipomoea purpurea. 1. Attempts to detect selection. Evolution 43: 563–572.

Rawlings, P., G. Davidson, R. K. Sakai, H. R. Rathor, M. Aslamkhan, and C. F. Curtis. 1981. Field measurement of the effective dominance of insecticide resistance in anopheline mosquitoes. Bull. WHO 59: 631–640.

Reeve, H. K., and P. W. Sherman. 1993. Adaptation and the goals of evolutionary research. Quart. Rev. Biol. 68: 1–32.

Reimer, O., and M. Tedengren. 1996. Phenotypical improvement of morphological defenses in the mussel Mytilus edulis induced by exposure to the predator Asterias rubens. Oikos 75: 383–390.

Reyer, H. U., G. Frei, and C. Som. 1999. Cryptic female choice: Frogs reduce clutch size when amplexed by undesired males. Proc. Royal Soc. Lond. Series B 266: 2101–2107.

Reynolds, J. D. 1996. Animal breeding systems. Tr. Ecol. Evol. 11: 68–72.

Reznick, D. N. 1982. The impact of predation on life history evolution in Trinidadian guppies: The genetic components of observed life history differences. Evolution 36: 1236–1250.

Reznick, D. N. 1996. Life history evolution in guppies: A model system for the empirical study of adaptation. Netherlands J. Zool. 46: 172–190.

Reznick, D. N. 1997. Life history evolution in Guppies *(Poecilia reticulata)*: Guppies as a model for studying the evolutionary biology of aging. Exp. Geront. 32: 245–258.

Reznick, D. N., and H. Bryga. 1987. Life-history evolution in guppies. 1. Phenotypic and genotypic changes in an introduction experiment. Evolution 41: 1370–1385.

Reznick, D. N., H. Bryga, and J. A. Endler. 1990. Experimentally induced life-history evolution in a natural population. Nature 346: 357–359.

Reznick, D. N., I. Butler M. J., F. H. Rodd, and P. Ross. 1996. Life history evolution in guppies *(Poecilia reticulata)*. 6. Differential mortality as a mechanism for natural selection. Evolution 50: 1651–1660.

Reznick, D. N., F. H. Shaw, F. H. Rodd, and R. G. Shaw. 1997. Evaluation of the rate of evolution in natural populations of guppies *(Poecilia reticulata)*. Science 275: 1934–1937.

Reznick, D. N., and J. Travis. 1996. The empirical study of adaptation in natural populations. Pp. 243–290 *in* M. R. Rose and G. V. Lauder (eds.), Adaptation. Academic Press, San Diego.

Rhen, T. 2000. Sex-limited mutations and the evolution of sexual dimorphism. Evolution 54: 37–43.

Rhen, T., and D. Crews. 1999. Embryonic temperature and gonadal sex organize male-typical sexual and aggressive behavior in a lizard with temperature-dependent sex determination. Endocrinology 140: 4501–4508.

Rhen, T., and J. W. Lang. 1994. Temperature-dependent sex determination in the snapping turtle: Manipulation of the embryonic sex steroid environment. Gen. Comp. Endocrin. 96: 243–254.

Rhen, T., and J. W. Lang. 1995. Phenotypic plasticity for growth in the common snapping turtle: Effects of incubation temperature, clutch, and their interaction. Am. Nat. 146: 726–747.

Rhen, T., and J. W. Lang. 1998. Among-family variation for environmental sex determination in reptiles. Evolution 52: 1514–1520.

Rhen, T., and J. W. Lang. 1999a. Temperature during embryonic and juvenile development influences growth in hatchling snapping turtles, *Chelydra serpentina*. J. Thermal Biol. 34: 33–41.

Rhen, T., and J. W. Lang. 1999b. Incubation temperature and sex affect mass and energy reserves of hatchling snapping turtles, *Chelydra serpentina*. Oikos 86: 311–319.

Rice, W. R. 1983. Parent-offspring pathogen transmission: A selective agent promoting sexual reproduction. Am. Nat. 121: 187–203.

Rice, W. R. 1984. Sex chromosomes and the evolution of sexual dimorphism. Evolution 38: 735–742.

Rice, W. R. 1996. Evolution of the Y sex chromosome in animals. BioScience 46: 331–343.

Richards, A. J. 1997. Plant breeding systems, 2nd ed. Chapman and Hall, London.

Ricklefs, R. E. 1976. The economy of nature. Chiron Press, Portland.

Ricklefs, R. E. 1998. Evolutionary theories of aging: Confirmation of a fundamental prediction, with implications for the genetic basis and evolution of life span. Am. Nat. 152: 24–44.

Rissing, S. W., G. B. Pollock, M. R. Higgins, R. H. Hagen, and D. R. Smith. 1989. Foraging specialization without relatedness or dominance among co-founding ant queens. Nature 338: 420–422.

Ritland, K. 1990. Inferences about inbreeding depression based on changes of the inbreeding coefficient. Evolution 44: 1230–1241.

Robertson, F. W., M. Shook, G. Takel, and H. Gaines. 1968. Observations on the biology and nutrition of *Drosophila disticha,* Hardy, and indigenous Hawaiian species. Studies in Genetics, IV. Res. Rep. 4: 279–299.

Roche, B. M., and R. S. Fritz. 1997. Genetics of resistance of *Salix sericea* to a diverse community of herbivores. Evolution 51: 1490–1498.

Roelke. M. E., J. S. Martenson, and S. J. O'Brien. 1993. The consequences of demographic reduction and genetic depletion in the endangered Florida panther. Curr. Biol. 3: 340–350.

Roff, D. A. 1980. Optimizing development time in a seasonal environment: The "ups and downs" of clinal variation. Oecologia 45: 202–208.

Roff, D. A. 1981. On being the right size. Am. Nat. 118: 405–422.

Roff, D. A. 1990. The evolution of flightlessness in insects. Ecol. Monogr. 60: 389–422.

Roff, D. A. 1992. The evolution of life histories: Theory and analysis. Chapman and Hall, New York.

Roff, D. A. 1996. The evolution of threshold traits in animals. Quart. Rev. Biol. 71: 3–35.

Roff, D. A. 1997. Evolutionary quantitative genetics. Chapman and Hall, New York.

Roff, D. A. 2000. Trade-offs between growth and reproduction: An analysis of the quantitative genetic evidence. J. Evol. Biol. 13: 434–445.

Roff, D. A., and D. J. Fairbairn. 1991. Wing dimorphisms and the evolution of migratory polymorphisms among the Insecta. Am. Zool. 31: 243–252.

Roff, D. A., G. Stirling, and D. J. Fairbairn. 1997. The evolution of threshold traits: A quantitative genetic analysis of the physiological and life-history correlates of wing dimorphism in the sand cricket. Evolution 51: 1910–1919.

Roff, D. A., J. Tucker, G. Stirling, and D. J. Fairbairn. 1999. The evolution of threshold traits: Effects of selection on fecundity and correlated response in wing dimorphism in the sand cricket. J. Evol. Biol. 12: 535–546.

Rohwer, S. 1982. The evolution of reliable and unreliable badges of fighting ability. Am. Zool. 22: 531–546.

Root, R. 1996. Herbivore pressure on goldenrods (Solidago altissima): Its variation and cumulative effects. Ecology 77: 1074–1087.

Rose, M. R., and T. J. Bradley. 1998. Evolutionary physiology of the cost of reproduction. Oikos 83: 443–451.

Rose, M. R., and B. Charlesworth. 1981. Genetics of life history in Drosophila melanogaster. 2. Exploratory selection experiments. Genetics 97: 187–196.

Rosenheim, J. A. 1999. The relative contributions of time and eggs to the cost of reproduction. Evolution 53: 376–385.

Rosenthal, G. A., C. Hughes, and D. Janzen. 1982. L-Canavanine, a dietary nitrogen source for the seed predator Caryedes brasiliensis (Bruchidae). Science 217: 353–355.

Roskam, J. C., and P. M. Brakefield. 1996. A comparison of temperature-induced polyphenism in African Bicyclus butterflies from a seasonal savannah-rainforest ecotone. Evolution 50: 2360–2372.

Roskam, J. C., and P. M. Brakefield. 1999. Seasonal polyphenism in Bicyclus (Lepidoptera: Satyridae) butterflies: Different climates need different cues. Biol. J. Linn. Soc. 66: 345–356.

Rosmoser, W. S., and J. G. Stoffolano, Jr. 1998. The science of entomology, 4th ed. McGraw-Hill, New York.

Rossi, L., A. Basset, and L. Nobile. 1983. A coadapted trophic niche in two species of crustacea (Isopoda): Acellus aquaticus (L.) and Proacellus coxalis Dolff. Evolution 37: 810–820.

Roush, R. T., and J. A. McKenzie. 1987. Ecological genetics of insecticide and acaricide resistance. Annu. Rev. Entomol. 32: 361–380.

Rowe, L., and D. Ludwig. 1991. Size and timing of metamorphosis in complex life cycles: Time constraints and variation. Ecology 72: 413–427.

Rowell-Rahier, M., and J. M. Pasteels. 1992. Third trophic level influences of plant allelochemicals. Pp. 243–278 in G. A. Rosenthal and M. R. Berenbaum (eds.), Herbivores: Their interactions with plant secondary metabolites. 2. Ecological and evolutionary processes. Academic Press, New York.

Rumrill, S. S. 1990. Natural mortality of marine invertebrate larvae. Ophelia 32: 163–198.

Ryan, M. J. 1985. The Túngara frog. University of Chicago Press, Chicago.

Ryan, M. J. 1997. Sexual selection and mate choice. Pp. 179–202 in J. R. Krebs and N. B. Davies (eds.), Behavioural ecology: An evolutionary approach, 4th ed. Blackwell, Oxford.

Ryan, M. J., J. H. Fox, W. Wilczynski, and A. S. Rand. 1990. Sexual selection for sensory exploitation in the frog Physalaemus pustulosus. Nature 343: 66–67.

Ryan, R. B. 1997. Before and after evaluation of biological control of the larch casebearer (Lepidoptera: Coleophoridae) in the Blue Mountains of Oregon and Washington, 1972–1995. Environ. Entomol. 26: 703–715.

Saccheri, I., M. Kuussaari, M. Kankare, P. Vikman, W. Fortelius, and I. Hanski. 1998. Inbreeding and extinction in a butterfly metapopulation. Nature 392: 491–494.

Sakai, A. K., and S. G. Weller. 1999. Gender and sexual dimorphism in flowering plants: A review of terminology, biogeographic patterns, ecological correlates, and phylogenetic approaches. Pp. 1–31 in M. A. Geber, T. E. Dawson, and L. F. Delph (eds.), Sexual and gender dimorphism in flowering plants. Springer-Verlag, Heidelberg.

Sakai, A. K., S. G. Weller, M.-L. Chen, S.-Y. Chou, and C. Tasanont. 1997. Evolution of gynodioecy and maintenance of females: The role of inbreeding depression, outcrossing rates, and resource allocation in Schiedea adamantis (Caryophyllaceae). Evolution 51: 724–736.

Sakaluk, S. K. 2000. Sensory exploitation as an evolutionary origin to nuptial food gifts in insects. Proc. R. Soc. Lond. B 267: 339–343.

Sakata, H. 1994. How an ant decides to prey on or to attend aphids. Res. Pop. Ecol. 36: 45–51.

Salomonson, A. 1996. Interactions between somatic mutations and plant development. Vegetatio 127: 71–75.

Saloniemi, I. 1993. A coevolutionary predator-prey model with quantitative characters. Am. Nat. 141: 880–896.

Sano, H., I. Kamada, S. Youssefian, M. Katsumi, and H. Wabiko. 1990. A single treatment of rice seedlings with 5-azacytidine induces heritable dwarfism and undermethylation of genomic DNA. Molecular and Genreral Genetics 220: 441–447.

Sapp, J. 1994. Evolution by association: A history of symbiosis. Oxford University Press, New York.

Sausman, K. A. 1984. Survival of captive-born *Ovis canadensis* in North American zoos. Zoo Biol. 3: 111–121.

Savalli, U. M. 1994a. Mate choice in the yellow-shouldered widowbird: correlates of male attractiveness. Behav. Ecol. Sociobiol. 35: 227–234.

Savalli, U. M. 1994b. Tail length affects territory ownership in the yellow-shouldered widowbird. Anim. Behav. 48: 105–111.

Savalli, U. M. 1995. The evolution of tail-length in widowbirds (Ploceidae): Tests of alternatives to sexual selection. Ibis 137: 389–395.

Savalli, U. M., and C. W. Fox. 1998. Sexual selection and the fitness consequences of male body size in the seed beetle *Stator limbatus*. Anim. Behav. 55: 473–483.

Schaffer, W. M. 1974a. Optimal reproductive effort in fluctuating environments. Am. Nat. 108: 783–790.

Schaffer, W. M. 1974b. Selection for optimal life histories: The effects of age structure. Ecology 55: 291–303.

Schaffer, W. M., and M. L. Rosenzweig. 1978. Homage to the Red Queen. 1. Coevolution of predators and their victims. Theoretical Population Biology 14: 135–157.

Scheiner, S. M. 1993. Genetics and evolution of phenotypic plasticity. Annu. Rev. Ecol. Sys. 24: 35–68.

Schierup, M. H., and F. B. Christiansen. 1996. Inbreeding depression and outbreeding depression in plants. Heredity 77: 461–468.

Schlichting, C. D. 1986. The evolution of phenotypic plasticity in plants. Annu. Rev. Ecol. Sys. 17: 667–693.

Schlichting, C. D. 1989. Phenotypic integration and environmental change. BioSci. 39: 460–464.

Schlichting, C. D., and M. Pigliucci. 1995. Gene regulation, quantitative genetics and the evolution of reaction norms. Evol. Ecol. 9: 154–168.

Schlichting, C. D., and M. Pigliucci. 1998. Phenotypic evolution, a reaction norm perspective. Sinauer, Sunderland, Mass.

Schluter, D. 1988. Estimating the form of natural selection on a quantitative trait. Evolution 42: 49–61.

Schluter, D. 1990. Species-for-species matching. Am. Nat. 136: 560–568.

Schluter, D. 1994. Experimental evidence that competition promotes divergence in adaptive radiation. Science 266: 798–801.

Schluter, D. 2000. The ecology of adaptive radiation. Oxford University Press, Oxford.

Schluter, D., and L. Gustafsson. 1993. Maternal inheritance of condition and clutch size in the collared flycatcher. Evolution 47: 658–667.

Schluter, D., and J. D. McPhail. 1992. Ecological character displacement and speciation in sticklebacks. Am. Nat. 140: 85–108.

Schluter, D., and D. Nychka. 1994. Exploring fitness surfaces. Am. Nat. 143: 597–616.

Schmalhausen, I. I. 1949. Factors of evolution. the theory of stabilizing selection. University of Chicago Press, Chicago.

Schmitt, J., S. Dudley, and M. Pigliucci. 1999. Manipulative approaches to testing adaptive plasticity: Phytochrome-mediated shade avoidance responses in plants. Am. Nat. 154: S43–S54.

Schneider, J. C. 1980. The role of parthenogenesis and female aptery in microgeographic ecological adaptation in the fall cankerworm, *Alsophila pometaria* Harris (Lepidoptera: Geometridae). Ecology 61: 1082–1090.

Schoener, T. W. 1970. Size patterns in West Indian *Anolis* lizards. 2. Correlations with the size of particular sympatric species—displacement and convergence. Am. Nat. 104: 155–174.

Schoener, T. W. 1983. Field experiments on interspecific competition. Am. Nat. 122: 240–285.

Schoener, T. W. 1987. A brief history of optimal foraging ecology. Pp. 5–67 *in* A. C. Kamil, J. R. Krebs, and H. R. Pulliam (eds.), Foraging behavior. Plenum Press, New York.

Schulke, B., and N. M. Waser. 2001. Long-distance pollinator flights and pollen dispersal between populations of *Delphinium nuttallianum*. Oecologia 127: 239–245.

Schwartz, J. 2000. Death of an altruist. Lingua Franca 10: 51–61

Searcy, W. A., and K. Yasukawa. 1989. Alternative models of territorial polygyny in birds. Am. Nat. 134: 323–343.

Searcy, W. A., and K. Yasukawa. 1995. Polygyny and sexual selection in red-winged blackbirds. Princeton University Press, Princeton, N.J.

Seeley, T. 1995. The wisdom of the hive. Harvard University Press, Cambridge.

Seger, J. 1992. Evolution of exploiter-victim relationships. Pp. 3–26 *in* M. Crawley (ed.), Natural enemies. Blackwell, Oxford.

Segraves, K. A., and J. N. Thompson. 1999. Plant polyploidy and pollination: Floral traits and insect visits to diploid and tetraploid *Heuchera grossulariifolia*. Evolution 53: 1114–1127.

Sgrò, C. M., and L. Partridge. 1999. A delayed wave of death from reproduction in *Drosophila*. Science 286: 2521–2524.

Sgrò, C. M., and L. Partridge. 2000. Evolutionary responses of the life history of wild-caught *Drosophila melanogaster* to two standard methods of laboratory culture. Am. Nat. 156: 341–353.

Shaw, R. F., and J. D. Mohler. 1953. The selective significance of the sex ratio. Am. Nat. 87: 337–342.

Sheffer, R. J., P. W. Hedrick, W. L. Minckley, and A. L. Velasco. 1997. Fitness in the endangered Gila topminnow. Cons. Biol. 11: 162–171.

Sheffer, R. J., P. W. Hedrick, and A. Velasco. 1999. Testing for inbreeding and outbreeding depression in the endangered Gila topminnow. Anim. Cons. 2: 121–129.

Sheldon, K. M., M. S. Sheldon, and R. Osbaldiston. 2000. Prosocial values, assortation, and group-selection in an N-person prisoner's dilemma. Human Nature 11: 387–404.

Shields, W. M. 1988. Sex and adaptation. Pp. 253–269 *in* R. E. Michod and B. R. Levin (eds.), The evolution of sex. Sinauer, Sunderland, Mass.

Shine, R. 1999. Why is sex determined by nest temperature in many reptiles? Trends Ecol. Evol. 14: 186–189.

Shine, R., and A. E. Greer. 1991. Why are clutch sizes more variable in some species than in others? Evolution 45: 1696–1706.

Shykoff, J. A., and P. Schmid-Hempel. 1991. Parasites and the advantage of genetic variability within social insect colonies. Proc. Roy. Soc. Lond. B 243: 55–58.

Sikes, R. S., and H. Ylönen. 1998. Considerations of optimal litter size in mammals. Oikos 83: 452–465.

Silbermann, R., and M. Tatar. 2000. Reproductive costs of heat shock protein in transgenic *Drosophila* melanogaster. Evolution 54: 2038–2045.

Simberloff, D., and W. Boecklen. 1981. Santa Rosalia reconsidered: Size ratios and competition. Evolution 35: 1206–1228.

Sinervo, B. 1999. Mechanistic analysis of natural selection and a refinement of Lack's and Williams's principles. Am. Nat. 154: S26–S42.

Slatkin, M. 1980. Ecological character displacement. Ecology 61: 163–177.

Slatkin, M. 1995. A measure of population subdivision based on microsatellite allele frequencies. Genetics 139: 457–462.

Smith, A. B, and C. H. Jeffery. 1998. Selectivity of extinction among sea urchins at the end of the Cretaceous period. Nature 392: 69–71.

Smith, C. C., and S. D. Fretwell. 1974. The optimal balance between size and number of offspring. Am. Nat. 108: 499–506.

Smith, F. A., and S. E. Smith. 1996. Mutualism and parasitism: Diversity in function and structure in the "arbuscular" (VA) mycorrhizal symbiosis. Adv. Bot. Res. 22: 1–43.

Smith, P., M. Berdoy, R. H. Smith, and D. W. MacDonald. 1993. A new aspect of warfarin resistance in wild rats: Benefits in the absence of the poison. Functional Ecol. 7: 190–194.

Smith, R. H., and C. M. Lessells. 1985. Oviposition, ovicide and larval competition in granivorous insects. Pp. 423–448 *in* R. M. Sibly and R. H. Smith (eds.), Behavioral ecology. Blackwell, Oxford.

Smith, T. B. 1993. Disruptive selection and the genetic basis of bill size polymorphism in the African finch *Pyrenestes*. Nature 363: 618–620.

Smith-Gill, S. J., and K. A. Berven. 1979. Predicting amphibian metamorphosis. Am. Nat. 113: 563–585.

Sober, E., and D. S. Wilson. 1998. Unto Others: The evolution and psychology of unselfish behavior. Harvard University Press, Cambridge.

Solbreck, C. 1986. Wing and flight muscle polymorphism in a lygaeid bug, *Horvathiolus gibbicollis*: Determinants and life history consequences. Ecol. Entomol. 11: 435–444.

Soltis, D. E., P. S. Soltis, and J. N. Thompson. 1992. Chloroplast DNA variation in *Lithophragma* (Saxifragaceae). Systematic Botany 17: 607–619.

Southwick, E. E. 1984. Photosynthate allocation to floral nectar: A neglected energy investment. Ecology 65: 1775–1779.

Southwood, T. R. E. 1962. Migration of terrestrial arthropods in relation to habitat. Biol. Rev. 37: 171–214.

Southwood, T. R. E. 1977. Habitat, the template for ecological strategies? J. Anim. Ecol. 46: 337–365.

Southwood, T. R. E., R. M. May, M. P. Hassell, and G. R. Conway. 1974. Ecological strategies and population parameters. Am. Nat. 108: 791–804.

Sponaugle, S., and R. K. Cowen. 1997. Early life history traits and recruitment patterns of Caribbean wrasses (Labridae). Ecol. Monog. 67: 177–202.

Stearns, S. C. 1992. The evolution of life histories. Oxford University Press, Oxford.

Stearns, S., G. de-Jong, and B. Newman. 1991. The effects of phenotypic plasticity on genetic correlations. Trends Ecol. Evol. 6: 122–126.

Stearns, S. C., and J. C. Koella. 1986. The evolution of phenotypic plasticity in life-history traits: Predictions of reaction norms for age and size at maturity. Evolution 40: 893–913.

Stephens, D. W., and S. R. Dunbar. 1991. Dimensional analysis in behavioral ecology. Behav. Ecol. 4: 172–183.

Stephens, D. W., and J. R. Krebs. 1986. Foraging theory. Princeton University Press, Princeton, N.J.

Stevens, L., C. J. Goodnight, and S. Kalisz. 1995. Multilevel selection in natural populations of *Impatiens capensis*. Am. Nat. 145: 513–526.

Stone, G., P. Willmer, and S. Nee. 1996. Daily partitioning of pollinators in an African *Acacia* community. Proc. Roy. Soc. Lond. B, Biol. Sci. 263: 1389–1393.

Stoner, D. S. 1990. Recruitment of a tropical colonial ascidian: Relative importance of pre-settlement vs. post-settlement processes. Ecology 71: 1682–1690.

Storfer, A., and A. Sih. 1998. Gene flow and ineffective antipredator behavior in a stream-breeding salamander. Evolution 52: 558–565.

Stowe, K. 1998. Experimental evolution of resistance in *Brassica rapa*: Correlated response of tolerance in lines selected for glucosinolate content. Evolution 52: 703–712.

Strathmann, R. R. 1985. Feeding and nonfeeding larval development and life-history evolution in marine invertebrates. Annu. Rev. Ecol. Syst. 16: 339–361.

Stratton, D. A. 1992. Components of selection in *Erigeron annuus*: 1. Phenotypic selection. Evolution 46: 92–106.

Strong, D. R., Jr., L. A. Szyska, and D. S. Simberloff. 1979. Tests of community-wide character displacement against null hypotheses. Evolution 33: 897–913.

Sultan, S. E. 1987. Evolutionary implications of phenotypic plasticity in plants. Evol. Biol. 21: 127–178.

Sun, J., and J. Tower. 1999. FLP/FRT-mediated induction of Cu/ZnSOD transgene expression can extend the life span of adult *Drosophila*. Mol. Cell. Biol. 19: 216–228.

Sutherland, W. J. 1996. From individual behaviour to population ecology. Oxford University Press, Oxford.

Swenson, W., D. S. Wilson, and R. Elias. 2000. Artificial ecosystem selection. Proc. Nat. Acad. Sci. 97: 9110–9114.

Taper, M., and T. J. Case. 1992. Models of character displacement and the theoretical robustness of taxon cycles. Evolution 46: 317–334.

Tatar, M., and J. R. Carey. 1995. Nutrition mediates reproductive trade-offs with age-specific mortality in the beetle *Callosobruchus maculatus*. Ecology 76: 2066–2073.

Tatar, M., D. W. Grey, and J. R. Carey. 1997a. Altitudinal variation in senescence in a *Mela-noplus* grasshopper species complex. Oecologia 111: 357–364.

Tatar, M., A. A. Khazaeli, and J. W. Curtsinger. 1997b. Chaperoning extended life. Nature 390: 30.

Taylor, C. E. 1986. Genetics and the evolution of resistance to insecticides. Biol. J. Linnean Soc. 27: 103–112.

Taylor, M., and R. Feyereisen. 1996. Molecular biology and evolution of resistance to toxicants. Mol. Biol. Evol. 13: 719–734.

Templeton, A. R. 1986. Coadaptation and outbreeding depression. Pp. 105–116 *in* M. E. Soulé (ed.), Conservation biology, the science of scarcity and diversity. Sinauer, Sunderland, Mass.

Templeton, A. R. 1998. Nested clade analyses of phylogeographic data: Testing hypotheses about gene flow and population history. Mol. Ecol. 7: 381–397.

Templeton, A. R., and B. Read. 1983. The elimination of inbreeding depression in a captive herd of Speke's gazelle. Pp. 241–261 *in* C. M. Schonewald-Cox, S. M. Chambers, B. MacBryde, and L. Thomas (eds.), Genetics and conservation. Benjamin/Cummings, Menlo Park, Calif.

Thébaud, C., and D. Simberloff. 2001. Are plants really larger in their introduced ranges? Amer. Nat. 157: 231–236.

Thompson, J. D. 1988. Evolutionary ecology of the relationship between oviposition preference and performance of offspring in phytophagous insects. Entomol. Exp. Appl. 47: 3–14.

Thompson, J. D. 1994. The coevolutionary process. University of Chicago Press, Chicago.

Thompson, J. N. 1982. Interaction and coevolution. Wiley, New York.

Thompson, J. N. 1988. Variation in interspecific interactions. Annu. Rev. Ecol. Syst. 19: 65–87.

Thompson, J. N. 1994. The coevolutionary process. University of Chicago Press, Chicago.

Thompson, J. N. 1996. The coevolutionary process. University of Chicago Press, Chicago.

Thompson, J. N. 1999. Specific hypotheses on the geographic mosaic of coevolution. Am. Nat. 153: S1–S14.

Thompson, J. N., B. M. Cunningham, K. A. Segraves, D. M. Althoff, and D. Wagner. 1997. Plant polyploidy and insect/plant interactions. Am. Nat. 150: 730–743.

Thompson, J. N., and O. Pellmyr. 1992. Mutualism with pollinating seed parasites amid copollinators: Constraints on specialization. Ecology 73: 1780–1791.

Thornhill, N. W. (ed.) 1993. The natural history of inbreeding and outbreeding: Theoretical and empirical perspectives. University Chicago Press, Chicago.

Thorson, G. 1950. Reproductive and larval ecology of marine bottom invertebrates. Biol. Rev. 25: 1–45.

Tousignant, A., and D. Crews. 1995. Incubation temperature and gonadal sex affect growth and physiology in the leopard gecko *(Eublepharis macularius)*, a lizard with temperature-dependent sex determination. J. Morphol. 224: 159–170.

Tousignant, A., B. Viets, D. Flores, and D. Crews. 1995. Ontogenetic and social factors affect the endocrinology and timing of reproduction in the female leopard gecko *(Eublepharis macularius)*. Horm. Behav. 29: 141–153.

Travis, J., M. G. McManus, and C. F. Baer. 1999. Sources of variation in physiological phenotypes and their evolutionary significance. Am. Zool. 39: 422–433.

Travis, J., and D. N. Reznick. 1998. Experimental approaches to the study of evolution. Pp. 310–352 *in* W. J. Resitarits and J. Bernardo (eds.), Issues and Perspectives in Experimental Ecology. Oxford University Press, New York.

Travis, J., and J. C. Trexler. 1987. Regional variation in habitat requirements of the sailfin molly, with special reference to the Florida Keys. Florida Game and Freshwater Fish Commission Nongame Wildlife Program Technical Report No. 3.

Trexler, J. C. 1988. Hierarchical organization of genetic variation in the sailfin molly, *Poecilia latipinna* (Pisces, Poeciliidae). Evolution 42: 1006–1017.

Trivers, R. L. 1971. The evolution of reciprocal altruism. Quart. Rev. Biol. 46: 35–57.

Trivers, R. L. 1972. Parental investment and sexual selection. Pp. 136–179 *in* B. Campbell (ed.), Sexual selection and the descent of man. Aldine, Chicago.

Trivers, R. L., and D. E. Willard. 1973. Natural selection of parental ability to vary the sex ratio of their offspring. Science 179: 90–92.

Tucic, B., V. Tomic, S. Avramov, and D. Pemac. 1998. Testing the adaptive plasticity of *Iris pumila* leaf traits to natural light conditions using phenotypic selection analysis. Acta Oecologica 19: 473–481.

Tuljapurkar, S. 1990. Population dynamics in variable environments. Springer-Verlag, Berlin.

Turmel, M., C. Otis, and C. Lemieux. 1999. The complete chloroplast DNA sequence of the green alga *Nephroselmis olicacea*: Insights into the architecture of ancestral chloroplast genomes. Proc. Natl. Acad. Sci. U.S.A. 96: 10248–10253.

Uri, N. D. 1998. Government policy and the development and use of biopesticides. Futures 30: 405–423.

Uyenoyama, M. K., K. E. Holsinger, and D. M. Waller. 1993. Ecological and genetic factors directing the evolution of self-fertilization. Oxford Surv. Evol. Biol. 9: 327–381.

Vacquier, V. D. 1995. Evolution of gamete recognition proteins. Science 281: 1995–1998.

Van Klinken, R. D. 1999. Is host specificity of biocontrol agents likely to evolve once they are released? P. 67 *in* Evaluating indirect ecological effects of biological control, Global IOBC Intern. Symp. IOBC wprs Bulletin. Montpellier, France.

van Noordwijk, A. J., and G. de Jong. 1986. Acquisition and allocation of resources: Their influence on variation in life history tactics. Am. Nat. 128: 137–142.

van Tienderen, P. H. 1991. Evolution of generalists and specialists in spatially heterogeneous environments. Evolution 45: 1317–1331.

Van Valen, L. 1973. A new evolutionary law. Evol. Theory 1: 1–30.

Varley, G. C., G. R. Gradwell, and M. P. Hassell. 1973. Insect population ecology. Blackwell, Oxford.

Vermeij, G. J. 1994. The evolutionary interaction among species: Selection, escalation, and coevolution. Annu. Rev. Ecol. Syst. 25: 219–236.

Via, S., and R. Lande. 1985. Genotype-environment interaction and the evolution of phenotypic plasticity. Evolution 39: 505–522.

Viets, B. E., A. Tousignant, M. A. Ewert, C. E. Nelson, and D. Crews. 1993. Temperature-dependent sex determination in the leopard gecko, *Eublepharis macularius*. J. Exp. Zool. 265: 679–683.

Vrijenhoek, R. C. 1994. Unisexual fish: Model systems for studying ecology and evolution. Annu. Rev. Ecol. Syst. 25: 71–96.

Vrijenhoek, R., M. E. Douglas, and G. K. Meffe. 1985. Conservation genetics of endangered fish population in Arizona. Science 229: 400–492.

Waage, J. 1990. Ecological theory and the selection of biological control agents. Pp. 135–158 *in* M. Mackauer, L. E. Ehler, and J. Roland (eds.), Critical issues in biological control. Intercept, Andover, UK.

Wade, M. J., and C. J. Goodnight. 1998. The theories of Fisher and Wright in the context of metapopulations: When nature does many small experiments. Evolution 52: 1537–1553.

Wagner, J. D., M. D. Glover, J. B. Moseley, and A. J. Moore. 1999. Heritability and fitness consequences of cannibalism in *Harmonia axyridis*. Evol. Ecol. Res. 1: 375–388.

Wagner, R. P. 1944. The nutrition of *Drosophila mulleri* and *D. aldrichi*: Growth of the larvae on cactus extract and the microrganisms found in cactus. University of Texas Publ. 4445: 104–28.

Wagner, W. L., and V. A. Funk. 1995. Hawaiian biogeography: Evolution on a hot spot archipelago. Smithsonian Press, Washington.

Wallace, A. R. 1858. On the tendency of varieties to depart indefinitely from the original type. J. Proc. Linn. Soc. (Zool.) 3: 53–62.

Walsh, B., and M. Lynch. 1998. Quantitative genetics. 2. Evolution and selection of quantitative traits. <http://nitro.biosci.arizona.edu/zbook/volume_2/chapters/vol2_14.html>.

Waples, R. S. 1995. Evolutionarily significant units and the conservation of biological diversity under the Endangered Species Act. Am. Fish. Soc. Symp. 1: 8–27.

Warner, R. R. 1975. The adaptive significance of sequential hermaproditism in animals. Am. Nat. 109: 61–82.

Warner, S. C., J. Travis, and W. A. Dunson. 1993. The effect of variation in pH level on the interspecific competition between two species of hylid tadpoles. Ecology 74: 183–194.

Waser, N. M. 1993. Population structure, optimal outbreeding, and assortative mating in Angiosperms. Pp. 173–199 *in* N. W. Thornhill (ed.), The natural history of inbreeding and outbreeding: Theoretical and empirical perspectives. University of Chicago Press, Chicago.

Waser, N. M. 1998. Pollination, angiosperm speciation, and the nature of species boundaries. Oikos 82: 198–201.

Waser, N. M., L. Chittka, M. V. Price, N. M. Williams, and J. Ollerton. 1996. Generalization in pollination systems, and why it matters. Ecology 77: 1043–1060.

Waser, N. M., and M. V. Price. 1994. Crossing distance effects in *Delphinium nelsonii*: Outbreeding and inbreeding depression in progeny fitness. Evolution 48: 842–852.

Waser, N. M., M. V. Price, and R. G. Shaw. 2000. Outbreeding depression varies among cohorts of *Ipomopsis aggregata* planted in nature. Evolution 54: 485–491.

Wattiaux, J. M. 1968. Cumulative parental age effects in *Drosophila subobscura*. Evolution 22: 406–421.

Weeks, S., and G. K. Meffe. 1996. Quantitative genetic and optimality analyses of life-history plasticity in the Eastern Mosquitofish, *Gambusia holbrooki*. Evolution 50: 1358–1365.

Weir, B. S., and C. C. Cockerham. 1984. Estimating F-statistics for the analysis of population structure. Evolution 38: 1358–1370.

Weis, A. E., and W. L. Gorman. 1990. Measuring selection on reaction norms: An exploration of the *Eurosta-Solidago* system. Evolution 44: 820–831.

Welch, D. M., and M. Meselson. 2000. Evidence for the evolution of bdelloid rotifers without sexual reproduction or genetic exchange. Science 288: 1211–1215.

Weldon, W. F. R. 1899. Presidential address. Report of the Sixty-Eighth Meeting of the British Association for the Advancement of Science, held at Bristol in September 1898. Murray, London.

Weller, S. G. 1976. Breeding system polymorphism in a heterostylous species. Evolution 30: 442–454.

Weller, S. G., M. Donoghue, and D. Charlesworth. 1995a. The evolution of self-incompatibility in flowering plants: A phylogenetic approach. Pp. 355–382 *in* P. C. Hoch and A. G. Stephenson (eds.), Experimental and molecular approaches to plant biosystematics. Monographs in systematic botany from the Missouri Botanical Garden 53.

Weller, S. G., and A. K. Sakai. 1999. Using phylogenetic approaches for the analysis of plant breeding system evolution. Annu. Rev. Ecol. Syst. 30: 167–199.

Weller, S. G., A. K. Sakai, A. E. Rankin, A. Golonka, B. Kutcher, and K. E. Ashby. 1998. Dioecy and the evolution of wind pollination in *Schiedea* and *Alsinidendron* (Caryophyllaceae: Alsinoideae) in the Hawaiian Islands. Am. J. Bot. 85: 1377–1388.

Weller, S. G., W. L. Wagner, and A. K. Sakai. 1995b. A phylogenetic analysis of *Schiedea* and *Alsinidendron* (Caryophyllaceae: Alsinoideae): Implications for the evolution of breeding systems. Systematic Botany 20: 315–337.

Wendt, D. E. 1998. Effect of larval swimming duration on growth and reproduction of *Bugula neritina* (Bryozoa) under field conditions. Biol. Bull. 195: 126–135.

West-Eberhard, M. J. 1989. Phenotypic plasticity and the origins of diversity. Annu. Rev. Ecol. Sys. 20: 249–278.

Westemeier, R. L., J. D. Brown, S. A. Simpson, T. L. Esker, R. W. Jansen, J. W. Walk, E. L. Kershner, J. L. Bouzat, and K. N. Paige. 1998. Tracking the long-term decline and recovery of an isolated population. Science 282: 1695–1698.

Whitlock, M. C., and N. H. Barton. 1997. The effective size of a subdivided population. Genetics 146: 427–441.

Widemo, F., and I. P. F. Owens. 1995. Lek size, male mating skew and the evaluation of lekking. Nature 373: 148–150.

Wiens, J. A. 1977. On competition and variable environments. Am. Sci. 65: 590–597.

Wikelski, M., and F. Trillmich. 1997. Body size and sexual size dimorphism in marine iguanas fluctuate as a result of opposing natural and sexual selection: An island comparison. Evolution 51: 922–936.

Wilbur, H. M., and J. P. Collins. 1973. Ecological aspects of amphibian metamorphosis. Science 182: 1305–1314.

Wiley, R. H., and J. Poston. 1996. Indirect mate choice, competition for mates, and coevolution of the sexes. Evolution 50: 1371–1381.

Williams, C. F., J. Ruvinsky, P. E. Scott, and D. W. Hews. 2001. Pollination, breeding system, and genetic structure in two sympatric Delphinium (Ranunculaceae) species. Am. J. Botany 88 (in press).

Williams, C. F., and N. M. Waser. 1999. Spatial genetic structure of Delphinium nuttallianum populations: Inferences about gene flow. Heredity 83: 541–550.

Williams, E. E. 1972. The origin of faunas: Evolution of lizard congeners in a complex island fauna: A trial analysis. Evol. Biol. 6: 47–89.

Williams, G. C. 1957. Pleiotropy, natural selection, and the evolution of senescence. Evolution 11: 398–411.

Williams, G. C. 1966. Adaptation and natural selection: A critique of some current evolutionary thought. Princeton University Press, Princeton, N.J.

Williams, G. C. 1975. Sex and evolution. Princeton University Press, Princeton, N.J.

Williams, G. C. 1992. Natural selection: Domains, levels and challenges. Oxford University Press, New York.

Williamson, D. H. 1993. Microbial mitochondrial genomes: Windows on other worlds. Pp. 73–106 in P. Broda, S. G. Oliver, and P. F. G. Sims (eds.), The eukaryotic genome: Organization and regulation. Cambridge University Press, Cambridge.

Willis, A., M. Thomas, and J. Lawton. 1999. Is the increased vigor of invasive weeds explained by a trade-off between growth and herbivore resistance? Oecologia 120: 632–640.

Willson, M. F. 1994. Sexual selection in plants: Perspective and overview. Am. Nat. 144: S13–S39.

Wilson, D. S. 1980. The natural selection of populations and communities. Benjamin/Cummings, Menlo Park, Calif.

Wilson, D. S. 1997a. Altruism and organism: Disentangling the themes of multilevel selection theory. Am. Nat. 150: S122–S134.

Wilson, D. S. 1997b. Biological communities as functionally organized units. Ecology 78: 2018–2024.

Wilson, D. S., and L. A. Dugatkin. 1997. Group selection and assortative interactions. Am. Nat. 149: 336–351.

Wilson, D. S., and K. M. Kniffin. 1999. Multilevel selection and the social transmission of behavior. Human Nature 10: 291–310.

Wilson, E. O. 1975. Sociobiology: The new synthesis. Harvard University Press, Cambridge.

Windig, J. J. 1994. Genetic correlations and reaction norms in wing pattern of the tropical butterfly Bicyclus anynana. Heredity 73: 459–470.

Winn, A. A. 1999. Is seasonal variation in leaf traits adaptive for the annual plant Dicerandra linearifolia? J. Evol. Biol. 12: 306–313.

Wolf, J. B., E. D. Brodie III, and A. J. Moore. 1999. Interacting phenotypes and the evolutionary process. 2. Selection resulting from social interactions. Am. Nat. 153: 254–266.

Woltereck, R. 1909. Weiterer experimentelle untersuchungen uber artveranderung, speziell uber das wessen quantitativer artunterschiede bei daphniden. Versuch. Deutsch Zool. Gesellschaft 19: 110–172.

Wright, S. 1922. Coefficients of inbreeding and relationship. Am. Nat. 56: 330–338.

Wright, S. 1931. Evolution in Mendelian populations. Genetics 16: 97–159.

Wright, S. 1943. Isolation by distance. Genetics 28: 114–138.

Wright, S. 1951. The genetical structure of populations. Ann. Eugenics 15: 323–354.

Wright, S. 1977. Evolution and the genetics of populations. Vol. 3. Experimental results and evolutionary deductions. University of Chicago Press, Chicago.

Wu, R. 1998. The detection of plasticity genes in heterogeneous environments. Evolution 52: 967–977.

Wynne-Edwards, V. C. 1962. Animal dispersion in relation to social behavior. Oliver and Boyd, Edinburgh.

Ydenberg, R. C., L.-A. Giraldeau, and D. L. Kramer. 1986. Interference competition, payoff asymmetries, and the social relationships of central place foragers. Theor. Pop. Biol. 30: 26–44.

Ydenberg, R. C., C. V. J. Welham, R. Schmid-Hempel, P. Schmid-Hempel, and G. Beauchamp. 1992. Time and energy constraints and the relationships between currencies in foraging theory. Behav. Ecol. 5: 28–34.

Zahavi, A. 1975. Mate selection—A selection for a handicap. J. Theor. Biol. 53: 205–214.

Zahavi, A. 1977. Reliability in communication systems and the evolution of altruism. Pp.

253–259 *in* B. Stonehouse and C. M. Perrins (eds.), Evolutionary ecology. Macmillan, London.

Zangerl, A. R., and M. R. Berenbaum. 1993. Plant chemistry and insect adaptation to plant chemistry as determinants of host plant utilization patterns. Ecology 74: 478–504.

Zera, A. J., and R. F. Denno. 1997. Physiology and ecology of dispersal polymorphism in insects. Annu. Rev. Entomol. 42: 207–230.

Zera, A. J., J. Sall, and K. Grudzinski. 1997. Flight-muscle polymorphism in the cricket *Gryllus firmus:* Muscle characteristics and their influence on the evolution of flightlessness. Physiol. Zool. 70: 519–529.

Zera, A. J., T. Sanger, and G. L. Cisper. 1998. Direct and correlated responses to selection on JHE activity in adult and juvenile *Gryllus assimilis:* Implications for stage-specific evolution of insect endocrine traits. Heredity 80: 300–309.

Zwölfer, H. 1973. Competition and coexistence in phytophagous insects attacking the heads of *Carduus nutans* L. Pp. 74–80 *in* Proc. II. Intern. Symp. Biological Control of Weeds.

Index